Lecture Notes in Control and Information Sciences

Edited by M. Thoma and A. Wyner

For information about Vols. 1–116 please contact your bookseller or Springer-Verlag

Lecture Notes
in Control and Information Sciences   184

Editors: M. Thoma and W. Wyner

T.E. Duncan, B. Pasik-Duncan (Eds.)

Stochastic Theory and Adaptive Control

Proceedings of a Workshop
held in Lawrence, Kansas, September 26 - 28, 1991

Springer-Verlag Berlin Heidelberg GmbH

Editor

Prof. T.E. Duncan
Prof. B. Pasik-Duncan

Department of Mathematics
University of Kansas
Lawrence, KS 66045
USA

ISBN 978-3-540-55962-7   ISBN 978-3-540-47327-5 (eBook)
DOI 10.1007/978-3-540-47327-5

© Springer-Verlag Berlin Heidelberg 1992
Originally published by Springer-Verlag Berlin Heidelberg New York in 1992.

Typesetting: Camera ready by authors
Offsetprinting: Mercedes-Druck, Berlin
61/3020 5 4 3 2 1 0 Printed on acid-free paper

PREFACE

This volume contains most of the papers that were presented at the Workshop on Stochastic Theory and Adaptive Control that was held at the University of Kansas, 26-28 September 1991. This workshop brought together many leading researchers on stochastic control and stochastic adaptive control to increase communication between these two subfields of stochastic analysis. Furthermore a number of graduate students in these two areas were invited to provide them with a broad exposure and additional education in these areas.

The workshop was supported by the National Science Foundation and the University of Kansas. We especially thank Dr. R. S. Baheti, Program Director of Engineering Systems and Dr. I. Peden, Director of the Division of Electrical Communications Systems at the National Science Foundation and C. Himmelberg, Chairman of the Mathematics Department at the University of Kansas, for their encouragement of the workshop. The workshop was also supported by the Kansas Center for Excellence in Computer Aided Systems Engineering. The following units of the University of Kansas supported the workshop: Office of Academic Affairs, Department of Mathematics, Center for Research, Inc., International Studies and Programs, Division of Continuing Education, College of Liberal Arts and Sciences and the Office of Research, Graduate Studies and Public Service. We thank all of the contributors for their essential support of this workshop. We thank a number of people at the University of Kansas for their generous help with the administration of the workshop. These include T. Atteberry, K. Coleman, M. Frei, P. Forio, S. Gumm, G. Prothe, S. Reed, H. H. Tsai and O. Zane.

We want to thank all of the participants of this workshop for making it very satisfying and memorable.

Lawrence, February 1992

T. E. Duncan
B. Pasik-Duncan

PARTICIPANTS

E. H. Abed
University of Maryland
College Park, MD 20742

A. Araposhathis
University of Texas at Austin
Austin, TX 78712

T. Atteberry
University of Kansas
Lawrence, KS 66045

J. Baillieul
Boston University
Boston, MA 02215

J. Balasiewicz
University of Colorado
Boulder, CO

J. Baras
University of Maryland
College Park, MD 20742

V. Benes
26 Taylor St.
Millburn, NJ 07041

P. Bertrand
Universite des Antilles et de la Guyanne
Pointe-A-Pitre Cedex, France

L. Brown
University of Illinois at Urbana-Champaign
Urbana, IL 61801

P. Caines
McGill University
Montreal, Quebec Canada H3A 2K6

H. F. Chen
Institute of Systems Science
Academia Sinica
Beijing, 100080, PR China

M. Deistler
Technical University of Vienna
Vienna, Austria

T. Duncan
University of Kansas
Lawrence, KS 66045

R. Elliott
University of Alberta
Edmonton, Alberta Canada T6G 2G1

E. Fernandez-Gaucherand
University of Arizona
Tucson, AZ 85721

J. Filar
University of Maryland Baltimore County
Baltimore, MD 21228

W. Fleming
Brown University
Providence, RI 02912

M. Frei
University of Kansas
Lawrence, KS 66045

L. Gerencsér
Computer and Automation Institute
Hungarian Academy of Sciences
Budapest, Hungary

M. Ghosh
University of Texas at Austin
Austin, TX 78712

A. J. Goldberg
Naval Research Laboratory
Washington, DC 20375-5000

R. E. Gover
Naval Research Laboratory
Washington, DC 20375-5000

A. Heinricher
University of Kentucky
Lexington, KY 40515

K. Helmes
University of Kentucky
Lexington, KY 40515

O. Hijab
Temple University
Philadelphia, PA 19122

C. J. Himmelberg
University of Kansas
Lawrence, KS 66045

I. Karatzas
Columbia University
New York, NY 10027

G. Kesidis
University of California at Berkeley
Berkeley, CA 94720

P. R. Kumar
University of Illinois at Urbana
Urbana, IL 61801

T. Lai
Stanford University
Stanford, CA 94305

I. Lasiecka
University of Virginia
Charlottesville, VA 22903

E. Lee
University of Minnesota
Minneapolis, MN 55455

A. Lindquist
Royal Institute of Technology
Stockholm, Sweden

L. Ljung
Linkoping University
Linkoping, Sweden

P. Mandl
Charles University
Prague, Czechoslovakia

S. Marcus
University of Maryland
College Park, MD 20742

C. Martin
Texas Tech University
Lubbock, TX 79409

W. McEneaney
Brown University
Providence, RI 02912

S. Meyn
University of Illinois at Urbana
Urbana, IL 61801

S. Naik
University of Illinois at Urbana
Urbana, IL 61801

A. Olbrot
Wayne State University
Detroit, MI 48202

B. Pasik-Duncan
University of Kansas
Lawrence, KS 66045

I. Peden
National Science Foundation
Washington, DC 20550

W. Ren
University of California at Berkeley
Berkeley, CA 94720

R. Rishel
University of Kentucky
Lexington, KY 40506

M. S. Radenkovic
University of Colorado
Boulder, CO

W.J. Runggaldier
University of Padova
Padova, Italy

M. Sain
University of Notre Dame
Notre Dame, IN 46556

S. Sethi
University of Toronto
Toronto, Ontario Canada M5S 1V4

A. Shwartz
Technion-Israel Institute of Technology
Haifa, Israel

V. Solo
Johns Hopkins University
Baltimore, MD 21218

M. Soner
Carnegie-Mellon University
Pittsburgh, PA 15213

L. Stettner
Institute of Mathematics
Polish Academy of Sciences
Warsaw, Poland

M. Taksar
SUNY at Stony Brook
Stony Brook, NY 11794

R. L. Tweedie
Colorado State University
Fort Collins, CO 80523

L. Valavani
Massachusetts Institute of Technology
Cambridge, MA 02139

F. VanVleck
University of Kansas
Lawrence, KS 66045

P. Varaiya
University of California at Berkeley
Berkeley, CA 94720

J. Walrand
University of California at Berkeley
Berkeley, CA 94720

Z. Wang
University of Kansas
Lawrence, KS 66045

W. S. Wong
AT&T Bell Laboratories
Holmdel, NJ 07733

X. Xue
Columbia University
New York, NY 10027

G. Yin
Wayne State University
Detroit, MI 48202

O. Zane
University of Kansas
Lawrence, KS 66045

Q. Zhang
University of Toronto
Toronto, Ontario Canada M5S 1V4

CONTENTS

Stochastic Analysis of Vaccination Strategies

L. Allen, T. Lewis, C. Martin, R. Carpio, M. Jones, M. Stamp, Texas
Tech University, G. Mundel, A. Way, Lubbock City Health Department

Abstract

In this paper data from a measles epidemic on a University campus is analyzed and is used to formulate a problem of the optimal distribution of vaccine during an epidemic. It is determined that the optimal strategy is to distribute the vaccine as soon as possible without regard to the spatial distribution between dormitory complexes. The result are obtained using extensive computer simulations.

1 Introduction

The number of cases of measles in the United States has risen dramatically in the last few years. In 1983, the number of measles cases was at an all time low, 1497 reported cases, but in 1988, there were 3411 cases and in 1989 there were approximately 17,850 cases (Centers for Disease Control 1990). The state of Texas reported the second highest number of cases in 1989 (3201 cases) and one of the worst outbreaks in the United States occurred in the Houston area (1802 cases reported from late 1988 to September 1989) (Centers for Disease Control 1990; Canfield 1989). A significant number of cases of the measles have occurred on university and college campuses and many of the epidemics have centered around public schools. The adult population in the United States is largely immune to the measles because of infection prior to the beginning of vaccination programs. The purpose of this paper is to present some data analysis and simulations associated with a recent outbreak of the measles on the campus of Texas Tech University.

Measles epidemics are very expensive both in the actual outlay of dollars and in the loss of productive time by the infected. There are a significant number of deaths and debilitating side effects associated with infection by the measles virus. The goal of this project at Texas Tech University is to develop vaccination strategies that will minimize the number of case of measles in Texas. However the object of this particular paper is to present some initial results on minimizing the number of cases of measles in an epidemic on a university campus. Typically the first vaccination is given 25 days after the first infection occurs. This is because there is 10 to 14 day latency period after the first person is infected. Often the first case is not recognized immediately and the second wave of the epidemic is in progress before it is recognized that there is an epidemic. The problem we consider is how should the vaccination be distributed on the campus in order to minimize the total number of cases. The ultimate problem is a nonlinear stochastic optimization problem with constraints. We have not made any attempt to attack the problem directly but have used extensive computer simulations of the epidemic to test a wide variety of vaccination strategies. These are contained in the master's thesis of Rosana Carpio and these result will be published in full detail elsewhere. Here we will describe the various strategies tested and summarize the results.

This increased incidence of measles in the general populations in recent years has resulted in new recommendations for measles vaccinations in the United States. The Immunization Practices Advisory Committee (ACIP) has recommended two doses of measles vaccine for all children, the first dose at 15 months of age and the second when the child is about to enter school at kindergarten or first grade (Immunization Practices Advisory Committee 1989). (Previously, only one vaccination at 15 months of age was recommended.) Ultimately, we hope that these new vaccination protocols will alleviate the need for the study of measles on university campuses.

Some theoretical studies have confirmed the need for a higher rate of immunity from measles than that achieved by a single vaccination given during the first two years of life (Hethcote 1983, 1988). These investigations are based on the analysis of epidemiological models of $SEIR$ type (S–Susceptible, E–Exposed, I–Infectious, R–Removed). Analyses of these models have provided a theoretical foundation for studying many questions related to the spread of an epidemic (Anderson and May 1979; Bailey 1975; Greenhalgh 1990; Hethcote 1976, 1983, 1988; Hethcote and Yorke 1984; Hoppensteadt 1975; May 1986; McLean and Anderson 1988; Waltman 1974).

In this report, *SEIR* models (stochastic and deterministic) are applied to a specific epidemic in a university setting. Data from a 1989 measles epidemic that occurred on the campus of Texas Tech University in Lubbock is thoroughly investigated. Calculations from the data and simulations of the models provide an estimate of the level of immunity that is necessary in this university population to prevent a measles epidemic.

2 Description of Data

An outbreak of measles (rubeola) occurred on the Texas Tech University campus and in the city of Lubbock in January, February, and March of 1989. The first case of measles was reported on January 13, 1989. Evidently, the student with the first reported case of measles infected a small group of students within his dormitory (dormitory 5) during spring semester enrollment. The epidemic continued until about the middle of March. After this point no more cases were reported at the university. Vaccination began on the university campus on February 1 (20 days after the first reported case). A total of 303 confirmed cases occurred in Lubbock; 198 of these were students at Texas Tech University which included 159 cases among students residing on campus (campus population size = 6104) and 39 cases among students residing off campus (off-campus population size = 17,396).

The data collected included onset date (first appearance of rash), age, race, sex, college of enrollment and living quarters, and were entered into one of the computer systems available at Texas Tech University. The dormitories at Texas Tech University were divided into ten complexes on the basis of common dining hall; students generally eat in the dining hall associated with their complex.

The data in Table 1 were used to estimate the level of immunity at the end of the epidemic. Two simplifying assumptions were made to estimate this level of immunity: (1) the susceptible proportion of the population equaled .06 before the epidemic began and (2) the individuals receiving vaccinations represented a random sample of the entire population. The infection rate was highest in dormitory 5 (where the epidemic began); it reached 5.8%. Therefore, an estimate of .06 for the susceptible proportion seems reasonable. (Estimates of .06 and .07 were used in the deterministic simulation; however, a susceptible proportion equal to .06 provided a closer fit to the data.) The high proportion of vaccine coverage (Table 1) is a good indication that individuals received vaccinations regardless of their vaccination history (ignorance of vaccination history or the possibility of a previous vaccine failure may have prompted many students to obtain the vaccine). Thus, based on our assumptions only 6% of the individuals receiving vaccinations were susceptible. We used the proportion vaccinated just during the first week of February as the proportion vaccinated during the epidemic, since they represented more than 90% of the total proportion vaccinated (Table 1) and the individuals vaccinated late in the epidemic probably did not affect the course of the epidemic. Thus, the level of immunity reached by the end of the epidemic is estimated as follows. The proportion infected (f_I) plus .06 of those individuals receiving vaccinations during the first week of February (f_V) equals the proportion of the susceptible class that became immune during the course of the epidemic. Adding this proportion to .94 (the proportion immune at the beginning of the epidemic) gives an estimate of the proportion immune or removed (f_R) at the end of the epidemic; i.e.,

$$f_I + .06 f_V + .94 = f_R$$

(Table 1). The average level of immunity is $\bar{f}_R$=.981. A 95% confidence interval was calculated for the average level of immunity, $\bar{f}_R$, at the end of the epidemic:

$$(.977, .985).$$

This estimate indicates (under the above assumptions) that a high level of immunity ($\approx$ 98%) must be achieved in this population for a measles epidemic to conclude. It also gives an indication of the level of immunity necessary to prevent an epidemic ($\approx$ 98%) in this population; the epidemic concluded due to the reduced proportion of the susceptibles in the population.

3 Logistic Regression Analysis

Logistic regression models are useful in problems where the dependent variable takes only two discrete values (Landwehr, Pregibon, and Shoemaker 1984). We consider the case in which the response is dichotomous (binary). The response variable can be coded as 1 if the student is infected (with probability p) and 0 otherwise (with probability $1 - p$). The logistic regression analysis will determine if any relationship exists between the probability of the disease susceptibility and the variables, namely, Dormitory (D), College (C), Sex (S), and Vaccination Record (V). (Age and race were not included. The ages were concentrated between 19 and 24 years of age; measles is not specific to a particular race.) The regression equation is

$$\text{Logit}(p) = \log\left(\frac{p}{1-p}\right) = F(D, C, S, V), \tag{1}$$

where p is the probability of susceptibility to the disease, D, C, S, and V are the explanatory variables, and F is a function to be determined. The data are not quantitative variables; therefore, all of the variables are transformed into dummy variables. There are 9 groups of dorms, 9 groups of colleges, 2 groups of gender and 3 groups of vaccination records (1–no vaccination; 2–vaccinated early(2/1/89–2/3/89); 3–vaccinated late (2/4/89–end). The specific equation is:

$$y = \log\left(\frac{p}{1-p}\right) = \beta_0 + \sum_{i=1}^{19} \beta_i x_i, \tag{2}$$

where x_i, $i = 1, \ldots, 8$ are the dummy variables for dorm, x_i, $i = 9, \ldots, 16$ are the dummy variables for college, x_{17} is the dummy variables for sex, and x_{18} and x_{19} are the dummy variables for vaccination record. All x_i, $i = 1, \ldots 19$ are non-random and all β_i, $i = 1, \ldots, 19$ are parameters to be determined.

A test of hypotheses was performed for each of the explanatory variables using the data analysis software package SAS. The null hypothesis (H_0) for testing the effect of one of the variables, D, C, S, or V, assumes the coefficients β_i associated with D, C, S, or, V are zero, whereas, the alternate hypothesis (H_a) assumes at least one of these β_i are not zero. The generalized likelihood-ratio test was used to test H_0 against H_a. The function $-2 \log \Lambda$ is distributed approximately as a Chi-square distribution with r degrees of freedom, where

$$\Lambda = \frac{\sup_{H_0} L(\beta; y)}{\sup_{H_a} L(\beta; y)}, \tag{3}$$

and $L(\beta; y)$ is the log-likelihood function (Gross and Clark 1975). Hypothesis H_0 is rejected at the αth significance level if and only if $-2 \log \Lambda > \chi^2_{1-\alpha}(r)$, where $\chi^2_{1-\alpha}$ is the $(1 - \alpha)$th quantile of the Chi-square distribution with r degrees of freedom. Out of the four tests of hypothesis that were performed, the null hypothesis was rejected at the $\alpha = .05$ level in two cases, in testing the effectiveness of dorm and of vaccination record (see Lo 1989). Therefore, in the following model, the population was subdivided according to place of residence; no other demographic variables were included. A brief summary of the model and the simulation results are presented in the next sections; more complete descriptions can be found in the references (Allen, Jones, and Martin 1991; Allen, Lewis, Martin, and Stamp 1990).

4 Model Formulation

The model is based on a discrete-time $SEIR$ model formulated by Rvachev and Longini (1985) for the global spread of influenza. The model consists of an $SEIR$ submodel for each dormitory complex and an additional one for those students who do not reside in a dormitory (off-campus students). Each of the states S, E, I, and R is defined below:

$S_i(t)$ = number of susceptible individuals on day t,

$E_i(\tau, t)$ = number of individuals on day t who were exposed on day $t - \tau$, $\tau = 0, \ldots, \tau_1$,

$I_i(\tau, t)$ = number of individuals that are infectious on day t who were exposed on day $t - \tau$, $\tau = 0, \ldots, \tau_2$,

$R_i(t)$ = number of individuals that are removed from the infection process by immunity or isolation on day t,

where $i = 1, 2, \ldots, 11$, representing the 10 dorms plus the off-campus students, τ_1 is the maximum length of the latent period (assumed to be 11 days) and τ_2 is the maximum length of the latent plus infectious period (assumed to be 14 days). The population size in each dormitory is assumed to be time invariant (no births or deaths), i.e.,

$$S_i(t) + \sum_{\tau=0}^{\tau_1} E_i(\tau, t) + \sum_{\tau=0}^{\tau_2} I_i(\tau, t) + R_i(t) = P_i, \tag{4}$$

where P_i is the population size within each dormitory, $i = 1, \ldots, 10$ (Table 1) and $P_{11} = 17,396$ is the off-campus population size.

Rvachev and Longini (1985) used an $SEIR$ submodel for each major city and linked the cities via a transportation operator. We do not use a transportation operator; in our model there are three sources of daily contact, within a dormitory, between dormitories, and within the entire student population and it is through these contacts the epidemic is spread. The basic time unit of the model is one day (reported number of cases are known on a daily basis).

A basic assumption in most $SEIR$ models is that the population mixes homogeneously. In a homogeneous-mixing model, it is assumed that the number of newly exposed individuals, $E(t)$, on day t is proportional to the product of the number of susceptible individuals, $S(t)$, and the number of infectious individuals, $I(t)$, on day t, i.e., $\alpha S(t) I(t)$. The constant of proportionality α is the contact rate, i.e., the average number of individuals with whom an infectious individual will make sufficient contact (to pass infection) in one day (Rvachev and Longini 1985).

Initially it was assumed that the population mixed homogeneously; however, agreement between observed and simulated results was poor. Therefore, it was assumed that α decreased with time. This is a reasonable assumption for many epidemics, since as the epidemic becomes more severe, individuals become more careful of their social contacts and avoid those individuals known to be infectious and those places where contagious individuals reside. In addition, this measles epidemic began at the start of spring semester, at a time when there is probably more socializing and a greater number of contacts than during midsemester when the epidemic concluded. We chose $\alpha = \bar{\alpha} S(t)$, where $\bar{\alpha}$ is a constant, α decreases with time since the susceptible group S decreases with time (Allen, Jones, and Martin 1991; Allen, Lewis, Martin, and Stamp 1990). The choice of α is specific to this population. At the beginning of the semester there was most likely a very high rate of contact (large $S(t)$) and as the semester progressed the rate of contact probably decreased to a constant level ($\bar{\alpha} S(t) \approx \alpha$). There are other reasonable forms that may be proposed for a decreasing contact rate (e.g., a contact rate dependent on $N - I(t)$) and further research is needed to test these various forms.

There are three parameters, $\bar{\alpha}_1$, $\bar{\alpha}_2$, and $\bar{\alpha}_3$ associated with the three sources of contact: within a dormitory ($\bar{\alpha}_1$), within the campus population, dorms 1-10, ($\bar{\alpha}_2$), within the entire student population, dorms 1-11, ($\bar{\alpha}_3$). Contact rates for measles (under the assumption of a homogeneous-mixing population) have been estimated for the United States, North America and Great Britain; they range from 12.5 to 15 (Hethcote 1983). However, depending on the structure of the population, the rates can be significantly different. Therefore, values for the parameters $\bar{\alpha}_i$, $i = 1, 2, 3$ were selected within a reasonable range of values to provide the best fit to data (minimize the absolute differences). After estimation of $\bar{\alpha}_i$ an average contact rate (corresponding to a homogeneous-mixing model) was calculated for the deterministic model, i.e., the average of $\sum_{i=1}^{3} \alpha_i$ over the course of the epidemic (60 days), and it was found to be ≈ 10, a value which agrees with previous estimates.

Measles is contagious from one to two days before the onset of symptoms (e.g., fever) and from three to five days before the appearance of a rash (Committee on Infectious Diseases 1988). It was assumed that an exposed individual remains in the exposed state from 9 to 11 days and that an infectious individual remains infectious for approximately three days. If the measles vaccine is given within 72 hours after exposure, it can still provide protection against infection (Committee on Infectious Diseases 1988). Therefore, it was assumed that if a susceptible individual is successfully vaccinated (95% efficacy (Immunization Practices Advisory Committee 1989)), or if an individual in the exposed class is successfully vaccinated within three (or four) days after exposure, that individual becomes immune, but if the vaccination is given four (or five) or more days after exposure, the vaccination has no effect. In the deterministic model four days were used and in the stochastic model five days were used.

The state equations for the deterministic model are defined below (similar equations hold for the stochastic model). The removed state, $R_i(t)$, can be determined from equation (??) and is omitted from the state equations.

$$S_i(t+1) = [1 - v_i(t)][S_i(t) - E_i(0, t+1)],$$

$$E_i(\tau+1, t+1) = \begin{cases} [1 - v_i(t)]E_i(\tau, t), & \tau = 0, \ldots, \beta_0, \\ k(\tau)E_i(\tau, t), & \tau = \beta_0 + 1, \ldots, \tau_1 - 1, \end{cases}$$

$$I_i(\tau+1, t+1) = \begin{cases} [1 - k(\tau)]E_i(\tau, t) + r(\tau)I_i(\tau, t), & \tau = 0, \ldots, \tau_1, \\ r(\tau)I_i(\tau, t), & \tau = \tau_1 + 1, \ldots, \tau_2 - 1, \end{cases}$$

where β_0 (3 or 4) is the number of days during the latent period in which the vaccine is effective and $v_i(t)$ is the fraction of susceptible students in dorm i who become immune due to vaccinations on day t (calculated from the data). The functions $k(\tau)$ and $r(\tau)$ are transition probabilities; $k(\tau)$ is the probability that an individual is latent on day $\tau + 1$ given that the individual was latent on day τ, and $r(\tau)$ is the probability that an individual is infectious on day $\tau + 1$ given that the individual was infectious on day τ. Specification of initial and boundary conditions completes the model formulation. The boundary conditions for $E_i(0, t+1)$, $t = 0, \ldots$, the number of newly exposed individuals on day $t + 1$, are specified via the contact rates (discussed previously). The boundary conditions for I_i are $I_i(0, t) = 0$, $t = 0, \ldots$, because it takes more than one day to become infectious. The initial conditions used in the simulations, $S_i(0)$, $E_i(\tau, 0)$, $I_i(\tau, 0)$, and $R_i(0)$, $\tau = 0, \ldots, \tau_2$ are specified in the next section. The model formulation and the functions v_i, k, and r are discussed in more detail in the references (Allen, Jones, and Martin 1991; Allen, Lewis, Martin, and Stamp 1990; Jones 1990; Longini 1986, 1988; Rvachev and Longini 1985).

5 Computer Simulations

For purposes of simulation it was assumed that 6% of the population was susceptible in each dormitory (5.8% infection rate was reported in dormitory 5). In dormitory 8 the fewest number of vaccinations were given (Table 1); therefore, we should expect a relatively high number of cases, but the reverse occurred. Therefore, the proportion susceptible was reduced to .03 in dormitory 8. (There is most likely some variation from the .06 level in all of the dormitories, but in dormitory 8 this variation was most evident.) Since the epidemic began in dormitory 5, the simulation was initiated (initial conditions) by putting four infected students in dormitory 5. Vaccination began on the twenty-fifth day of the simulation (twentieth day since the first reported cases of measles). The three parameters $\bar{\alpha}_1$, $\bar{\alpha}_2$, and $\bar{\alpha}_3$ were estimated by trial and error until the simulated results were in close agreement with the observed results. In the deterministic model the estimates were chosen to minimize the absolute difference between the simulated and the reported results at 80 days (Allen, Jones, and Martin 1991). The results of the simulations (the average of 50 simulations in the stochastic model) are compared with the reported results in Figure 1. (The solid curves represent the simulated results and the dotted curve the reported results.)

The simulations were used to estimate the level of immunity necessary to prevent an epidemic (herd immunity). Vaccinations were eliminated and the rate of immunity was varied from .95 to .99 (all other parameters were kept fixed). The results for each model, number of cases at each rate of immunity (95% to 99%), are given in Table 2 (only one simulation output at each rate of immunity in the stochastic model). An immunization rate above 98% does not generate any new cases in the stochastic model (four students were initially infected); in the deterministic model two cases accumulate over a period of 80 days. This slight discrepancy in the two models is due to the fact that the stochastic output generates integer values, whereas, the deterministic output generates real values. The epidemic continues in the deterministic model with only a small infected population (< 1), but this is impossible in the stochastic model since values < 1 are zero (signifying the end of the epidemic). The simulation results indicate a high level of immunity ($> 98\%$) is required to end the epidemic. This estimate agrees with the estimate obtained from the raw data.

6 Vaccination

The optimal strategy would be to vaccinate all of the students on the first hint of an epidemic. Unfortunately this is not possible because of the expense of the vaccination and because of the difficulty of getting the student to be vaccinated. There are also real limitations as to the number of personnel available to give the vaccine, the number of doses available on any given day and the total number of doses available in the state at any given time. In this project theses limitations are strictly considered.

For this paper the following aspects will be considered:

I. The distribution of the vaccines. In order to analyze this aspect we will keep constant the day when the vaccination begins. It will be the same as in the real epidemic, that is on the 25th day of the simulation. The number of vaccines to be given per dormitory complex will be changed. The following strategies will be tested, considering the aspect above.

a) No vaccination will be given in the dormitory complex where the epidemic started (host dorm) since it spread rapidly through this dormitory complex. Concentrate the vaccines in other dormitories, following the schedule that was actually used.

b) No vaccination will be given in the host dorm. Concentrate the vaccines in the dormitory complexes with fewer cases of infected people.

c) No vaccination will be given in the host dorm. Concentrate the vaccines in the dormitory complexes that have a larger percentage of infected people.

d) Some vaccines will be given in the host dorm and for the other dormitory complexes, strategies a,b, or c will be considered, depending on which one gives the fewest number of total cases.

II. The time when to vaccinate. In order to analyze this aspect, the same strategies a,b,c, and d will be considered as in I, but the vaccination will begin on different days. The following strategies will be tested, considering this second aspect:

a) Begin the vaccination on the 15th day of the simulation, that is ten days before the real case. This is after the first wave of cases of infections. Notice that this strategy will be considered four times (i.e., with each one of the strategies in I).

b) Begin the vaccination after the first cases of infection appear, that is, on the 4th day of the simulation. This strategy will also be considered with each one of the strategies in I.

c) Begin the vaccination after the first cases of infection appear, that is, on the 4th day of the simulation. Increase the number of vaccines for dormitory complex number 2 with respect to strategy d in I.

III. The number of vaccines available. The following strategies will be tested considering this aspect:

a) Consider the 1st day of the vaccination to be as in the real case, the 25th day of the simulation. For this day use the vaccines available for the actual first three days, shift the others up.

b) Consider the strategy that gives the least number of total cases in II. Use the vaccines available for the first three days and shift the others up.

The idea is to measure the effect of each vaccination strategy by considering the total number of cases obtained after the vaccination.

In order to test each vaccination strategy, fifty runs of the simulation program were made for each. An average of the output was taken and used to compare the vaccination strategies.

An interpretation of the results obtained for each one of the vaccination strategies tested by using the simulation program is summarized here.

For Strategy Ia through Id we want to test in which dormitory complexes to concentrate the vaccines. The day the vaccination program starts will be constant and the number of vaccines to be given per dormitory was changed according to the vaccination strategy. It is assumed that slightly more than 94% of the population was immune at the beginning of the epidemic. Thus 6% of the population was susceptible to the measles.

In Strategies Ia, Ib, and Ic, no vaccination was given to dormitory complex 2. This is because the 6% susceptible implies that there were approximately 81 susceptible students in dormitory complex 2. From these 81 susceptible students approximately 67 students were already infected when the vaccination program started.

Strategy Ia. The vaccination program starts on the 25th day of the simulation, that is 25 days after the epidemic began. By this time the epidemic has already spread. In the dormitory complex where the epidemic started (dorm 2), it spread rapidly. Thus, no vaccines are given to dormitory 2 and instead they are used for other dormitories but following the schedule that was actually used.

Using all the vaccines available (5895), a total of 356 students became immune due to the vaccination, slightly more than 25% of the susceptible students. And there was a total of 189 cases (see Table 3.1), which is 13.4% of the susceptible students.

In order to reduce the cost of the epidemic, the number of cases must decrease. The worst case had 217 cases and the best had 156 cases.

Strategy Ib. As stated before, the starting day for the vaccination program is the same, the 25th day of the simulation. No vaccines are given to dormitory complex 2. The vaccines are concentrated in the dormitories with fewer numbers of cases.

Using all the vaccines available (5895 vaccines) with this vaccination strategy, 303 students became immune due to this vaccination, that is 21.49% of the susceptible students. But more important for the purposes of this study is the number of cases that needs to be reduced. Using this strategy, there were 187 cases, which corresponds to a 13.26% of the susceptible students.

Using this strategy we obtain slightly better results than the real case, even though the worst case of the fifty runs of the simulation (i.e., the one with more number of cases) falls above the real case, with 218 cases.

Strategy Ic. The vaccination program also starts on the 25th day of the simulation. No vaccines are given to dormitory complex 2. The vaccines are distributed proportionally to the percentage of cases in each dormitory complex, except for dormitory complex 2.

Using the 5895 vaccines with this vaccination strategy, a total of 302 students became immune due to the vaccination, that is 21.42% of the susceptible students, and a total of 173 cases, which is 12.27% of the susceptible students.

The best case with 143 cases and the worst with 207 cases.

Comparing Strategies Ia, Ib, and Ic, Strategy Ic is the one that gives the fewest number of cases.

Strategy Id. For this vaccination strategy the same criterion will be used for the distribution of the vaccines as in Strategy Ic, since it gives the fewest number of cases. The only difference will be that this strategy will consider the distribution of some vaccines to dormitory complex 2 (where the epidemic started).

It will be interesting to see if it is beneficial to give some vaccines in the dormitory where the epidemic started, especially considering that by the time the vaccination program started, approximately 83% of the susceptible students (i.e, 67 out of 81 students) were already infected.

The day when the vaccination program starts stays constant. Some vaccines are given to the host dorm and the distribution given in Strategy Ic will be considered for the other dormitory complexes (i.e., more vaccines to the dormitories with higher percentage of infected people).

278 vaccines were given in dormitory complex 2 and only 2 students out of the 14 susceptible became immune due to that vaccination. There was a total of 75 cases in dormitory complex 2, only 2 less than the number of cases considering no vaccination for that dormitory.

In order to give some vaccines in dormitory complex 2, the number of vaccines in other dormitories needed to be reduced. This fact made the number of cases increase in some dormitory complexes.

Considering the totals, there were 175 cases, that is 12.41% of the susceptible students. Of the fifty runs of the simulation, the best case was the one with 134 cases and the worst with 205 cases.

It seems that giving some vaccines to dormitory complex 2 after the epidemic is already spread, may reduce the number of cases in that dormitory (in a very small percentage) but increases the number of cases in other dormitories. In general, it increases the total number of cases.

If the number of vaccines to be given in dormitory complex 2 increases, then the number of vaccines available for the other dormitories decreases, increasing in this way the total number of cases.

We see that if the number of vaccines given in dormitory complex 2 increases (from 278 to 408, that is 47%) the number of cases in that dormitory stays at 75. But the number of cases in the dormitories where the vaccines were reduced increases; thus the total number of cases increases. This is what actually happened on the Texas Tech University campus. More vaccines were given in dormitory complex 2 (Table 1.1) and so the total number of cases was 198.

Strategies IIa, IIb, IIc, IId. The same vaccine distribution used in Strategies Ia, Ib, Ic and Id will be used for Strategies IIa, IIb, IIc, and IId, respectively. The only difference will be the day when the vaccination program begins. Instead of beginning on the 25th day of the simulation program, the vaccination program will begin on the 15th day of the simulation. This is after the first wave of cases.

Comparing the results for these strategies with the results for Strategies Ia, Ib, Ic and Id it can be seen very clearly that the number of cases has been reduced. Notice that Strategy IIa considers no vaccination in dormitory complex 2 and for the other dormitories uses the same distribution used for the real epidemic. Strategy IIb considers no vaccination in dormitory complex 2 and concentrates the vaccines in the dormitory complexes with fewer cases. Strategy IIc considers no vaccination in dormitory complex 2 and concentrates the vaccines in the dormitory complexes with more percentage of infected people. Finally, Strategy IId considers some vaccines in dormitory complex 2 and for the other dormitories uses the same vaccine distribution as in Strategy IIc. For all these four strategies the day when the vaccination program starts is on the 15th day of the simulation.

Using Strategy IId, the total number of cases is reduced by a 48% (i.e., from 198 cases to 102 cases, (see Table 3.9)). Notice that if the vaccination program starts earlier, then it is beneficial to give some vaccines in dormitory complex 2.

Using Strategy IId, 278 vaccines were given in dormitory complex 2 (the same number of vaccines were given when using Strategy Id) and the number of cases decreased by 12% with respect to the number of cases obtained between Strategy Id and IId is the day when the vaccination program begins.

Strategies IIIa, IIIb, IIIc, IIId. The same vaccine distribution used in Strategies IIa, IIb, IIc, and IId will be used for Strategies IIIa, IIIb, IIIc, and IIId, respectively. The only difference is the day when the vaccination program starts. This time, the vaccination program begins on the fourth day of the simulation, that is, after the first cases of infection appear.

The distribution of the vaccines, the results obtained after testing each strategy and the corresponding graphs are presented in figures and tables at the end of this chapter and in the Appendix.

If the day when the vaccination program begins is early enough, like for these strategies on the fourth day of the simulation of the epidemic (21 days before than in the real case) then the effect of the vaccines given in dormitory complex 2 is better. Also, it is better for the other dormitories.

Using Strategy IIId, (the one that considers giving some vaccines in dormitory complex 2 and for the other dormitory complexes it concentrates the vaccines in the dormitories with a higher percentage of infected people), the total number of cases decreases from 198 cases to 71 cases, that is 64%.

Strategy IV. From the results obtained by using Strategy IIId, it seems that if the number of vaccines to be given in dormitory complex 2 increases, then the number of cases will decrease even more.

Strategy IV considers the beginning of the vaccination program on the fourth day of the simulation. The number of vaccines to be given in dormitory complex 2 will be increased from 278 (considered in Strategy IIId) to 408.

Using this strategy, the total number of cases is reduced to 60 cases, with respect to the real epidemic, that is 70%. The effect of the vaccine is better, the sooner the vaccination program begins.

Strategy V. This strategy considers the first day of the vaccination to be on the 25th day of the simulation. For this day it uses the vaccines available for the actual first three days, that is, in stated of 1399 vaccines, 3526 vaccines will be distributed among the students on the day the vaccination program begins. On the second day of the vaccination, the vaccines that were available on the fourth day of the vaccination for the real case will be used. In other words, the other vaccines will be shifted up.

The same distribution as in the real epidemic is used. With this strategy, the possibility of decreasing the total number of cases by increasing the number of vaccines to be used on the first day of the vaccination will be tested.

These results were analyzed using SAS for significance and it was concluded that the only statistically significant differences occurred when the vaccinations were begun early. In fact, every day that the vaccination is moved forward makes a very significant difference in the total number of cases in the epidemic.

7 Summary

It appears that the present level of immunity obtained with one vaccination is insufficient to prevent measles epidemics. Hopefully, the recommendation by the ACIP (Immunization Practices Advisory Committee 1989) of two measles vaccines before entry into school will increase the percentage immune to a sufficiently high level to prevent measles outbreaks. An estimate of herd immunity in various population settings will be useful in determining whether the present strategy (if achieved) will eliminate measles epidemics in these settings.

A number of simplifying assumptions were made in this measles model. Although many of the assumptions appear reasonable, they require further validation and verification from other investigations. Our analysis raises some questions. Is the assumption of nonhomogeneous-mixing a valid one for a measles epidemic in a university setting? Is behavior modification (or other factors) an important consideration in modeling social contacts? How does the structure of the population affect the spread of the epidemic? These and other important questions which have been addressed in other model simulations (Ackerman, Elveback, and Fox 1984) require further investigation to justify the model assumptions and the conclusions derived from them.

References

[1] Ackerman, E., Elveback, L. R., and Fox, J. P. (1984), *Simulation of Infectious Disease Epidemics*, Springfield, Illinois: Charles C. Thomas Publisher.

[2] Allen, L. J. S., Jones, M. A., and Martin, C. F. (1991), "A Discrete-Time Model with Vaccination For a Measles Epidemic," *Mathematical Biosciences*, 105, 111-131.

[3] Allen, L., Lewis, T., Martin, C. F., and Stamp, M. (1990), "A Mathematical Analysis and Simulation of a Localized Measles Epidemic," *Applied Mathematics and Computation*, 39, 61-77.

[4] Anderson, R. M., and May, R. M. (1979), "Population Biology of Infectious Diseases: Part I," *Nature*, 280, 361-367.

[5] Bailey, N. J. T. (1975), *The Mathematical Theory of Infectious Diseases*, (2nd ed.) New York: Macmillan.

[6] Canfield, M. (1989), "Measles Epidemic in Houston/Harris County, 1988-89: A Perspective From the Harris County Health Department," *Texas Preventable Disease News*, October 21, 1989, 49, 1-4.

[7] Carpio, R. "Intervention stratigies for local epidemics", Masters Thesis, Texas Tech University, 1991.

[8] Centers for Disease Control. (1990), "Measles–United States, 1989 and First 20 Weeks 1990," *Morbidity and Mortality Weekly Report*, 39, 353-363.

[9] Committee on Infectious Diseases. (1988), *Report of the Committee on Infectious Diseases*, 21st ed., American Academy of Pediatrics.

[10] Greenhalgh, D. (1990), "Vaccination Campaigns for Common Childhood Diseases," *Mathematical Biosciences*, 100, 201-240.

[11] Gross, A. J., and Clark, V. A. (1975), *Survival Distributions: Reliability Applications in the Biomedical Sciences*, New York: John Wiley & Sons.

[12] Hethcote, H. W. (1976), "Qualitative Analyses for Communicable Disease Models," *Mathematical Biosciences*, 28, 335-356.

[13] Hethcote, H. W. (1983), "Measles and Rubella in the United States," *American Journal of Epidemiology*, 117, 2-13.

[14] Hethcote, H. W. (1988), "Optimal Ages of Vaccination for Measles," *Mathematical Biosciences*, 89, 29-52.

[15] Hethcote, H. W., and Yorke, J. A. (1984), *Gonorrhea Transmission Dynamics and Control*, Lecture Notes in Biomathematics, Berlin: Springer-Verlag.

[16] Hoppensteadt, F. (1975), *Mathematical Theories of Populations: Demographics, Genetics and Epidemics*, Regional Conference Series in Applied Mathematics, Philadelphia, Penn.: SIAM.

[17] Immunization Practices Advisory Committee. (1989), "Measles Prevention: Recommendations of the Immunization Practices Advisory Committee," *Morbidity and Mortality Weekly Report*, 38(no.S-9).

[18] Jones, M. A. (1990), "A Deterministic Mathematical Model of a Measles Epidemic, " Master's Report, Department of Mathematics, Texas Tech University.

[19] Landwehr, J. M., Pregibon, D., and Shoemaker, A. C. (1984), "Graphical methods for assessing logistic regression models," *Journal of American Statistical Association*, 79, 61-74.

[20] Lo, C. K. (1989), "The Statistical Data Analysis of the 1989 Measles Outbreak on the Texas Tech Campus," Master's Report, Department of Mathematics, Texas Tech University.

[21] Longini, Jr., I. M. (1986), "The Generalized Discrete-Time Epidemic Model with Immunity: a Synthesis," *Mathematical Biosciences*, 82, 19-41.

[22] Longini, Jr., I. M. (1988), "A Mathematical Model for Predicting the Geographic Spread of New Infectious Agents," *Mathematical Biosciences*, 90, 367-383.

[23] May, R. M. (1986), "Population Biology of Microparasitic Infections", In: *Biomathematics*, Vol. 17, T. G. Hallam and S. A. Levin (Eds.) pp. 405-442. Berlin: Springer-Verlag.

[24] McLean, A. R., and Anderson, R. M. (1988), "Measles in Developing Countries. Part II. The Predicted Impact of Mass Vaccination," *Epidemiol. Infect.* , 100, 419-442.

[25] Rvachev, L. A., and Longini, Jr, I. M. (1985) "A Mathematical Model for the Global Spread of Influenza," *Mathematical Biosciences*, 75, 3-22.

[26] Waltman, P. (1974), *Deterministic Threshold Models in the Theory of Epidemics*, Lecture Notes in Biomathematics, Berlin: Springer-Verlag.

Dorm	Population Size	Proportion Infected f_I	Proportion Vaccinated f_V (2/1/89-2/7/89)	Total Proportion Vaccinated	Proportion Removed f_R
1	628	.022	.443	.467	.989
2	420	.019	.488	.507	.988
3	354	.025	.381	.395	.988
4	477	.010	.413	.470	.975
5	1355	.058	.409	.422	——
6	670	.010	.481	.500	.984
7	633	.027	.395	.425	.991
8	867	.008	.249	.273	.963
9	144	.007	.514	.542	.978
10	55	.018	.354	.379	.979

Table 1: Estimate Of The Proportion Immune At The End Of The Epidemic (Dormitory 5 is not included since most of the cases in dormitory 5 occurred prior to the commencement of the vaccination program.)

Percent Immunity	Stochastic		Deterministic	
	Total	On Campus	Total	On Campus
95	252	223	276	234
96	160	135	168	151
97	55	53	73	69
98	19	17	23	22
99	4	4	6	6

Table 2: Number Of Cases After 80 Days For Each of The Models With Various Immunity Rates And No Vaccinations

Self-organizing Behavior in a Simple Controlled Dynamical System

J. Baillieul *

Abstract: A standard paradigm in control theory involves the use of feedback to change the dynamics of a system in some significant way. In the language of Willems ([9]), this paradigm prescribes the use of feedback to create desirable "behavior" in the system. There is growing interest (see e.g. [8]) in exploring an alternative paradigm applied to the control of systems in which there is a set of various behaviors pre-existing within the natural (uncontrolled) dynamics of the system, and wherein control acts in a minimalistic way to entrain a mode of behavior chosen from this set. We shall explore the latter in the context of some mechanical systems in which the control is only allowed to act intermittently. The systems we look at involve the controlled one dimensional scattering of a certain number of particles. In the absence of control, the systems are similar to the Toda lattices that have been considered by Moser ([7]) and others. We introduce boundary controls and confine our analysis to two classes of open loop controls—roughly corresponding to constant and periodic forcing. For the constant controls, the set of possible behaviors is easily described using fixed point analysis. For periodic forcing, on the other hand, the behavior set is very rich, and is modeled as the dynamics of an iterated 2-d mapping. Results on the stability and bifurcations of periodic orbits are given.

1 Introduction

This paper describes the dynamics of a system of colliding particles whose motions will be controlled by varying the position, velocity, and inertia of a "racquet" at the boundary of a domain to which the particles are confined. While the system is extremely simple, the dynamics are nevertheless nonlinear, and there is a rich set of natural behaviors which can be produced by appropriate control actions.

The motivation for studying this system comes in part from a toy recently shown to the author by one of his children. The toy contains several hard elastic spheres which are constrained to move so

---

*The author gratefully acknowledges the support of the U. S. Air Force through grant AFOSR-90-0226. Thanks are also extended to Professor S.K. Mitter for hospitality shown during the author's recent sabbatical visit to MIT.

as to guarantee collisions. The object of play is to produce any of several immediately perceivable patterns of collisions. A broader motivation is to understand the general problem of controlling complex systems for which any controlling agent must learn how to influence the internal dynamics so as to elicit desired modes of behavior. The objective is non-classical in the sense that we wish to use feedback only to change the observed behavior of the system but not the qualitative nature of the system itself. For the system of colliding particles studied below, we find a large number of possible modes of behavior, but our understanding of how these depend on the control variables is not yet complete. Before a mature behavior-oriented approach to nonlinear control theory can be applied to such systems, it will be necessary to develop a more complete stability and bifurcation theory of the open-loop dynamics of prototypical systems. These objectives are currently being pursued in conjunction with controlling the gait of walking robots (c.f. [3]), controlling juggling ([5] and [6]), and controlling kinematic chains in which the number of degrees of freedom exceeds the number of actuators which are directly controlled. (See [1] and [2].)

While there is not presently a large body of research literature on the behavior-oriented control of nonlinear systems (c.f. the linear theory proposed in [9]), recent work by Ott *et al.* ([8]) has been directed somewhat along these lines. We believe that the existence or non-existence of chaos in the dynamics we are trying to control is perhaps a red herring, however, since we only need a rich set of natural dynamics, not necessarily fully developed chaos, to make the behavior oriented approach to control interesting. Also, we wish to avoid chaotic transients in switching from one stable mode of behavior to another.

The paper is organized as follows. In the next section, we review some elementary facts regarding the dynamics of elastic collisions. Section 3 describes the control problem to be studied, and discusses the system's response to steady state forcing. Section 4 presents preliminary results on the system's response to periodic forcing. Concluding remarks are given which summarize simulations showing the co-existence of distinct stable modes of behavior. While it must be pointed out that we only treat the response of the system to open loop forcing, simulations indicate that the set of possible responses is extremely rich, and further study of prescriptive control strategies seems warranted.

2 Preliminaries on the dynamics of elastic collisions

Before describing the system in detail, it is useful to recall some elementary facts about elastic collisions, and in so doing, it is important to distinguish between collisions involving two and collisions involving simultaneously more than two particles. Consider first two particles of mass m_1 and m_2 constrained to move without friction on a line. In an elastic collision, both total momentum and energy are conserved. Suppose the particles in our system have velocities v_1^i and v_2^i just prior to colliding and v_1^f and v_2^f just after colliding. Because momentum $\mathcal{M}$ and energy $\mathcal{E}$ are conserved, both the initial velocity pair, (v_1^i, v_2^i), and final velocity pair, (v_1^f, v_2^f), simultaneously satisfy the equations

$$m_1 v_1 + m_2 v_2 = \mathcal{M} \tag{1}$$

$$m_1 v_1^2 + m_2 v_2^2 = 2\mathcal{E} \qquad (2)$$

It is easy to see that there are precisely two points in the (v_1, v_2)-plane at which the momentum locus intersects the energy locus. Thus if the initial velocity pair (v_1^i, v_2^i) is given, it defines values $\mathcal{M}$ and $\mathcal{E}$, and from (1)-(2) we find the post collision velocity pair (v_1^f, v_2^f). On the other hand, if the same pair of particles were to collide with initial velocities (v_1^f, v_2^f), the post collision velocity pair would be (v_1^i, v_2^i). The implication of this observation is that the pairs $\vec{v}^i = (v_1^i, v_2^i)$ and $\vec{v}^f = (v_1^f, v_2^f)$ are related by an idempotent matrix. I.e., there is a 2×2 matrix A such that

$$\vec{v}^f = A\vec{v}^i, \qquad (3)$$

and $A^2 = I$. We have the following explicit characterization.

Lemma 2.1 *Suppose that two particles of mass m_1 and m_2 slide without friction along an infinitely long linear tract in the absence of exogenous forces. If the particles undergo an elastic collision and if the pre- and post- collision velocities are $\vec{v}^i = (v_1^i, v_2^i)$ and $\vec{v}^f = (v_1^f, v_2^f)$ respectively, then these are related by (3) where*

$$A = \begin{pmatrix} \frac{m_1-m_2}{m_1+m_2} & \frac{2m_2}{m_1+m_2} \\ \frac{2m_1}{m_1+m_2} & \frac{m_2-m_1}{m_1+m_2} \end{pmatrix}.$$

Proof: That such a matrix A exists follows from our above remarks. That it has this particular form follows from a simple calculation involving the conservation laws (1)-(2). $\square$

Conservation of momentum and energy does not suffice to characterize elastic collisions simultaneously involving three or more particles. The post collision distribution of velocities in such a collision will depend not only on the pre-collision velocities but also on the relative amounts of time each particle spends in contact with the others. Such collisions are not easily analyzed, and the velocity transition must generally be determined by integrating the actual equations of motion.

3 The control of colliding particles in 1-dimension

The dynamics of the system to be studied are described in terms of Figure 1. It consists of a frictionless line or track along which n particles of unit mass may slide. The particles undergo elastic collisions, and they are confined to move between two barriers with which elastic collisions also occur. We assume that the left hand barrier is fixed, but that the right hand barrier moves and functions as a racquet which strikes the particles to influence their motions. The effect of the racquet striking a particle is described by a scattering law of the form (3). More specifically, suppose the racquet has mass M and velocity v_r^i just prior to striking the particle which has unit mass and

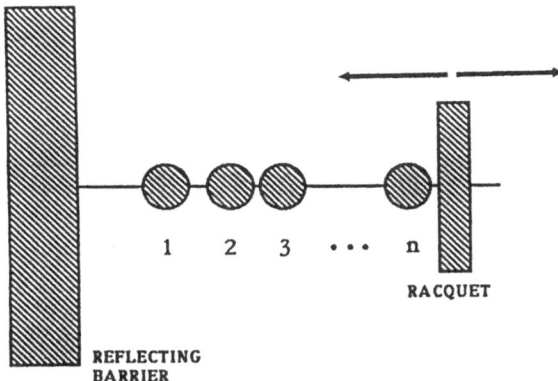

Figure 1: Particles of unit mass slide without friction in one dimension between a reflecting barrier and a racquet which may be moved left and right.

is moving at pre-collision velocity v_p^i. Then the post collision velocity of the particle and racquet are respectively v_p^f and v_r^f given by the formula

$$\begin{pmatrix} v_p^f \\ v_r^f \end{pmatrix} = \begin{pmatrix} \frac{1-M}{M+1} & \frac{2M}{M+1} \\ \frac{2}{M+1} & \frac{M-1}{M+1} \end{pmatrix} \begin{pmatrix} v_p^i \\ v_r^i \end{pmatrix}. \tag{4}$$

The control objective which we pursue for this system is to program sequences of racquet strikes to elicit prescribed stable patterns of motion among the particles. The principal results of this and the next section of the paper are to characterize several achievable patterns. Since we assume control actions are effected by moving the racquet left and right, we must in principle incorporate the dynamics of the racquet into any control strategy. We shall ignore the details of the racquet's dynamics, however, and assume that it may be moved as fast as necessary to any point of the line to strike with velocity v_r. In the next section, we shall also study the dynamics of the racquet and particles system under the assumption that the racquet inertia M may be varied for the purpose of controlling the system. This type of control is used by tennis players who change the effective inertia of a tennis racquet by adjusting their grip to be a greater or lesser distance from the racquet head.

Virtual Particles. If there is more than one particle in our system as depicted in Figure 1, then there are two equivalent ways to view the motions. Physically, the particles remain ordered from left to right, and they constantly exchange velocities by means of elastic collisions. In effect, however, because the particles all have unit mass, each pairwise collision simply results in an exact exchange of velocities. Hence, to describe the overall motion of the system, it is possible to label the particles according to their velocities. From this point of view, the particles "pass through" each other whenever a pairwise collision occurs. We shall refer to these velocity-labeled particles as *virtual particles*. It is convenient to study the motions of the virtual particles since their velocities only change at the reflecting barrier and when they are struck by the racquet.

As mentioned in the preceding section, velocity exchanges resulting from collisions of three or more particles being struck simultaneously by the racquet do not obey a simple algebraic scattering law of the form (3). The following definition thus highlights an important distinction.

Definition 3.1 Collisions simultaneously involving two particles are called *simple*. A racquet strike involving one particle is called a *simple racquet strike*. Collisions simultaneously involving more than two particles and racquet strikes simultaneously involving more than one particle are called *non-simple*.

The first set of controlled behaviors which we wish to study are responses to what may be thought of as constant inputs for the system. Suppose the n particles have initial velocity distribution $v_1^0, \ldots, v_n^0$. We choose coordinates, denoted, say, by x, for the axis along which the racquet and particles move such that the reflecting barrier is located at $x = 0$. Although the racquet may be moved left or right to strike the left-most particle at any point along its trajectory, the nominal rest position of the racquet will be $x = 1$ in our chosen coordinate system. We implement the following simple

CONTROL LAW: the rightmost particle is struck by the racquet moving at velocity $-v$ each time the particle crosses the position $x = 1$ moving in a left to right (positive) direction.

The following characterizes the behavior of the particle motions under this law.

Theorem 3.1 *Assuming that only simple racquet strikes and collisions occur, each virtual particle which is initially in motion will approach a steady state speed Mv under the above control law.*

Proof: Clearly the theorem will hold if it holds for a single particle system. We keep track of the particle's motion by listing its speed just prior to each racquet strike. Under the law (4), the evolution of these speeds is given by

$$v_p^{k+1} = \left(\frac{M-1}{M+1}\right)v_p^k + \frac{2M}{M+1}v.$$

This mapping has a fixed point: $v_p = Mv$, and this proves the theorem. □

Remark 3.1 In steady state under the above control law, there is an exact momentum exchange between the racquet and particle at each strike.

The following theorem shows that generically only simple collisions and racquet strikes occur.

Theorem 3.2 *For a generic set of initial particle velocities $v_1^0, \ldots, v_n^0$ and racquet inertias M, each virtual particle which is initially in motion will approach a steady state speed Mv under the control law of the previous theorem.*

Proof: We prove this theorem in two parts. First we show that if the speeds evolve to within a prescribed threshold of their claimed steady state values, then unless the initial velocities and M have a functional dependency, no two particles approach each other closer than a certain positive distance. It will then be noted that for a generic choice of initial velocities and racquet inertias, no non-simple collisions or racquet strikes occur before the particle speeds have gotten to within the necessary threshold of their steady state values.

To carry out the argument, we shall want to analyze the trajectories of a typical pair of particles. It will be shown that the pair remains separated by a positive distance asymptotically except for regular encounters as directions are reversed at the reflecting barrier and racquet. To eliminate such irrelevant encounters from consideration, we unfold particle motions to occur on a doubly infinite line which is subdivided into subintervals of length two. We lift the dynamics to this unfolded domain by stipulating that all particle motions will be from left to right on the infinite line. The effect of the racquet is modeled by having the particles undergo velocity transitions according to the law

$$v_p \mapsto \left(\frac{M-1}{M+1}\right)v_p - \frac{2M}{M+1}v$$

each time the particle transits one of the length two subintervals.

Consider now the motion of two particles. Assume, without loss of generality, that the left hand particle is slower than the right. (Because the particles are actually moving in a compact interval we may always arrange them such that the slower is on the left in the "unfolded" model.) Suppose also that the initial conditions are such that $x_1(0)$ is a transit point: that is to say the velocity of the left-hand particle has just undergone a transition. Let T denote the amount of time required for the faster particle to subsequently cross the point of velocity transition. In the instants of time immediately after the faster particle has undergone a transition, the positions of the two particles are given by

$$x_1(t) = v_1 \cdot (t + T) + x_1(0), \quad \text{and}$$

$$x_2(t) = \frac{M-1}{M+1}v_2 - \frac{2M}{M+1}v + v_2 \cdot T + x_2(0).$$

The next velocity transition for the slow particle occurs at t satisfying $v_1 \cdot (t + T) = -2$. At this point, we have

$$x_2(t) = \left(-\frac{2}{v_1} - T\right)\left(\frac{M-1}{M+1}v_2 - \frac{2M}{M+1}v\right) + v_2 T + x_2(0).$$

Since T may be expressed in terms of $x_1(0)$, $x_2(0)$, and v_2, we may rewrite this formula as

$$x_2(t) = (\frac{-2}{v_1} - \frac{x_1(0) - x_2(0)}{v_2})(\frac{M-1}{M+1}v_2 - \frac{2M}{M+1}v) + x_1(0).$$

Since $x_1(t) = x_1(0) - 2$, the relative distance between x_2 and x_1 at the next velocity transition for the slow particle, x_1, is

$$x_2(t) - x_1(t) = (\frac{-2}{v_1} - \frac{x_1(0) - x_2(0)}{v_2})(\frac{M-1}{M+1}v_2 - \frac{2M}{M+1}v) + 2.$$

Stated in slightly different notation, suppose x_1^k and x_2^k are the respective positions of the particles when the k-th velocity transition of the slow particle, x_1, occurs, then

$$x_2^{k+1} - x_1^{k+1} = 2(\frac{v_1^k - v_2^{k+1}}{v_1^k}) + \frac{v_2^{k+1}}{v_2^k}(x_2^k - x_1^k),$$

where

$$v_i^{k+1} = (\frac{M-1}{M+1})v_i^k - \frac{2M}{M+1}v.$$

From this we may write

$$x_2^{k+1} - x_1^{k+1} = 2\left[\frac{v_2^{k+1}}{v_2^{k+1}}\left(\frac{v_1^k - v_2^{k+1}}{v_1^k}\right) + \frac{v_2^{k+1}}{v_2^k}\left(\frac{v_1^{k-1} - v_2^k}{v_1^{k-1}}\right) + \frac{v_2^{k+1}}{v_2^{k-1}}\left(\frac{v_1^{k-2} - v_2^{k-1}}{v_1^{k-2}}\right) + \cdots + \frac{v_2^{k+1}}{v_2^1}\left(\frac{v_1^0 - v_2^1}{v_1^0}\right)\right]$$

$$+ \frac{v_2^{k+1}}{v_2^0}(x_2^0 - x_1^0). \tag{5}$$

Now barring multiple collisions, we have $lim_{k\to\infty} v_i^k = -Mv$. Hence suppose we take as our initial velocities $v_1^0 = -Mv + \epsilon + \delta$ and $v_2^0 = -Mv + \epsilon$, where ϵ and δ are small positive numbers. A typical term in the above sum is

$$v_2^{k+1}\left(\frac{v_1^{j-1} - v_2^j}{v_2^j v_1^{j-1}}\right).$$

Now if we write $v_i^j = -Mv + \eta$, then

$$v_i^{j+1} = (\frac{M-1}{M+1})v_i^j - \frac{2M}{M+1}v = -Mv + \frac{M-1}{M+1}\eta.$$

Hence a typical term in the sum is the product of

$$-Mv + (\frac{M-1}{M+1})^{k+1}\epsilon$$

and

$$\frac{(\frac{M-1}{M+1})^{j-1}(\delta + \frac{2}{M+1}\epsilon)}{(-mv + (\frac{M-1}{M+1})^j\epsilon)(-mv + (\frac{M-1}{M+1})^{j-1}(\epsilon + \delta))}.$$

These sums are bounded and monotonic and thus clearly form a convergent sequence as k tends to ∞. Hence the relative distance (5) approaches a finite limit d as $k \to \infty$ and this will generally not be equal to zero.

It remains to note that our assumption that the initial velocities were close to the limit $-Mv$ is consistent with the evolution of the system from a generic set of initial velocities. This follows from our discussion of single particle dynamics and the observation that the only way the particles could fail to approach this limit would be if non-simple racquet strikes repeatedly occurred. But the occurrence of a non-simple racquet strike of two or more particles imposes algebraic conditions on the initial position and velocity data and system parameters. Thus, we find it is generic that non-simple racquet strikes do not occur in any pre-specified finite interval of time. Since without non-simple racquet strikes the velocities will get close to the limiting value $-Mv$ in a finite time (which is easily computed for any initial conditions), we see that our assumption holds generically. $\square$

Remark 3.2 Since (generically) any virtual particle having nonzero initial velocity will tend toward steady state speed Mv, while any virtual particle which is initially at rest will remain at rest, it follows from the previous theorem that there are precisely $n+1$ constant-velocity behaviors for our system. These may be indexed by either the number k of particles in motion or the complementary number $n-k$ which are at rest. We note that with even the smallest initial velocity, a virtual particle will eventually approach steady state speed Mv, and hence the only constant-velocity steady state behavior which is stable is the one in which all particles are moving with speed Mv.

Remark 3.3 Suppose we slightly deform the track in our racquet and particle system so that it has a parabolic shape with minimum at $x = 1/2$. This adds a gravitational force which is felt by the moving particles and changes the stability characteristics described above. All virtual particles which are initially sufficiently close to $x = 1/2$ and which have initial velocities too small to make it up the potential well to the strike-point $x = 1$ will remain near the minimum of the potential $x = 1/2$ for all time. On the other hand, any virtual particle which is moving fast enough initially to get to $x = 1$ will be forced toward the steady state speed Mv. Thus for this modified system, there are $n+1$ constant-velocity motions counted, as in the previous remark, according to the number of virtual particles which are in motion. Each of these motions is stable in the sense that if the rest particles (point masses lying at the minimum of the potential) are perturbed slightly and the velocities deviate slightly from 0, the phase portrait of each rest particle remains in a neighborhood of the point $(x, v_p) = (1/2, 0)$, and those particles moving near the speed Mv tend toward this speed as time evolves.

Remark 3.4 The coexistence of a number of stable modes of behavior as described in the previous remark is an important feature which will also be noted for the periodic racquet motions discussed in the next section. Adding a small amount of friction to the particle motions will result in the steady state motions we have described being asymptotically stable. By making slight planned deviations from the control we have prescribed, it is possible to use the racquet strikes to move the system among the various domains of attraction. Feedback control strategies along these lines will be discussed elsewhere.

4 The response to a periodically moving racquet

Under the control law studied in the preceding section, the dynamics of the individual virtual particles are decoupled, and hence our investigation is reduced to the study of certain iterated scalar mappings. By contrast, even when there is only a single particle in our system, the response to periodic motions of the racquet is governed by two dimensional dynamics. This is because characterization of the response must be given not only in terms of the velocity transitions that occur but also how these are synchronized with the period of the racquet. We again explicitly describe the open loop control law (=periodic racquet motion) we shall study.

PERIODIC CONTROL LAW: The racquet moves back and forth between the positions $x = 1$ and $x = 2$ in the saw-tooth wave form:

$$v(t) = \begin{cases} v & 0 \leq t < h \\ -v & h \leq t < 2h \\ v(t) = v(t \pm 2h) & \text{otherwise} \end{cases}.$$

Note that because we have fixed the amplitude of the racquet's motion, the period and velocity are related by $vh = 1$. The remainder of this section will be devoted to obtaining an understanding of the dynamical response of our system of particles to this forcing.

For the moment, assume there is a single particle. As it moves, it will eventually collide with the racquet when it is moving either left or right. The velocity is changed according to the scattering law (4). To keep track of the particle dynamics over the course of many racquet strikes, as in the preceding section, is is useful to record the particle velocities just prior to each racquet strike. The rules describing the evolution of these quantities will depend on the particle path as it leaves the racquet:

TRANSITION RULES

Type (i): $v_p \mapsto \left(\frac{1-M}{M+1}\right)v_p + \frac{2m}{M+1}v$

  if the particle meets the racquet moving to the right and will again be struck by the racquet on its next left-stroke;

Type (ii): $v_p \mapsto \left(\frac{M-1}{M+1}\right)v_p - \frac{2m}{M+1}v$

  if the particle meets the racquet moving to the right and will next encounter the reflecting barrier before being struck by the racquet again;

Type (iii): $v_p \mapsto (\frac{M-1}{M+1})v_p + \frac{2m}{M+1}v$

if the particle meets the racquet when it is moving left.

For fixed M and v it is somewhat more complicated than what was done in the preceding section to keep track of the evolution of these velocities.

As in the preceding section, the only control variables which may be employed to modify the system's behavior within the structure of the assumed law are the racquet inertia M and the racquet velocity v. In Section 3, varying either M or v could change the steady state velocity of the system, but the dynamics remained qualitatively unchanged for all choices of M and v (both positive). It is not difficult to see that the qualitative response to the periodic control law described above is also unchanged as we vary v. This follows from writing out the explicit dependence of the particle velocity in terms of successive velocity transitions of the above form. As the number of transitions becomes large it is clear that the particle velocity v_p ceases to depend on the initial velocity, and under any change in racquet speed $v \mapsto \alpha v$, we shall have the particle velocity at any instant in time scaled by the same factor. Since the period of the racquet motion will be scaled inversely (i.e. $h \mapsto h/\alpha$ because we keep the amplitude of the racquet trajectory fixed), it follows that the motions of the racquet and particle remain unchanged except for the combined system speeding up or slowing down. Qualitative changes in the particle motion are produced by varying M, however, and it is these changes which we now summarize.

Theorem 4.1 *For $1 < M < 2$, the periodic control law above leads to an asymptotically stable periodic orbit of period $2h$ for the particle.*

Proof: The proof uses a construction which will be more broadly useful than this theorem alone. We note that given any time $0 \le t_0 < 2h$ and particle velocity v_p measured just prior to a racquet strike at t_0, it is possible to write down the next recorded velocity $\bar{v}_p$ (in terms of the velocity transition mappings above) and the time t when the next racquet strike occurs. This defines a function $\bar{F}(v_p, t_o) = (\bar{v}_p, t)$. This is the key to our proof, since we show there is an orbit as claimed by showing that the function $\bar{F}(v_p, t_0) - \begin{pmatrix} 0 \\ 2h \end{pmatrix}$ has a stable fixed point. It is not difficult to show that the only possible velocity value that could be a component of this fixed point is itself a fixed point of the third velocity transition

$$v_p \mapsto (\frac{M-1}{M+1})v_p + \frac{2M}{M+1}v,$$

which is obviously Mv. The t-component of $\bar{F}(v_p, t_0)$ may be explicitly written in this case as

$$t = \frac{(M-1)(v_p + v)t_0 + 8(M+1)}{(M-1)v_p + (3M+1)v}$$

Substituting the steady state velocity Mv for v_p, we wish to solve

$$t = (\frac{M-1}{M+1})t_0 + \frac{8}{M+1)v}$$

and

$$t = t_0 + 2h = t_0 + 2/v$$

for t_0 in the interval $h \leq t_0 < 2h$. Solving, we find $t_0 = \frac{3-M}{v}$, and this will be in the desired range precisely when $1 < M \leq 2$. This shows the period $2h$ orbit exists under the conditions of the formula. To show that it is stable, we need to evaluate the derivative of $\bar{F}$ at the fixed point. This may be done by the obvious explicit calculation in this case, and we find that this derivative has repeated eigenvalues $\frac{M-1}{M+1}$, $\frac{M-1}{M+1}$ at the fixed point. Since these are both strictly less than 1 in absolute value, the fixed point is stable, and this proves the theorem. $\square$

Remark 4.1 It turns out that this orbit is stable precisely when it exists. Generally, this will be true of all the velocity orbits we study for this system.

Our study of the particle dynamics produced by the periodic racquet motions we have described may be reduced in general to studying the iterated function dynamics of $F : V \times [0, 2h] \to V \times [0, 2h]$, where V is the set of velocities "recorded" just prior to each racquet strike ($V = [-v, \infty)$), and F maps points in this velocity-time space as follows:

If $0 \leq t_0 < h$ then

$$(v_0, t_0) \overset{F}{\longmapsto} \begin{cases} (\frac{1-M}{M+1}v_0 + \frac{2M}{M+1}, \frac{(M-1)(v-v_0)t_0+2(M+1)}{(1-M)v_0+(3M+1)v}) & v < v_0 \leq \frac{4Mv+(1-M)v^2t_0}{(M-1)(2-vt_0)} \\ ((\frac{M-1}{M+1}v_0 - \frac{2M}{M+1}, s) & v_0 > \frac{4Mv+(1-M)v^2t_0}{(M-1)(2-vt_0)} \end{cases}$$

If $h \leq t_0 < 2h$ then

$$(v_0, t_0) \overset{F}{\longmapsto} \begin{cases} (\frac{M-1}{M+1}v_0 + \frac{2M}{M+1}, \frac{(M-1)(v+v_0)t_0+8(M+1)}{(M-1)v_0+(3M+1)v} - \frac{2}{v}) & -v \leq v_0 < \frac{(M-1)v^2t_0-(M-5)v}{(M-1)(3-vt_0)} \\ ((\frac{M-1}{M+1}v_0 + \frac{2M}{M+1}, \frac{(M-1)(v+v_0)t_0+2(M+1)}{(M-1)(v_0+v)} - \frac{2}{v}) & \frac{(M-1)v^2t_0-(M-5)v}{(M-1)(3-vt_0)} \leq v_0 \leq \frac{(M-1)v^2t_0+4v}{(M-1)(2-vt_0)} \\ (\frac{M-1}{M+1}v_0 + \frac{2M}{M+1}, \frac{(M-1)(v+v_0)t_0+6(M+1)}{(M-1)v_0+(3M+1)v} - \frac{2}{v}) & \frac{(M-1)v^2t_0+4v}{(M-1)(2-vt_0)} < v_0 \end{cases}$$,

where s denotes the time (mod($2h$)) at which the next collision between the racquet and the particle occurs. (We omit the explicit expression for s, since it is in principle straightforward to calculate but roughly doubles the complexity of the explicit formula for $F(v_0, t_0)$.) The idea here is that this function describes simultaneously the velocity transitions of our system together with the sequence of times *modulo the basic period* $2h$ at which the racquet strikes the particle. While a complete characterization of the iterated function dynamics for F cannot be given here, we are able to carry out some elementary calculations which illustrate important qualitative features of the system dynamics.

Thinking of the three basic types of velocity transitions in the above table as letters in an alphabet, we may uniquely identify any trajectory obtained from iterating the function F by the sequence of velocity transitions which it defines. Conversely, if any sequence of velocity transitions is written down, it will define a trajectory of iterates of F provided an appropriate sequence of transition times can be given.

To illustrate what is involved in finding a trajectory which realizes a prescribed sequence of velocity transitions, we shall investigate the existence of several periodic trajectories. Call a trajectory in which there is a repeated pattern of velocity transitions consisting of a type (i) transition followed by $k-1$ type (iii) transitions a *type 1 velocity cycle*. We shall show that type 1 velocity cycles may or may not exist.

Proposition 4.1 *There is no type 1 velocity cycle of length 2.*

Proof: A velocity cycle of length 2 would alternate velocity transitions of type (i) and type (ii). One can explicitly solve for the particle velocities comprising this cycle:

$$v_1 = \frac{2M^2}{(M+1)^2}v \quad \text{and} \quad v_2 = \frac{2M}{(M+1)^2}v.$$

The corresponding 2-cycle of collision times t_1, t_2 would be related according to the above definition of F by the formulas

$$t_2 = \frac{(M-1)(v-v_1)t_1 + 2(M+1)}{(1-M)v_1 + (3M+1)v}$$

and

$$t_1 = \frac{(M-1)(v_2+v)t_2 + 2(M+1)}{(M-1)(v_2+v)} - 2/v.$$

In order for a type 1 2-cycle to exist, we must be able to solve these equations simultaneously for t_1, t_2 in the respective intervals $0 \le t_1 < h$ and $h \le t_2 < 2h$. Solving the equations simultaneously for t_1 yields

$$t_1 = \frac{M^6 + 6M^5 + 15M^4 + 56M^3 + 39M^2 + 10M + 1}{(M-1)(M+1)^3(M^2+4M+1)}h.$$

(Recall that $h = 1/v$.) It is not difficult to show that on the interval $1 < M < \infty$, t_1 is monotonically decreasing and always greater than h. Since we therefore cannot have $t_1 < h$, we have shown that no 2-cycle of type 1 exists. $\square$

A tedious but elementary calculation of this type shows that a length 3 type 1 velocity cycle exists with

$$(v_1, v_2, v_3) = \left(\tfrac{3M^2+1}{M^2+3}v, \ \tfrac{-M^2+4M+1}{M^2+3}v, \ \tfrac{M^2+4M-1}{M^2+3}v \right)$$

provided we can find a corresponding time cycle (t_1, t_2, t_3) with $0 \leq t_1 < h$ and $h \leq t_2, t_3 < 2h$. Elementary but lengthy arguments along the lines described in proving the proposition show that all required inequalities are satisfied provided

$$3M^2 - 8M + 1 > 0$$

and

$$M^3 - 11M^2 + 3M - 1 < 0.$$

These will simultaneously hold for $2.53518\ldots < M < 10.7291\ldots.$

In principle, the same type of elementary argument may be used to confirm or rule out the existence of an orbit of any type, but the complexity of the formal manipulations places practical limits what can be determined in this way.

We conclude with some remarks based on simulation. Although an asymptotically stable length 3 type 1 cycle exists for the range of inertias M we have indicated, the corresponding domain of attraction may be quite small. Other types of orbits are found to coexist for various values of M, and the observed behavior will depend sensitively on initial conditions. (For instance, for $M \approx 5.4$, a length 3 velocity cycle consisting of a type (ii) and two type (iii) transitions seems to dominate the dynamics.) The possible multiplicities of coexisting stable periodic orbits and the ways in which their geometry may vary with the racquet inertia remains open at the present time. It is precisely by understanding the dependence of the system's dynamics on such control parameters that we hope to develop a control theory for systems of this type.

References

[1] J. BAILLIEUL, 1991. "The Behavior of Super-articulated Mechanisms Subject to Periodic Forcing," in *Analysis of Controlled Dynamical Systems*, B. Bonnard, B. Bride, J.P. Gauthier, I. Kupka, Eds., Birkhašer, pp. 35-50.

[2] J. BAILLIEUL, 1991. "The Behavior of Single-Input Super-articulated Mechanisms," Proceedings *1991 American Control Conf.*, Boston, June 26-28, pp. 1622-1626.

[3] J.K. HODGINS, 1991. "Biped Gait Transitions," Proceeding of *1991 IEEE International Conference on Robotics and Automation*, Sacramento, April, pp. 2092-2097.

[4] F.C. HOPPENSTEADT, 1986. *An Introduction to the Mathematics of Neurons*, Cambridge University Press, Cambridge, England.

[5] M. BÜHLER, D.E. KODITSCHEK, & P.J. KINDLMANN, 1990. "A Family of Robot Control Strategies for Intermittent Dynamical Environments," *IEEE Control Systems Magazine*, February, 1990, pp. 16-22.

[6] M. BÜHLER & D.E. KODITSCHEK, 1990. "From Stable to Chaotic Juggling: Theory, Simulation, and Experiments," Proceedings of *1990 IEEE International Conf. on Robotics and Automation*, May, Cincnnati, pp. 1976-1981.

[7] J. MOSER, 1974. "Finitely Many Mass Points on the Line Under the Influence of an Exponential Potential—An Integrable System," *Dynamical Systems Theory and Applications: Battelles Seattle 1974 Rencontres*, Springer Lecture Notes in Physics, vol. 38, pp. 467-497.

[8] E. OTT, C. GREBOGI, & J.A. YORKE, 1990. "Controlling Chaos," *Phys. Rev. Lett.*, #11, pp. 1196-1199.

[9] J.C. WILLEMS, 1991. "Paradigms and Puzzles in the Theory of Dynamical Systems," *IEEE Trans. Aut. Control*, vol. 36, pp. 259-294.

Aerospace/Mechanical Engineering
Boston University
Boston, MA 02215

CONSISTENT ESTIMATION
OF THE ORDER OF HIDDEN MARKOV CHAINS [1]

John S. Baras

Electrical Engineering Department

and Systems Research Center

University of Maryland at College Park

and

Lorenzo Finesso

CNR Ladseb

Corso Stati Uniti, 4

35020 Padova, Italy

Abstract

The structural parameters of many statistical models can be estimated maximizing a penalized version of the likelihood function. We use this idea to construct strongly consistent estimators of the order of Hidden Markov Chain models. The specification of the penalty term requires precise information on the rate of growth of the maximized likelihood ratio. We find an upper bound to the rate using results from Information Theory. We give sufficient conditions on the penalty term to avoid overestimation and underestimation of the order. Examples of penalty terms that generate strongly consistent estimators are also given.

1. Introduction

Let $\{Y_t, t \epsilon Z\}$ be a stationary finitely valued stochastic process that admits a representation of the form $Y_t = f(X_t)$ where $\{X_t, t \epsilon Z\}$ is a finite Markov chain and f is a many-to-one function. We call such a process a Hidden Markov Chain (HMC).

Under well known conditions on f a HMC inherits the Markov property of X_t and becomes a finite Markov chain itself, but this case is non-generic. In general a HMC need not be a Markov chain of any finite order and will therefore exhibit long-range dependencies of some kind. This fact means that the class of HMC's is a very rich one and it comes to no surprise that it is extensively present in many applications.

We can find HMC's appear under various disguises in such diverse fields as: engineering (stochastic automata, speech recognition), biosciences (in medicine to study neurotransmission), economics (stock market predictions), and many others.

On the theoretical side the same fact (lack of the Markov property) makes the class of HMC's difficult to work with. The general methods developed for the study of stationary processes apply but being non-specific they will not give the best results. Theoretical work on the specific class of HMC's has proceeded along two main lines.

The early contributions, inspired by the work of Blackwell and Koopman [4], concentrated on the probabilistic aspects. The basic question was the characterization of HMC's. More specifically the problem analyzed was: *among all finitely valued stationary processes Y_t characterize those*

---

[1] This research was supported by National Science Foundation grant NSFD CDR 8803012, under the Engineering Research Centers Program.

that admit a HMC representation. This problem was solved by Heller [11] in 1965. To some extent Heller's result is not quite satisfactory since his methods are non-constructive. Even if Y_t is known to be representable as a HMC, no algorithm has been devised to produce a Markov chain X_t and a function f such that $Y_t = f(X_t)$ or at least $Y_t \sim f(X_t)$ (i.e. they have the same laws). In recent years the problem has attracted the attention of workers in the area of Stochastic Realization Theory, and while some of the issues have been clarified a constructive algorithm is still missing.

The first contributions dealing with statistical aspects were made in the late sixties. Baum and Petrie [3] studied maximum likelihood estimation of the parameters of a HMC proving consistency and asymptotic normality of the MLE. They also provided an algorithm for the numerical computation of the MLE (of course there is little hope for an explicit solution in a non-Markovian setting) basically inventing the EM algorithm that became popular only later thanks to the work of Dempster, Laird and Rubin [7]. After the mid seventies HMC's made only sporadic appearances in the statistical literature. In 1975 HMC's were proposed by Baker [2] as models for automatic speech recognition (ASR) and ever since they have been adopted as one of the models of choice in this field. Computational aspects became very important and much work was done on the implementation of Baum's algorithm. A good survey of this area of research is [12] which also includes an extensive bibliography.

Although much work has been dedicated to parameter estimation for HMC's only very recently the order estimation problem received some attention. The order of an HMC Y_t is the minimum integer q for which there exists a q-valued Markov chain X_t such that $Y_t = f(X_t)$ for some f. The knowledge of the order of an observed HMC Y_t allows the construction of the *most economical* representations $f(X_t)$ in the sense that the number of parameters (the transition probabilities of X_t) is minimized. The order cannot be estimated using the classical maximum likelihood because increasing the parameter q automatically increases the likelihood. This is the typical behavior of the likelihood function when the parameter is *structural* i.e. the parameter (usually integer valued) indexes the complexity of the model. As another example of structural parameter we mention the order of a Markov chain i.e. the smallest integer m such that:

$$P(X_t \mid X_1^{t-1}) = P(X_t \mid X_{t-m}^{t-1}) \quad \forall t > m + 1, \ \forall X_1^t.$$

Again the maximum likelihood technique fails when applied to the estimation of the parameter m.

In this paper we describe our recent results on the problem of order estimation for hidden Markov chains. The detailed proofs can be found in [8]. The technique we adopt is based on the compensation of the likelihood function. A penalty term, decreasing in q (or m), is added to the maximum likelihood and the resulting compensated likelihood is maximized with respect to q (or m). Proper choice of the penalty term allows the strongly consistent estimation of the structural parameter. Accurate information on the almost sure asymptotic behavior of the maximum likelihood is of critical importance for the correct choice of the penalty term and the Law of the Iterated Logarithm (LIL) is therefore the best tool for this study.

The technique that we have just (roughly) described and the same probabilistic tools have been used for the estimation of the structural parameters of ARMA processes (see e.g. [1], [10]), but we are not aware of any previous work that employs this approach for hidden Markov chains. The behavior of the maximum likelihood is difficult to evaluate because no explicit expressions for the estimators are available. The LIL works for one special case, but we must use other methods to evaluate the asymptotics. We resort to a result from Information Theory to get the necessary asymptotics of the maximum likelihood.

2. Towards a Realization Theory for HMC's

There are many equivalent ways of defining HMC's. We particularly like the definition that originated in Realization Theory [16] and we will borrow it.

Definition 2.1 (SFSS): A pair $\{X_t, Y_t\ t\epsilon N\}$ of stochastic processes defined on a probability space $(\Omega, \mathcal{F}, \mathcal{P})$ and taking values in the finite set $\mathcal{X} \times \mathcal{Y}$ is said to be a stationary finite stochastic system (SFSS) if the following conditions are met:

(i) (X_t, Y_t) are jointly stationary

(ii) $P(Y_{t+1} = y_{t+1}, X_{t+1} = x_{t+1} \mid Y_1^t = y_1^t, X_1^t = x_1^t) = P(Y_{t+1} = y_{t+1}, X_{t+1} = x_{t+1} \mid X_t = x_t)$

The processes X_t and Y_t are called respectively the *state* and the *output* of the SFSS. The cardinality of $\mathcal{X}$ will be called the *size* of the SFSS.

Definition 2.2 (HMC): A stochastic process Y_t with values in the finite set $\mathcal{Y}$ is a Hidden Markov Chain (HMC) if it is equivalent to the output of a SFSS.

Recall that two stochastic processes are said to be equivalent if their laws coincide. Definition 2.2 has therefore to be interpreted as follows: the process Y_t is a HMC if its probability distribution function $P_Y(y_1^n) := Pr[Y_1^n = y_1^n]$ can be represented as $P_Y(y_1^n) = P(\tilde{Y}_1^n = y_1^n)$ where $\tilde{Y}_t$ takes value in $\mathcal{Y}$ and is the output of a SFSS. Observe that we do not require $\tilde{Y}_t$ to be defined on the same probability space $(\Omega, \mathcal{F}, P)$ as Y_t; they can be completely different objects but they are indistinguishable from observation. From now on when we refer to Y_t as a HMC we will actually refer to any process $\tilde{Y}_t$ in the same equivalence class. We will refer to any $SFSS$ $(X_t, \tilde{Y}_t)$ with $\tilde{Y}_t$ equivalent to Y_t as a *representation* of the HMC Y_t.

In the introduction we referred to HMC's as stationary processes of the form $Y_t = f(X_t)$ where X_t is a stationary Markov Chain, but this is equivalent to Definition 2.2. Clearly, if $Y_t = f(X_t)$ with X_t stationary Markov, the pair (X_t, Y_t) will be a SFSS and Y_t a HMC according to Definition 2.2. Conversely, let Y_t be a HMC according to Definition 2.2 and X_t be the state process of a SFSS associated with Y_t. If we sum (ii) of Definition 2.1 over y_{t+1} we get $P(X_{t+1} = x_{t+1} \mid X_1^t = x_1^t, Y_1^t = y_1^t) = P(X_{t+1} = x_{t+1} \mid X_t = x_t)$ and after taking conditional expectations with respect to X_1^t we have $P(X_{t+1} = x_{t+1} \mid X_1^t = x_1^t) = P(X_{t+1} = x_{t+1} \mid X_t = x_t)$. Therefore X_t is a Markov Chain. As a direct consequence of Definition 2.1 (ii) we also have that the process

$S_t = (X_t, Y_t)$ is a Markov Chain. Taking $f : \mathcal{X} \times \mathcal{Y} \to \mathcal{Y}$ to be the projection map on the second component i.e., $f(x, y) = y$ we get the representation $Y_t = f(S_t)$ as desired.

In general HMCs do not have finite memory. Nevertheless their laws are completely specified by a finite number of parameters. In fact to specify the laws of a SFSS it is sufficient to specify the finite set of matrices $\{M(y), y \epsilon \mathcal{Y}\}$ whose elements are: $m_{ij}(y) := P(Y_{t+1} = y, X_{t+1} = j \mid X_t = i)$, $i, j = 1, 2, \cdots \mid \mathcal{X} \mid$. Observe that the matrix $A := \sum_y M(y)$ is the transition matrix of the Markov Chain X_t. If to the matrices $M(y)$ we add an initial distribution vector π such that $\pi = \pi A$ (stationarity) then we have a complete specification of the laws of the $SFSS$.

Very often in the literature the following "factorization" hypothesis is made:

$$P(Y_{t+1} = y, X_{t+1} = j \mid X_t = i) = P(Y_{t+1} = y \mid X_{t+1} = j)P(X_{t+1} = j \mid X_t = i)$$

Since the factorization hypothesis always holds for the process $S_t = (X_t, Y_t)$ we will assume it without loss of generality. Let $b_{iy} := P(Y_t = y \mid X_t = i)$, B the $\mid \mathcal{X} \mid \times \mid \mathcal{Y} \mid$ matrix of the b_{iy}'s, and $B_y := \text{diag} \{b_{1y}, b_{2y}, \cdots b_{cy}\}$ (where $c := \mid \mathcal{X} \mid$). The factorization hypothesis now gives: $M(y) = AB_y$.

In [11] Heller characterized the finite valued stationary processes Y_t that are HMC's. Let $\mathcal{Y}$ denote a finite set, $\mathcal{Y}^*$ the set of finite words from $\mathcal{Y}$, and C^* the set of probability distributions on $\mathcal{Y}^*$. C^* is convex. A convex subset $C \subset C^*$ is polyhedral if $C = \text{conv} \{q_1(\cdot), \cdots, q_c(\cdot)\}$ i.e. C is generated by finitely many distributions $q_i(\cdot) \epsilon C^*$. A convex polyhedral subset $C \subset C^*$ is stable if $C = \text{conv} \{q_1(\cdot), \cdots, q_c(\cdot)\}$ and for $1 \leq i \leq c$ and $\forall y \epsilon \mathcal{Y}$ the conditional distributions

$$q_i(\cdot \mid y) := \frac{q_i(y \cdot)}{q_i(y)} \epsilon C$$

Then

Theorem 2.1 (Heller [11]): $P_Y(\cdot)$ is the *pdf* of a HMC *iff* the set $C_Y := conv\{P_Y(\cdot \mid u) \quad u \epsilon \mathcal{Y}*\}$ is contained in a polyhedral stable subset of C^*.

Consider now a HMC Y_t with the set of parameters $\mathcal{M} := \{c, M(y), \pi\}$ where $c = \mid \mathcal{X} \mid$. It is natural to identify a representation of the HMC Y_t with the set $\mathcal{M}$. When clear from the context we will omit c from the list of parameters. Two questions now arise naturally.

The first question is: can the parameters of a representation be determined directly from $P_Y(\cdot)$?

Such a representation of Y_t is inherently non-unique and we would like to find the "simplest" one. Take $\mid \mathcal{X} \mid$ as a measure of complexity, and for a given HMC Y_t define its *order* as the minimum of $\mid \mathcal{X} \mid$ among all representations. A representation for which $\mid \mathcal{X} \mid$ equals the order is said to be a *minimal* representation.

The second question is: can the order be determined?

Past work contains some partial answers. Unless otherwise noted the following summary is derived from the works of Gilbert [9], Carlyle [5] and Paz [14]. Let $p(\cdot)$ be an arbitrary *pdf* (not necessarily HMC), and $v_1 \cdots v_n \ v_1' \cdots v_n'$, $2n$ arbitrary words from $\mathcal{Y}^*$. The compound sequence matrix (c.s.m.) $P(v_1 \cdots v_n, \ v_1' \cdots v_n')$ is the $n \times n$ matrix with i,j element $p(v_i v_j')$. The *rank* of $p(\cdot)$ is defined as the maximum of the ranks of all possible c.s.m. if such maximum exists or $+\infty$ otherwise. Suppose now that $p(\cdot)$ is the *pdf* of a HMC which admits a representation $\mathcal{M} := \{c, M(\cdot), \pi\}$ of size c. Then we have that: $P(v_1 \cdots v_n v_1' \cdots v_n') = G(v_1 \cdots v_n) H(v_1' \cdots v_n')$ where G, H are $n \times c$ and $c \times n$ matrices respectively, the i-th row of G is $g(v_i)$, the j-th column of H is $h(v_j')$ and $p(v_j v_j') = \pi M(v_i v_j')e = \pi M(v_i) M(v_j')e = g(v_i) h(v_j')$.

It clearly follows that the rank of a HMC cannot exceed the size of any of its representations and therefore in particular:

The rank of a HMC is a lower bound to its order.

It is important to note that the concept of the rank of a *pdf* is only loosely related to the HMC property because there are examples of *pdf's* with finite rank that do not correspond to HMC's. Also there are examples of HMC's whose order is strictly greater than their rank.

A representation $\mathcal{M} = \{c, M(\cdot), \pi\}$ of size c is *regular* if the rank of the corresponding *pdf* equals c. It is not difficult to establish that regular representations are minimal. As it was just noted not all HMC's admit regular representations, but the following two results will justify our interest in them. The first result states that it is "easy" to check regularity. Or more precisely : *A finite number of operations is sufficient to determine the regularity of a given representation* $\mathcal{M} = \{c, M(\cdot), \pi\}$. The second result states that almost all representations are regular. Let Γ be the set of all $\mathcal{M} := \{c, M(\cdot), \pi\}$ of size c. Γ is a compact set in $\mathcal{R}^k$ for some k depending on c. Then: *The non-regular elements of* Γ *are a closed subset of* $\mathcal{R}^k$*-Lebesgue measure zero.*

3. Families of HMC's

In this section we introduce the families of HMC's that will be used as model classes. From now on $\mathcal{Y}$ will be a fixed finite set with $\mid \mathcal{Y} \mid = r$. The family Θ of all HMC's of all orders (taking values in $\mathcal{Y}$) can be identified with the family of all $\theta := \{c_\theta, M_\theta(y), \pi_\theta\}$ with $c\epsilon N$. For $\theta\epsilon\Theta$ define $P_\theta(y_1^n) := \pi_\theta M_\theta(y_1^n)e_{c_\theta}$; we will often drop the subscripts and simply write $P_\theta(y_1^n) = \pi M(y_1^n)e)$.

Define $\Theta_q := \{\theta\epsilon\Theta; c_\theta = q\}$. Note that $\forall \ q \forall \theta\epsilon\Theta_q \ \exists\bar{\theta}\epsilon\Theta_{q+1}$ such that $P_{\bar{\theta}}(\cdot) = P_\theta(\cdot)$ or, abusing the notation, $\Theta_q \subset \Theta_{q+1}$. Statisticians refer to families having the last property as *nested families*.

A few considerations about the identifiability of Θ are now in order. A point $\theta\epsilon\Theta_q$ is *identifiable* in Θ_q if for any $\theta' \neq \theta(\theta'\epsilon\Theta_q)P_\theta(\cdot) \neq P_{\theta'}(\cdot)$ i.e. for at least one word w, $P_\theta(w) \neq P_{\theta'}(w)$. This definition is too strong and it would give no identifiable points in any Θ_q. In fact for a given θ at least the (finitely many) points θ' obtained by permutations of the rows and columns of $M(y)$ and π give $P_{\theta'}(\cdot) = P_\theta(\cdot)$. We will say that $\theta\epsilon\Theta_q$ is *identifiable modulo permutations* (i.m.p.) if the only points $\theta'\epsilon\Theta_q$ with $P_\theta(\cdot) = P_{\theta'}(\cdot)$ are obtained by permutation as described above. Regular points $\theta\epsilon\Theta_q$ (i.e. points for which rank $P_\theta = q$) are good candidates for being i.m.p. but a few (mild) extra conditions must be added. We have adapted to our case the following theorem from Petrie [15] on identifiability.

Definition 3.1: $\theta = \{q, M(y), \pi\}$ is a Petrie point if: θ is regular, $M(y)$ is invertible $\forall y$, and $\exists y \epsilon \mathcal{Y}$ such that b_{iy}, $(i = 1, 2, \cdots q)$, are distinct.

Theorem 3.1(Petrie [15] adapted): The Petrie points of Θ_q are identifiable modulo permutation.

Theorem 3.2[8]:The set of Petrie points is open and of full Lebesgue measure in Θ_q.

It will often be convenient to somewhat restrict the family Θ in order to simplify statistical considerations. To this end we have the **Definition 3.2:**for $0 < \delta < 1/q$ define:
$$\Theta_q^\delta := \{\theta \epsilon \Theta_q; a_{ij} \geq \delta, b_{jy} \geq \delta, \quad \forall i, j, y\}.$$

With the abuse of notation introduced earlier we have:

$$\Theta_q^\delta \subset \Theta_q^{\delta/2}$$

This nested property will be essential later.

4. HMC's as Models of Stationary Processes

The consistency of the Maximum Likelihood Estimator (MLE) for HMC's was established in [3] under the assumption that the true distribution of the observations comes from a HMC. In our work we have shown that, if Y_t is stationary and ergodic, the MLE taken on a class of HMC's converges to the model closest to the true distribution in the divergence sense. The result in [3] is therefore a special case of ours. In the course of this work we have also obtained a slightly generalized version of the Shannon-McMillan-Breiman theorem.

Suppose a given series of observations $\{y_1, y_2 \cdots y_n\}$ is to be modeled for some specific reason. For example we might want to predict y_{n+1} or compress $\{y_1 \cdots y_n\}$ for storage. Confronted with this problem a statistician would most likely set up a related parameter estimation problem as follows. First assume that the sample is generated by some unknown stochastic mechanism, let us say $y_k = g_k(\omega), 1 \leq k \leq n$. The observed data sample is now interpreted as the initial segment of a realization of an unknown stochastic process. Based on prior information, insight, and mathematical tractability, a class of models would then be selected. The models in the class will be denoted $\{f_k(\cdot, \theta), \theta \epsilon \Theta\}$ where $\{f_k(\cdot, \theta)\}_{k \geq 1}$ is a stochastic process whose probability law is completely specified by the parameter θ. The modeling problem is now reduced to an estimation problem. According to some specified criterion of optimality the statistician selects a model, i.e. estimates the θ, that best fits the data. Let us call the estimator based on n observations $\hat{\theta}_n$.

How are we to judge the quality of $\hat{\theta}_n$? Ideally we should compare $f_k(\cdot, \hat{\theta}_n)$ to $g_k(\cdot)$ but the latter is unknown. There are two possible solutions. The classical one is to assume that the unknown process g_k is actually a member of the selected class i.e. $g_k(\cdot) = f_k(\cdot, \theta_0)$ for some true (but unknown) θ_0. The estimator $\hat{\theta}_n$ is then judged to be good if it behaves well, uniformly with respect to $\theta_0 \epsilon \Theta$. Based on this idea a great deal of statistical theory has been developed on the asymptotic properties of various estimators.

The second approach (which we prefer) does not rely on the existence of a true parameter θ_0 in Θ. After all the class of models was chosen more or less arbitrarily, why should g_k belong to it? The problem is transformed into one of best approximation. A distance $d(\cdot, \cdot)$ between probability measures is introduced and θ_* is defined as $d(P_g, P_{\theta_*}) = \min_\theta d(P_g, P_\theta)$. The estimator $\hat{\theta}_n$ is judged to be good if it is close to θ_*. In the statistical literature this is known as the misspecified model approach.

In this section we introduce our first statistical result involving HMC's. We observe the process Y_t with values in the finite set $\mathcal{Y}$. The only assumptions on Y_t are stationarity and ergodicity. Denote by Q the probability distribution on $\mathcal{Y}^*$ induced by Y_t. The class of models for Y_t will be $\Psi := \Theta_q^\delta$ with q and δ fixed. Notice that we do not assume a priori that $Q = P_{\theta_0}$ for some $\theta_0 \epsilon \Psi$. Instead we are adopting the misspecified model approach.

Our goal is to establish the analog of the consistency of the maximum likelihood estimator in this set up. Toward this end define:

$$h_n(\theta, Y) := \frac{1}{n}\log P_\theta(Y_1^n)$$

Following the terminology from [13] we define the *quasi-maximum likelihood estimator* $\hat{\theta}(n)$ as:

$$\hat{\theta}(n) := \{\tilde{\theta}\epsilon\Psi; \ h_n(\tilde{\theta}, Y) = \sup_{\theta\epsilon\Psi} h_n(\theta, Y)\}$$

Note that $\hat{\theta}(n)$ is defined as a set because no uniqueness is guaranteed for this class of models. It is easy to see that in the last equation the sup can be replaced by a max.

We need a notion of "distance" between Q and the P_θ's. A reasonable choice justified by its widespread use in statistics and engineering would be the divergence rate:

$$D(Q \parallel P_\theta) := \lim_{n\to\infty} \frac{1}{n} E_Q \left[\log \frac{Q(Y_1^n)}{P_\theta(Y_1^n)}\right]$$

It can also be shown that:

$$D(Q \parallel P_\theta) = H_Q - H_Q(\theta) \geq 0$$

where $H_Q := E_Q[\log Q(Y_0 \mid Y_{-\infty}^{-1})]$ is minus the entropy of Y_t under Q, and $H_Q(\theta) := E_Q[\log P_\theta(Y_0 \mid Y_{-\infty}^{-1})]$ is a well-defined and continuous function of $\theta\epsilon\Psi$.

Next define the *quasi-true parameter* set as:

$$\mathcal{N} := \{\tilde{\theta}\epsilon\Psi; \quad D(Q \parallel P_{\tilde{\theta}}) = \min_{\theta\epsilon\Psi} D(Q \parallel P_\theta)\}$$

An equivalent description is

$$\mathcal{N} = \{\tilde{\theta}\epsilon\Psi; \ H_Q(\tilde{\theta}) = \max_{\theta\epsilon\Psi} H_Q(\theta)\}$$

For the proof of Theorem 4.1 below we need the following result, established in [].

$$h_n(\theta, Y) \to H_Q(\theta) \quad a.s. \ Q, \ uniformly \ in \ \theta.$$

We recall the notion of a.s. set convergence that will be used. For any subset $\mathcal{E} \subset \Psi$ define the ε-fattened set $\mathcal{E}_\varepsilon := \{\theta\epsilon\Psi; \ \rho(\theta, \mathcal{E}) < \varepsilon\}$, where ρ is the euclidean distance. Then $\hat{\theta}(n) \to \mathcal{N}$ a.s. Q if $\forall \varepsilon > 0 \ \exists \ N(\varepsilon, \omega)$ such that $\forall n \geq N(\varepsilon, \omega), \ \hat{\theta}(n) \subset \mathcal{N}_\varepsilon$.

We are now ready to state our result:

Theorem 4.1[8]:

$$\hat{\theta}(n) \to \mathcal{N} \quad a.s. \ Q$$

This proof [8] is even simpler than the one given by Baum and Petrie [3] for the case of perfect modeling (i.e. $Q = P_{\theta_0}$ for some $\theta_0 \epsilon \Psi$) because it uses the uniform convergence of $h_n(\theta, Y)$.

We now present a slightly generalized version of the Shannon-McMillan-Breiman (SMB). The SMB theorem, first introduced by Shannon in 1948, has already a rich history of extensions and generalizations vestiges of which are found in its very name. The classic version of the theorem is the following:

Theorem 4.2: Let Y_t be a finitely valued stationary ergodic process with probability distribution $Q(\cdot)$. Then:

$$\frac{1}{n} \log Q(Y_1^n) \to E_Q[\log Q(Y_0 \mid Y_{-\infty}^{-1})] \quad a.e. \ and \ in \ L_1$$

In this form the theorem has direct application in Information Theory because it allows the estimation of the entropy rate of a finite alphabet stationary ergodic source. Generalizations of Theorem 4.2 have appeared for the case of real valued processes. Our result generalizes Theorem 4.2 to reference measures M of the HMC type but it applies only to finitely valued processes.

Theorem 4.3[8]: Let Y_t be a process with values in the finite set $\mathcal{Y}$. Assume Y_t to be stationary ergodic under the probability distribution Q and a HMC under the alternative distribution $P \epsilon \Theta_q^\delta$ for some fixed q and δ. Let $q(Y_1^k) = Q(Y_1^k)/P(Y_1^k)$ and define:

$$D_1(Q \parallel P) := \lim_k E_Q[\log q(Y_k \mid Y_0^{k-1})]$$

Then D_1 is well defined and moreover:

$$\frac{1}{n} \log \frac{Q(Y_1^n)}{P(Y_1^n)} \to D_1(Q \parallel P) \quad a.e. \ Q$$

5. Estimation of the Order of a Hidden Markov Chain

The technique that was employed in [8, Chapter 3] for the estimation of the order of a Markov chain will now be adapted to the estimation of the order of a HMC. As we have seen [8] in the Markov case, the crucial step is the evaluation of the rate of growth of the maximized likelihood ratio (MLR). For Markov chains we evaluated this rate to be $O_{a.s.}(\log \log n)$ and we also had very precise results for the $\overline{\lim}$ and the $\underline{\lim}$ of the MLR. For HMC's we will be able to get the rate $O_{a.s.}(\log \log n)$ only in special cases. For the general case we get $O_{a.s.}(\log n)$.

At first the problem of estimating the rate of the MLR for HMC seems easy to solve. For any y_1^n write: $P_\theta(y_1^n) = \sum_{x_1^n} P_\theta(y_1^n, x_1^n) = \sum_{x_1^n} P_\theta(s_1^n)$ where the process $S_t = (X_t, Y_t)$ is a Markov chain.

Clearly $\max_\theta P_\theta(y_1^n) \leq \sum_{x_1^n} \max_\theta P_\theta(s_1^n)$.

Since S_t is a Markov chain we know from [8, Theorem 3.3.2] that:

$$\frac{\max_\theta P_\theta(s_1^n)}{P_{\theta_0}(s_1^n)} = e^{\alpha_n}$$

where $\alpha_n = O_{a.s.}(\log \log n)$

Substituting in the previous inequality we find:

$$\max_\theta P_\theta(y_1^n) \leq \sum_{x_1^n} e^{\alpha_n} P_{\theta_0}(s_1^n) = e^{\alpha_n} \sum_{x_1^n} P_{\theta_0}(s_1^n) = e^{\alpha_n} P_{\theta_0}(y_1^n)$$

From this we immediately get the desired rate:

$$\log \frac{\max_\theta P_\theta(y_1^n)}{P_{\theta_0}(y_1^n)} = O_{a.s.}(\log \log n)$$

This idea, or variations of it, has appeared in the literature, but unfortunately it is wrong. The problem is that Theorem 3.3.2 of [8] does *not* state that $\alpha_n = O_{a.s.}(\log \log n)$ *uniformly with respect to the realization ω.*

In Section 2 we defined the order of a HMC Y_t as the minimum integer q for which there exists a representation of Y_t with $| \mathcal{X} | = q$. We would like to construct a consistent estimator of the order based on the compensated maximum likelihood. The HMC case is complicated by the fact that our knowledge of the set of equivalent representations is only partial. To cope with this difficulty we have to impose restrictions on the observed process Y_t thus limiting the applicability of the results. Fortunately all of the assumptions are satisfied by a generic HMC and therefore the results are still widely applicable.

Assumption 5.1:

The observed process Y_t is a HMC taking values in $\{1, 2, \cdots r\}$, of unknown order q_0. One representation of Y_t is given by $\theta_0 = \{q_0, A_0, B_0\}$ where θ_0 is a Petrie point of $\Theta_{q_0}^\delta$ for some $\delta > 0$.

The class of parametric models that will be used is

$$\Theta := \cup_{q \geq 1} \Theta_q^\delta.$$

The results of Section 2 guarantee that Θ_q^δ contains no point equivalent to θ_0 if $q < q_0$ and a finite number of points equivalent to θ_0 if $q = q_0$. For $q > q_0$ there are infinitely many points in Θ_q^δ equivalent to θ_0. The compensated maximum log-likelihood is defined as:

$$C(q, n) := -L_n(\hat{\theta}_q(n)) + \delta_n(q)$$

where:

$\hat{\theta}_q(n)$ is the MLE of $\theta \epsilon \Theta_q^\delta$ based on n observations

$L_n(\hat{\theta}_q(n)) := \frac{1}{n} \log P_{\hat{\theta}_q(n)}(Y_1^n)$

$\delta_n(q)$ is a positive increasing function of q and n to be determined.

The estimator of the order is defined by:

$$\hat{q}(n) := \min\{\arg \min_{q \geq 1} C(q, n)\}$$

The problem of order estimation can now be posed as follows.

Problem:

The HMC Y_t satisfying Assumption 5.1 is observed. Find a compensator sequence $\delta_n(q)$ such that the estimator $\hat{q}(n)$ is strongly consistent i.e. $\hat{q} \to q_0$ a.s. P_{θ_0}.

The analog of Theorem 3.4.2 of [8] is valid and we can easily give a sufficient condition on $\delta_n(q)$ that avoids underestimation.

Theorem 5.1 (Compensators avoiding underestimation)[8]: Let Y_t be a process satisfying Assumption 5.1. If $\lim_{n \to \infty} \delta_n(q) = 0$ $(\forall q)$, then $\underline{\lim}_{n \to \infty} \hat{q}(n) \geq q_0$ $P_{\theta_0} - a.s.$

To estimate the rate of convergence in $\Theta_{q_0}^\delta$, we study next the rate of growth of the maximized log-likelihood ratio (MLR)

$$\log \frac{P_{\hat{\theta}_{q_0}(n)}(y_1^n)}{P_{\theta_0}(y_1^n)}$$

Since q_0 is fixed, $\hat{\theta}_{q_0}(n)$ will be denoted $\hat{\theta}_n$. We need one extra assumption on the HMC Y_t which will be in force through this section.

Assumption 5.2:

$$-\frac{\partial^2}{\partial \theta^2} H_{\theta_0}(\theta) \mid_{\theta_0} > 0$$

Recall that: $H_{\theta_0}(\theta) := E_{\theta_0}[\log P_\theta(Y_0 \mid Y_{-\infty}^{-1})]$.

After giving two preliminary results we will prove that the MLR is $0_{a.s}(\log \log n)$. Recall that:

$$\hat{\theta}_n = \{\hat{\theta}\epsilon\Theta_{q_0}^\delta \; ; \; P_{\hat{\theta}}(y_1^n) = \max_{\theta} P_\theta(y_1^n)\}$$

and that in general $\hat{\theta}_n$ is not a singleton. Our first preliminary result shows that it is always possible to choose a convergent sequence $\hat{\theta}_n' \epsilon \hat{\theta}_n$.

The second preliminary result establishes the following bound needed for the application of the Law of Iterated Logarithm.

For some finite C, $\forall k$, $\forall l$, $\forall \theta$:

$$| \frac{\partial}{\partial \theta_l} \log P_\theta(y_k \mid y_1^{k-1}) | \leq C \quad a.s. \; P_{\theta_0}$$

We are now ready to study the rate of convergence.

Theorem 5.2[8]:

$$\frac{1}{n} \log P_{\hat{\theta}_n}(y_1^n) = \frac{1}{n} \log P_{\theta_0}(y_1^n) + O_{a.s.}\left(\frac{\log\log n}{n}\right)$$

We next use a result from Information Theory to get a useful bound on the MLR valid for all values of q. Recall that by $P_{\hat{\theta}_q(n)}(y_1^n)$ we denoted the maximized probability $P_\theta(y_1^n)$ for P_θ a HMC with $\theta\epsilon\Theta_q^\delta$. We denote by $P_{ML_q}(Y_1^n)$ the corresponding maximized probability when $\theta\epsilon\Theta_q$. The next result is crucial. A complete proof is to be found in Csiszar [6].

Theorem 5.3: There exists a probability measure Q on $\mathcal{Y}^\infty$ such that

$$\log \frac{P_{ML_q}(y_1^n)}{Q(y_1^n)} \leq \frac{d(q)}{2} \log n - c \quad for \; all \; n \; and \; y_1^n$$

where c is a constant and $d(q) := q(q + r - 2)$ As a sketch of the proof we observe first that:

$$P_{ML_q}(y_1^n) = \max_{\theta\epsilon\Theta_q} P_\theta(y_1^n) = \max_{\theta\epsilon\Theta_q} \sum_{x_1^n} P_\theta(y_1^n \mid x_1^n) P_\theta(x_1^n)$$

$$\leq \sum_{x_1^n} \max_{\theta} P_\theta(y_1^n \mid x_1^n) \cdot \max_{\theta} P_\theta(x_1^n)$$

The proof proceeds by showing the existence of probability measures Q_1 and Q_2 such that:

$$\max_{\theta} P_\theta(y_1^n \mid x_1^n) \leq Q_1(y_1^n \mid x_1^n) n^{q(r-1)/2}$$

$$\max_{\theta} P_\theta(x_1^n) \leq Q_2(x_1^n) n^{q(q-1)/2}$$

Clearly $Q(y_1^n) := \sum_{x_1^n} Q_1(y_1^n \mid x_1^n) Q_2(x_1^n)$ is a probability measure on $\mathcal{Y}^\infty$ and substituting into completes the proof. The existence of Q_1 and Q_2 is proved directly by actually constructing measures Q_1 and Q_2 that satisfy the above.

The following Theorem, based on Theorem 5.3, will be essential to finding estimators of the order that avoid overestimation.

Theorem 5.4[8]:

$$\overline{\lim} \, (\log n)^{-1} \log \frac{P_{\hat{\theta}_q(n)}(y_1^n)}{P_{\theta_0}(y_1^n)} \leq \frac{d(q)}{2} + 2 \quad a.s. \, P_{\theta_0}$$

We are finally able to give a set of sufficient conditions on the compensators of the maximized likelihood (the sequences $\delta_n(q)$) to avoid overestimation of the order. Theorem 5.5 is complementary to Theorem 5.1: together they allow us to construct compensators $\delta_n(q)$ that guarantee strong consistency of the order estimator $\hat{q}(n)$.

Theorem 5.5(Compensators avoiding overestimation)[8]: Let Y_t be a process satisfying Assumptions 5.1 and 5.2. If the compensator is of the form:

$$\delta_n(q) := \varphi(n)h(q)$$

where the function φ satisfies:

$$\underline{\lim} \left(\frac{\log n}{n} \right)^{-1} \varphi(n) > 1$$

and the function h satisfies:

$$h(q') - h(q) \geq \frac{d(q')}{2} + 2 \quad \forall q' > q \geq 1$$

Then:

$$\overline{\lim} \, \hat{q}(n) \leq q_0 \quad a.s. \, P_{\theta_0}$$

The existence of a strongly consistent estimator $\hat{q}(n)$ of the order q_0 will be established by giving examples of functions $h(\cdot)$ and $\varphi(\cdot)$ satisfying both the conditions imposed by Theorem 5.1 and Theorem 5.5.

Theorem 5.6[8]: The compensator

$$\delta_n(q) := 2d^2(q)\frac{\log n}{n}$$

produces a strongly consistent estimator $\hat{q}(n)$ of q_0.

Proof: Clearly $\lim \delta_n(q) = 0 \; \forall q$ thus satisfying the conditions of Theorem 5.1. The function $\varphi(n) := 2(\log n/n)$ is such that $\underline{\lim} \left(\frac{\log n}{n} \right)^{-1}\varphi(n) = 2 > 1$ and therefore satisfies the condition imposed by Theorem 5.5. For the function $h(q) := d^2(q)$ we must check the condition:

$$h(\hat{q}) - h(q) \geq \frac{d(\hat{q})}{2} + 2 \quad \forall \, \hat{q} > q \geq 1$$

Recall that $d(q) := q(q + r - 2)$. The condition to be verified is equivalent to:

$$\hat{q}(\hat{q} + r - 2)[\hat{q}(\hat{q} + r - 2) - \frac{1}{2}] \geq q^2(q + r - 2)^2 + 2$$

for all $\hat{q} > q \geq 1$. This is easily established observing that the left-hand side is increasing in $\hat{q}$ and that for $\hat{q} = q + 1$ the inequality is verified.

6. References

[1] Azencott, R. and Dacunha-Castelle, D., *Series of Irregular Observations*, New York: Springer Verlag, 1986.

[2] Baker, J.K., "Stochastic Modeling for Automatic Speech Understanding", in *Speech Recognition*, Reddy, R. ed., New York: Academic Press, 1975.

[3] Baum, L.E. and Petrie, T., "Statistical Inference for Probabilistic Functions of Finite State Markov Chains", *Ann. Math. Stat.*, 37 (1966), 1554-63.

[4] Blackwell, D. and Koopmans, L., "On the Identifability Problem for Functions of Finite Markov Chains", *Ann. Math. Stat.*, 28 (1957), 1011-15.

[5] Carylye, J.W., "Stochastic Finite-State System Theory", in *System Theory*, Zadeh, L.A., Polak, E. eds., New York: McGraw-Hill, 1969.

[6] Csiszar, I., *Information Theoretic Methods in Statistics*, Notes for course ENEE 728F, University of Maryland, Spring 1990.

[7] Dempster, A.P., Laird, N.M. and Rubin, D.B., "Maximum Likelihood from Incomplete Data via the EM Algorithm", *J. Roy. Statist. Soc., Ser. B*, 39 (1977), 1-38.

[8] Finesso, L. "Consistent Estimation of the Order for Markov and Hidden Markov Chains", Ph. D. Thesis Electrical Engin. Department, Technical Report Ph.D. 91-1, Systems Research Center, University of Maryland, College Park, 1991.

[9] Gilbert, E.J., "On the Identifiability Problem for Functions of Finite Markov Chains", *Ann. Math. Stat.*, 30 (1959), 688-697.

[10] Hannan, E.J. and Deistler, M., *The Statistical Theory of Linear Systems*, New York: Wiley, 1988.

[11] Heller, A., "On Stochastic Processes Derived from Markov Chains", *Ann. Math. Stat.*, 36, (1965), 1286-91.

[12] Levinson. S.E., Rabiner, L.R. and Sondhi, M.M., "An Introduction to the Application of the Theory of Probabilistic Functions of a Markov Process to Automatic Speech Recognition", *Bell Syst. Tech. J.*, 62, (1983), 1035-74.

[13] Nishii, R., "Maximum Likelihood Principle and Model Selection when the True Model is Unspecified", *J. Multiv. Anal.*, 27 (1988), 392-403.

[14] Paz, A., *Introduction to Probabilistic Automata*, New York: Academic Press 1971.

[15] Petrie, T., "Probabilistic Functions of Finite State Markov Chains", *Ann. Math. Stat.*, 40 (1969), 97-115.

[16] Picci, G., "On the Internal Structure of Finite State Stochastic Processes", in *Recent Developments in Variable Structure Systems*, New York: Springer Verlag (Lecture Notes in Economics and Math. Systems, Vol. 162) 1978.

Adaptive control of partially observed linear systems, the scalar case

Pierre BERTRAND
Département de Mathématiques
Université des Antilles et de la Guyane
97159 Pointe a Pitre Cedex
FRANCE

Abstract: We study scalar adaptive control. The control problem is linear quadratic with partial observation. The drift term depends affinely from an unknown parameter α. We show that the maximum likelihood is strongly consistent. At each time, we replace the true parameter α_0 by the estimate α_t to compute the feedback law and we prove that the associate cost converges to the optimal cost, almost surely.

Key Words: Stochastic control, Adaptive control, partial observation, identification, Maximum likelihood estimate.

0) Introduction

The problem of adaptive control with full information has been studied extensively. A good reference for the continuous case is Duncan & Pasik-Duncan [DP]. The adaptive control with partial observation has almost not been studied. In the continuous case, the only reference is Duncan [D], but he assumes the existence of a consistent estimate and the stability of the estimate filter when we prove them; he just proves the convergence in average of the estimate cost to the optimal cost.

As a first step, we study the linear control problem partially observed in the scalar case. The stochastic systems are described by linear stochastic differential equations, the cost is linear quadratic and the unknown parameter appears affinely in the drift term of the state equation. We adapt the method of [DP]: we separate the identification and the control problem.

This paper is organized as follow: In the first section, we describe the problem and give a first simplification. In the second section, we recall the consistency result of Borkar & Bagchi [BB] for the Maximum Likelihood Estimate and deduce, in the linear case, the strong consistency. In the third section, we build the adaptive control policy, using the estimate α_t of the parameter instead the true parameter α_0 in the optimal feedback law. We also use an approximate filter, replacing α_0 by α_t in the Kalman equation. We show the stability of the approximate filter and the convergence in average to the true filter, almost surely. In the last section, we prove that the observed cost converges in average to the optimal cost, almost surely.

1) Description of the problem and simplification.

The partially observed stochastic system for the adaptive control problem is:

(1) $\qquad dX_t^\alpha = (A_\alpha X_t^\alpha + U_t)\, dt + dW_t \qquad X^\alpha(0) = h$

(2) $\qquad dY_t^\alpha = H X_t^\alpha\, dt + dB_t \qquad Y(0) = 0$

The cost function $C(t)$ is defined as

(3) $\qquad C(t) = \int_0^t |X_s|^2 + N\,|U_s|^2\, ds$

We consider the scalar case: the state X_t, the control U_t and the observation Y_t are real. W_t and B_t are independent real brownian motions. The drift term depends linearly of the unknown parameter α: $A_\alpha = F_0 + \alpha F_1$ with $\alpha, F_0, F_1 \in \mathbf{R}$. We suppose that the parameter α belongs to a bounded interval K of $\mathbf{R}$. It is usual to assume that the operator A_α is stable, uniformly on K; in this case we can simplify the problem and replace (1) by the equation (1') with the hypothesis (H1):

(1') $dX_t^\alpha = (\alpha X_t^\alpha + U_t) \, dt + dW_t$ $X^\alpha(0) = h$

(H1) $\alpha \in [a, b]$ with $b < 0$ and $H \neq 0$

2)   Estimation of the unknown parameter.

We follow the idea of Duncan & Pasik-Duncan [DP] in the case with full information: we first prove the consistency of the estimate for a large class of control (verifying (H2)). After we will use the optimal feedback law with the estimate parameter and show it verifies the hypothesis (H2).

(H2) $\forall \; T{>}0 \;\; U \in L^2([0, T] \times \Omega, \mathbb{R}^n)$ and

$$\frac{1}{t} \int_0^t |u(s, \omega)|^2 \, ds \leq K < +\infty \quad \text{a.s.}$$

We suppose that the system has reached the steady state, and recall the consistency results given by Borkar & Bagchi [BB]. Let $P_{\alpha,t}$ and P_α denote the probability measures induced on $C([0, t], \mathbb{R}^p)$ and $C([0, \infty[, \mathbb{R}^p)$ by Y_t, we have:

Proposition 1:

If (H1) and (H2) hold then $P_{\alpha,t}$ is absolutely continuous with respect to $P_{\alpha_0,t}$ and

$$\frac{dP_{\alpha,t}}{dP_{\alpha_0,t}} = ML(\alpha, t) = \exp \; \{ -\frac{1}{2} \int_0^t |H \; \hat{X}(\alpha, s) - H \; \hat{X}(\alpha_0, s)|^2 \, ds$$

$$+ \int_0^t <H \; \hat{X}(\alpha, s) - H \; \hat{X}(\alpha_0, s), dZ_s> \} \quad P_{\alpha_0} \text{ a.s.}$$

with Z_t, $\hat{X}(\alpha_0, s)$ and Π_α defined by

(4) $Z_t = Y_t - \int_0^t H \; \hat{X}(\alpha_0, s) \, ds$

(5) $\qquad d\hat{X}(\alpha, s) = [\alpha \, \hat{X}(\alpha, s) + U_s] \, ds + \Pi_\alpha \, H \, [dY_s - H \, \hat{X}(\alpha, s) \, ds]$

(6) $\qquad \Pi_\alpha = \dfrac{\alpha + \sqrt{\alpha^2 + C^2}}{C^2}$

Moreover, Z_t is a real brownian motion adapted to $\mathcal{F}(Y_s, s \leq t)$.

Proof: This is proved by [BB, thm 2.1, p. 204] for the case without control, with the notation $(L(\alpha).dY)(t) = H \, \hat{X}(\alpha, s)$. Since Π_α is the positive solution of the Riccati equation $2 \, \alpha \, \Pi_\alpha - H^2 \alpha^2 + 1 = 0$, we have (6). The paper [BB] adapts a previous result of Balakrishnan [Bala] with a control U_t verifying (H2) and it is easy to combine the two results to obtain proposition 1.

Remark 1: By the hypothesis (H2), $\hat{X}(\alpha, s)$ is well defined and $\hat{X}(\alpha, s) \in L^2([0, T], \mathbb{R}^n)$ a.s.

At each time t, we choose α_t which maximises $ML(\alpha, t)$. We give another condition on α_t to obtain α_t dependent only from the observation (not from the unknown parameter α_0).

<u>Proposition 2</u>:

We have $f(\alpha_t, t) = \underset{\alpha \in [a,b]}{max} \, f(\alpha, t) \; P_{\alpha_0}$ a.s. where

$$f(\alpha, t) = \int_0^t <H \, \hat{X}(\alpha, s), \, dY_s> - \frac{1}{2} \int_0^t |H \, \hat{X}(\alpha, s)|^2 \, ds \qquad P_{\alpha_0} \text{ a.s.}$$

Proof: Since $dZ = dY - H \, \hat{X}(\alpha_0, s) \, dt$, we have

$$\int_0^t <H \, \hat{X}(\alpha, s) - H \, \hat{X}(\alpha_0, s), \, dZ_s> - \frac{1}{2} \int_0^t |H \, \hat{X}(\alpha, s) - H \, \hat{X}(\alpha_0, s)|^2 \, ds$$

$$= \int_0^t <H \, \hat{X}(\alpha, s), \, dY_s> - \frac{1}{2} \int_0^t |H \, \hat{X}(\alpha, s)|^2 \, ds$$

$$- \int_0^t <H \hat{X}(\alpha_0, s), dY_s> + \frac{1}{2} \int_0^t |H \hat{X}(\alpha_0, s)|^2 \, ds.$$

From now on, all probabilitic statements will be made with respect to P_{α_0}. We recall the "weak" consistency result [BB, cor. 3.1]:

Proposition 3:

If (H1) and (H2) hold, then $\alpha_t \to \alpha_1$ when $t \to +\infty$ a.s. with α_1 verifying $\hat{X}(\alpha_1, t) = \hat{X}(\alpha_0, t) \; \forall \, t > 0$ a.s.

We deduce from the above proposition the strong consistency of the ML estimate:

Theorem 1:

If (H1) and (H2) hold, then $\alpha_t \to \alpha_0$ when $t \to +\infty$ a.s.

Proof: From (5) and proposition 3, we have

(7) $$0 = (\alpha_1 - \alpha_0) \int_0^t \hat{X}(\alpha_0, s) \, ds + (\Pi_{\alpha_1} - \Pi_{\alpha_0}) H Z(t) \quad \text{a.s.}$$

Assume $\alpha_1 \neq \alpha_0$ then $(\Pi_{\alpha_1} - \Pi_{\alpha_0}) \neq 0$, since $Z(t)$ is a real brownian motion (7) is absurde.

3)   Adaptive control and stability result.

In the non adaptive case, the linear control problem (1'), (2) with partial observation (3) is well known. The stationary optimal control is $U_t^0 = -\frac{1}{N} P_0 \hat{X}_t^0$ where $\hat{X}_t^0$ satisfies the Kalman equation:

(8) $$d\hat{X}_t^0 = [\alpha \hat{X}_t^0 + U_t^0] \, dt + \Pi_{\alpha_0} H [dY_t^0 - H\hat{X}_t^0 \, dt]$$

where Y_t^0 is the observation corresponding to the control U_t^0 and P_0 the positive solution of the Riccati equation (with $\alpha = \alpha_0$):

(9) $$2 P A_\alpha - \frac{1}{N} P^2 + 1 = 0$$

If P_α is solution of (9) then $P_\alpha = N[\alpha + \sqrt{\alpha_t^2 + \frac{1}{N}}\,]$

Thus $\hat{X}_t^o$ satisfies

(8') $d\hat{X}_t^o = [\alpha_0 - \frac{P_0}{N} - \Pi_0 H^2]\, \hat{X}_t^o\, dt + \Pi_0 H\, dY_t^o$

In the adaptive case, it is usual to replace P_0 by P_t solution of (9) with $\alpha = \alpha_t$ (the estimate of the parameter at the time t) and $\hat{X}_t^o$ by $\hat{X}(\alpha_t, t)$ to obtain the control policy. Since $\hat{X}(\alpha_t, s)$ is solution of a anticipative stochastic equation, we prefer to form another approximation $\hat{X}_t$ of the Kalman filter, by solving

(10) $d\hat{X}_t = -[\alpha_t + \sqrt{\alpha_t^2 + \frac{1}{N}} + \sqrt{\alpha_t^2 + H^2}\,]\, \hat{X}_t\, dt + \Pi_t H\, dY_t^u$

We denote Y_t^u the observation corresponding to the control $U_t = -\frac{1}{N} P_t \hat{X}_t$. The real Kalman filter corresponding to the control U_t is $\hat{X}_t^u$ and it satisfies

(11) $d\hat{X}_t^u = [\alpha_0 \hat{X}_t^u + U_t]\, dt + \Pi_0 H\, dY_t^u$

In this section, we will prove that the system (10), (11) is stable and deduce that $\hat{X}_t$ converges in average to $\hat{X}_t^u$ almost surely.

Proposition 4:

Let $\xi_t = (\hat{X}_t^u, \hat{X}_t)$, then ξ_t satisfies the stochastic equation

(12) $d\xi_t = \mathcal{M}_{\alpha_t} \xi_t\, dt + \mathcal{P}_t\, dZ_t$ $\xi(0) = 0$

with $\mathcal{M}_\alpha = \begin{pmatrix} \alpha_0 & -(\alpha + \sqrt{\alpha^2 + \frac{1}{N}}\,) \\[2ex] \alpha + \sqrt{\alpha^2 + H^2} & -(\alpha + \sqrt{\alpha^2 + H^2} + \sqrt{\alpha^2 + \frac{1}{N}}\,) \end{pmatrix}$

and $\mathcal{P}_t = \begin{pmatrix} \Pi_0 & C \\ \Pi_t & C \end{pmatrix}$

Proof: We use $dY_t = dZ_t + H \hat{X}_t^u$ in (10) and (11).

<u>Proposition 5</u>:

If (H1) holds then there is a $K > 0$ such that

$$\limsup_{t \to +\infty} \frac{1}{t} \int_0^t |\xi_s|^2 \, ds \leq K \quad \text{a.s.}$$

Proof: The eigenvalues of the limit matrix $\mathcal{M}_{\alpha_0}$ are $-\sqrt{\alpha_0^2 + H^2}$ and $-\sqrt{\alpha_t^2 + \frac{1}{N}}$ thus $\mathcal{M}_{\alpha_0} = P_0^{-1} \Delta_0 P_0$ with Δ_0 a diagonal matrix. We denote $\Delta(\alpha) = P_0 \mathcal{M}_\alpha P_0^{-1}$, we have $\Delta(\alpha) \to \Delta_0$ as $\alpha \to \alpha_0$ and $\Delta(\alpha)$ is a C^1 function of α. We apply the Taylor formula to $\Delta(\alpha)$ and we obtain:

$$\Delta(\alpha) = \Delta_0 + (\alpha - \alpha_0) \frac{d}{d\alpha} \Delta(\theta) \text{ with } \theta \in [a, b]$$

But $\frac{d}{d\alpha} \Delta(\theta) = P_0^{-1} \frac{d}{d\alpha} \mathcal{M}(\theta) P_0$ and $\frac{d}{d\alpha} \mathcal{M}(\theta)$ is uniformly bounded on [a, b], we deduce (for $\|M\|_2$ the euclidian matrix norm):

$$\exists \, K_1 > 0 \text{ such that } \forall \, \theta \in [a, b] \quad \|\frac{d}{d\alpha} \Delta(\theta)\|_2 \leq K_1$$

Let ζ_s be a $\mathbb{R}^2$ stochastic process, we have

$$E \zeta_s^* \Delta(\alpha_s) \zeta_s = E \zeta_s^* \Delta_0 \zeta_s + E (\alpha_s - \alpha_0) \zeta_s^* \frac{d}{d\alpha} \Delta(\theta) \zeta_s$$

but

$$E \zeta_s^* \Delta_0 \zeta_s \leq - \sqrt{\alpha_0^2 + (H^2 \wedge \frac{1}{N})} \; E|\zeta_s|^2$$

and

$$E (\alpha_s - \alpha_0) \zeta_s^* \frac{d}{d\alpha} \Delta(\theta) \zeta_s \leq [E (\alpha_s - \alpha_0)^2]^{1/2} \, K_1 \, E|\zeta_s|^2$$

Since $\alpha_s \to \alpha_0$ when $s \to +\infty$ a.s. and α_s is bounded we deduce $E (\alpha_s - \alpha_0)^2 \to 0$ when $s \to +\infty$. We have proved that:

$$\exists \, T_0 > 0 \text{ such that } E \, \zeta_s^* \Delta(\alpha_s) \zeta_s \leq -\mu_1 \, E \, |\zeta_s|^2 \text{ with } \mu_1 = -\frac{1}{2} \sqrt{\alpha_0^2 + (H^2 \wedge \frac{1}{N})}$$

The process $\zeta_t = P_0\, \xi_t$ is solution of the equation:

$$d\zeta_t = \Delta(\alpha_t)\, \zeta_t\, dt + P_0\, \mathcal{P}_t\, dZ_t$$

We apply the Itô formula and we obtain

(13) $\qquad |\zeta_t|^2 - 2 \int_0^t \zeta_s^* \Delta(\alpha_s)\, \zeta_s\, ds = |\zeta_{T_0}|^2 + \int_0^t |P_0 \mathcal{P}_s|^2\, ds + 2 \int_0^t \zeta_s^* P_0 \mathcal{P}_s\, dZ_s$

Since $|P_0 \mathcal{P}_s| \le K_2$ we have

$$E\, |\zeta_t|^2 + 2\mu_1 \int_0^t E\, |\zeta_s|^2\, ds \le E\, |\zeta_{T_0}|^2 + \; + K_2\, (t - T_0)$$

We deduce $\displaystyle \limsup_{t \to +\infty} \frac{1}{t} \int_0^t E\, |\zeta_s|^2\, ds \le \frac{K_3}{2\mu_1}$

But $M_t = \displaystyle\int_0^t \zeta_s^* P_0 \mathcal{P}_s\, dZ_s$ is a martingale and by the above majoration, we can apply the strong law of large numbers; we deduce $\frac{1}{t}\, M_t \to 0$ when $t \to +\infty$, almost surely. But for almost every ω

$$\exists\, T(\omega) > 0 \text{ such that } \forall\, s > T(\omega),\ \forall \zeta \in \mathbf{R}^2,\quad \zeta^* \Delta(\alpha_s)\, \zeta \le \mu_1\, |\zeta|^2$$

We deduce from (13) that

$$\limsup_{t \to +\infty} \frac{1}{t} \int_0^t E\, |\zeta_s|^2\, ds \le \frac{K_3}{2\mu_1} \qquad \text{a.s.}$$

Using $|\xi_s| \le \|P_0\|_2\, |\zeta_s|$ we deduce the result of the proposition 5. This insures that (H2) holds, thus we can omit this hypothesis in the assumptions of the proposition 5.

Remark 2: In the proof of the above proposition, we do not use $\alpha < 0$. The limit matrix has negative eigenvalues even for $\alpha > 0$. The hypothesis (H1) is useful only to obtain the consistency result of theorem 1. We now prove the convergence of $\hat{X}_t$ to the true Kalman filter $\hat{X}_t^u$.

Proposition 6:

If (H1) holds, we have $\displaystyle\lim_{t\to+\infty} \frac{1}{t} \int_0^t |\hat{X}_s - \hat{X}_t^u|^2 \, ds = 0$ a.s.

Proof: We denote $r\hat{X}_t^u = \hat{X}_t - \hat{X}_t^u$ and we obtain

$$d r\hat{X}_t^u = -\sqrt{\alpha_t^2 + H^2} \;\; r\hat{X}_t^u \, dt + f(t) \, \hat{X}_t^u \, dt + (\Pi_t - \Pi_0) \, H \, dZ_t \qquad r\hat{X}_t^u(0) = 0$$

with $f(t) = \sqrt{\alpha_0^2 + H^2} - \sqrt{\alpha_t^2 + H^2} + (\Pi_t - \Pi_0) \, H$

By the Itô formula, we have

$$|r\hat{X}_t^u|^2 + 2 \int_0^t \sqrt{\alpha_s^2 + H^2} \;\; |r\hat{X}_t^u|^2 \, ds \leq 2 \int_0^t f(s) \;\; r\hat{X}_t^u \; \hat{X}_t^u \, ds$$

$$+ 2 \int_0^t (\Pi_s - \Pi_0) \, H \, r\hat{X}_t^u \, dZ_s + \int_0^t |(\Pi_s - \Pi_0) \, H|^2 \, ds$$

But $\alpha_s^2 \geq b^2$ gives

(14) $\qquad 2\sqrt{b^2 + H^2} \;\; |r\hat{X}_t^u|^2_{L^2(0,t)} \leq 2 \, |f(s) \, \hat{X}_t^u|_{L^2(0,t)} \;\; |r\hat{X}_t^u|_{L^2(0,t)} + M_t$

with $M_t = 2 \displaystyle\int_0^t (\Pi_s - \Pi_0) \, H \, r\hat{X}_t^u \, dZ_s + \int_0^t |(\Pi_s - \Pi_0) \, H|^2 \, ds$

Since α is bounded we have $E \, M_t \leq K \, t$ and

$$2\sqrt{b^2 + H^2} \;\; E|r\hat{X}_t^u|^2_{L^2(0,t)} \leq 2 \, E|f(s) \, \hat{X}_t^u|_{L^2(0,t)} \;\; E|r\hat{X}_t^u|_{L^2(0,t)} + K \, t$$

We also have

$$\frac{1}{t} \, E|f(s) \, \hat{X}_t^u|_{L^2(0,t)} \leq \frac{1}{t} \, E|f(s)|_{L^2(0,t)} \;\; E|\hat{X}_t^u|_{L^2(0,t)}$$

From $f(t) \to 0$ when $t \to +\infty$ a.s. and $f(t)$ bounded, we deduce $\displaystyle\lim_{t\to+\infty} \frac{1}{t} \, E|f(s)|^2_{L^2(0,t)} = 0$ and from proposition 5 $\frac{1}{t} \, E|\hat{X}_t^u|_{L^2(0,t)}$ is bounded. This insures $\frac{1}{t} \, E|r\hat{X}_t^u|^2_{L^2(0,t)}$ bounded.

Thus we can apply the strong law for large numbers to $\int_0^t (\Pi_s - \Pi_0) \, H \, r\hat{X}_t^u \, dZ_s$ and

deduce $\frac{1}{t} M_t \to 0$ a.s. From $f(t) \to 0$ when $t \to +\infty$ a.s., $f(t)$ bounded and proposition 5 we

deduce $\frac{1}{t} |f(s) \hat{X}_t^u|_{L^2(0,t)} \to 0$ when $t \to +\infty$ a.s. With (14) we deduce the result.

We easily deduce the following corollary from the above proposition:

Corollary 1:

If (H1) holds, we have $\lim_{+\infty} \frac{1}{t} \int_0^t \hat{X}(s)^2 - \hat{X}^u(s)^2 \, ds = 0$ a.s.

4) Convergence of the average cost.

In this last section, we will show that the observed cost $\hat{C}(t)$ converges (a.s. in average) to the optimal cost $C_0(t)$ for the stationary control problem with partial observation (1'), (2), (3). We have

$$\hat{C}(t) = \int_0^t |\hat{X}_s|^2 + \frac{1}{N} |P_s \hat{X}_s|^2 \, ds \qquad \text{and}$$

$$C(t) = \int_0^t |\hat{X}_s^0|^2 + \frac{1}{N} |P_0 \hat{X}_s^0|^2 \, ds$$

$\hat{X}_t^0$ is the Kalman filter corresponding to the optimal control for the non adaptive problem (1'), (2), (3): when α_0 is known. $\hat{X}_t^0$ is given by the equations:

$$(15) \qquad d\hat{X}_t^0 = -[\alpha_t + \sqrt{\alpha_t^2 + \frac{1}{N}} + \sqrt{\alpha_t^2 + H^2}] + \Pi_t \, H \, dY_t^0$$

$$(16) \qquad dX_t^0 = (A_{\alpha_0} X_t^0 - \frac{1}{N} P_0 \hat{X}_t^0) \, dt + dW_t$$

$$(17) \qquad dY_t^0 = H X_t^0 \, dt + dB_t, \qquad Y^0(0) = 0$$

We denote $rX_t = X_t - X_t^0$ and $r\hat{X}_t^0 = \hat{X}_t - \hat{X}_t^0$ where $\hat{X}_t$ is the solution of (10) and X_t the solution of (1') corresponding to the control $U_t = -\frac{1}{N} P_t \hat{X}_t$. We have

(18) $\quad drX_t = (\alpha_0 \ rX_t - \frac{1}{N} \ P_0 \ r\hat{X}_t^0) \ dt - \frac{1}{N} \ (P_t - P_0) \ \hat{X}_t \ dt$

(19) $\quad dr\hat{X}_t^0 = -v(\alpha_0) \ r\hat{X}_t^0 \ dt + \Pi_0 \ H^2 \ rX_t \ dt$

$$+ \ (v(\alpha_0) - v(\alpha_t)) \ \hat{X}_t \ dt + (\Pi_t - \Pi_0) \ H^2 \ \hat{X}_t^u \ dt + (\Pi_t - \Pi_0) \ H \ dZ_t$$

with $v(\alpha) = \alpha + \sqrt{\alpha^2 + \frac{1}{N}} + \sqrt{\alpha^2 + H^2}$

The system (18), (19) is a linear system with matrix $\mathcal{M}_{\alpha_0}$

Proposition 7:

If (H1) holds, we have

$$\lim_{t \to +\infty} \frac{1}{t} \int_0^t E|r\hat{X}_s^0|^2 + E|rX_s|^2 \ ds = 0 \qquad \text{and}$$

$$\lim_{t \to +\infty} \frac{1}{t} \int_0^t |r\hat{X}_s^0|^2 + |rX_s|^2 \ ds = 0 \qquad \text{a.s.}$$

Proof: We adapt the proof of the proposition 5, this is more simple since we work directly with the limit matrix $\mathcal{M}_{\alpha_0}$. This time the second member depends from $v(\alpha_0) - v(\alpha_t)$, $(P_t - P_0)$ and $(\Pi_t - \Pi_0)$; but v, P and Π are continuous functions of α and $\alpha_t \to \alpha_0$ when $t \to +\infty$ a.s. insure $\lim \frac{1}{t} \int_0^t \varphi(s) \ ds = 0$ for $\varphi(t) = v(\alpha_0) - v(\alpha_t)$ or $\varphi(t) = (P_t - P_0)$ or $\varphi(t) = (\Pi_t - \Pi_0)$. This gives the results. We finally deduce the theorem:

Theorem 2:

If (H1) holds, we have $\lim_{+\infty} \frac{1}{t} [\hat{C}(t) - \hat{C}_0(t)] = 0$ a.s.

and $\qquad \lim_{t \to +\infty} \frac{1}{t} [E\hat{C}(t) - E\hat{C}_0(t)] = 0$

Proof: $\qquad \hat{C}(t) - \hat{C}_0(t) = (1 + \frac{1}{N} \ P_0^2) \int_0^t |\hat{X}_s|^2 - |\hat{X}_s^0|^2 \ ds + \frac{1}{N} \int_0^t (P_s^2 - P_0^2) \ |\hat{X}_s|^2 \ ds$

Since $(P_s^2 - P_0^2) \to 0$ when $s \to +\infty$ and $(P_t^2 - P_0^2)$, $\frac{1}{t} \int_0^t |\hat{X}_s|^2 \, ds$ bounded we have

$\frac{1}{t} \int_0^t (P_s^2 - P_0^2) |\hat{X}_s|^2 \, ds \to 0$ when $t \to +\infty$ a.s. It remains to prove $\frac{1}{t} \int_0^t |\hat{X}_s|^2 - |\hat{X}_s^0|^2 \, ds \to 0$

a.s. This results from

$$\int_0^t |\hat{X}_s|^2 - |\hat{X}_s^0|^2 \, ds \leq |r\hat{X}|_{L^2(0,t)}^2 + |r\hat{X}|_{L^2(0,t,)} \, |\hat{X}|_{L^2(0,t)}$$

$$\lim \frac{1}{t} |r\hat{X}|_{L^2(0,t)}^2 = 0 \text{ a.s. and } \frac{1}{t} |\hat{X}|_{L^2(0,t)}^2 \text{ bounded.}$$

Conclusion:

We have proved the strong consistency of the maximum likelihood estimate and built an approximate of the Kalman filter which converges to the true filter, almost surely. We have also proved that the associate cost converges in average to the optimal cost, almost surely and in mean. We only assume that the system is stable, this is an usual assumption. We use an estimate which is not recursive, it should be nice to find another estimate which would converge slowly but would be recursive.

REFERENCES

[BB]   A. Bagchi & V. Borkar, *"Parameter identification in infinite dimensional linear systems"*, Stochastics 1984, vol. 12, pp. 201-213.

[Bala]   A. V. Balakrishnan, *"Stochastic Differential Systems"*, Springer Verlag, Berlin, 1973.

[D]   T. E. Duncan, *"Adaptive control of some partially observed linear stochastic systems"*, Lecture Notes in Control and Info. Sci. 126, 1989, pp. 102-114.

[DP]   T. E. Duncan & B. Pasik-Duncan, *"Adaptive control of continuous time linear systems"*, M.C.S.S. 3, 1990, pp. 45-60.

System and Control Theory Perspectives of the IMAGE Greenhouse Model

Roger D. Braddock
Australian Environmental Studies
Griffith University, Nathan
Queensland, Australia

Jerzy A. Filar
Department of Mathematics (and Statistical Policy Branch)
University of Maryland Baltimore County (and EPA)
Baltimore, Maryland, USA

Radoslaw Zapert
Department of Mathematics and Statistics
University of Maryland Baltimore County
Baltimore, Maryland, USA

January 30, 1992

Introduction

The issues surrounding the anticipated impacts of the **enhanced greenhouse effect** are likely to form a significant part of of scientific activities in the next decade. However, the emphasis in the previous sentence should properly be placed on the words *anticipated impacts*. The understanding of both the power and the limitations of the tools by which these anticipated impacts are derived will be absolutely essential both to the scientists and for government regulatory agencies. The majority of these tools are mathematical/computer models which simulate certain mathematical models of the underlying atmospheric chemistry and physics. The outputs of these models are then linked into further mathematical models of socio-economic factors that in turn simulate the above mentioned "anticipated impacts".

This paper is really only a "progress report" on a line of research intended to demonstrate that some of the best known greenhouse models can, and perhaps should, be formulated and analyzed within the framework of modern control theory. Thus the paper is intended for applied mathematicians and engineers with interest in dynamical systems and control theory on the one hand, and environmental problems on the other hand. We hope to achieve the following objectives:

1. To demonstrate that the system/control theoretic framework is well suited for the modeling and the analysis of some important environmental phenomena,

2. Communicate the importance, in the environmental systems context, of a few of the most basic concepts of control,

3. To provide a large scale "case study" illustration of the control theoretic approach to environmental modeling.

A key element, and an essential starting point, in our research is IMAGE; The Integrated Model for the Assessment of the Greenhouse Effect (see Rotmans [9]). IMAGE is a parameterized scientifically based simulation model developed for the calculation of historical and future effects of emissions of greenhouse gases on the global temperature, sea level rise, ecological, and socio-economic interests etc.. IMAGE has been developed at The National Institute of Public Health and Environmental Protection (RIVM) for the Dutch government, and is regarded as one of the best models of this kind in the world.

The entire paper is motivated by our belief that the perception of certain environmental phenomena as stochastic, dynamical systems "controlled" by one or more controllers will prove to have useful applications that reach beyond the scope of this research. Traditionally, we are brought up to think of atmospheric phenomena as "acts of God" beyond human control. It is our opinion that the notion that we can somehow "control" the climate, is still sufficiently foreign to the general population, even in the industrialized countries, as to diminish public support for (potentially expensive) environmental policies. We hope that the control theoretic perspective can "filter down" to the larger, nonscientific, community and help to increase the understanding of the dependence of the phenomena such as the greenhouse effect on human interference/"controls".

Qualification

It must be emphasized that the mathematical analyses carried out in the subsequent sections are quite elementary. However, they are performed on an existing "large scale" computer simulation model of the greenhouse effect that has been calibrated with real data, and as such this paper may constitute one of the first attempts to apply system/control theoretic techniques to a global environmental problem.

1. Formulation and Structure

As mentioned in the Introduction the model IMAGE can be regarded as a controlled dynamical system. We refer the reader to Braddock et. al. [2] and Rotmans [9] for the complete construction of the system equations.

In view of the fact that the latter is a system of 159 ordinary differential equations connecting many physical phenomena a complete description is rather involved. Below, we give only a sketch of the system's structure[1].

We define state variables as:

$\mathbf{x} = (x_1, \ldots, x_{12})$, the carbon contents in the 12 ocean layers of an ocean box diffusion model, in Giga tones of carbon (GtC) (Braddock and Filar [1]).

$\mathbf{y} = y$, the concentration of carbon as carbon dioxide (CO_2) in the atmosphere, in parts per million by volume, $ppmv$.

$\mathbf{z} = (z_1, \ldots, z_{49})$, the carbon content in the seven levels of the seven categorized ecosystems, in GtC (Goudriaan and Ketner [10] and Rotmans [9]).

$\mathbf{s} = (s_1, \ldots, s_7)$, the total area of land in each of the seven categorized ecosystems, in hectares (ha) (Rotmans [9]).

$\mathbf{S} = (S_1, \ldots, S_{22})$, the nonzero entries of the 7×7 land use transfer matrix S:

---

[1]System equations encoded in MATLAB are available upon request

$$S = \begin{bmatrix} 0 & 0 & 0 & S_1 & S_2 & 0 & 0 \\ S_3 & 0 & 0 & S_4 & S_5 & 0 & S_6 \\ 0 & 0 & 0 & 0 & 0 & 0 & 0 \\ S_7 & S_8 & S_9 & 0 & S_{10} & 0 & 0 \\ S_{11} & S_{12} & 0 & 0 & 0 & 0 & S_{13} \\ S_{14} & S_{15} & S_{16} & S_{17} & S_{18} & 0 & 0 \\ S_{19} & S_{20} & 0 & S_{21} & S_{22} & 0 & 0 \end{bmatrix}$$

whose (k, l)-th entry denotes the amount of land per year transferred from ecosystem k to ecosystem l.

$u = (u_1, \ldots, u_{49})$, the change in temperature since 1900, in the 49 layers of the ocean under the ocean mixed layer, in degrees Celsius (Rotmans [9]).

$v = v$, the change in temperature since 1900, in the mixed surface layer of the ocean, in degrees Celsius (Wigley and Schlesinger [11]).

$w = (w_1, \ldots, w_{18})$, the concentration of various gases with CFC's $w_1, \cdots, w_{15}$ in $pptv$, and of carbon monoxide, (w_{16}) in $ppmv$, methane, (w_{17}) in $ppmv$, and nitrous oxide, (w_{18}) in $ppbv$, in the atmosphere (Den Elzen et. al., [4], [5]).

The vector state variable $X(t)$, where t represents time, for the full system is a function of time, contains the above scalar and column vector components in the given order, and is a 159 element column vector which is partitioned, and represented by

$$X(t) = [x^T, y^T, z^T, s^T, S^T, u^T, v^T, w^T]^T$$

where T denotes the matrix transpose.

The model IMAGE[2] incorporates a modular structure whereby dynamic interactions between the state variables are permitted using models for interactions between various components of the state vector. These dynamic interactions can be represented as

$$\frac{dX(t)}{dt} = AX(t) + N(X(t)) + U(t), \tag{1}$$

together with the initial condition

$$X_0 = X(1990)$$

---

[2]We attempt to capture only the most important dynamical processes modeled by IMAGE. The full IMAGE model also includes many additional input and output processes.

where A is a 159×159 matrix which is partitioned in the same manner as $X(t)$, and X_0 is the initial value of the state vector $X(t)$. Thus,

$$A = (A_{IJ}), \quad I, J = x, y, z, s, S, u, v, w$$

where the partitioned submatrices A_{IJ} take their structure from $X(t)$. Thus A_{zs} is a 49×7 submatrix relating the linear effects of the area of land allocated to various ecosystems, to the carbon content in those ecosystems. The vector function $N(X(t))_{159 \times 1}$, is a nonlinear vector function of the state variables, while $U(t)_{159 \times 1}$ is a vector function of time, and represents the forcing terms such as the emission of the various gases as functions of time and the rate of changes of S. The vectors:

$$N = (N_I) \ and \ U = (U_I) \qquad I = x, y, z, s, S, u, v, w$$

have the same partitioned structure as $X(t)$. The partitioned submatrices A_{IJ}, and subvectors N_I and U_I describe the operations of the constituent modules of IMAGE and the interactions between the variables; sometimes referred to as *feedbacks*, (Lashoff [7]). Note that some interactions do not occur, e.g., the carbon content x of the ocean compartments and the area of land allocated to the ecosystems s, are not related, and thus the corresponding parts of A, N and U are zero. Indeed, there is a high level of sparsity in the system, that can be seen if (1) is written in the matrix form:

$$\frac{d}{dt}
\begin{bmatrix} x \\ y \\ z \\ s \\ S \\ u \\ v \\ w \end{bmatrix}
=
\begin{bmatrix}
A_{xx} & 0 & 0 & 0 & 0 & 0 & 0 & 0 \\
A_{yx} & A_{yy} & A_{yz} & 0 & 0 & 0 & 0 & 0 \\
0 & 0 & A_{zz} & 0 & 0 & 0 & 0 & 0 \\
0 & 0 & 0 & A_{sS} & 0 & 0 & 0 & 0 \\
0 & 0 & 0 & 0 & 0 & 0 & 0 & 0 \\
0 & 0 & 0 & 0 & 0 & A_{uu} & A_{uv} & 0 \\
0 & 0 & 0 & 0 & 0 & A_{vu} & A_{vv} & A_{vw} \\
0 & 0 & 0 & 0 & 0 & 0 & A_{wv} & A_{ww}
\end{bmatrix}
\begin{bmatrix} x \\ y \\ z \\ s \\ S \\ u \\ v \\ w \end{bmatrix}
+
\begin{bmatrix} N_x \\ N_y \\ N_z \\ 0 \\ 0 \\ 0 \\ N_v \\ N_w \end{bmatrix}
+
\begin{bmatrix} 0 \\ U_y \\ 0 \\ 0 \\ U_S \\ 0 \\ 0 \\ U_w \end{bmatrix}
\qquad (2)$$

The initial condition is taken at the year 1990. A simplified system diagram for our system is shown in Figure 1, and illustrates the major interactions between the state variables. The two main linkages in the system are the atmospheric concentration y of CO_2, and the temperature of the ocean mixed layer v. Carbon cycles through the ocean, the biosphere, and the atmosphere, with the atmosphere serving as the link mechanism for all three reservoirs. The CO_2 in the atmosphere also links to the temperature v, of the ocean mixed layer through changes in the radiation patterns in the atmosphere. These radiation patterns are also affected by the other atmospheric gases, w, and v also links to the temperature of the deeper ocean. The temperature v as well interacts with the biosphere and the ocean carbon content since the uptake of CO_2 by these components is temperature dependent.

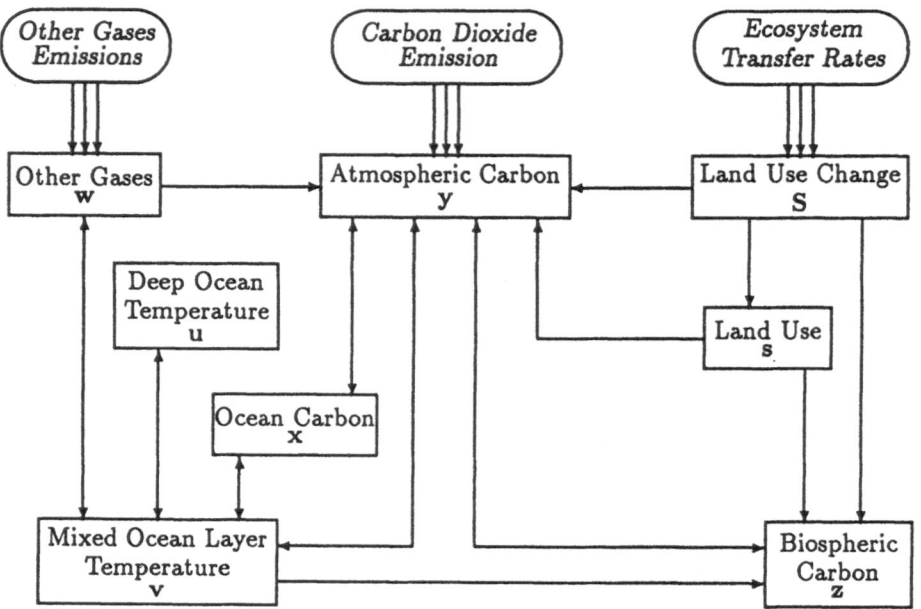

Figure 1: *Systems diagram for IMAGE indicating interactions between the state variables.*

The two main differences between our dynamical system, and that of Braddock et. al. [2] are in the carbon cycle. They are described in the following remarks.

Remark 1 In the ocean carbon circulation model the equilibrium carbon contents of the two surface layers were dropped as state variables. Whenever they appear in the formulation given in [2], they were replaced by an analytical expression derived from the expression for the equilibrium phase between CO_2 in the atmosphere and the ocean

$$\frac{dx_{eq}}{x_{eq}} = \frac{1 - .04v}{4.05\ln(0.033y)} \cdot \frac{dy}{y}$$

where x_{eq} stands for the equilibrium carbon content of a given surface ocean layer. The above equation is a slight modification of (3.11) in Rotmans [9]. The "Net Primary Production" variables were treated in the similar manner, with equation (3.15) in Rotmans [9] providing a basis for replacing these variables with algebraic expressions in y

and v. The above reductions eliminated all the "conceptual variables" from the state vector $\mathbf{X}(t)$. The remaining state variables represent actual physical quantities some of which, nonetheless, may be hard to measure.

Remark 2 We are now treating all of the nonzero entries of the land use transfer matrix S as 22 new state variables. Note, that the corresponding differential equations in (1) have zero linear and nonlinear terms, and hence we are implicitly treating the rates of change of the land use transfers as "control" variables. This is conceptually reasonable because these transfers are a product of human interference, rather than of natural processes. In addition, we have combined the areas belonging to a given ecosystem in Africa, Asia and South America into seven state variables corresponding to the total areas of the seven ecosystems.

2. Equilibrium and Stability

The first important concept related to the dynamical system (1) is that of an *equilibrium*. We shall say that $\mathbf{X}_{eq}$ is an *equilibrium point* of the system corresponding to a constant control $\mathbf{U}^c$ if:

$$0 = \mathbf{AX}_{eq} + \mathbf{N}(\mathbf{X}_{eq}) + \mathbf{U}^c \tag{3}$$

For an environmental model the knowledge of an equilibrium point (if one exists) is very important because it represents the eventual or "sustainable" state of the system when the external (usually human) interference levels off at the constant amount contained in the vector $\mathbf{U}^c$. Thus if $\mathbf{X}_{eq}$ is somehow attained, the environmental variables of interest are no longer changing as long as the control remains equal to $\mathbf{U}^c$. The technical problem implied by (3) is the solution of large number of nonlinear equations. Alternatively, a reasonable heuristic is to initiate a numerical algorithm for simulating (1) (with $\mathbf{U}(t) = \mathbf{U}^c$ for all t), observe the resulting trajectory $\mathbf{X}(t)$, and to terminate the algorithm when the norm of the right hand side of (3) (evaluated at $\mathbf{X}(t)$) is sufficiently near 0.

In Braddock et. al. [2] the first of the above approaches was used. Since the nonlinear equations analogous to (3) were badly conditioned, a fixed point iteration algorithm was successfully used to compute $\bar{\mathbf{X}}_{eq}$. In view of Remarks 1 and 2 our system of equations is slightly different than that used in [2]. Consequently $\bar{\mathbf{X}}_{eq}$ was no longer an equilibrium point of (1). However, it provided a good starting point for the simulation heuristic mentioned above. Thus a fourth order Runge-Kutta algorithm initiated at $\bar{\mathbf{X}}_{eq}$ was used to approximately compute $\mathbf{X}_{eq}$, an equilibrium of (1). The simulation was stopped when

$$\|\mathbf{AX}(t) + \mathbf{N}(\mathbf{X}(t)) + \mathbf{U}^c\| \leq 10^{-6}.$$

Of course, $\mathbf{X}_{eq}$ depends essentially on the constant input/control vector $\mathbf{U}^c$. In particular, constant $\mathbf{U}^c$ implies that the areas belonging to the seven ecosystems also remain constant, and hence that the land use transfer matrix $S \equiv 0$. It is now interesting to note how some of the more important variables in $\mathbf{X}_{eq}$ vary with the constant inputs $\mathbf{U}^c$. Figures 2-3 indicate some interesting relationships that can be discovered in this manner.

In particular, in the computations of the graphs in these figures we fixed the levels of all emissions other than that of CO_2 at constant levels corresponding roughly to the average global emission levels in the year 2100. The CO_2 emission levels were allowed to range from 8 to 50 GtC corresponding roughly to the present global emission levels through to those forecasted by IMAGE's "business as usual" or "unrestricted trends" scenario, in the year 2100. The first graph, Figure 2 below shows that equilibrium concentration of CO_2 (the y entry of $\mathbf{X}_{eq}$) grows linearly with the above CO_2 emissions.

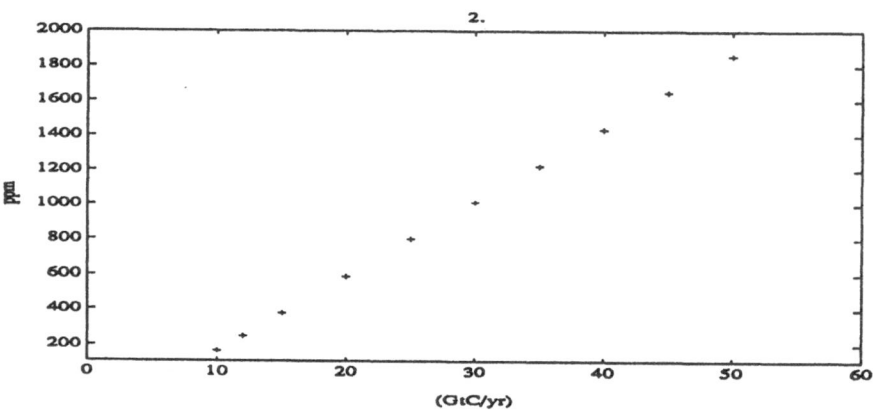

Figure 2: *Equilibrium Atmospheric CO_2 Concentration as a Function of a Constant CO_2 Emission Level*

However, the equilibrium temperature change of the mixed ocean layer with respect to the year 1900 (the v entry of $\mathbf{X}_{eq}$)[3] appears to be growing only logarithmically with the fixed CO_2 emission levels. This can be seen from Figure 3A, which also shows that if CO_2 were held fixed at the current level of approximately 8 GtC per year then the equilibrium temperature change of the ocean's mixed layer will actually be negative! Of course, some of the response times reported later in this section indicate that the

---

[3]This variable plays a vital role in the IMAGE model, since the ocean's mixed surface layer has a strong moderating effect on the entire system

periods of time required to reach these equilibria can be multiples of more than 500 years. Nonetheless, this graph says something about the system's (planet's?) ability to stabilize itself in the long-run. Finally, Figure 3B shows that the equilibrium sea level rise, due to thermal expansion, also seems to grow logarithmically with the (fixed) CO_2 emission levels.

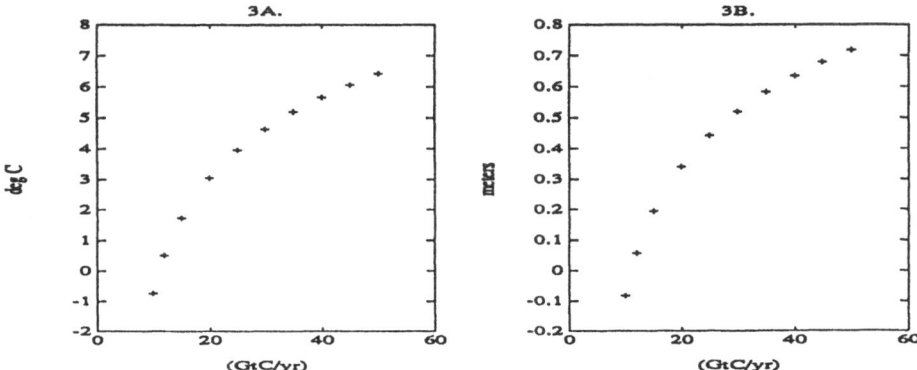

Figure 3: *Equilibrium Temperature Change of the Mixed Ocean Layer (3A) and the Equilibrium Thermal Expansion of the Ocean (3B) as a Function of CO_2 Emission Level*

It is important to note that the system (1) simplifies greatly if the nonlinear part does not appear in the equation, that is, $N(X(t)) = 0$. In such a case the more powerful theory of *linear dynamical control systems* can be utilized. Even when $N(X(t)) \neq 0$ it is possible to locally approximate (1) by a linear system, under quite general conditions. In particular, in a neighborhood of a point $\tilde{X} = X(\tilde{t})$ (for some $t \geq \tilde{t}$) the nonlinear term of (1) can be expressed as:

$$
\begin{aligned}
N(X(t)) &\approx N(\tilde{X}) + \nabla N(\tilde{X})[X(t) - \tilde{X}] \\
&= \nabla N(\tilde{X})X(t) + [N(\tilde{X}) - \nabla N(\tilde{X})X] \\
&= \nabla N(\tilde{X})X(t) + \tilde{U}
\end{aligned}
\tag{4}
$$

where $\nabla N(\tilde{X})$ is the gradient matrix of partial derivatives of $N(X(t))$ with respect to the state variables, evaluated at $\tilde{X}$. Now combining (2) with (1) yields a linearized system that is locally a good approximation of (1) and is of the form:

$$
\begin{aligned}
\frac{d}{dt}X(t) &= (A + \nabla N(\tilde{X}))X(t) + U(t) + \tilde{U} \\
&= \tilde{A}X(t) + \tilde{U}(t)
\end{aligned}
\tag{5}
$$

$$\mathbf{X}(t_o) \;=\; \check{\mathbf{X}}$$

where $\tilde{\mathbf{A}} = \mathbf{A} + \nabla \mathbf{N}(\check{\mathbf{X}})$ and $\tilde{\mathbf{U}}(t) = \mathbf{U}(t) + \tilde{\mathbf{U}}$.

Now with $\check{\mathbf{X}} = \mathbf{X}_{eq}$ the equilibrium point calculated above, we can investigate the stability of the linearized system (5). In particular, that system has been checked to be stable since all of the eigenvalues of $\tilde{\mathbf{A}}$ (of course, with s and S blocks deleted) lie in the left half plane.

The remaining eigenvalues correspond closely to those reported in Braddock et al. [2]. There are six complex pairs of eigenvalues, and the real parts range from -0.002 to -39.96. Thus, the corresponding *response times* (defined as $\frac{1}{|Re(\lambda)|}$), for an eigenvalue λ in the left half-plane, range from 500 years down to approximately 9 days.

Remark 3 It is, perhaps, remarkable that our version of IMAGE is stable. Often, models of climatic phenomena are inherently chaotic. We conjecture that the stability of (5) (about its equilibrium) can be explained in one of three ways: 1. The underlying physical system is indeed stable, perhaps, due to the great amount of global averaging that is part of the computation of IMAGE's variables that induces stability in the "mean trend" of these variables, 2. The deterministic nature of IMAGE's equations ((1) and (5)) neglects the stochastic aspect of the relevant physical processes, and the usual chaotic behavior is due to that stochasticity, or 3. The formulation of IMAGE does not model some important physical processes that, otherwise, would have induced chaotic behavior.

Remark 4 Another interesting observation is that the linearized system (5) appears to be a rather good approximation of (1) for substantial time intervals. We compared a simulated trajectory $\mathbf{X}_1(t)$ of (1) obtained by the fourth order Runge - Kutta method with a simulated trajectory $\mathbf{X}_2(t)$ of (5). When the linearization was around the current state $\mathbf{X}_c = \mathbf{X}_c(t_c)$ we observed that:

$$\frac{\|\mathbf{X}_1(t) - \mathbf{X}_2(t)\|}{\|\mathbf{X}_1(t)\|} \le 0.05$$

for $t_c \le t \le t_c + 8$, with t measured in years.

3. Controllability and Observability

In the case of a nonlinear system such as (1) the simplest approach to the problem of controllability is via the linearized system (5), which can be written in the more general form:

$$\frac{d}{dt}\mathbf{X}(t) \;=\; \tilde{\mathbf{A}}\mathbf{X}(t) + \mathbf{B}\tilde{\mathbf{U}}(t)$$
$$\mathbf{X}(t_o) \;=\; \mathbf{X}_o \tag{6}$$

where **B** is a matrix that may combine (or exclude) some of the components of the control/input. It is very well known (e.g., see [3] and [8]) that the system (6) is completely controllable if and only if:

$$rank(\mathcal{C}_n) = n \qquad (7)$$

where n is the dimension of the system matrix $\tilde{\mathbf{A}}$ and $\mathcal{C}_i$ denotes i-controllability matrix:

$$\mathcal{C}_i = [\mathbf{B}|\tilde{\mathbf{A}}\mathbf{B}|\tilde{\mathbf{A}}^2\mathbf{B}|...|\tilde{\mathbf{A}}^{i-1}\mathbf{B}], \qquad i = 1...n \qquad (8)$$

Of course, in the case of our application the seemingly trivial task of checking the rank of a matrix is complicated by the fact that $n = 159$. However, an iterative algorithm can be easily devised whereby the rank of $\mathcal{C}_1 = [\mathbf{B}]$ is checked, then the rank of $\mathcal{C}_2 = [\mathbf{B}|\tilde{\mathbf{A}}\mathcal{C}_1] = [\mathbf{B}|\tilde{\mathbf{A}}\mathbf{B}]$ etc. The algorithm stops either when $i = n$ or the condition: $rank(\mathcal{C}_i) = rank(\mathcal{C}_{i+1})$ is satisfied for the first time.

A preliminary investigation of the controllability properties of our version of IM-AGE, tested the rank of $\mathcal{C}_n$ where $\mathcal{C}_n$ was computed for the linearized system (6) with $\tilde{\mathbf{X}} = \mathbf{X}(1990), \mathbf{X}(2000), \mathbf{X}(2050), \mathbf{X}(21000)$ respectively. The results are summarized in Table 1. below, and show that each of these linearized systems is completely controllable if the input matrix **B** selects sufficiently many controls.

Of course, it can be argued that the controls required to drive the system to many desirable states in sufficiently short time will be infeasible from a practical standpoint. The computation of the standard feedback controls required to reach such target states is part of ongoing research.

Table 1: *Controllability of the linearized system (YES means controllable)*

Full System (6)					Reduced System (9)				
Variable	1990	2000	2050	2100	Variable	1990	2000	2050	2100
y, S, w	YES	YES	YES	YES	y, w	YES	YES	YES	YES
S, w	YES	YES	YES	YES	y	NO	NO	NO	NO
y, w	NO	NO	NO	NO	w	YES	YES	YES	YES

Further, in line with the emissions scenarios developed in Rotmans [9] we will now suppose that the land use variables (which are driven by population growth) are not to be regarded as controls. Then we can still consider the controllability of a reduced linearized system:

$$\begin{aligned}\frac{d}{dt}\hat{\mathbf{X}}(t) &= \hat{\mathbf{A}}\hat{\mathbf{X}}(t) + \hat{\mathbf{B}}\hat{\mathbf{U}}(t) \\ \hat{\mathbf{X}}(t_o) &= \hat{\mathbf{X}}_o\end{aligned} \qquad (9)$$

that is obtained from (6) by deleting the s and S variables from $X(t)$, and by replacing these variables with their appropriate constant values. The controllability of the reduced system (9) tells us whether that system is controllable only by controlling emissions of greenhouse gases. These results are also given in Table 1 above.

It is clear that in applications to environmental systems both the observation and reconstruction problems are of great importance. Accurate estimation of the state of the environmental system at a given time is important irrespective of whether this estimation is done using historical measurements of some indicator/output variables, or whether it is to be recovered from measurements to be taken in the future. It is important to mention that the linearized system (5) augmented by the output vector:

$$Y(t) = CX(t) \qquad (10)$$

is completely observable if

$$rank(\mathcal{O}_n) = n \qquad (11)$$

where n is the dimension of the system matrix $\tilde{A}$ and $\mathcal{O}_i$ denotes i-observability matrix:

$$\mathcal{O}_i = [C^T|\tilde{A}^T C^T|(\tilde{A}^T)^2 C^T|...|(\tilde{A}^T)^{i-1}C^T]^T, \qquad i = 1...n \qquad (12)$$

The data on the state variables of our greenhouse model indicate, that, at best, only the blocks of variables y, s, S, v, w could be regarded as observable[4]. Consequently, we have considered three alternative selections of these blocks of variables as constituting matrix C. In each case we tested the rank of the n-observability matrix given in (10). The results are summarized in Table 2.

Table 2: *Observability of the linearized system (YES means observable)*

Variable	1990	2000	2050	2100
y, s, S, v,w	YES	YES	YES	YES
y, w	YES	YES	YES	YES
w	NO	NO	NO	NO

4. Stochasticity/Variability

Our formulation of IMAGE as a controlled dynamical system (1) is purely deterministic. Nonetheless, it is clear that there are many uncertainties in this model. Consequently,

---

[4]For instance, while there are estimates of the area covered by, say, closed tropical forest the reliability of these estimates is unclear.

the actual trajectory $X(t)$ should, perhaps, be regarded as a realization of a complex stochastic process. The technical difficulty in designing a tractable stochastic control formulation of IMAGE lies in the modeling of the many distinct sources of variability.

In particular, the parameters of the system equations in (1) are all known only within some range of accuracy. Rotmans [9] gives "intervals of uncertainty" for many of the important parameters. For instance, an important diffusivity parameter, *Diff*, in the ocean model is given a range of [3700,6000] with the value of *Diff* = 4000 being the one used in IMAGE. However, the above interval merely captures the range of values considered by the scientists modeling the ocean, and has no statistical interpretation. In designing a stochastic model one could assume that *Diff* is uniformly distributed on [3700,6000], or that it is normally distributed with mean 4000 and standard deviation being equal to some percentage of the mean, or for that matter anyone of a number of other distributional assumptions could be made. Perhaps, an empirical distribution should be fitted to some indirect measurements/estimates of *Diff*?

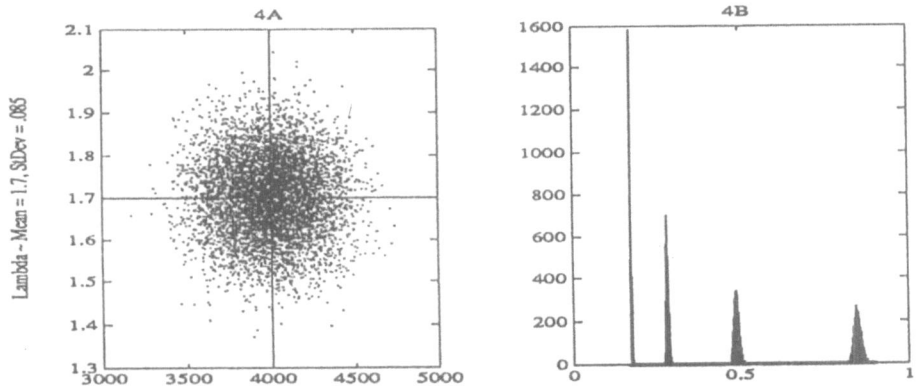

Diffusivity: Mean = 4000, StDev = 200Diff.: Mean=4000,StD.=200; Lambda: Mean=1.7,StD=.085

Figure 4: *Expected Sea level Rise in: 2000, 2020, 2050, 2100*

Furthermore, the initial state vector X_o should also be treated as a random vector with the same difficulties concerning its distributional properties. However, the fact that IMAGE uses the year $t_o = 1900$ as the initial simulation time just about eliminates the possibility of fitting a good empirical distribution to X_o.

In addition, the usual introduction of an additive noise term into (1) entails still more assumptions about the dependence/independence of errors and the manner in which these errors grow over time. Nonetheless, we believe that it is important to formulate and analyze a stochastic version of (1), with the tools of stochastic control.

So far, we have performed only some exploratory experiments of the robustness of IMAGE under stochastic perturbations.

In the case of stochastic uncertainty of the parameters we assumed normal distributions with means equal to the values used by IMAGE and standard deviations taken as some percentage of the means. We then sampled randomly from these distributions, ran IMAGE for each set of sampled values, and constructed histograms of relevant variables. Figure 4A shows a sample of 10,000 from a bivariate normal distribution with mean $\mu = (4000, 1.7)$ and variance-covariance matrix $V = diag(40000, 0.085)$. These were treated as randomly selected pairs of realizations of two important parameters: $Diff$, the ocean diffusivity; and λ, the climate sensitivity parameter[5]. In Figure 4B, histograms of expected sea level rises in years 2000, 2020, 2050, and 2100 are plotted for the case of $Diff \approx N(4000, 40000)$, and $\lambda \approx N(1.7, 0.0072)$.

Figure 5: *Distribution of the Temperature Change for Four Scenarios, in 2000, 2010, ..., 2100*

---

[5]The climate sensitivity parameter $\lambda = 1.7$ used by IMAGE plays a major role in the climate (surface temperature) module.

Remark 5 It is worth observing the trend of increasing dispersion over time. However, it is also worth pointing out that these empirical results indicate remarkable robustness of IMAGE with respect to stochastic perturbations in these parameters. For instance, the entire span of the histogram corresponding to the year 2100 in Figure 4B is only about 10 centimeters.

It seems, however, that IMAGE is more sensitive to stochastic uncertainty in the knowledge of the initial conditions (X_o). In Figure 5. we present the plots of histograms of expected temperature change from runs of IMAGE with initial (in the year 1900) amounts of carbon in the mixed layers of the ocean treated as normal variables. Again, the means were those values that were used by IMAGE and standard deviations were taken as 10% of the means. The four "mountain ridges"' going from bottom left to the right represent the histograms corresponding to the four emission scenarios considered in Rotmans [9]. While these scenarios are drastically different in terms of economic and environmental control assumptions, we see that the histograms of the three "conservative" scenarios overlap significantly. The time step (along the horizontal) is taken as 10 years starting with the year 2000.

We may conclude from the above that input from environmental scientists that would narrow down the uncertainty in the initial conditions would considerably decrease the variability in IMAGE's outputs.

References

[1] BRADDOCK, R.D., FILAR, J.A., *Response Times of the Oceans and Greenhouse Forcing*, submitted to Climatic Change, 1990.

[2] BRADDOCK, R.D., FILAR, J.A., den ELZEN, M.G., ROTMANS, J., ZAPERT, R. *Mathematical Formulation of the IMAGE Greenhouse Model*, Techical Report Australian Environmental Studies, Griffith University, Nathan, Queensland, Australia, 1991.

[3] CASTI, J.L, *Linear Dynamical Systems*, Academic Press, Orlando, 1987

[4] den ELZEN, M.G., ROTMANS, J., SWART R.J., *The Greenhouse Effect of Halocarbons After the Montreal Protocol*, submitted to International Environmental Affairs, 1990.

[5] den ELZEN, M.G.J., ROTMANS, J., VLOEDBELD, M. *Modeling Climate Related Feedback Processes*, RIVM Report nr. 222901007 1991

[6] den ELZEN, M.G.J., KREILEMAN, E., ROTMANS, J., *Documented Version of the IMAGE Model (v.1.0)*, RIVM Report nr. *******, 1991

[7] LASHOF, D. A., *The Dynamic Greenhouse, Feedback Processes that may Influence Future Concentrations of Atmospheric Trace Gases and Climatic Change*, Climatic Change, Vol. 14, pp. 213-242.

[8] NIJMEYER, Van der SCHAFT, *Nonlinear Dynamical Systems*, Springer Verlag, Amsterdam, 1991

[9] ROTMANS, J., *IMAGE: An Integrated Model to Assess the Greenhouse Effect*, Kluwer, Dovdvecht, The Netherland, 1990.

[10] GOUDRIAAN, J. KETNER, P., *A Simulation Study for the Global Carbon Cycle Including Main Impact on the Biosphere*, Climatic Change, Vol. 6 pp. 167–192, 1984.

[11] WIGLEY, T.M., SCHLESINGER, M.E., *Analytical Solution for the Effect of Increasing CO_2 on Global Mean Temperature*, Nature, Vol. 315, pp. 649–652, 1985

[12] *Matlab Mathworks v.3.1*, 1989

Identification of Linear Systems
Using Rational Approximation Techniques [++]

Man-ping Cai [+] and E. B. Lee [+]

Abstract

Given measured frequency response data from a linear system (amplitude ratio and phase shift) at a significant number of frequencies, one can use a recently developed rational approximation algorithm to find a reasonable model of the system (in the transfer function form) with H_∞-norm error estimates. The H_∞-norm error can be made arbitrarily small by increasing the order of the approximant and therefore leads to models which are sufficient for robust controller designs. The algorithm uses FFT type calculations which make it convenient to use and computationally well behaved. The use of the algorithm for system identification along with some aids developed to make it easier to use will be described.

I> Introduction

It is unusual to find a system which is finite dimensional. Thus, there is always a need to say what can safely be ignored[4]. This is done in most engineering projects by using performance specifications for the various subsystems which are used to compose the overall system. These error tolerances ensure that everything will fit together so as to perform the given task.

The use of various error tolerances in the design of controlled systems continues to be a major theme of control systems research. Finding a close fit to frequency response data is commonly a first step in feedback control system design (see reference [10] section 9-9 for example). Once an adequate model is known, Feedback Control Theory [3] can then be used in synthesizing the controller to meet various design specifications such as stability and robust optimal performance.

We shall describe here a recently developed algorithm (based on a Fourier-Laguerre theory) for finding rational approximations [5] and [6] ; also see [8] and [15]. While the algorithm was developed to find finite dimensional approximations for infinite dimensional systems, it has found application in the finite dimensional cases as a reasonable way to get a reduced order model with known H_∞-norm error estimates [1]. Also, since it can be directly applied to numerical frequency response data, it can be used to replace the curve fitting procedures now used to find state or transfer function models [10]. It has also found applications in finding adequate finite dimensional controllers for infinite dimensional systems, [11] and [12], and combined with other robust controller synthesis techniques provides an H_∞-norm approximant model sufficient for robust stabilization [2]. Some recent comparisons of the Fourier-Laguerre series based techniques conclude that

---

+ Center for Control Sciences and Dynamical Systems, University of Minnesota, 200 Union Street, Minneapolis, Minnesota 55455.
++ The research reported on here was supported by NSF grant DMS 9002919.

while the convergence rate is not optimal they are usually easily computable, see for example [13] or [14].

In the next section II, a brief description of the rational approximation algorithm based on the Fourier-Laguerre series techniques will be presented. In the section III, we describe various ways to select the parameters of the algorithm; these have to do with the Laguerre parameter λ, the number of points M of the DFT, and the number N of Laguerre functions used in the approximation procedure. Also the error estimate depends on the reduction of the Laguerre model to reasonable order, say n, by balanced model reduction. Examples illustrate the tradeoffs between these parameters which then becomes a trial and error interaction with the computer program (the MATLAB software with control toolbox and μ-synthesis option were used on a PC computer in our study).

It should be noted that to obtain frequency response data for an unstable system requires special care. First there is the need to do the experimentation only after the system in included in a stabilizing feedback situation. Second, sensor noise or other influences can affect the quality of the frequency response data. We assume that there are only a finite number of unstable modes and that reasonably accurate frequency response data is available.

II> Rational Approximation Algorithm and Model Reduction Algorithm

Assume that the system is linear and has transfer function model $T(s)$. We seek a finite dimensional approximant $\tilde{T}(s)$ such that

$$\left\| T(s) - \tilde{T}(s) \right\|_\infty$$

is sufficiently small. We now describe the algorithm of Gu et al. [5, 6] for doing this approximation.

One way to obtain a finite-dimensional approximation is to use a partial fraction expansion to decompose $T(s)$ into $T_s(s) + T_u(s)$, in which $T_s(s)$ and $T_u(-s)$ are both analytic in the open right half plane, and then find a finite dimensional approximation of $T_s(s)$ and $T_u(-s)$. It is a fact that for any Laplace transformable function $H(s)$, we can write

$$H(s) = \sum_{k=0}^{\infty} H_k \, \phi_k(s) \tag{2.1}$$

where

$$\phi_k(s) = \frac{\sqrt{2\lambda}}{\lambda + s}\left(\frac{\lambda - s}{\lambda + s}\right)^k \quad , \quad H_k \in R^{m \times p}$$

This is then a Fourier-Laguerre series for $H(s)$, where the Fourier-Laguerre coefficients H_k are defined in usual way (see theorem 2 below).

Defining

$$H(s) = \frac{\sqrt{2\lambda}}{\lambda + s} T(s)$$

then

$$T_s(s) = \sum_{k=0}^{\infty} H_k \left(\frac{\lambda - s}{\lambda + s} \right)^k = \sum_{k=0}^{\infty} H_k E_k \tag{2.2}$$

Because $T_u(-s)$ is also analytic in the open right half plane,

$$T_u(s) = \sum_{k=1}^{\infty} H_{-k} \left(\frac{\lambda - s}{\lambda + s} \right)^{-k} \tag{2.3}$$

Thus,

$$T(s) = T_s(s) + T_u(s) = \sum_{k=-\infty}^{\infty} H_k \left(\frac{\lambda - s}{\lambda + s} \right)^k \tag{2.4}$$

According to the bilinear transformation

$$s := \lambda \frac{1-z}{1+z}, \quad \text{or} \quad z := \frac{\lambda - s}{\lambda + s} \tag{2.5}$$

which is a conformal mapping from the open right half plane into the unit disk, we can get

$$F(z) := T(\lambda \frac{1-z}{1+z})$$

$$= \sum_{k=-\infty}^{\infty} f_k z^k = F_s(z) + F_u(z)$$

where

$$F_u(z) = \sum_{k=1}^{\infty} f_{-k} z^{-k} \quad \text{and} \quad F_s(z) = \sum_{k=0}^{\infty} f_k z^k.$$

Obviously, $F_u(z)$ corresponds to the unstable part $T_u(s)$ and $F_s(z)$ corresponds to the stable part $T_s(s)$.

Because for the stable part

$$F_s(z) = \sum_{k=0}^{\infty} H_k z^k \approx \sum_{k=0}^{N-1} H_k z^k =: F_{NS}$$

$F_s(z)$ can be approximated if the finite number of coefficients H_k ($= f_k$) for $k = 0, 1, 2, \ldots, N-1$ are known. Further, according to the definition of the FFT, we have

$$H_M(k) = \frac{1}{M} \sum_{r=0}^{M-1} F(W_{2M}^r) W_{2M}^{-rk}, \quad k = 0, \cdots, M-1.$$

where

$$W_{2M} = \exp\left(\frac{j\pi}{M}\right),$$

We can easily obtain thereby approximates of H_k. Then, using the bilinear transformation on the approximation of $F_s(z)$, we get the approximation of $T_s(s)$. In the reference [5], the convergence of such an approximation scheme is developed.

Theorem 1: [5]
    Let $T_s(s)$ and $\{f_k\}$ be as above. If dT_s/ds is continuous, then $\{ \|f_k\| \}$ is an absolutely summable sequence, where $\| * \|$ is any matrix norm. In this case, the partial sum

$$F_{NS} := \sum_{k=0}^{N} f_k z^k$$

converges to $F_s(z)$ (or $T_s(s)$) in H_∞-norm as N approaches ∞.

    An error bound was also obtained in reference [5] as follows.

Theorem 2:
    Let $T_s(s)$, $\{f_k\}$ and F_{NS} be as defined above. Suppose $(1-s)dT_s/ds$ is an L_2-function, then partial sum F_{NS} converges to $F_s(z)$ in H_∞-norm as N approaches ∞. Further,

$$\| T_s - T_{NS} \|_\infty = \left\| T_s - \sum_{k=0}^{N} H_k E_k \right\|_\infty \leq \sqrt{\frac{1}{N}} \left(\sum_{k=N+1}^{\infty} \overline{\sigma}^2(kH_k) \right)^{1/2}$$

$$= \frac{1}{\sqrt{2\lambda N}} \left(\sum_{k=N+1}^{\infty} \overline{\sigma}^2(\overline{H}_k) \right)^{1/2}$$

where $\overline{\sigma}(G)$ is largest singular value of operator G, the Fourier coefficients are

$$\overline{H}_k \equiv \left\langle (\lambda - s)\frac{dT_s}{ds}, \phi_k \right\rangle = \frac{1}{2\pi j} \int_{-j\infty}^{j\infty} (\lambda - s)\frac{dT_s}{ds} \phi_k(-s) ds$$

and T_{NS} is partial sum

$$\sum_{k=0}^{N} H_k \left(\frac{\lambda - s}{\lambda + s} \right)^k .$$

    Now, we focus on approximation of the unstable part $F_u(z)$. Clearly, $F_u(z)$ and $T_u(s)$ have the same McMillan degree which by assumption is finite. And, since the bilinear transformation does not change the Hankel singular values of the original transfer function, the Hankel singular values of $F_u(z)$ are exactly the same as those of $T_u(s)$. But, how can $F_u(z)$ be computed approximately without actually doing the partial fraction expansion?

According to the **Stone-Weirestrass Theorem**, any continuous function on unit circle, $\{|z|=1\}$, can be uniformly approximated by trigonometric polynomials. Assume $T(s)$ is continuous on the imaginary axis, for

$$T\left(\lambda\frac{1-z}{1+z}\right) = F(z)$$

$F(z)$ can be approximated as

$$F(z) \approx \sum_{k=-N}^{N-1} f_k z^k$$

where

$$F_s(z) \approx \sum_{k=0}^{N-1} f_k z^k \quad \text{and} \quad F_u(z) \approx \sum_{k=-N}^{-1} f_k z^k .$$

Using 2M-point IFFT, we have

$$f_M(k) = \frac{1}{2M}\sum_{r=-M}^{M-1} F\left(W_{2M}^r\right)W_{2M}^{-rk}, \qquad k=-M,-M+1,\cdots,0,1,2,\cdots,M-1.$$

For the convergence, the following theorems are established in reference [6].

Theorem 3:
Let $F(z)$ be defined as before and let the McMillan degree of $F_u(z)$ be n. Suppose $dF(e^{jw})/d(e^{jw})$ in $L_2[0,2\pi]$. Then,

(i) $\{\|f_k\|\}$ is an absolutely summable sequence and

(ii) $f_M(k) = \sum_{L=-\infty}^{\infty} f_{2LM+k}$,

where f_k and $f_M(k)$ are defined as before, and L is an integer.

Theorem 4:
Let $F(z)$ be defined as before and $F_u(z)$ have McMillan degree n. Define

$$S_N^M(z) \equiv \sum_{k=1}^{N} f_M(-k)z^{-k}$$

where $f_M(k)$ are as before, and $M > N$. Suppose that $dF(e^{jw})/d(e^{jw})$ in $L_2[0,2\pi]$. Then,

$$\lim_{\sqrt{M}\geq N\to\infty}\left\|F_u - F_{u;n}^{M;N}\right\|_\infty = 0,$$

where $F_{u;n}^{M;N}(z)$ is an n-th order approximant function of $S_N^M(z)$ obtained using the balanced truncation scheme (or the optimal Hankel approximation), see below for balancing.

However, the McMillan degree n of $F_u(z)$ may not be known in advance. From the approximate function of $S_N^M(z)$, we know that the first n Hankel singular values of $S_N^M(z)$ converges to the true Hankel singular values of $F_u(z)$ and the rest of the Hankel singular values of $S_N^M(z)$ converges to zero as (M, N) approaches to (∞, ∞) with M > N. Otherwise $S_N^M(z)$ can't be the approximate function of $F_u(z)$. Thus, as M, N are both large, a gap between $\sigma_n(S_N^M)$ and $\sigma_{n+1}(S_N^M)$ would be significant if $\sigma_n(F_u)$ is not too small. In this case, the McMillan degree of $F_u(z)$ can be identified in the approximation process.

Since the order of the model obtained by the Fourier-Laguerre series method is quite large (typically forty or fifty, maybe even a hundred series coefficients are needed when a model of that order is obtained), a model reduction step must be used to get a realistic order model for controller synthesis. A suggested model reduction approach is now described. It will provide Hankel singular values which can be used to determine how far to reduce., see [5] or [6] for more details.

In single-input/single-output case, according to the FFT algorithm, there exists

$$F(z) \approx \sum_{k=-N}^{N-1} f_k z^k$$

where

$$f_M(k) = \frac{1}{2M} \sum_{r=-M}^{M-1} F\left(W_{2M}^r\right) W_{2M}^{-rk}, \qquad k = -M, \cdots, M-1;$$

is used as approximate value of f_k. For the stable part, since

$$F_s(z) = T_N^M\left(\lambda \frac{1-z}{1+z}\right) = \sum_{k=0}^{N-1} f_k z^k$$

select a simple realization (A, B, C, D) for G(z) as

$$A_N = \begin{bmatrix} 0 & 0 & \cdots & 0 \\ 1 & 0 & \cdots & 0 \\ \vdots & \ddots & \ddots & \vdots \\ 0 & \cdots & 1 & 0 \end{bmatrix}, \quad B_N = \begin{bmatrix} 1 \\ 0 \\ \vdots \\ 0 \end{bmatrix}, \quad C_N^T = \begin{bmatrix} f_1 \\ f_2 \\ \vdots \\ f_N \end{bmatrix}, \quad D = f_0.$$

$$(2.6)$$

Then the controllability and observability Gramians W_c and W_o for this system are, respectively,

$$W_o = \sum_{n=0}^{\infty} (A^T)^n C^T C A^n, \quad \text{and} \quad W_c = \sum_{n=0}^{\infty} A^n B B^T (A^T)^n$$

As described in [6], and also see [7] and [9], W_c and W_o are uniquely determined by the Lyapunov equations

$$A W_c + W_c A^T + B B^T = 0$$
$$A^T W_o + W_o A + C^T C = 0 \qquad (2.7)$$

Substituting A, B and C from (2.6) to (2.7), we obtain that $W_c = I_N$ and $P = \{P_{ij}\}$ where P_{ij} is the (i, j)-th element of P. And P_{ij} can be obtained by the following recursive formulas:

$$P_{k,N} = P_{N,k}^T = (C^T C)_{k,N}, \qquad k = 1,2,\ldots,N$$

and

$$P_{ij} = P_{i+1,j+1} + (C^T C)_{i,j}, \qquad (1,1) \le (i,j) \le (N,N)$$

According to the algorithm of model reduction which is described above, the normalized (balanced) system can then be obtained as follows:

$$\hat{A} = S^{-1} A S \qquad \hat{B} = S^{-1} B \qquad \hat{C} = C S \qquad \hat{D} = D$$

Here, S is a nonunique orthogonal matrix. So a reduced model of order n is obtained by direct truncation:

$$A_r = \begin{bmatrix} I_n & 0 \end{bmatrix} \hat{A} \begin{bmatrix} I_n \\ 0 \end{bmatrix}, \quad B_r = \begin{bmatrix} I_n & 0 \end{bmatrix} \hat{B}, \quad C_r = \hat{C} \begin{bmatrix} I_n \\ 0 \end{bmatrix}, \quad D_r = \hat{D}$$

Obviously, n < N. Using (2.5), a reduced-order model (A_c, B_c, C_c, D_c) for $T_N^M(s)$ is

$$A_c = \lambda (I - A_r)(I + A_r)^{-1}, \quad B_c = \sqrt{2\lambda} (I + A_r)^{-1} B_r$$
$$C_c = \sqrt{2\lambda} C_r (I + A_r)^{-1}, \quad D_c = D_r - C_r (I + A_r)^{-1} B_r$$

Based on the periodic property of FFT,

$$\sum_{k=-N}^{N-1} f_k z^k = \sum_{k=0}^{2N-1} f_k z^k = \sum_{k=0}^{N-1} f_k z^k + \sum_{k=N}^{2N-1} f_k z^k = F_s(z) + F_u(z)$$

and

$$\sum_{k=-N}^{-1} f_k z^k = \sum_{k=N}^{2N-1} f_k z^k = F_u(z)$$

Thus, for the stable part and the unstable part, we can use the same method to do the approximation.

In the multi-input/multi-output case, the theorems are no different. And in the algorithm, just approximate each element of the transfer function matrix separately. The algorithm for doing the approximation is described in more detail in the next section.

III> Algorithm and Examples

Based on the summary given in section II, the suggested algorithm is as follows:

Step 1: For a given possibly unstable transfer function T(s), verify first if

$$(\lambda - jw)\frac{dT(jw)}{djw} \in L_2[-\infty, \infty]$$

and choose $\lambda > 0$ to find F(z) in section II;

Step 2: Use 2M-point FFT algorithm to compute $f_M(k)$ as defined in section II.

Step 3: Computer Hankel singular values of

$$F_N := \sum_{k=0}^{N} f_k z^k$$
$$S_N^M(z) \equiv \sum_{k=1}^{N} f_M(-k)z^{-k} \quad ,$$

where $N^2 \leq M$. Determine n_1, the number of stable poles of T(s) or the error involved in taking a prescribed number, and n_2, the number of unstable poles of T(s).

Step 4: Use bilinear transform to obtain the stable part approximation and unstable part approximation of T(s), respectively.

Step 5: Add stable part approximation and unstable part approximation of T(s) together. The total approximation of T(s) is then obtained.

Simulations of the approximation algorithm can be done by using **manage.m** program[+]. The **procedure** for running the program **manage.m** on a personal computer is available in report form.

We now describe some of the aids available in the manage.m program.

First there is the question of what Laguerre parameter λ to choose. In work of Partington (see reference [4] and references cited there) it is suggested that λ should be n, the order of the approximant. Here (as suggested in [5] and [6]) we use the bandwidth of $T(s)|_{s=jw}$ as λ. Thus in the manage.m program the frequency response $|T(jw)|$ is calculated at a sufficient number of points w_k so that the bandwidth can be approximately selected.

Next the number of points M in the FFT algorithm is selected; usually 1024, or 2048 point FFT is used in MATLAB software. Then the number of terms for truncation of

---

Note : * means that (return) is a key on the keyboard.

+ manage.m is the result of a senior design project at the University of Minnesota; see acknowledge below.

the Fourier-Laguerre series is to be selected, both for unstable as well as stable parts. The program manage.m helps one do this by displaying the discrete time signal for the stable as well as unstable parts. See figure A for a typical plot of this. Note periodic property means for the stable part we look at the first part of the 1024 point spectrum (stable part corresponds to phase lag, while unstable part to phase lead). In figure A after about 50 the data is approximately zero so we select $N = 50$ to do an initial approximation.

To select the order n of the unstable part (we want same number of poles in right half plane for our model as the system has) we look for gaps in the Hankel singular values. Manage.m program plots these as an aid to find n_2 (see figure C).

To show more details of the algorithm and the aids available in the program manage.m we now consider a couple of examples. In each case the frequency response data was generated by creating a file from a given (analytical) transfer function. Many other examples have been tried with very promising results.

Example 3.1 Consider a system with transfer function $T(s) = \dfrac{(5 - s)}{(s+1)(s+8)}$. manage.m computes the system bandwidth to be approximately 1.8, so select $\lambda = 2$. Next compute the IFFT coefficients $f_m(k)$ and display (see Figures a and b); select $N_1 = 10$ as order of Fourier-Laguerre series truncations for stable part and $N_2 = 0$ for unstable part. Figure c is a plot of Hankel singular values for stable part versus order n_1. Big gap is between 2 and 3, but for this example select $n_1 = 4$. Figures d, e, and f show magnitude, phase and error comparisons for the approximation.

Example 3.2 Consider a system with transfer function

$$T(s) = \frac{20(6e^{-2s}+2e^{-s}-6)}{6s^2 + (6e^{-2s}-2e^{-s}-66)s - (2e^{-3s}+30e^{-2s}-12e^{-s}-180)} \quad .$$

In reference [6], λ was selected to be 10. If one uses manage.m program the bandwidth is computed to be approximately 2.2. Thus we select $\lambda = 3$ for our first calculations. (Also the selection of $\lambda = 5$ will be carried along.)

After computing the IFFT coefficients $f_m(k)$ (see equation 3.4 in reference [6]) these are displayed in manage.m program to select the order of the Fourier-Laguerre series truncation for both stable and unstable parts. Figure A shows the first 200 values in graphical form, while Figure B shows the last 200 values in graphical form. From these figures, we select $N_1 = 50$ for the stable part and $N_2 = 20$ for the unstable part. Next manage.m displays the plot of Hankel singular values, (unstable) see Figure C (also they appear in reference [6] in table 4.2). The gap between the second and third suggest two unstable poles, so select $n_2 = 2$ for the unstable part reduced order model. Just for comparison with $\lambda = 3$, we have taken $N_2 = 20$, and $n_2 = 5$ for the unstable part, and $N_1 = 50$ and $n_1 = 10$ for the stable part. Figures D, E, and F display the magnitude, phase and error of the approximant and the original model.

For further comparison with $\lambda = 5$, we again take $N_2 = 20$ and $n_2 = 5$ for the unstable part and $N_1 = 75$ and $n_1 = 20$ for the stable part. Figure α shows the first 200 points of IFFT coefficients, while Figure β shows the last 200 values. Figure γ is the same as figure C, but for the stable Hankel singular values, Figures δ, ε and ζ are the magnitude, phase and error.

Using a slightly more accurate computation yields the figures A, B and Γ for this same transfer function with $\lambda = 10$, $N_1 = 75$, $n_1 = 15$, $N_2 = 9$, and $n_2 = 2$.

References

1. H. Beke, Transfer Function Fitting to Gain-Phase Constraints, Ph.D. Thesis, University of Minnesota, 1992.

2. R. F. Curtain and K. Glover, Robust Stabilization of infinite-dimensional systems by finite-dimensional controllers, Syst. Contr. Lett. 7 (1986), pp. 41-47.

3. J. C. Doyle, B. A. Francis, and A. R. Tannenbaum, Feedback Control Theory, Macmillan Publishing Co., New York, 1992.

4. P. A. Fuhrmann, Linear Systems and Operators in Hilbert Spaces, McGraw Hill Inc., New York, 1981.

5. G. Gu, P. Khargonekar, and E. B. Lee, Approximation of infinite-dimensional systems, IEEE Trans. on Automat. Contr. 34 (1989) 610-618.

6. G. Gu, P. Khargonekar, E. B. Lee, and P. Misra, Rational approximations of unstable infinite dimensional systems, SIAM J. Contr. and Opt., 1992.

7. A. J. Laub, "Computation of System Balancing Transformations," Proceedings of 25th Conference on Decision and Control, Dec. 1986, pp. 548-553.

8. P. Mäkilä, "Approximation of stable systems by Laguerre filters", Automatica, 26, 1990, 333-345.

9. B. C. Moore, "Principal Component Analysis in Linear Systems: Controllability, Observability, and Model Reduction," in IEEE trans. on automat contr., AC-26 (1981), pp.17-31.

10. K. Ogata, Modern Control Engineering, Prentice Hall Inc., New Jersey, 1970.

11. H. Özbay, "H$^\infty$ optimal controller design for a class of distributed parameter systems", EE Dept. preprint, Ohio State Univ., 1991.

12. H. Özbay, "Controller reduction in the two block H$^\bullet$-optimal design for distributed plants", Int. J. Control, 54 (1991) 1291-1308.

13. J. Partington, "Approximation of delay systems by Fourier-Laquerre Series", Automatica, 27, (1991), 569-572.

14. B. Wahlberg, System Identification Using Laguerre Models, IEEE Trans. on Automat. Contr. 36 (1991), 551-562.

15. Eva Wu and G. Gu, "Discrete fourier transform and H^2 approximation", IEEE Trans. Automat. Contr. 35 (1990) 1044-1046.

Acknowledgement The authors are indebted to the senior design group: Mark Buer, Tamara Irwin, Linda Nowak, Kevin Sarkinen, and Craig Anderson, who implemented the algorithm and provided many of the aids which make it easy to use in the MATLAB framework.

79

$$T(s) = \frac{(5-s)}{(s+1)(s+8)} \qquad \lambda = 2 \qquad N_1 = 10 \qquad N_2 = 0 \qquad n_1 = 4 \qquad n_2 = 0$$

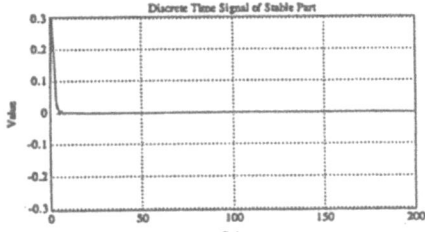

Figure a

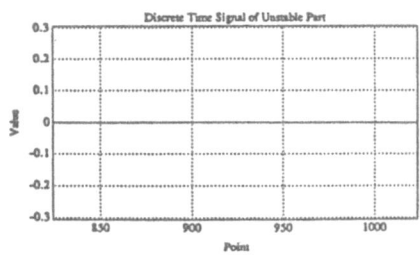

Figure b

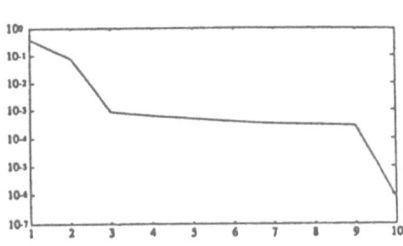

Figure c

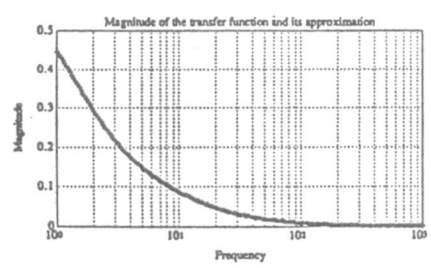

Figure d

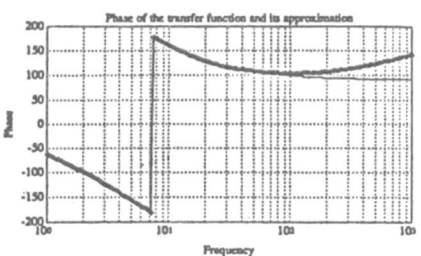

Figure e

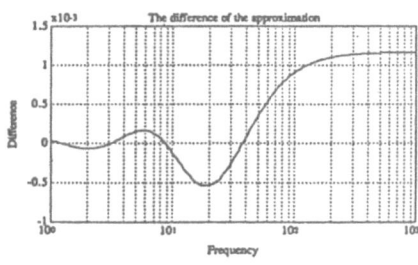

Figure f

$$T(s) = \frac{20(6e^{-2s}+2e^{-s}-6)}{6s^2 + (6e^{-2s}-2e^{-s}-66)s - (2e^{-3s}+30e^{-2s}-12e^{-s}-180)}$$

$\lambda = 3$ $N_1 = 50$ $N_2 = 20$ $n_1 = 10$ $n_2 = 5$

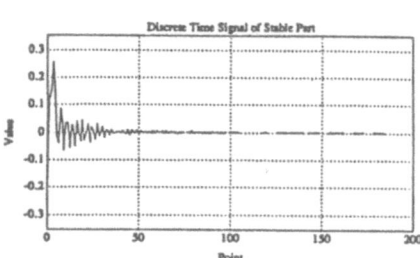

Figure A

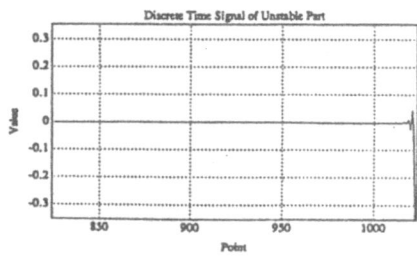

Figure B

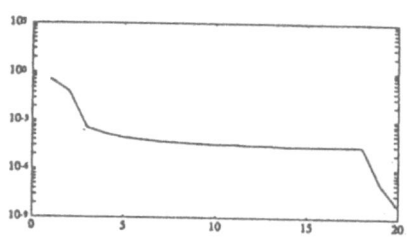

Figure C

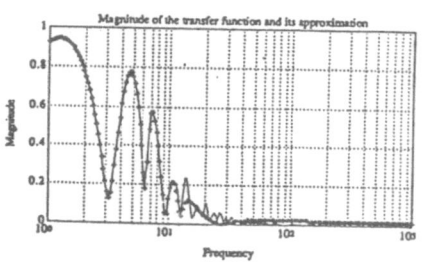

·Figure D

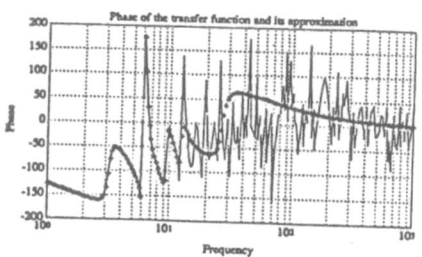

Figure E

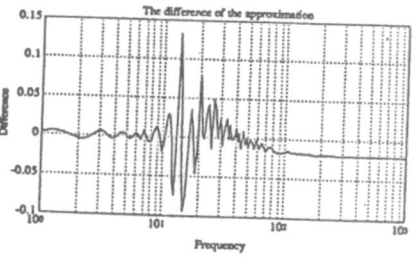

Figure F

$\lambda = 5$ $\qquad$ $N_1 = 75$ $\qquad$ $n_1 = 20$ $\qquad$ $N_2 = 20$ $\qquad$ $n_2 = 5$

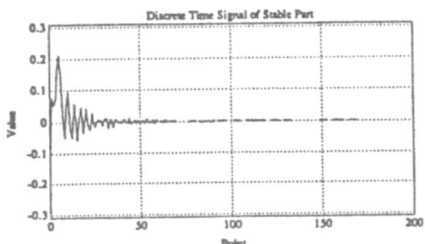

Figure α

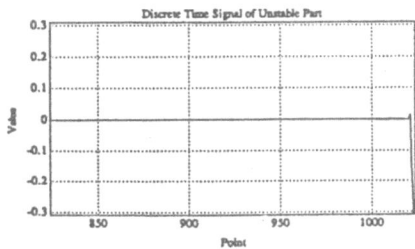

Figure β

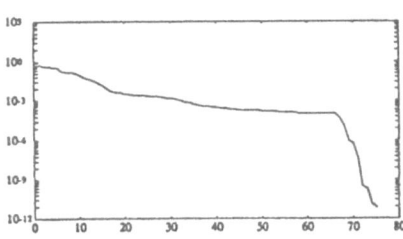

Figure γ

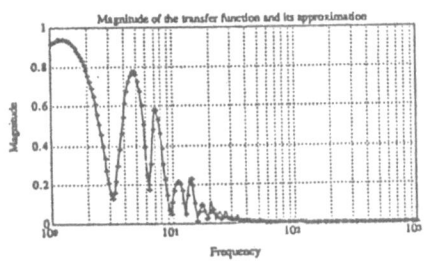

Figure δ

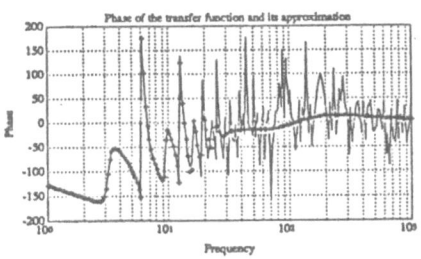

Figure ε

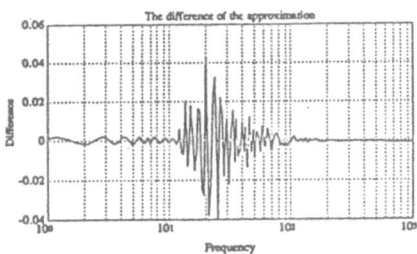

Figure ζ

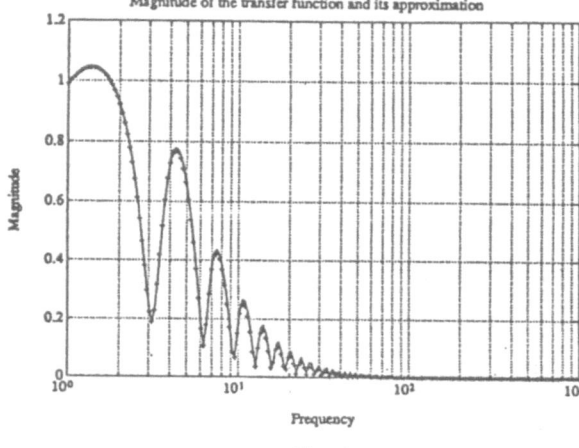

Figure A

$$\lambda = 10$$
$$N_1 = 75$$
$$N_2 = 9$$
$$n_1 = 15$$
$$n_2 = 2$$

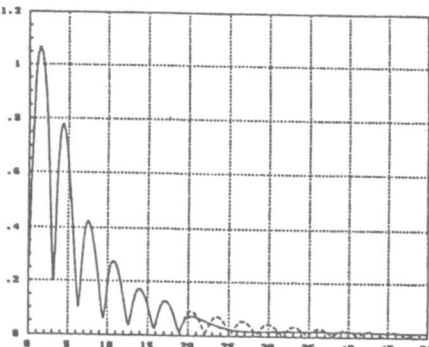

Figure B : Frequency response of $\hat{T}(s)$ and $T(s)$

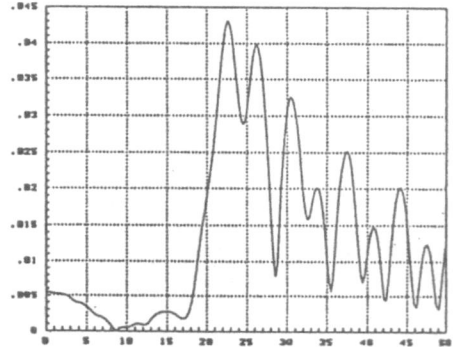

Figure Γ: Magnitude frequency response of $T(s) - \hat{T}(s)$.

Recent Results in Stochastic Adaptive Control: Non-explosion, ϵ-Consistency and Stability

P.E. Caines *

Dept. of Electrical Engineering, McGill University,
3480 University St., Montreal, P.Q. H3A 2A7, Canada,
and
Canadian Institute for Advanced Research
peterc@moe.mcrcim.mcgill.edu

Abstract

For completely observed continuous time constant parameter stochastic linear systems, an indirect adaptive control law is presented which, subject principally to a weak location hypothesis concerning the true parameter, and a persistent excitation hypothesis, generates ϵ-consistent recursive least squares parameter estimates and ensures the system is mean square sample path stable. The adaptive control algorithm entails (i) recursively calculating the least squares estimate of the system parameter, and (ii) recursively generating a stabilizing LQR feedback gain matrix. The gain matrix is projected onto the closed convex hull $\mathrm{ch}\Gamma$ of a set of matrix gains Γ known to contain a stabilizing gain and the resulting projection is used as the adaptive gain in the linear feedback control law. The a.s. non-explosion of the system is a direct consequence of this construction.

Keywords: stochastic control, adaptive control, stochastic stability

*Research partly supported by Canadian NSERC grant A 1329 and by INRIA, France.

1   Introduction

In this paper we give a non-explosion and adaptive stabilization result for completely observed continuous time parameter stochastic linear systems. Results in this area for partially observed systems have been obtained by Gevers, Goodwin and Wertz [1987], Goodwin and Mayne [1990] and Goodwin, Gevers, Mayne and Wertz [1991]; these authors employ modified forms of the standard estimation algorithms together with various hypotheses including those implying certain a-prior knowledge on the location of the parameters. There are also the robustness results of Chen and Guo [1990], who, in particular, assume the existence of strong solutions to the closed loop system stochastic differential equations. We adopt a hypothesis implying knowledge of an (arbitrarily large) region containing the true system parameter and a persistent excitation hypothesis. When the controlled system is subject to a certain class of linearly bounded indirect adaptive control laws, which use unconstrained recursive least squares parameter estimates, we prove (i) the global existence of solutions to the system equations, (ii) the parameter estimates in the closed loop system are ε-consistent and (iii) the mean square sample path stability of the system.

An aspect of our approach that we believe to be of interest is that one obtains a relatively simple proof of adaptive stabilization at the cost of a strong but easy to comprehend persistent excitation hypothesis. Furthermore, this hypothesis lends itself to further investigation; in particular, a question that merits study is whether it holds when the adaptive control signal u contains a dither processes independent of the system noise w (see e.g. Caines [Section 2, Chapter 12, 1988]).

To extend our results to the partially observed case, we may adopt—in common with the previously cited authors—a version of the familiar "least squares positive real" condition (SPR) for the disturbance transfer function. This will be described in a subsequent paper.

2   The System and the Adaptive Control Law

The class of system models Σ will be a subset of the completely observed time invariant real linear finite dimensional systems described by Itô equations of the form

$$dx_t = Fx_t dt + Gu_t dt + C dw_t, \quad t \geq 0, \tag{2.1}$$

where x_t, u_t, w_t take values in $\mathbb{R}^n$, $\mathbb{R}^m$, $\mathbb{R}^r$ respectively, w is a standard r-dimensional Wiener process on a sufficiently large underlying probability space $(\Omega, \mathcal{B}, P)$ such that w is measurable w.r.t. $\mathcal{F}^w \triangleq \{\mathcal{F}_t^w; t \geq 0\}$ and x_0 is independent of $\mathcal{F}_\infty^w$.

In order to specify the adaptive control law we first make a geometric construction which shall be used to express certain a priori knowledge on the true value of the system parameter.

Let $\theta^T = [F, G] \epsilon \mathbb{R}^{n(n+m)}$ and let V_θ denote the solution of the corresponding Riccati equation with identity weighting matrices of appropriate dimension, namely:

$$V_\theta F_\theta + F_\theta^T V_\theta - V_\theta G_\theta G_\theta^T V_\theta + I = 0. \tag{2.2}$$

As is well known, whenever $[F_\theta, G_\theta]$ is stabilizable, (2.2) possesses a positive solution; furthermore, because $[I, F_\theta]$ is observable, this solution is strictly positive, unique among positive solutions of (2.2) and yields $F_\theta - G_\theta G_\theta^T V_\theta$ asymptotically stable. The control law $u_t = -K(\theta)x_t$, where $K(\theta) \triangleq G_\theta^T V_\theta$, minimizes $\int_0^\infty (\|x_t\|^2 + \|u_t\|^2)dt$ for the system $\dot{x}_t = F_\theta x_t + G_\theta u_t$, for any $x_0 = x(0)\epsilon\mathbb{R}^n$, with respect to all feedback control laws $u_t = u(x_t), t \geq 0$, for which unique solutions to the system ODE are defined locally on $[0, t^*)$ for some $0 < t^* \leq \infty$. If $[F, G]$ is not stabilizable there will exist initial conditions for which the cost is unbounded and correspondingly solutions to (2.2) will not exist (see below).

Let $D \subset \mathbb{R}^{n(n+m)}$ denote the open subset of $\mathbb{R}^{n(n+m)}$ where $[F_\theta, G_\theta]$ is stabilizable. Now let $K : \mathbb{R}^{n(n+m)} \rightarrow \mathbb{R}^{nm}$ be given by $K(\theta) = G_\theta^T V_\theta$. By Delchamps' Lemma (Delchamps [1980]), this map is analytic on the set D. Next let $\{\Theta_\gamma; \gamma \geq 0\}$ be a sequence of compact subsets of D which are increasing on D, i.e. $\Theta_\gamma \uparrow D$ as $\gamma \rightarrow \infty$, and let $\Gamma \triangleq K(\Theta_\gamma)$. We note that Γ is compact and we shall denote by $ch\Gamma$ the compact set which is the convex hull of Γ.

We may now phrase our first hypothesis:

H1: The system in the class Σ is parameterized by (F_θ, G_θ, C) and γ is such that (i) $\theta\epsilon\Theta_\gamma$, and (ii) $\theta_0 = [0, I]\epsilon\Theta_\gamma$. □

We shall set $\bar{\gamma} = \sup\{(\text{Tr}M^T M)^{\frac{1}{2}}; M \epsilon ch\Gamma \equiv chK(\Theta_\gamma)\}$. We may also now state our hypothesis concerning the information dependence of the adaptive control function u.

HID: Let $\mathcal{B}^n$ denote the Borel σ-field on $\mathbb{R}^n$, C_n^t the set of continuous functions from $[0, t]$ to $\mathbb{R}^n$ and let $\mathcal{F}_t^x$ denote the σ-field generated by the trajectories $\{x_s; 0 \leq s \leq t\}$. Then the measurability constraint on $u(\cdot, \cdot) \equiv u.(\cdot)$ is given by

$$u.(\cdot) : \mathbb{R}^1 \times C_n^t \rightarrow \mathbb{R}^m; u : (t, x_0^t) \mapsto u_t(x_0^t)$$

with $u.(\cdot, \cdot)\epsilon\mathcal{B}^1 \otimes \mathcal{F}_t^x$. □

The estimation algorithm we shall employ is the *unconstrained* least squares algorithm given below wherein it is assumed that u satisfies HID.

First we rewrite the basic system equation (2.1) in the convenient transposed form:

$$dx_t^T = [x_t^T u_t^T]\begin{bmatrix} F^T \\ G^T \end{bmatrix}dt + dw_t^T C^T \tag{2.3a}$$

$$= \phi_t^T \theta dt + dw_t^T C^T, \tag{2.3b}$$

where $\phi_t^T \triangleq [x_t^T, u_t^T], \theta^T = [F, G], x_0 = x(0)$ and $w_0 = 0$. Then we set

$$d\theta_t = R_t\phi_t[dx_t^T - d\hat{x}_t^T]$$
$$\equiv -R_t\phi_t\phi_t^T\theta_t dt + R_t\phi_t dx_t^T, \tag{2.4}$$

where

$$dR_t^{-1} = \phi_t\phi_t^T dt, \tag{2.5a}$$

i.e.

$$dR_t = -R_t \phi_t \phi_t^T R_t dt, \tag{2.5b}$$

where $(\theta_0, R_0) = ([0, I], I)$. And so the equation for the parameter estimation error $\tilde{\theta}_t = \theta_t - \theta$ is

$$
\begin{aligned}
d\tilde{\theta}_t &= d\theta_t - d\theta \\
&= -R_t \phi_t \phi_t^T \theta_t dt + R_t \phi_t [\phi_t^T \theta dt + dw_t^T C^T] \\
&= -R_t \phi_t \phi_t^T \tilde{\theta}_t dt + R_t \phi_t dw_t^T C^T,
\end{aligned} \tag{2.6}
$$

where $\tilde{\theta}_0 = -\theta$.

The process $\{\theta_t; t \geq 0\}$ generated by (2.3), through (2.6) satisfies the measurability relation

$$\theta.(\cdot, \cdot) : \mathbb{R}^1 \times C_n^t \times C_m^t \to \mathbb{R}^{n(m+n)} : \theta(t, x_0^t, u_0^t) \mapsto \hat{\theta}_t$$

with $\theta.(\cdot, \cdot) \epsilon \, \mathcal{B}1^x \otimes \mathcal{F}_t^x \otimes \mathcal{F}_t^u$, over any interval for which a solution exists. Furthermore, whenever the dependence of u on x_0^t in (2.3) is only via (θ_t, x_t), the required measurability property HID on u is necessarily satisfied by the scheme (2.3) - (2.6).

Finally the adaptive control law u is defined as follows: assume θ is such that (2.2) has a solution V_θ (a sufficient condition for this being $\theta \epsilon D$) and let $K(\theta_t) = G^T(\theta_t) V(\theta_t)$ denote the corresponding feedback gain. Then the feedback gain matrix $K^P(\theta_t)$ to be used in the adaptive state feedback control law is given by

$$u_t = -K_t^P x_t \triangleq -K^P(\theta_t) x_t, \tag{2.7a}$$

where

$$K^P(\theta) \triangleq \begin{cases} \arg \min_{M \epsilon ch\Gamma} \|M - K(\theta)\| & \text{if } \theta \epsilon D \\ K(\theta_0) & \text{if } \theta \epsilon D^c \end{cases} \tag{2.7b}$$

and where $K^P(\theta)$ is evidently uniquely defined by the closed convex nature of $ch\Gamma$. The equations (2.3) - (2.7) constitute the complete set of equations for the closed loop system with state vector $(x_t, \tilde{\theta}_t, R_t)$. We shall now define a solution to (2.2) - (2.6), together with a piece-wise continuous solution to (2.7), on the half-open interval $[0, \tau_\theta)$, where τ_θ is an explosion time for θ_t.

Assume that a solution to (2.5a), (2.6) exists on $[0, t^*)$ with $\lim \sup_{t \to t^*} \|\theta_t\| < \infty$ (i.e. no escape at t^*), that $\theta_t \epsilon D$ for $t \epsilon [0, t^*)$ and $\theta_t \to \theta_{t^*} \epsilon D^c$ as $t \to t^*$. Recall that at $t = 0, \theta_0 = [0, I] \epsilon D$. Then a solution to (2.3) - (2.7) may be constructed on a half-open interval $[0, t^{**}) \supset [0, t^*]$ in the following manner. By construction, $[F_{\theta_{t^*}}, G_{\theta_{t^*}}]$ is not stabilizable. From (2.2)

$$
\begin{aligned}
V(\theta_t) = \int_0^\infty &\{\exp[F(\theta_t) - G(\theta_t) G^T(\theta_t) V(\theta_t)]s\} \\
&\times \{V(\theta_t) G(\theta_t) G^T(\theta_t) V^T(\theta_t) + I\} \\
&\times \{\exp[F(\theta_t) - G(\theta_t) G^T(\theta_t) V(\theta_t)]^T s\} ds.
\end{aligned}
$$

Consequently, when $F^K(\theta) \triangleq F(\theta) - G(\theta)G^T(\theta)V(\theta)$, (2.2) gives

$$\liminf_{t \to t^*} \mathrm{Tr}\,V(\theta_t) \geq \int_0^\infty \liminf_{t \to t^*} \mathrm{Tr}\{\exp[F^K(\theta_t)]s \exp[F^K(\theta_t)]^T s\} ds$$
$$= \mathrm{Tr} \int_0^\infty \{\exp[F^K(\theta_{t^*})]s \exp[F^K(\theta_{t^*})]^T s\} ds$$
$$= \infty.$$

It follows that solutions $V(\theta_t)$ to (2.2) exist at each $\theta = \theta_t$ on $[0, t^*)$ but not on $[0, t^*]$. However, for $0 \leq t < t^*$, $K^P(\theta_t)$ is given as the unique continuous solution to the scheme (2.2), (2.7b) in the compact set $ch\Gamma$, and this implies in particular that the limit $K^P(\theta_{t^*}) = \lim_{t \to t^*} K^P(\theta_t)$ exists in $ch\Gamma$, although at t^* the control law (2.7) employs the gain $K(\theta_0)$.

Hence at t^* the quantities $x_{t^*} = \lim_{t \to t^*} x_t$, $\theta_{t^*} = \lim_{t \to t^*} \theta_t \epsilon D^c$, $K_{t^*} = K(\theta_0)$, $R_{t^*} = \lim_{t \to t^*} R_t$, all exist a.s. and are bounded a.s. (in the case of x this follows from the uniform boundedness of the coefficients of x in (2.3)). Consequently there exists an interval $[t^*, t^{**})$ on which a solution to (2.3) - (2.7) exists.

It is clear that $K(\theta_t)$ may have only a piece-wise continuous trajectory on the interval $[0, t^{**})$. However, a unique solution (x_t, θ_t, R_t) to (2.3) - (2.6), together with a uniquely specified $K^P(\theta_t)$, is defined on some $[0, t^{**}) \supset [0, t^*]$ and hence such a solution may be extended throughout $[0, \tau_\theta)$. Finally, for all $t \geq \tau_\theta$, we shall set $K_t^P \triangleq \lim_{s \uparrow \tau_\theta} K^P(\theta_s)$.

For all θ in the compact set Θ_γ, let $\{P_\theta; \theta\epsilon\Theta_\gamma\}$ denote the compact set of (unique, strictly positive) solutions to the Lyapunov equation

$$P_\theta(F_\theta - G_\theta K^P(\theta)) + (F_\theta - G_\theta K^P(\theta))^T P_\theta = -I, \qquad \theta\epsilon\Theta_\gamma,$$

and let δ_γ be such that

$$4\delta_\gamma \left(\max_{\theta\epsilon\Theta_\gamma} \|P_\theta G_\theta\| \right) < 1.$$

Such a δ_γ clearly exists. Then take ϵ_γ such that

$$\epsilon_\gamma \triangleq \sup\{\epsilon : \|\theta - \theta'\| < \epsilon \implies \|K(\theta) - K^P(\theta')\| < \delta_\gamma; \forall\theta, \theta', \theta\epsilon\Theta_\gamma, \theta'\epsilon\mathbb{R}^{n(n+m)}\};$$

again such an ϵ_γ obviously exists.

Having defined the quantity ϵ, we may now frame the following persistent excitation hypothesis:

(H2)$_\epsilon$

$$\liminf_{t \to \infty} \frac{1}{t}\lambda_{\min}(R_t^{-1}) > \epsilon^{-2}4 \max_{\theta\epsilon\Theta_\gamma, M\epsilon ch\Gamma} \|F_\theta - G_\theta M\| \quad \text{a.s.}$$

□

The hypothesis (H2)$^\gamma$ is taken to be the statement that (H2)$_\epsilon$ holds for $\epsilon = \epsilon_\gamma$.

One method for attempting to ensure that (H2)$_\epsilon$ holds is to add an independent dither process v_t to the control as follows:

$$u_t = = -K_t^P x_t + v_t,$$
$$dv_t = \Lambda v_t dt + dn_t,$$

(2.8)

where n is an m-dimensional Brownian motion process independent of w and Λ is an asymptotically stable matrix.

3 ϵ-Consistency and Stability

Theorem 3.1 Subject to the hypothesis H1, the set of equations (2.3) - (2.7), with the given initial conditions, has no finite time of escape and R_t is positive and non-singular for all $t \geq 0$.

Proof For convenience we take $(x, R, \tilde{\theta})$ as the state of the system of equations (2.3) - (2.7). The coefficients on the right hand side of the state equations are piece-wise continuously differentiable with respect to the state. It follows that in any compact ball in the state space a uniform Lipschitz condition holds for the vector of drift and diffusion coefficients. Consequently by Theorem 2.5, Chapter 4, Ikeda and Watanabe [1989] there exists a unique pathwise solution up until a random explosion time $\tau(\omega) \leq \infty$.

Evidently the solution $R_t, t \geq 0$, to (2.5b) is defined up until the time of explosion $\tau_\phi(\omega)$ of ϕ. Let $\tau_\theta(\omega)$ be the time of explosion of $\tilde{\theta}_t$ and hence θ_t.

For all sample paths such that $\tau_\theta(\omega) \leq \tau_\phi(\omega)$, the projection rule (2.7b) yields the control law

$$u_t = -K^P_{\tau_\phi} x_t \, \Delta - \left[\lim_{t \to \tau_\phi} K^P(\theta_t) \right] x_t \tag{3.1}$$

for $\tau_\theta \leq t \leq \tau_\phi$.

However, since for $0 \leq t < \tau_\phi$ the solutions ϕ_t and R_t exist, we have the explicit solution for θ_t given by

$$\theta_t = R_t [\int_0^t \phi_s . dx_s + \theta_0], \qquad t < \tau_\phi. \tag{3.2}$$

This shows $\tau_\phi < \tau_\theta$ and hence $\tau_\theta = \tau_\phi$.

For all $t\epsilon[0, \tau_\phi(\omega))$, (2.3) is given by

$$dx_t = [F - GK^P_t] x_t dt + C dw_t, \tag{3.3}$$

and since this is a linear stochastic differential equation with uniformly bounded random coefficients defined for all $t \geq 0$, the solution x is defined globally, i.e. $\tau_\phi = \infty$ a.s., and hence $\tau = \infty$ a.s.

It follows that (2.5) has a decreasing solution for all $t \geq 0$. Furthermore, since the growth of R_t^{-1} is governed by the growth of ϕ_t, and hence of x_t, and since $\|x_t\|$ is an (at most) exponential function of time, we see that $\det (R_t) > 0$ for all $t \geq 0$. $\qquad \square$

Theorem 3.2 Subject to the hypotheses (H1) and (H2), the system of equations (2.3) - (2.7), with the given initial conditions, generates a strongly ϵ-consistent parameter estimate, that is to say

$$\limsup_{t \to \infty} \|\tilde{\theta}_t\| < \epsilon \qquad \text{a.s.}$$

Proof As in Chen and Moore [1987] and Chen and Guo [1990], let us differentiate $\frac{1}{2}\mathrm{Tr}\tilde{\theta}_t^T R_t^{-1}\tilde{\theta}_t$ along a solution curve to obtain

$$d(\frac{1}{2}\mathrm{Tr}\tilde{\theta}_t^T R_t^{-1}\tilde{\theta}_t)$$

$$= \frac{2}{2}\mathrm{Tr}[\tilde{\theta}_t^T R_t^{-1}\{-R_t\phi_t\phi_t^T\tilde{\theta}_t dt + R_t\phi_t dw_t^T C^T\}]$$

$$+\frac{1}{2}\{\frac{2}{2}\mathrm{Tr}\phi_t^T R_t\phi_t C^T C dt\} \tag{3.4a}$$

$$+\frac{1}{2}\mathrm{Tr}\tilde{\theta}_t^T \phi_t\phi_t^T\tilde{\theta}_t dt \tag{3.4b}$$

$$= -\frac{1}{2}\phi_t^T\tilde{\theta}_t\tilde{\theta}_t^T \phi_t dt + \frac{1}{2}\phi_t^T R_t\phi_t (\mathrm{Tr}C^T C) dt$$

$$+\mathrm{Tr}[\tilde{\theta}_t^T \phi_t dw^T C^T] \tag{3.4c}$$

Proof We now apply the result of Christopeit [1986] to estimate the third term on the right hand side of (3.4c), the hypotheses of Lemma 4 of the cited paper being satisfied here. This yields

$$\frac{1}{2}\mathrm{Tr}\tilde{\theta}_t^T R_t^{-1}\tilde{\theta}_t \le \frac{1}{2}\mathrm{Tr}\tilde{\theta}_0 R_0^{-1}\tilde{\theta}_0$$

$$-\frac{1}{2}\int_0^t \|\phi_s^T\tilde{\theta}_s\|^2 ds + \frac{1}{2}(\mathrm{Tr}C^T C)\log(\det R_t^{-1})$$

$$+ o(\int_0^t \|\tilde{\theta}_s^T\phi_s\|^2 ds)^{1/2+\eta} + 0(1), \tag{3.5}$$

where $0 < \eta < \frac{1}{2}$.

The following fundamental inequality follows from (3.5):

$$\frac{1}{2}\mathrm{Tr}\tilde{\theta}_t^T\tilde{\theta}_t \le \frac{1}{2}\mathrm{Tr}\tilde{\theta}_t^T[\frac{1}{\lambda_t}R_t^{-1}]\tilde{\theta}_t$$

$$\le \frac{1}{\lambda_t}(\frac{1}{2}\mathrm{Tr}\tilde{\theta}_0^T R_0^{-1}\tilde{\theta}_0 + O(1))$$

$$-\frac{1}{2}\frac{1}{\lambda_t}\int_0^t \|\tilde{\theta}_s^T\phi_s\|^2 ds \tag{3.6}$$

$$+\frac{1}{\lambda_t}o(\int_0^t \|\tilde{\theta}_s^T\phi_s\|^2 ds)^{1/2+\eta}$$

$$+\frac{1}{2}\frac{1}{\lambda_t}(\mathrm{Tr}C^T C)\log(\det R_t^{-1}),$$

for all $t \ge 0$, where $\lambda_t = \lambda_{\min}(R_t^{-1})$.

Hypothesis (H2) implies $\lambda_t \to \infty$ a.s. as $t \to \infty$, and clearly

$$\limsup_{t\to\infty}\left[-\frac{1}{2}\int_0^t \|\tilde{\theta}_s^T\phi_s\|^2 ds + o\left(\int_0^t \|\tilde{\theta}_s^T\phi_s\|^2 ds\right)^{1/2+\eta}\right] < \infty \text{ a.s.} \tag{3.7}$$

Hence

$$\limsup_{t\to\infty}\frac{1}{2}\mathrm{Tr}\widetilde{\theta}_t^T\widetilde{\theta}_t \le \limsup_{t\to\infty}\frac{1}{2\lambda_t}\log(\det R_t^{-1}).\qquad(3.8)$$

To obtain an estimate for the right hand side of (3.11) we observe that

$$R_t^{-1} = I + \int_0^t \phi_s\phi_s^T ds = I + \int_0^t \begin{bmatrix} I \\ -K_s^P \end{bmatrix} x_s x_s^T[I,-K_s^{P^T}]ds$$

yields

$$\mathrm{Tr}R_t^{-1} \le n + \int_0^t \mathrm{Tr}\{[I + K_s^{P^T}K_s^P]x_s x_s^T\}dr$$

$$\le n + (n + \sup_{0\le s\le t}\{\|K_s^P\|^2\})\int_0^t \|x_s\|^2 ds \qquad(3.9)$$

$$\le n + (n + \overline{\gamma}^2)\left[\int_0^t \|\Psi_s x_0 + \int_0^s \Psi_{s,\tau}C dw_\tau\|^2 ds\right]$$

$$\le n + 2(n + \overline{\gamma}^2)\left[\int_0^t [\|\psi_{s,0}x_0\|^2 + \|\int_0^s \Psi_{s,\tau}C dw_\tau\|^2]ds\right]$$

where

$$\frac{d\Psi_{s,\tau}}{ds} = [F - GK_s^P]\Psi_{s,\tau}, \quad \Psi_{\tau,\tau} = I, \quad s,\tau \ge 0. \qquad(3.10)$$

By construction, $\|F - GK_s^P\| \le \zeta_\gamma$, for all $s \ge 0$, where $\zeta_\gamma \triangleq \max_{\theta\in\Theta_\gamma, M\in\Gamma}\|F_\theta - G_\theta M\|$, and hence

$$\|\Psi_{s,\tau}\| \le \exp\zeta_\gamma(s - \tau). \qquad(3.11)$$

Employing the estimate of Christopeit once more, (3.10) and (3.11) yield, for some $\eta, 0 < \eta < 1/2$,

$$\mathrm{Tr}R_t^{-1} \le n + 2(n + \overline{\gamma}^2)\Bigg\{\int_0^t \|x_0\|^2(\exp 2\zeta_\gamma s)ds$$

$$+ \int_0^t \left[o\left(\int_0^s \|C\|^2(\exp 2\zeta_\gamma(s - \tau))d\tau\right)^{1/2+\eta} + 0(1)\right]ds\Bigg\}$$

$$\le n + 2(n + \overline{\gamma}^2)\left\{\frac{\|x_0\|^2}{2\zeta_\gamma}\exp 2\zeta_\gamma t + k^{(3)}\frac{\|C\|^2}{2\zeta_\gamma}\int_0^t(\exp 2\zeta_\gamma s)ds + k^{(4)}t\right\}$$

$$\le n + 2(n + \overline{\gamma}^2)\left\{\frac{\|x_0\|^2}{2\zeta_\gamma}\exp 2\zeta_\gamma t + k^{(3)}\frac{\|C\|^2}{(2\zeta_\gamma)^2}\exp 2\zeta_\gamma t + k^{(4)}t\right\}$$

for positive constants $k^{(3)}, k^{(4)}$ and for all t sufficiently large.

Invoking the hypothesis (H2)$_\epsilon$ gives

$$\limsup_{t\to\infty}\mathrm{Tr}\widetilde{\theta}_t^T\widetilde{\theta}_t \le \limsup_{t\to\infty}\frac{1}{\lambda_t}\log(\det R_t^{-1})$$

$$\le \limsup_{t\to\infty}\left\{\frac{1}{\lambda_t}\log\frac{(n + \overline{\gamma}^2\|x_0\|^2)}{\zeta_\gamma}\exp 2\zeta_\gamma t\right.$$

$$+ \quad \frac{1}{\lambda_t} \log \frac{k^{(3)}(n + \overline{\gamma}^2)}{2\zeta_\gamma^2} \|C\|^2 \exp 2\zeta_\gamma t$$

$$+ \frac{1}{\lambda_t} \log 2k^{(4)}(n + \overline{\gamma}^2)t\Big\}$$

$$= \limsup_{t \to \infty} \Big\{ \frac{4}{\lambda_t} \zeta_\gamma t \Big\} < \epsilon^2$$

and hence the required result. $\qquad\qquad\qquad\qquad\qquad\qquad\qquad\qquad\qquad\qquad$ □

Theorem 2.3 The closed loop system (2.3)-(2.6), with the given initial conditions, subject to the hypotheses (H1) and (H2)$^\gamma$, is sample mean square stable in the sense that

$$\limsup_{t \to \infty} \frac{1}{t} \int_0^t \|x_s\|^2 ds < \infty \qquad \text{a.s.}$$

Proof By Theorem 2.2 with the hypothesis (H2)$^\gamma$ in force, $\theta_t, t \geq 0$, is a strongly ϵ_γ-consistent process of estimates of θ, the true system parameter. By hypothesis (H1), $\theta \epsilon \Theta_\gamma$ and so by ϵ_γ-consistency $\limsup_{t \to \infty} \|K(\theta_t) - K^P(\theta)\| < \delta_\gamma$ a.s. Let $\Delta K_t \triangleq K(\theta) - K^P(\theta_t)$, and let P_θ satisfy

$$P_\theta(F_\theta - G_\theta K(\theta)) + (F_\theta - G_\theta K(\theta))^T P_\theta = -I.$$

Then

$$\begin{aligned}
d(x_t^T P_\theta x_t) = \ &x_t^T [P_\theta \tilde{F}_\theta + \tilde{F}_\theta^T P_\theta] x_t dt \\
&+ 2x_t^T P_\theta G_\theta \Delta K_t x_t dt + 2x_t^T P_\theta C dw_t \\
&+ \text{Tr} P_\theta C C^T dt,
\end{aligned}$$

where $\tilde{F}_\theta = F_\theta - G_\theta K(\theta)$. Hence, using Christopeit's estimate again with $0 < \eta < 1/2$,

$$\begin{aligned}
x_t^T P_\theta x_t \leq \ &x_{t_0}^T P_\theta x_{t_0} - \int_{t_0}^t \|x_s\|^2 ds \\
&+ 2 \limsup_{t \geq t_\omega} \|\Delta K_t\| \|P_\theta G_\theta\| \int_{t_\omega}^t \|x_s\|^2 ds \\
&+ 2 \int_{t_0}^{t_\omega} x_s^T P_\theta G_\theta \Delta K_s x_s ds \\
&+ o\big(\int_{t_0}^t \|x_s\|^2 ds\big)^{1/2+\eta} \\
&+ \text{Tr}(P_\theta C C^T)(t - t_0) + 0(1).
\end{aligned}$$

By ϵ_γ-consistency, we may take t_ω sufficiently large that

$$(-1 + 2 \limsup_{t \geq t_\omega} \|\Delta K_t\| \|P_\theta G_\theta\|) < -1/2,$$

and then we obtain

$$\limsup_{t \to \infty} \frac{1}{t} \int_{t_0}^t \|x_s\|^2 ds < \infty \qquad \text{a.s.}$$

as required. $\qquad\qquad\qquad\qquad\qquad\qquad\qquad\qquad\qquad\qquad\qquad\qquad\qquad\qquad$ □

The analysis of the persistent excitation condition in the presence of an appropriate dither and the extension of these results to partially observed systems will be presented in a subsequent paper. Futhermore, we shall examine the construction of continuous $K^P(\theta_t)$ functions via an analysis of conditions implying that $\theta_t \in D^c$ for only a countable set of instants t.

Acknowledgements

The author gratefully acknowledges conversations about this work with H.F. Chen, E. Pardoux and J.C. Willems.

References

Caines, P.E. *Linear Stochastic Systems*, John Wiley, NYC, 1988.

Chen, H.F. and J.B. Moore Convergence Rates of Continuous-time Stochastic ELS Parameter Estimation. *IEEE Trans-Automatic Control*, Vol AC 32, 1987, pp 267-269.

Chen, H.F. and L. Guo "Continuous Time Stochastic Adaptive Tracking—Robustness and Asymptotic Properties", *SIAM J. Control and Opt.*, Vol.28, No.3, May 1990, pp.513-527.

Christopeit, N. "Quasi-least-squares Estimation in Semi-martingale Regression Models", *Stochastics*, Vol. 16, 1986, pp.255-278.

Delchamps D. "A Note on the Analyticity of the Riccati Metric" in *Algebraic and Geometric Methods in Linear System Theory* Lectures in Applied Mathematics, Vol. 18, AMS, Providence, Rhode Island, 1980.

Gevers, M., G.C. Goodwin and V. Wertz "Continuous Time Stochastic Adaptive Control", *SIAM J. Control and Opt.* Vol. 29, No.2, March 1991, pp. 264-282.

Goodwin, G.C. and D.Q. Mayne, "Continuous Time Stochastic Model Reference Adaptive Control", Technical Report, Department of Electrical and Computer Engineering, University of Newcastle, March 1990.

Goodwin, G.C., M. Gevers, D.Q. Mayne and V. Wertz. "Stochastic Adaptive Control: Results and Perspectives" in *Topics in Stochastic Systems: Modelling Estimation and Adaptive Control* L. Gerencsér and P.E. Caines, Eds: LNCIS, Springer Verlag, Heidelberg, 1991.

Has'minskii, R.Z., *Stochastic Stability of Differential Equations*, Sijthoff and Nordhoff, Amsterdam, 1980.

Ikeda, N. and S. Watanabe, *Stochastic Differential Equations and Diffusion Processes*, North Holland/Kodansha, Amsterdam and Tokyo, 1989.

Stochastic Adaptive Control*

Han-Fu CHEN

Institute of Systems Science, Academia Sinica
Beijing 100080, P. R. China

Abstract.

Some results in stochastic adaptive control for the ARMAX system and its time-varying analogue are surveyed. Including one or other result in this paper is to demonstrate the theoretical possibility of stochastic adaptive control, rather than to mean its importance in applications. The emphasis is on the convergence aspect of the problem. Proofs for some results are outlined where appropriate.

The paper first presents convergence and consistency results on parameter estimates, given by the stochastic gradient (SG) and the extended least squares (ELS) algorithms, which are the basis of all adaptive controls considered in the paper. Then the paper discusses the following adaptive control problems: adaptive tracking, including both SG- and ELS-based trackers; adaptive model reference control; adaptive stabilization and adaptive LQ control. Special attention is paid to achieving both optimality of control performance and consistency of parameter estimates under minimal conditions imposed on the system. Finally, robustness analysis and systems with time-varying parameters are concerned.

1 Introduction

Some results in stochastic adaptive control for the ARMAX system and its time-varying analogue are surveyed. The results are selected for the paper in order to demonstrate the theoretical possibility of adaptive control, rather than to mean their importance in applications. No attempt has been made to provide a complete list of papers in the field. The emphasis is on the convergence aspect of the problem.

The ARMAX system is a stochastic difference equation

$$
\begin{aligned}
y_n + A_1 y_{n-1} + \ldots + A_p y_{n-p} &= b_1 u_{n-d} + B_2 u_{n-d-1} + \ldots + B_q u_{n-d-q+1} \\
&\quad + w_n + C_1 w_{n-1} + \ldots + C_r w_{n-r}, \\
y_n = 0, \quad u_n = 0, \quad w_n &= 0, \quad \forall n < 0,
\end{aligned}
\tag{1.1}
$$

---

* Project supported by the National Natural Science Foundation of China.

or compactly,

$$A(z)y_n = B(z)u_{n-d} + C(z)w_n,$$

where

$$A(z) = I + A_1z + ... + A_pz^p, \tag{1.2}$$

$$B(z) = B_1z + ... + B_qz^{q-1}, \tag{1.3}$$

$$C(z) = I + C_1z + ... + C_rz^r, \quad \text{with} \quad zy_n = y_{n-1}. \tag{1.4}$$

In system (1.1) y_n, u_n and w_n are the m-dimensional output, l-dimensional input and m-dimensional driven noise, respectively. $\{w_n, \mathcal{F}_n\}$ is assumed to be a martingale difference sequence with respect to a family $\{\mathcal{F}_n\}$ of nondecreasing σ-algebras, d is the system delay, while (p, q, r) are system orders. u_n is of feedback feature. Thus u_n is $\mathcal{F}_n$-measurable.

The problem we deal with is that the feedback control u_n is required to design so that a performance index is minimized or the closed-loop system possesses some prescribed properties, but the system delay d, orders (p, q, r) and the system coefficient θ

$$\theta = [-A_1 \ ... \ -A_p \quad B_1 ... B_q \quad C_1 ... C_q]^r \tag{1.5}$$

are unknown.

So, in solving adaptive control problems it is natural first to estimate unknown parameters. In the case, where the estimation error is small, the certainly-equivalency-control (CEC) will, hopefully, give satisfactory results. Otherwise, if estimates are far from their true values, we have no reason to expect that the CEC is acceptable. On the other hand, a diminishing excitation technique, as will be explained later on, will lead to consistency of estimates if the system is already stabilized. Thus we are in a cyclic position: a good estimation implies a good system behavior, which incorporating with a dither technique, in turn leads to a good estimation. This is one of the major difficulties in adaptive control. To overcome this, we shall find that a diminishing excitation technique and a switch technique, which consists in switching the control at stopping times in accordance with the estimation accuracy, will play an important role.

In Section 2 we give convergence and consistency results on parameter estimates, given by the stochastic gradient (SG) and the extended least squares (ELS) algorithm, which are the basis of all adaptive controls considered in the paper. In Section 3 we discuss adaptive tracking, including both SG- and ELS-based trackers and adaptive model reference control. Section 4 is on adaptive stabilization under minimal conditions imposed on $A(z)$ and $B(z)$, while Section 5 concerns the adaptive LQ control. Special attention is paid to achieving both optimality of control performance and consistency of parameter estimates. In Section 6, some results on the robustness analysis and on time-varying systems are presented.

2 Parameter Estimation

For System (1.1) with (p, q, r) fixed and with $d = 1$ the commonly used in applications method for estimating unknown coefficient θ is the ELS algorithm which is defined as follows:

$$\theta_{n+1} = \theta_n + a_n P_n \varphi_n (y_{n+1}^\tau - \varphi_n^\tau \theta_n), \tag{2.1}$$

$$P_{n+1} = P_n - a_n P_n \varphi_n \varphi_n^\tau P_n, \quad a_n = (1 + \varphi_n^\tau P_n \varphi_n)^{-1} \tag{2.2}$$

with arbitrary initial values θ_0 and $P_0 > 0$. For convenience, we set $P_0 = \alpha_0 I$, $0 < \alpha_0 < \frac{1}{e}$, where

$$\varphi_n^\tau = [y_n^\tau \cdots y_{n-p+1}^\tau \quad u_n^\tau \cdots u_{n-q+1}^\tau \quad \hat{w}_n^\tau \cdots \hat{w}_{n-r+1}^\tau], \tag{2.3}$$

$$\hat{w}_n = y_n - \theta_n^\tau \varphi_{n-1}. \tag{2.4}$$

In the case $C(z) = I$, (2.3) turns to be

$$\varphi_n^\tau = [y_n^\tau \cdots y_{n-p+1}^\tau \quad u_n^\tau \cdots u_{n-q+1}^\tau]$$

and the ELS coincides with the well-known LS estimate.

Sometimes, for convenience of analysis the ELS algorithm is simplified to a so-called SG algorithm which is described as follows:

$$\theta_{n+1} = \theta_n + \frac{a\psi_n}{r_n}(y_n^\tau - \psi_n \theta_n^\tau), \quad 0 < a < 1, \tag{2.5}$$

where ψ_n denotes φ_n defined by (2.3) with $\hat{w}_n$ replaced by

$$\hat{w}_i' = y_i - \theta_{i-1}^\tau \psi_{i-1}, \quad i = n, \ldots, n - r + 1$$

and

$$r_n = 1 + \sum_{i=1}^{n} \|\psi_i\|^2. \tag{2.6}$$

For ELS it is easy to verify that

$$P_n = \left(\sum_{i=0}^{n-1} \varphi_i \varphi_i^\tau + \frac{1}{\alpha_0} I \right)^{-1}. \tag{2.7}$$

Let us denote by $\lambda_{\min}^{(n)}$ and $\lambda_{\max}^{(n)}$ the minimum and the maximum eigenvalue of P_n^{-1} respectively. In general, $\lambda_{\min}(X)$ denotes the minimum eigenvalue of a matrix X.

The following is a rough estimate for estimation error

$$\tilde{\theta}_n = \theta - \theta_n, \tag{2.8}$$

produced by LS, but it is valid for a general class of system noise ξ_n. To be specific, temporarily, let the system be

$$A(z)y_n = B(z)u_{n-1} + \xi_n, \tag{2.9}$$

where ξ_n is the system noise not necessarily equal to $C(z)w_n'$.

Theorem 2.1. *For System (2.9) if*

$$\limsup_{n\to\infty} \frac{1}{n} \sum_{i=1}^{n} \|\xi_i\|^2 < \infty \quad a.s., \tag{2.10}$$

then for LS

$$\|\theta - \theta_n\|^2 = O\left(\frac{n}{\lambda_{\min}^{(n)}}\right) \tag{2.11}$$

Proof. It is straightforward to verify that

$$tr\tilde{\theta}_{k+1}^\tau P_{k+1}^{-1}\tilde{\theta}_{k+1} \le tr\tilde{\theta}_k^\tau P_k^{-1}\tilde{\theta}_k - \|\tilde{\theta}_{k+1}^\tau\varphi_k\|^2 - 2\xi_{k+1}^\tau\tilde{\theta}_{k+1}^\tau\varphi_k. \tag{2.12}$$

From this by (2.10) we derive

$$tr\tilde{\theta}_{n+1}^\tau P_{n+1}^{-1}\tilde{\theta}_{n+1} = O(n), \tag{2.13}$$

which implies (2.11).

$\square$

The physical meaning of (2.11) is quite clear: In the case where $\lambda_{\min}^{(n)}$ diverges faster than n, the system noise satisfying (2.10) is relatively small in comparison with input-output data or the system is asymptotically equivalent to the deterministic one, then the LS algorithm provides a consistent estimate.

For System (1.1) we have more precise result[9,43,44].

Theorem 2.2. *For System (2.1) assume that*

1. $\sup_{n\ge 0} E[\|w_{n+1}\|^\beta | \mathcal{F}_n] < \infty$ *a.s.*, $\beta \ge 2$
2. $C^{-1}(e^{i\lambda}) + C^{-\tau}(e^{i\lambda}) - I > 0$, $\forall\lambda \in [0, 2\pi]$, *i.e.* $C^{-1}(z) - \frac{1}{2}I$ *is SPR.*

Then the ELS algorithm leads to

$$\|\theta_{n+1} - \theta\|^2 = O\left(\frac{\log\lambda_{\max}^{(n)}(\log\log\lambda_{\max}^{(n)})^{\delta(\beta-2)}}{\lambda_{\min}^{(n)}}\right) \quad a.s. \tag{2.14}$$

where

$$\delta(x) = \begin{cases} 0, & \text{if } x \ne 0 \\ c,, & \text{if } x = 0, \end{cases} \text{ with arbitrary } c > 1.$$

Proof. We give an outline of the proof.

Comparing (2.14) with (2.11), (2.13) we find that it suffices to prove

$$tr\tilde\theta^{\tau}_{n+1}P^{-1}_{n+1}\tilde\theta_{n+1} = O\big(\log\lambda^{(n)}_{\max}(\log\log\lambda^{(n)}_{\max})^{\delta(\beta-2)}\big). \tag{2.15}$$

Corresponding to (2.12) we now have

$$tr\tilde\theta^{\tau}_{n+1}P^{-1}_{n+1}\tilde\theta_{n+1} = O(1) - k_0\sum_{i=0}^{n}\|\tilde\theta^{\tau}_{i+1}\varphi_i\|^2 - 2\sum_{i=0}^{n}w_{i+1}\tilde\theta^{\tau}_{i+1}\varphi_i \tag{2.16}$$

where $k_0 > 0$ is a small positive constant coming from using the SPR condition.

The right-hand side of (2.16) can be estimated by $O(1) + \sum_{i=0}^{n}(a_i\varphi^{\tau}_iP_i\varphi_i)^2\|w_{i+1}\|^2$. The key step is to show that this term is dominated by the right-hand side of (2.15).

□

The method for analysing the ELS algorithm is a probabilistic one, essentially based on the estimation for weighted sum of martingale difference sequence, while for the SG algorithm the method is partly algebraic. This is understandable, because from (2.5)

$$\tilde\theta_{n+1} = \Big(I - \frac{a\psi_n\psi^{\tau}_n}{r_n}\Big)\tilde\theta_n + \frac{a\psi_n\varphi^{\zeta}_n}{r_n}\theta - \frac{a\psi_n}{r_n}w^{\tau}_{n+1}, \tag{2.17}$$

and the behavior of $\tilde\theta_n$ virtually depends on the transition matrix

$$\Phi(n+1,i) = \Big(I - \frac{a\psi_n\psi^{\tau}_n}{r_n}\Big)\Phi(n,i), \quad \Phi(i,i) = I, \tag{2.18}$$

where

$$\varphi^{\zeta}_n = [0...0 \quad \zeta^{\tau}_n \cdots \zeta^{\tau}_{n-r+1}]^{\tau}, \quad \zeta_n = y_n - w_n - \theta^{\tau}_{n-1}\psi_{n-1}.$$

Theorem 2.3. *Assume that for System (1.1) the following conditions hold:*

1. $\quad \sup_{n\geq 0} E[\|w_{n+1}\|^2|\mathcal{F}_n] < \infty$ *a.s.*

2. $\quad C(z) - \frac{a}{2}I$ *is SPR, i.e.* $C(e^{i\lambda}) + C^{\tau}(e^{i\lambda}) - aI > 0, \forall\lambda \in [0,2\pi].$

Then θ_n given by the SG algorithm (2.5) converges to θ on the set

$$S \triangleq \{\omega : \Phi(n,0) \xrightarrow[n\to\infty]{} 0\}.$$

In the special case $r = 0$ the converse is also true, i.e. $\{\omega : \Phi(n,0) \xrightarrow[n\to\infty]{} 0\} = \{\omega : \theta_n \xrightarrow[n\to\infty]{} \theta\}$ *for any θ_0.*

It is clear that the solution of the homogeneous equation corresponding to (2.17) tends to zero if and only if $\{\omega : \Phi(n,0) \xrightarrow[n\to\infty]{} 0\}$. So the main task is to analyse the last two terms of (2.16). For proof we refer to [11, 13].

It is worth noting that to guarantee $\{\omega : \Phi(n,0) \xrightarrow[n\to\infty]{} 0\}$ the persistent excitation (PE) condition is not necessary, in fact, $\lambda^{(n)}_{\max}/\lambda^{(n)}_{\min}$ may be allowed to diverge at a rate of $(\log r_n)^{\frac{1}{4}-\delta}$, $\delta \in [0,\frac{1}{4})$.

In order to guarantee strong consistency of parameter estimates, conditions, presented in the previous Theorems, are not easy to be verified for a feedback control system and, in general, not satisfied by an adaptive control system. In [44] an occasional excitation technique is used to get consistency of parameter estimates in adaptive control systems. For the same purpose we here present a different approach, the diminishing excitation technique, which consists in disturbing the desired control by an excitation signal tending to zero so that it makes $\lambda_{\min}^{(n)}$ to diverge at a sufficiently large rate in order to guarantee strong consistency, but it goes to zero and hence has no effect on the control performance.

Let $\{\varepsilon_i\}$ be a sequence of l-dimensional independent and identically distributed random vectors with continuous distribution and let $\{\varepsilon_i\}$ be independent of $\{w_i\}$ with $E\varepsilon_i = 0$, $E\varepsilon_i\varepsilon_i^\tau = I$ and $\|\varepsilon_i\| \leq \delta$, $\delta > 0$. Define

$$v_n = \varepsilon_n/n^{\varepsilon/2}, \quad \varepsilon \in \left[0, \frac{1}{2(t+1)}\right), \tag{2.19}$$

$$t = \max(p,q,r) + mp - 1. \tag{2.20}$$

In what follows $\{v_n\}$ given by (2.19) will serve as the diminishing excitation signal when the ELS algorithm is used, while for the SG algorithm instead of (2.19) we define the excitation as

$$v_n = \frac{\varepsilon_n}{\log^{\varepsilon/2} n}, \quad \varepsilon \in \left(0, \frac{1}{4s(m+2)}\right), \tag{2.21}$$

$$s = \max(p,q,r+1). \tag{2.22}$$

Without loss of generality we may assume $\{\mathcal{F}_n\}$ is sufficiently rich such that w_n and v_n are $\mathcal{F}_n$-measurable, otherwise we need only to extend $\mathcal{F}_n$ appropriately.

Set

$$\mathcal{F}_n' = \sigma\{w_i,\ 0 \leq i \leq n,\ v_j,\ 0 \leq j \leq n-1\}.$$

Let $\mathcal{F}_n'$-measurable u_n^s be the desired control at time n. The diminishing excitation technique suggests to take control as

$$u_n = u_n^s + v_n. \tag{2.23}$$

Theorem 2.4. *Suppose that for System (1.1) the following conditions are satisfied.*

i).
$$\sup_{n \geq 0} E[\|w_{n+1}\|^\beta | \mathcal{F}_n] < \infty \ a.s., \ \beta \geq 2, \tag{2.24}$$

$$\liminf_{n \to \infty} \lambda_{\min}\left(\frac{1}{n}\sum_{i=0}^n w_i w_i^\tau\right) > 0 \quad and \quad \limsup_{n \to \infty} \frac{1}{n}\sum_{i=0}^n \|w_i\|^2 < \infty \ a.s. \tag{2.25}$$

ii). *$A(z)$, $B(z)$, $C(z)$ have no common left factor and $[A_p, B_q, C_r]$ is of row-left-rank.*

iii). *Control (2.23) is applied with*

$$\frac{1}{n}\sum_{i=0}^n \|u_i^s\|^2 = O(n^\delta), \quad \delta \in \left[0, \frac{1 - 2\varepsilon(t+1)}{2t+3}\right), \tag{2.26}$$

where ε and t are defined by (2.19) and (2.20).

Then the following assertions are true:

i). *There are constant $c_0 > 0$ and integer n_0 such that*

$$\lambda_{\min}\left(\sum_{i=0}^{n} \varphi_i^0 \varphi_i^{0r}\right) \geq c_0 n^\alpha, \quad n \geq n_0 \tag{2.27}$$

for any $\alpha \in \left(\frac{1}{2}(1+\delta), 1 - (t+1)(\varepsilon+\delta)\right]$, where φ_i^0 denotes φ_i with $\hat{w}_j$ replaced by w_j, $j = i$, ..., $i - r + 1$.

ii). *If $C^{-1}(z) - \frac{1}{2}I$ is SPR and*

$$\|y_n\|^2 = O(n^b) \quad \text{for some } b > 0, \tag{2.28}$$

then the ELS estimate is strongly consistent with convergence rate

$$\|\theta_n - \theta\|^2 = O\left(\frac{\log n (\log \log n)^{\delta(\beta-2)}}{n^\alpha}\right) \tag{2.29}$$

It is not difficult to conclude (2.29) from Theorem 2.2 and (2.27). So the crucial part of the theorem is to prove (2.27), which can be shown by successively applying estimates for weighted sums of martingale difference sequence[9].

If the desired control u_n^* and the system output y_n are bounded, then (2.26), (2.28) are obviously satisfied, and under the rest conditions of the theorem we have the convergence rate indicated in (2.29).

For the SG algorithm we can also apply the diminishing excitation technique to get consistency of the estimate if conditions of Theorem 2.4 are satisfied with (2.26), (2.28) replaced by[11]

$$\frac{1}{n}\sum_{i=1}^{n}(\|u_i^*\|^2 + \|y_i\|^2) = O((\log n)^\delta) \quad \text{for some small } \delta > 0.$$

The order and time-delay estimation problem for a feedback control system is different from that considered in the time series analysis[34]. Under the condition that the upper bounds for unknown orders of a stochastic feedback control system are available, the consistent order estimate is first given in [12, 29], where the ELS algorithm is used for coefficient estimation, while in [35] a predicted least squares method is used. Further, the upper bound assumption has been removed in [37] by using regression vectors with increasing length. The consistent time-delay estimate is given in [16].

For feedback control systems it is important to consider system parameter estimation in the case where the data $\{u_n\}$, $\{y_n\}$ are corrupted by the non-Gaussian noise. It is interesting to remove all condition on $C(z)$ and to estimate coefficients of $A(z)$ and $B(z)$ only. It is also of interests to get a recursive formula for estimating orders. These problems belong to further study.

3 Adaptive Tracking and Model Reference Control

Let $\{y_n^*\}$ be a given bounded reference signal. The control purpose is to minimize

$$\limsup_{n\to\infty} \frac{1}{n}\sum_{i=1}^{n} \|y_i - y_i^*\|^2. \tag{3.1}$$

Assume $d = 1$ and (p, q, r) are the upper bounds for true orders. It is well-known that the optimal control is defined from the following equation

$$\theta^\tau \varphi_n = y_{n+1}^*. \tag{3.2}$$

Let us exclude B_1 and u_n from θ and φ_n respectively, and denote the results by $\bar\theta$ and $\overline{\varphi}_n$. Then the solution of (3.2) is expressed by

$$u_n = B_1^{-1}(y_{n+1}^* - \bar\theta^\tau \overline{\varphi}_n) \tag{3.3}$$

whenever $\det B_1 \neq 0$, $m = l$.

For the case where $\bar\theta$ is unknown and $y_n^* \equiv 0$ Åström and Wittenmark in [1] replaced $\bar\theta$ by its ELS estimate $\bar\theta_n$ and called the resulting adaptive control system as the self-tuning regulator. They proved that it is optimal if $\bar\theta_n$ converges (not necessarily to the true value).

Åström-Wittenmark adaptive trackers have got a great success in applications, and naturally have drawn much attention from control theorists. The optimality of the adaptive tracker was first proved for the one based on the SG algorithm[33].

Theorem 3.1. *Assume that*

i). $\quad \sup_{n\geq 0} E[\|w_{n+1}\|^\beta | \mathcal{F}_n] < \infty$ *a.s.,* $\beta \geq 2$,

$$\lim_{n\to\infty} \frac{1}{n}\sum_{i=0}^{n} w_i w_i^\tau = R;$$

ii). $\quad d = 1$, $m = l$, $\det B(z) \neq 0$, $|z| \leq 1$;

iii). $C(z) - \frac{a}{2}I$ *is SPR for some* $a \in (0, 1]$.

Then the certainty-equivalency-control defined from

$$\theta_n^\tau \psi_n = y_{n+1}^*. \tag{3.4}$$

leads to

$$\limsup_{n\to\infty} \frac{1}{n}\sum_{i=0}^{n} (\|u_i\|^2 + \|y_i\|^2) < \infty \quad a.s. \tag{3.5}$$

and

$$\lim_{n\to\infty} \frac{1}{n}\sum_{i=0}^{n} (y_i - y_i^*)(y_i - y_i^*)^\tau = R \tag{3.6}$$

where θ_n is given by the SG algorithm (2.5).

The theorem is proved by using a stochastic Lyapunov function. The key step is to show that

$$\sum_{i=0}^{\infty} \frac{\|y_{i+1} - w_{i+1} - \theta_i^r \psi_i\|^2}{r_i} < \infty \quad a.s.$$

With this done, we then have

$$\frac{1}{r_n} \sum_{i=0}^{n} \|y_{i+1} - w_{i+1} - \theta_i^r \psi_i\|^2 \to 0,$$

from which, by the minimum phase condition ii), it is easy to conclude (3.6) and (3.5).

It is instructive to note that although the SG-based adaptive tracker is optimal, the estimate θ_n used in the tracker is not necessarily consistent.

In case $y_n^* \equiv 0$, it is shown in [2] that $\theta_n \to \gamma\theta$ for some γ.

For the case $d \geq 1$, optimality and convergence of an SG-based tracker are proved in [13, 14].

The great progress made in studying SG-based trackers has stimulated research on ELS-based adaptive trackers, which are applied in practice and supposed to have a better convergence rate.

In [27] and [45] an SG-based adaptive tracker is used for a finite time period to stabilize the system. After the system having been stabilized the tracker is switched on the ELS algorithm. It is proved that such a complicated tracker asymptotically behaves as the ELS-based adaptive tracker. Under a Gaussian assumption in [39] it is shown that the LS-based adaptive tracker is optimal except a set of measure zero in the parameter space.

Let us consider the Åström-Wittenmark adaptive tracker (1.1) with

$$u_n = B_1^{-1}(y_{n+1}^* - \overline{\theta}_n^r \overline{\varphi}_n) \tag{3.7}$$

where $\overline{\theta}_n$ is the ELS estimate for $\overline{\theta}$:

$$\overline{\theta}_{n+1} = \overline{\theta}_n + \overline{a}_n \overline{P}_n \overline{\varphi}_n (y_{n+1} - B_1 u_n - \overline{\theta}_n^r \overline{\varphi}_n)^r, \tag{3.8}$$

$$\overline{P}_{n+1} = \overline{P}_n - \overline{a}_n \overline{P}_n \overline{\varphi}_n \overline{\varphi}_n^r \overline{P}_n, \ \overline{a}_n = (1 + \overline{\varphi}_n^r \overline{P}_n \overline{\varphi}_n)^{-1} \tag{3.9}$$

$$\overline{\varphi}_n = [y_n^r ... y_{n-p+1}^r \ u_n^r ... u_{n-q+1}^r \ \overline{w}_n^r ... \overline{w}_{n-r+1}^r]^r, \tag{3.10}$$

$$\overline{w}_n = y_n - B_1 u_{n-1} - \overline{\theta}_n^r \overline{\varphi}_{n-1}, \quad \overline{w}_n = 0, \ n < 0 \tag{3.11}$$

with arbitrary $\overline{\theta}_0, \overline{\varphi}_0 \neq 0$ and $\overline{P}_0 > 0$.

Theorem 3.2.[28] *Assume that*

i). $\quad \sup_{n \geq 0} E[\|w_{n+1}\|^{\beta} | \mathcal{F}_n] < \infty \ a.s., \ \beta \geq 2 \ and$

$$\lim_{n \to \infty} \frac{1}{n} \sum_{i=0}^{n} w_i w_i^r = R > 0;$$

ii). $C^{-1}(z) - \frac{1}{2}I$ *is SPR;*

iii). $\quad d = 1, m = l, \det B(z) \neq 0, \forall z : |z| \leq 1.$

Then the Åström-Wittenmark adaptive tracker (1.1), (3.7)-(3.11) is stable and optimal:

$$\limsup_{n \to \infty} \frac{1}{n} \sum_{i=0}^{n} (\|u_i\|^2 + \|y_i\|^2) < \infty \quad a.s. \tag{3.12}$$

$$\lim_{n \to \infty} \frac{1}{n} \sum_{i=0}^{n} (y_i - y_i^*)(y_i - y_i^*)^{\tau} = R \quad a.s. \tag{3.13}$$

Further, if $\|w_n\|^2 = O(d_n)$ *with* d_n *nondecreasing and* $\sup_n \dfrac{d_{n+1}}{d_n} < c,$ *then*

$$\sum_{i=0}^{n} \|y_i - y_i^* - w_i\|^2 = O(n^\varepsilon d_n) \quad a.s. \quad \forall \varepsilon > 0.$$

Proof. We prove (3.12), (3.13) for the simplest case $C(z) = I$, $\|w_n\|^2 \leq W$.
We have

$$y_{k+1} = B_1 u_k + \bar{\theta}^{\tau} \overline{\varphi}_k + w_{k+1} = \tilde{\bar{\theta}}_k^{\tau} \overline{\varphi}_k + y_{k+1}^* + w_{k+1}, \quad \tilde{\bar{\theta}}_k = \bar{\theta}_k - \bar{\theta}, \tag{3.14}$$

and hence

$$\frac{1}{n} \sum_{i=0}^{n} (y_{i+1} - y_{i+1}^*)(y_{i+1} - y_{i+1}^*)^{\tau} = R(1 + o(1)) + O\left(\frac{1}{n} \sum_{i=0}^{n} \|\tilde{\bar{\theta}}_i^{\tau} \overline{\varphi}_i\|^2\right). \tag{3.15}$$

So it suffices to show that the last term tends to zero as $n \to \infty$.
Set

$$s_n = e + \sum_{i=0}^{n} \|\overline{\varphi}_i\|^2, \quad \alpha_i = \frac{\|\tilde{\bar{\theta}}_i^{\tau} \overline{\varphi}_i\|^2}{1 + \overline{\varphi}_i^{\tau} \overline{P}_i \overline{\varphi}_i}, \quad \overline{P}_{n+1}^{-1} = \overline{P}_0^{-1} + \sum_{i=0}^{n} \overline{\varphi}_i \overline{\varphi}_i^{\tau}.$$

From the properties of the ELS estimate we have

$$\sum_{i=0}^{n} \alpha_i = O(\log s_n). \tag{3.16}$$

Then from (3.14), (3.16) it follows that

$$\|y_{k+1}\|^2 \leq 2\|\tilde{\bar{\theta}}_k^{\tau} \overline{\varphi}_k\|^2 + O(1) = 2\alpha_k[1 + \overline{\varphi}_k^{\tau} \overline{P}_k \overline{\varphi}_k + \overline{\varphi}_k^{\tau}(\overline{P}_k - \overline{P}_{k+1})\overline{\varphi}_k] + O(1)$$

$$\leq 2\alpha_k \delta_k \|\overline{\varphi}_k\|^2 + O(\log s_k),$$

where $\delta_k = tr(\overline{P}_k - \overline{P}_{k+1}) \geq 0$, $\sum_{i=0}^{\infty} \delta_i < \infty$.
Using the minimum phase condition we then have

$$\|y_{k+1}\|^2 \leq M \alpha_k \delta_k \sum_{i=0}^{k} \lambda^{k-i} \|y_i\|^2 + O(\log s_k), \quad \lambda \in (0, 1)$$

or

$$h_{k+1} \leq (\lambda + M\alpha_k\delta_k)h_k + O(\log s_k), \qquad (3.17)$$

where $h_k = \sum_{i=0}^{k} \lambda^{k-i}\|y_i\|^2$.

From (3.17) we obtain

$$
\begin{aligned}
h_{k+1} &\leq O\left(\sum_{i=0}^{k-1} \lambda^{k-i} \prod_{j=i+1}^{k} \left(1 + \frac{M}{\lambda}\alpha_j\delta_j\right)\log s_i\right) + O(\log s_k) \\
&= O\left(\sum_{i=0}^{k-1} \lambda^{k-i}\log s_i e^{\frac{M}{\lambda}\sum_{j=i+1}^{k}\alpha_j\delta_j}\right) + O(\log s_k). \qquad (3.18)
\end{aligned}
$$

For any $\varepsilon > 0$, there exists i_0 such that

$$\frac{M}{\lambda}\sum_{j=i+1}^{k}\alpha_j\delta_j \leq \frac{M}{\lambda}\log s_k \sum_{j=i+1}^{k}\delta_j \leq \varepsilon\log s_k, \quad \forall k \geq i \geq i_0.$$

Therefore, (3.18) implies that

$$
\begin{aligned}
h_{k+1} &\leq O\left(\sum_{i=0}^{k-1}\lambda^{k-i}\log s_i e^{\varepsilon\log s_k}\right) + O(\log s_k) \\
&= O(s_k^\varepsilon), \quad \forall\varepsilon > 0,
\end{aligned}
$$

which means $\|\overline{\varphi}_k\|^2 = O(s_k^\varepsilon)$.

Consequently, we have

$$
\begin{aligned}
\sum_{i=0}^{n}\|\widetilde{\overline{\theta}}_i^r\overline{\varphi}_i\|^2 + O(1) &= \sum_{i=0}^{n}\alpha_i(1 + \overline{\varphi}_i^r\overline{P}_i\overline{\varphi}_i) = O(\log s_n) + O(s_n^\varepsilon\log s_n) \\
&= O(s_n^\varepsilon), \quad \forall\varepsilon > 0
\end{aligned}
$$

and $\sum_{i=0}^{n}\|y_{i+1}\|^2 = O(s_n^\varepsilon) + O(n)$.

Hence, we find that

$$s_n = O\left(\sum_{k=0}^{n}\sum_{i=0}^{k}\lambda^{k-i}\|y_i\|^2\right) + O(n) = O(s_n^\varepsilon) + O(n) = O(n)$$

and

$$\frac{1}{n}\sum_{i=0}^{n}\|\widetilde{\overline{\theta}}_i^r\overline{\varphi}_i\|^2 = O(s_n^\varepsilon/n) = o(1).$$

$\square$

We now deal with the case where B_1 is unknown. We use ELS algorithm (2.1) and (2.2) for estimating θ defined by (1.5).

Write θ_n in the block form

$$\theta_n^r = [-A_{1n} \cdots -A_{pn} \quad B_{1n} \cdots B_{qn} \quad C_{1n} \cdots C_{rn}]. \qquad (3.19)$$

Similar to (3.4) we define the certainty-equivalency-control from

$$\theta_n^\tau \varphi_n = y_{n+1}^*, \tag{3.20}$$

i.e.

$$u_n = B_{1n}^{-1}\{y_{n+1}^* + (B_{1n}u_n - \theta_n^\tau \varphi_n)\} \tag{3.21}$$

whenever $\det B_{1n} \neq 0$. However, $P\{\det B_{1n} = 0\}$ may be greater than 0. Because of this we modify B_{1n} to $\hat{B}_{1n}$ as follows. Let the singular value decomposition for B_{1n} be

$$B_{1n} = V_n \begin{bmatrix} \Lambda_n & 0 \\ 0 & 0 \end{bmatrix} U_n^\tau \tag{3.22}$$

where V_n and U_n are orthogonal matrices and $\Lambda_n > 0$ is a diagonal matrix. Define

$$\hat{B}_{1n} \triangleq \begin{cases} B_{1n}, & \text{if } B_{1n}^\tau B_{1n} \geq \dfrac{1}{\log^\mu s_{n-1}} I \\ B_{1n} + V_n U_n^\tau \dfrac{1}{\log^{\mu/2} s_{n-1}}, & \text{otherwise,} \end{cases} \tag{3.23}$$

and the adaptive control

$$u_n = \hat{B}_{1n}^{-1}\{y_{n+1}^* + (B_{1n}u_n - \theta_n^\tau \varphi_n)\} \tag{3.24}$$

where $\mu \geq 1$ is an integer.

The system (1.1) with control (3.21) or (3.24) and estimate θ_n given by (2.1) and (2.2) is called the ELS-based adaptive tracker.

Theorem 3.3. *Assume Conditions of Theorem 3.2 hold. Then the adaptive tracker with control defined by (3.24) is stable and optimal in the sense that (3.12), (3.13) hold. Moreover, $\|y_n\|^2 + \|u_n\|^2 = o(n^\epsilon d_n)$ a.s. $\forall \epsilon > 0$ where $\{d_n\}$ is defined in Theorem 3.2.*

The idea of the proof is similar to that for Theorem 3.2. For details we refer to [28]. The modification (3.23) is introduced just for some technical reason. It is believed that the theorem holds true without modification. For case $d > 1$, to analyse the ELS-based adaptive tracker is an open problem.

As mentioned above, the parameter estimate in an optimal adaptive tracker is not necessarily consistent. However, if the reference signal $\{y_n^*\}$ satisfies the sufficient richness condition, the SG estimate is consistent[40]. We introduce a decaying richness condition:

$$\liminf_{n \to \infty} \frac{\log^{1/4} n}{n} \lambda_{\min}\left(\sum_{i=1}^n y_i^{\prime *} y_i^{\prime * \tau}\right) \neq 0 \tag{3.25}$$

with

$$y_i^{\prime *} = [y_i^{* \tau} \ \cdots \ y_{i-p-l(q-1)}^{* \tau}]^\tau$$

which, obviously, is weaker than the sufficient richness condition.

Theorem 3.4.[10] *Let Conditions i), ii), iii) of Theorem 3.1 be satisfied. Assume that B_1 and B_q are of full rank, $A(z)$ and $B(z)$ are left-coprime and $\{w_n\}$ are mutually independent with mutually independent components having continuous distribution. If the initial value for (2.5) is selected so that $\det B_{10} \neq 0$, then u_n can be defined from (3.4), and (3.5), (3.6) hold. Further, if the decaying richness condition (3.25) is satisfied, then $\theta_n \xrightarrow[n \to \infty]{} \theta$ a.s. where θ_n is given by (2.5).*

The proof is completed by 3 steps: First, it is to show that

$$\liminf_{n \to \infty} \frac{\log^{1/4} n}{n} \lambda_{\min} \left(\sum_{i=1}^{n} \varphi_i' \varphi_i'^{\tau} \right) \neq 0 \tag{3.26}$$

where

$$\begin{aligned}
\varphi_i' &= [(w_i + y_i^*)^\tau \ \dots \ (w_{i-p+1} + y_{i-p+1}^*)^\tau \\
&\quad (u_i - B(z)^{-1}A(z)\varsigma_{i+1})^\tau \ \dots \ (u_{i-q+1} - B(z)^{-1}A(z)\varsigma_{i-p+2})^\tau \\
&\quad w_i^\tau \ \dots \ w_{i-r+1}^\tau]^\tau,
\end{aligned}$$

with $\varsigma_i = y_i - w_i - y_i^*$.

Second, it is proved that (3.26) implies

$$\Phi'(n, 0) \xrightarrow[n \to \infty]{} 0, \tag{3.27}$$

where $\Phi'(n, 0)$ is defined as

$$\Phi'(n+1, i) = \left(I - \frac{a\varphi_n'\varphi_n'^{\tau}}{r_n'} \right) \Phi'(n, i), \quad \Phi'(i, i) = I,$$

$$r_n' = 1 + \sum_{i=0}^{n} \|\varphi_i'\|^2.$$

Finally, by Theorem 2.3 to complete the proof it suffices to show that (3.27) yields $\Phi(n, 0) \to 0$.

For general $\{y_n^*\}$, by using the diminishing excitation technique we can get optimal tracking and consistent estimating simultaneously.

Define

$$u_n^* \triangleq \hat{B}_{1n}^{-1}\{y_{n+1}^* + (B_{1n}u_n - \theta_n^\tau \varphi_n)\}, \tag{3.28}$$

$$u_n = u_n^* + v_n, \quad v_n = \varepsilon_n / s_{n-1}^{\epsilon/2}, \tag{3.29}$$

$$\varepsilon \in \left(0, \frac{1}{2(t+1)} \right), \quad t = \max(p, q, r) + mp - 1, \tag{3.30}$$

where $\{\varepsilon_n\}$ is defined in Section 2 and $s_n = tr P_{n+1}^{-1} > e + \sum_{i=0}^{n} \|\varphi_i\|^2$.

Theorem 3.5. *Suppose that Conditions of Theorem 3.2 are satisfied, $A(z)$, $B(z)$ and $C(z)$ have no common left factor and $[A_p, B_q, C_r]$ is of row-full-rank. Then the ELS-based adaptive tracker (1.1), (2.1), (2.2),(3.29) leads to*

$$\|\theta_n - \theta\|^2 = O\left(\frac{\log n}{n^{1-(t+1)\epsilon}}\right) \quad a.s.$$

and

$$\sum_{i=1}^{n}\|y_i - y_i^* - w_i\|^2 = O(n^{1-\epsilon}) + O(d_n) \quad a.s.$$

The proof is based on Theorems 2.4 and 3.3.

Adaptive tracking discussed above for System (1.1) is a special case of adaptive model reference control for more general CARIMA systems considered by [20, 52]:

$$A(z)y_n = B(z)u_{n-d} + e_n, \tag{3.31}$$

$$D(z)e_n = C(z)w_n, \quad y_i = w_i = 0, \ u_i = 0, \ i < 0, \tag{3.32}$$

where $A(z)$, $B(z)$ and $C(z)$, $\{w_n\}$ are the same as for System (1.1), while $D(z)$ is a scalar polynomial with known coefficients

$$D(z) = 1 + \alpha_1 z + ... + \alpha_s z^s. \tag{3.33}$$

Obviously, if $D(z) = 1$, then System (3.31), (3.32) reduces to (1.1).

Let $A^0(z)$ and $B^0(z)$ be given matrix polynomials, and let $\{y_i^*\}$ be a given signal. The reference model is

$$A^0(z)y_n = B^0(z)y_n^*, \tag{3.34}$$

and the control purpose is to lead System (3.31), (3.32) to (3.34) as close as possible, or more general, to minimize

$$\limsup_{n \to \infty} \frac{1}{n}\sum_{i=0}^{n}\|A^0(z)y_i - B^0(z)y_i^* + Q(z)D(z)u_{i-d}\|^2, \tag{3.35}$$

where $Q(z)$ is an arbitrary but given polynomial.

By using an implicit approach, in [15] the optimal adaptive control minimizing (3.35) is given, while in [16] not only the index (3.35) is minimized, but also the coefficient θ, the time-delay d and the orders are consistently estimated.

4 Adaptive Stabilization

In adaptive control to stabilize a system is of primary importance. When we solved the adaptive tracking problem we assumed $B(z)$ was of minimum-phase. It is easy to understand that we might assume stability of $A(z)$ to replace minimum-phase condition in order to get same conclusions. When the adaptive stabilization problem is concerned, it is not desirable to impose stability either on $A(z)$ or on $B(z)$. For stabilizing the system the standard condition is the coprimeness of $A(z)$ and $B(z)$.

Let us consider System (1.1) with $m = l = 1$, $d = 1$ and with (p, q) known.

If $A(z)$ and $zB(z)$ are coprime, then there exist two polynomials

$$G(z) = 1 + \sum_{j=1}^{q-1} g_j z^j, \quad H(z) = \sum_{j=0}^{p-1} h_j z^j \tag{4.1}$$

such that

$$A(z)G(z) - zB(z)H(z) = 1. \tag{4.2}$$

By (4.2) and (1.1) we have the following expressions

$$
\begin{aligned}
y_n &= G(z)C(z)w_n + zB(z)[G(z)u_n - H(z)y_n] \\
u_n &= H(z)C(z)w_n + A(z)[G(z)u_n - H(z)y_n].
\end{aligned}
$$

Therefore, the control defined by

$$G(z)u_n - H(z)y_n = 0. \tag{4.3}$$

leads to

$$\limsup_{n \to \infty} \frac{1}{n} \sum_{i=1}^{n} (y_i^2 + u_i^2) < \infty,$$

if $\sum_{i=0}^{n} w_i^2 = O(n)$.

It is natural to ask whether or not the coprimeness condition is sufficient for adaptive stabilization.

The system is adaptively stabilized, but in addition to coprimeness in [49, 19] the coefficient θ and parameters in the controller are assumed to be in a known region, in [30, 31, 48] a lower bound for the coprimeness degree is assumed to be known, and in [22] a positive constant $\delta > 0$ is required to be available so that

$$|\det[B_n \ A_n B_n \ \dots \ A_n^{s-1} B_n]| > \delta > 0,$$

where

$$
A_n = \begin{bmatrix}
A_{1n} & 1 & 0 & \dots & 0 \\
\vdots & 0 & \ddots & \ddots & \vdots \\
\vdots & \vdots & & \ddots & 0 \\
\vdots & \vdots & & & 1 \\
A_{sn} & 0 & \dots & \dots & 0
\end{bmatrix}, \quad
B_n = \begin{bmatrix}
B_{1n} \\
\vdots \\
B_{sn}
\end{bmatrix}, \quad s = \max(p, q)
$$

and $\theta_n = [A_{1n} \ldots A_{pn} \quad B_{1n} \ldots B_{qn}]^r$ is the estimate at time n for unknown parameter θ, $A_{in} = 0$, $B_{jn} = 0$ for $i > p$, $j > q$.

In [17, 18] under the coprimeness condition only, the system is adaptively stabilized.

For the special case $w_n \equiv 0$, it is shown[17] that the problem can be reduced to a nonadaptive one if $\max(p,q)$ is known. So, if $w_n \equiv 0$ the effort is devoted to the case where both p and q are unknown. In this case the stochastic adaptive control is given so that the system input and output tend to zero exponentially fast. Further, if the unknown parameters are also required to be estimated, then an adaptive control is designed so that the input and output tend to zero at a rate of $|y_n| + |u_n| = O\left(\frac{1}{n^{\epsilon/2}}\right)$, $\epsilon \in \left(0, \frac{1}{2}\right)$, the orders are consistently estimated and the coefficient θ is also consistently estimated with convergence rate $O\left(\frac{1}{n^{\alpha}}\right)$, $\alpha > 0$.

Let us consider the stochastic case but with $C(z) = 1$. Assume $l = m = d = 1$ and (p, q) known.

It is easy to find that coefficients

$$\psi^r = [1 \ g_1 \ldots g_{q-1} \quad h_0 \ldots h_{p-1}]$$

of $G(z)$ and $H(z)$ can explicitly be expressed via a matrix M consisting of coefficients A_i, B_j and 0, 1 so that

$$\psi = M^{-1}e, \quad e^r = [1 \ 0 \ \ldots \ 0]_{1 \times (p+q)}.$$

Replacing θ by θ_n in M we denote the result by M_n, which may be degenerate.

When $\det M_n \neq 0$, we then get an estimate $M_n^{-1}e$ for ψ, and $G_n(z)$ and $H_n(z)$ for $G(z)$ and $H(z)$ respectively. In this case it is easy to verify that

$$A_n(z)G_n(z) - zB_n(z)H_n(z) = 1. \tag{4.4}$$

The certainty-equivalency-control u_n is then defined from

$$G_n(z)u_n - H_n(z)y_n = 0. \tag{4.5}$$

Because the coprimeness of $A(z)$ and $B(z)$ is equivalent to non-degeneracy of M, M_n will be nondegenerate, and the certainty-equivalency-control will be close to the control defined from (4.3) if $\tilde{\theta}_n$ is small.

However, in general, the closeness of θ_n to θ is not guaranteed. For approaching θ we apply an explosive signal $\{u'_n\}$ which is defined as follows. Let $\{\varepsilon_n\}$ be the iid sequence defined in Section 2, and let $D(z) = 1 + \sum_{i=1}^{p+q} d_i z^i$ be an unstable polynomial of degree $p + q$, i.e. $D(z) \neq 0$, $|z| \geq 1$ and $d_{p+q} \neq 0$.

Define $\{u'_n\}$ from

$$D(z)u'_n = \varepsilon_n. \tag{4.6}$$

With the help of Theorem 2 of [43] it is not difficult to show that

$$\liminf_{n\to\infty} \frac{\lambda_{\min}^{(n)}}{a^n} \stackrel{\Delta}{=} c > 0 \quad a.s. \tag{4.7}$$

for some constant $a > 1$, if $A(z)$ and $zB(z)$ are coprime, $\lim_{n\to\infty} \frac{1}{n}\log\sum_{i=0}^{n} w_i^2 = 0$ and $u_n = u_n'$.
Therefore, by Theorem 2.1, in this case $\theta_n \xrightarrow[n\to\infty]{} \theta$ a.s.

Thus, the explosive input leads to consistency of the LS estimate on the one hand, and makes the input and output in closed-loop exponentially divergent on the other hand. The latter fact directly contradicts the stabilization purpose. So, the explosive input can only be applied for a finite period. Once the estimation accuracy becomes appropriate, one should switch on the certainty-equivalency-control. Thus, it is important to define stopping times $0 \stackrel{\Delta}{=} \tau_0 < \sigma_1 < \tau_1 < \sigma_2 < \tau_2 < \ldots.$ at which we switch on one or other control law:

$$\sigma_i = \min\left\{ n > \tau_{i-1} : \sum_{j=0}^{n-1} \varphi_j\varphi_j^\tau - (\log s_n)^3\delta_n^{-4}I > 0; \quad \det M_n \neq 0; \right.$$

$$\|G_n(z)\|^2 + \|H_n(z)\|^2 \leq \frac{1}{\max(p,q)\delta_n};$$

$$\left. \left| \frac{1}{s_n}\sum_{j=0}^{n-1}(y_j - \varphi_{j-1}^\tau\theta_n)^2 - \frac{1}{s_n}\sum_{j=0}^{n-1}(y_j - \varphi_{j-1}^\tau\theta_j)^2 \right| \leq 2\delta_n^2 \right\} \tag{4.8}$$

$$\tau_i = \min\left\{ n > \sigma_i : \left| \frac{1}{s_n}\sum_{j=0}^{n-1}(y_j - \varphi_{j-1}^\tau\theta_{\sigma_i})^2 - \frac{1}{s_n}\sum_{j=0}^{n-1}(y_j - \varphi_{j-1}^\tau\theta_j)^2 \right| > \delta_{\sigma_i}^2 + \delta_n^2 \right\}. \tag{4.9}$$

where $\{\delta_n\}$ is a sequence of real numbers with $0 < \delta_n < 1$, $\delta_n \to 0$, $\delta_n n^\alpha \to \infty$ for some $\alpha \in \left(0, \frac{1}{4}\right)$ and s_n is defined by $s_0 = 1$,

$$s_n = n \times \max\left\{ 1, \frac{1}{k}\sum_{j=0}^{k-1}(y_j^2 + u_j^2), \quad k = 1, \ldots, n \right\}, \quad \forall n \geq 1. \tag{4.10}$$

We now define adaptive control as

$$u_n = \begin{cases} u_n', & \text{if } n \in [\tau_i, \sigma_{i+1}) \text{ for some } i \geq 0, \\ H_{\sigma_i}(z)y_n - (G_{\sigma_i}(z) - 1)u_n + \frac{\varepsilon_n}{(n+1)^\varepsilon}, & \text{if } n \in [\sigma_i, \tau_i) \text{ for some } i \geq 1. \end{cases} \tag{4.11}$$

where $\varepsilon \in \left(0, \frac{1}{4(p+q)}\right)$.

From definitions, roughly speaking, σ_i is the time, after that the estimation accuracy is acceptable, while τ_i is the time starting from that the difference between θ_{σ_i} and the current LS estimate is not negligible. Therefore, u_n' is applied starting from τ_i in order to improve estimation quality, and a diminishingly excited certainty-equivalency-control is used starting from σ_i.

Theorem 4.1. *If $A(z)$ and $zB(z)$ are coprime, $C(z) \equiv 1$, $m = l = d = 1$ and $\sum_{i=0}^{n} w_i^2 = O(n)$, then adaptive control (4.11) stabilizes the closed-loop system and the LS estimate is strongly consistent, precisely,*

$$\limsup_{n \to \infty} \frac{1}{n} \sum_{j=0}^{n} (y_j^2 + u_j^2) < \infty$$

$$\|\theta_n - \theta\|^2 = O\left(\frac{\log n (\log \log n)^\gamma}{n^{1-2\varepsilon(p+q)}}\right), \quad \gamma > 1.$$

The key step of the proof is to show that the number of switches is finite and after that the diminishingly excited certainty-equivalency-control will be used for ever. The assertion on parameter estimate is the consequence of Theorem 2.4. For details we refer to [18].

5 Adaptive LQ Control

We now consider System (1.1) with $d = 1$ and an associated quadratic loss function

$$J(u) = \limsup_{n \to \infty} \frac{1}{n} \sum_{i=0}^{n-1} [(y_i - y_i^*)^\tau Q_1 (y_i - y_i^*) + u_i^\tau Q_2 u_i], \tag{5.1}$$

where $Q_1 \geq 0$, $Q_2 > 0$ and $\{y_i^*\}$ is a bounded deterministic reference signal.

Setting

$$A = \begin{bmatrix} A_1 & 1 & 0 & \ldots & 0 \\ \vdots & 0 & \ddots & \ddots & \vdots \\ \vdots & \vdots & & \ddots & 0 \\ \vdots & \vdots & & & 1 \\ A_s & 0 & \ldots & \ldots & 0 \end{bmatrix}, \quad B = \begin{bmatrix} B_1 \\ \vdots \\ \vdots \\ B_s \end{bmatrix}, \quad C = \begin{bmatrix} I \\ C_1 \\ \vdots \\ C_{s-1} \end{bmatrix}, \quad H^\tau = \begin{bmatrix} I \\ 0 \\ \vdots \\ 0 \end{bmatrix}_{ms \times m} \tag{5.2}$$

$$s = \max(p, q, r+1) \quad \text{and } A_i = 0, \, B_j = 0 \, C_k = 0 \text{ for } i > p, \, j > q, \, k > r,$$

we have

$$y_k = H x_k, \quad x_0^\tau = [y_0^\tau \; 0 \; \ldots \; 0] \tag{5.3}$$

$$x_{k+1} = A x_k + B u_k + C w_{k+1}. \tag{5.4}$$

Under certain conditions the Riccati equation

$$S = A^\tau S A - A^\tau S B (Q_2 + B^\tau S B)^{-1} B^\tau S A + H^\tau Q_1 H \tag{5.5}$$

has a unique solution $S > 0$.

In an appropriately selected set U of admissible controls the optimal control is

$$u_n^0 = Lx_n + d_n, \tag{5.6}$$

where

$$L = -(Q_2 + B^r SB)^{-1}B^r SA, \tag{5.7}$$

$$d_n = -(Q_2 + B^r SB)^{-1}B^r b_{n+1}, \tag{5.8}$$

$$b_i = -\sum_{j=0}^{\infty} F^{jr} H^r Q_1 y_{i+j}^* = F^r b_{i+1} - H^r Q_1 y_i^*, \tag{5.9}$$

$$F = A - B(Q_2 + B^r SB)^{-1}B^r SA, \tag{5.10}$$

and

$$\min_{u \in U} J(u) = J(u^0) = \limsup_{n \to \infty} \frac{1}{n} \sum_{i=0}^{n-1} [y_i^{*r} Q_1 y_i^* - b_{i+1}^r B(Q_2 + B^r SB)^{-1}B^r b_{i+1}]$$
$$+ tr SCRC^r, \tag{5.11}$$

where existence of

$$R = \limsup_{n \to \infty} \frac{1}{n} \sum_{i=0}^{n} w_i w_i^r \quad \text{with} \quad R > 0$$

is assumed.

When θ is unknown, the certainty-equivalency-control may not work. This is the difficulty of stochastic adaptive control. In [38, 41, 36, 5] solutions are given for the case where the unknown parameters are valued in a finite set, while in [53] the bounded disturbances are treated. Recently, the problem is solved for infinite dimensional case in [21].

We now present a solution of adaptive LQ control without any condition imposed on input-output data. The conditions needed are as follows:

i). $\sup_n E[\|w_{n+1}\|^\beta | \mathcal{F}_n] < \infty$, $\beta > 2$, $\lim_{n \to \infty} \frac{1}{n} \sum_{i=0}^{n} w_i w_i^r = R > 0$;

ii). $d = 1$, $m = l$, $p \geq q$, $\det B(z) \neq 0$, $\forall |z| \leq 1$;

iii). $C^{-1}(z) - \frac{1}{2}I$ is SPR;

iv). (A, B, D) is controllable and observable with D satisfying $D^r D = H^r Q_1 H$;

v). $[A_p, B_q, C_r]$ is of row-full-rank.

Let θ_n be the ELS estimate for θ and $A(n)$, $B(n)$ and $C(n)$ be estimates given by θ_n for A, B, C respectively. Define estimates for S recursively by

$$S_n = A^r(n)S_{n-1}A(n) - A^r(n)S_{n-1}B(n)(Q_2 + B^r(n)S_{n-1}B(n))^{-1}$$
$$\cdot B^r(n)S_{n-1}A(n) + H^r Q_1 H, \quad \text{with arbitrary} \quad S_0 \geq 0. \tag{5.12}$$

The x_k is estimated by an adaptive filter:

$$\hat{x}_{n+1} = A(n)\hat{x}_n + B(n)u_n + C(n)(y_{n+1} - HA(n)\hat{x}_n - HB(n)u_n) \tag{5.13}$$

$$\hat{x}_0 = [y_0^r \ 0 \ ... \ 0].$$

Putting these estimates into (5.7)-(5.10) we correspondingly obtain estimates L_n, $\hat{d}_n$, $\hat{b}_n$ and $F(n)$. Finally, from (5.6) we get the certainty-equivalency-control

$$\hat{u}_n^0 = L_n \hat{x}_n + \hat{d}_n. \tag{5.14}$$

Similar to the situation in Section 4, $\hat{u}_n^0$ may be unacceptable, if the estimation accuracy is not high enough. This is indicated by the stopping time σ_k:

$$\sigma_k = \sup \left\{ t > \tau_k : \sum_{i=\tau_k}^{j-1} \|y_i\|^2 \le (j-1)^{1+\delta} + \|y_{\tau_k}\|^2, \quad \forall j \in (\tau_k, t] \right\}, \tag{5.15}$$

because by Theorem 2.4 the diminishingly excited control leads to consistency of estimates, if growth rates of input and output are within a certain bound. After σ_k it is natural to apply the adaptive tracking control (3.24) with $y_n^* \equiv 0$ in order to stabilize the system. To avoid ambiguity let us denote it by u_n', i.e.

$$u_n' = \hat{B}_{1n}^{-1}(B_{1n}u_n - \theta_n^\tau \varphi_n). \tag{5.16}$$

We now define control

$$u_n = u_n^s + v_n, \tag{5.17}$$

where v_n is given by (3.29) and

$$u_n^s = \begin{cases} 0, & \text{if } n \in [\tau_k, \sigma_k) \cap \Lambda^c \text{ for some } k, \\ \hat{u}_n^0, & \text{if } n \in [\tau_k, \sigma_k) \cap \Lambda \text{ for some } k, \\ u_n', & \text{if } n \in [\sigma_k, \tau_{k+1}) \text{ for some } k, \end{cases} \tag{5.18}$$

where

$$\Lambda = \{i : \|\hat{u}_i^0\| \le i^{1+\delta}, i \ge 1\}, \tag{5.19}$$
$$\delta \in \left[0, \frac{1-2\varepsilon(t+1)}{2t+3} \right), \quad \varepsilon \in \left[0, \frac{1}{2(t+1)} \right),$$
$$t = \max(p, q, r) + mp - 1$$

and

$$\tau_{k+1} = \inf \left\{ t > \sigma_k : \sum_{i=\tau_k}^{\sigma_k - 1} \|y_i\|^2 \le \frac{t^{1+\delta}}{2^k}; \quad \sum_{i=\sigma_k}^{t} \|y_i\|^2 \le \frac{t^{1+\delta}}{2^k}; \right.$$
$$\left. \sum_{i=0}^{t} \|\hat{x}_i\|^2 \le t^{1+\delta/2} \right\}. \tag{5.20}$$

Theorem 5.1. *Assume that Conditions i)-v) listed above are satisfied, then the adaptive control (5.17) leads to*

$$\lim_{n \to \infty} \frac{1}{n} \sum_{i=0}^{n-1} [(y_i - y_i^*)^\tau Q_1(y_i - y_i^*) + u_i^\tau Q_2 u_i] = \min$$

and

$$\|\theta_n - \theta\|^2 = O\left(\frac{\log n}{n^{1-(l+1)\epsilon}}\right) \quad a.s.$$

The proof is completed by three steps. First, the estimate $\frac{1}{n^{1+\delta}}\sum_{i=1}^{n}\|y_i\|^2 = O(1)$ is verified separately for three cases: 1) $\tau_i < \infty$, $\sigma_i = \infty$ for some i; 2) $\tau_i < \infty$, $\sigma_i < \infty$ for all i and 3) $\sigma_i < \infty$ and $\tau_i = \infty$ for some i. Hence, by Theorems 2.4 and 3.5 θ_n is strongly consistent. Second, it is shown that there is k such that $\tau_k < \infty$ and $\sigma_k = \infty$. Third, Λ is shown to contain all integers starting from some sufficiently large n_k. Therefore, the certainty-equivalency-control is applied after n_k. This together with consistency of θ_n yields all assertions of the theorem. For details we refer to [13].

6 Deviation From Constant Parameters

A real system can rarely be modeled by an exact linear deterministic or stochastic system with constant parameters. In dealing with this situation one way is to model the system as a linear system with time-varying parameters, the other way is to analyse the influence of unmodeled dynamics contained in a system, which is modeled as a system with constant parameter, upon the behavior of adaptive control systems. The latter is the so-called robustness analysis.

In recent years much attention has been given to the issue of robust adaptive control in deterministic case. Research on robust stochastic adaptive controllers is relatively new [7, 8, 50].

In [8] it is assumed that the true system is described by (1.1) added by an unmodelled dynamics η_n which is assumed to be $\mathcal{F}_n$-measurable and dominated by

$$\|\eta_n\| \leq \epsilon \sum_{i=0}^{n} a^{n-i}(\|y_i\| + \|u_i\| + \|w_i\| + 1)$$

with $a \in (0,1)$, $\epsilon > 0$. A robustness analysis is given for the stochastic adaptive tracker designed for model (1.1) (with η_n ignored) and with control disturbed by a dither with constant variance σ^2. It is shown that under an appropriate set of conditions the system can still be stabilized, the tracking error differs from its ideal minimum by two terms, one proportional to ϵ^2 and the other proportional to σ^2, and the ELS estimate θ_n deviates from θ by $c\epsilon$ with c being a constant. In [50] normalized signals are used for parameter estimation algorithm and the adaptive tracking control is shown to be optimal in an ideal case. Further, it is also shown that the adaptive controller is robust in the sense that if mean-square stability holds for an ideal system Π_0, it will continue to hold for all systems in an open graph topological neighborhood of Π_0. In [51] a modified version of SG algorithm is used for parameter estimation. The authors developed a

Lyapunov method by which the authors show that the adaptive tracking system is stable and the tracking error deviates from the ideal minimum by a quantity proportional to the size of unmodeled dynamics. These results have been obtained without requiring SPR condition on the noise model and persistent excitation condition on the regressor.

To model systems as with time-varying parameters is another approach to deal with the deviation from constant parameters in (1.1). Let us rewrite (1.1) as

$$y_k = \varphi_k^\tau \theta + v_k \tag{6.1}$$

in the case $m = l = d = 1$, where

$$\varphi_k^\tau = [y_{k-1} \ldots y_{k-p} \quad u_{k-1} \ldots u_{k-p}], \tag{6.2}$$

$$\theta^\tau = [A_1 \ldots A_p \quad B_1 \ldots B_q], \quad v_k = C(z)w_k. \tag{6.3}$$

When θ is time-varying the corresponding system is expressed as

$$y_k = \varphi_k^\tau \theta_k + v_k \tag{6.4}$$

where v_k denotes the system noise, not necessarily being $C(z)w_n$. To estimate θ_k is the topic which has being drawn much attention from researchers in signal processing (see e.g. [3, 46]). But they usually need conditions which are hardly satisfied by feedback control systems.

For estimating θ_k in (6.4) the Kalman filter is the one of usually used methods. A precise analysis is given in [26], showing that the averaged estimation error is proportional to the averaged system noise and parameter variation. Further, these results are improved in [54]. It is worth noting that there no condition like stationarity or independence is used in the analysis. A condition called conditional richness

$$E\left[\sum_{k=m+1}^{m+h} \frac{\varphi_k \varphi_k^\tau}{1 + \|\varphi_k\|^2} \Big| \mathcal{F}_m\right] \geq \frac{1}{\alpha_m} I \quad a.s. \quad \forall m \geq 0 \tag{6.5}$$

is crucial in the analysis where α_n possibly tends to infinity. It is also shown that Condition (6.5) is satisfied by a large class of processes, for example, ϕ-mixing process, the output process of a output-controllable stochastic system and so on.

The problem of adaptive control for time-varying systems is difficult. Some results have been obtained in adaptive stabilization. For a simple system in [47] it is shown that

$$\lim_{n \to \infty} E y_n^2 = \lim_{n \to \infty} \frac{1}{n} \sum_{i=0}^{n} y_i^2 = \int y^2 d\pi, \tag{6.6}$$

where π is an invariant probability on the state-space generating the output y. Further, in [23] it is proved that adaptive control yielding (6.6) is robust in a certain sense and that stability remains valid if Gaussianality assumption used in [47] is removed.

We now present a result given in [24].

Theorem 6.1. Assume that i) $q = 1$ in (6.4), ii) the system noise $\{v_k, \mathcal{F}_k\}$ satisfies

$$E\{\exp[\varepsilon|v_{k+1}|^2|\mathcal{F}_{k-r}]\} \leq \exp\{M_v\} \quad a.s. \quad \forall k \geq 0$$

for some integer $r \geq 0$ and some constants ε and M_v, iii) $\{\theta_k, \mathcal{F}_k\}$ satisfies

$$E\{\exp[M|\theta_{k+1}|^2|\mathcal{F}_{k-m}]\} \leq \exp\{M_\theta\} \quad a.s. \quad \forall k \geq 0$$

$$E\{\exp[M|\Delta_{k+1}|^2|\mathcal{F}_{k-m}]\} \leq \exp\{\delta_\theta\} \quad a.s. \quad \forall k \geq 0$$

where $\Delta_{k+1} = \theta_{k+1} - \theta_k$, $m \geq 0$ is an integer, M, M_θ, $\delta_\theta < 1$ are positive constants.

The parameter estimates are given by

$$\widehat{\theta}_{k+1} = \pi_d \left\{ \widehat{\theta}_k + \frac{\varphi_k}{d + \|\varphi_k\|^2}(y_{k+1} - u_k - \varphi_k^\tau \widehat{\theta}_k) \right\},$$

where

$$D = \{x = (x_1 \ldots x_p) \in \mathbb{R}, |x_i| \leq L, 1 \leq i \leq p\}$$

and $\pi_D(x)$ is the nearest point in D from x.

Then the adaptive stabilization control $u_k = -\varphi_k^\tau \widehat{\theta}_k$ leads to

$$\limsup_{n \to \infty} E\{|y_n|^\beta + |u_n|^\beta\} < \infty$$

$$\limsup_{N \to \infty} \frac{1}{N} \sum_{n=0}^{N} \{|y_n|^2 + |u_n|^2\} < \infty \quad a.s.$$

if M and δ_θ are suitably large and small respectively, and L and d are taken sufficiently large.

7 Concluding Remarks

1). In stochastic adaptive control there has been made a considerable progress for discrete-time linear stochastic systems with constant parameters in recent years, but a series of problems are still open. In continuous-time case the existence of solution for adaptive control systems is of primary importance[25].

2). One of the initial motivations of adaptive control is to find controllers that work satisfactorily for systems with time-varying parameters or with unmodeled dynamics. Adaptive control theory developed for systems with constant parameters cannot completely meet demands from applications. Therefore, research on adaptive control for systems with time-varying parameters and with unmodeled dynamics is stimulated not only by theoretical interests but also by practical applications. However, at present, only a few precise results are available in this field.

References

[1]. K.J. Åström and B. Wittenmark, On self-tuning regulators, *Automatica*, 9(1973), 195-199.

[2]. A. Becker, P.R. Kumar and C.Z. Wei, Adaptive control with the stochastic approximation algorithm: geometry and convergence, *IEEE Trans. Autom. Control*, 30(1985), 4, 330-338.

[3]. A. Benveniste, M. Métivier and P. Priouret, *Adaptive Algorithms and Stochastic Approximations*, Springer, 1990.

[4]. P.E. Caines, *Linear Stochastic Systems*, Wiley, New York, 1988.

[5]. P. E. Caines and H. F. Chen, Optimal adaptive LQG control for system with finite state process parameter, *IEEE Trans. on Automatic Control*, AC-30, 1985, 185-189.

[6]. H. F. Chen, *Recursive Estimation and Control for Stochastic Systems*, Wiley, New York, 1985.

[7]. H. F. Chen and L. Guo, Robustness analysis of identification and adaptive control for stochastic system, *Systems & Control Letters*, 9, 1987, 131-140.

[8]. H. F. Chen and L. Guo, A robust stochastic adaptive controller, *IEEE Trans. on Automatic Control*, AC-33, No.11, 1988 1035-1043.

[9]. H.F. Chen and L. Guo, Convergence rate of least squares identification and adaptive control for stochastic systems, *Int. J. Control*, 44(1986), 5, 1459-1476.

[10]. H.F. Chen and L. Guo, Strong consistence of parameter estimates in optimal adaptive tracking systems, *Scientia Sinica (Series A)*, 29(1986), 11, 1145-1156.

[11]. H. F. Chen and L. Guo, Adaptive control with recursive identification for stochastic linear systems, in *Control and Dynamic Systems* (ed. C.T. Leondes), Vol.26(2), 1987, Academic Press, INC.

[12]. H.F. Chen and L. Guo, Consistent estimation of the order of the stochastic control systems, *IEEE Trans. Autom. Control*, 32(1987), 6, 531-535.

[13]. H.F. Chen and L. Guo, *Identification and Stochastic Adaptive Control*, Birkhäuser, Boston, 1991.

[14]. H. F. Chen and J.F. Zhang, Convergence rate in stochastic adaptive tracking, *Int. J. Control*, 49(1989), 6, 1915-1935.

[15]. H. F. Chen and J.F. Zhang, Stochastic adaptive control for systems with noise being an ARMA process, *Syst. Sci. & Math. Scis.*, **2**(1989), 1, 40-53.

[16]. H. F. Chen and J.F. Zhang, Identification and adaptive control for systems with unknown orders, time-delay and coefficients (Uncorrelated noise case), *IEEE Trans. Autom. Control*, **35**(1990), 8, 866-877.

[17]. H. F. Chen and J.F. Zhang, Adaptive regulation for deterministic systems, *Acta Math. Appl. Sinica*, **7**(1991), 4, 332-343.

[18]. H. F. Chen and J.F. Zhang, Adaptive stabilization of unstable and nonminimum-phase stochastic systems, submitted for publication, 1991.

[19]. H. M. J. Cantalloube, C.E. Nahum and P.E. Caines, Robust adaptive control: a direct factorization approach, *Tech. Report*, Dept. of Electrical Engineering, McGill University, Montreal, Canada, 1990.

[20]. D. W. Clarke, C. Mohtadi and P. S. Tuffs, Generalized predictive control Part I and Part II, *Automatica*, **23**(1987), 137-160.

[21]. T. E. Duncan, B. Pasik-Duncan and B. Goldys, Adaptive control of linear stochastic evolution systems, *Stochastics and Stochastic Reports*, Vol.**36**, No.2, 1991, 71-90.

[22]. B. Egardt and C. Samson, Stable adaptive control of non-minimum phase systems, *Systems & Control Letters*, **2**(1982), No.3.

[23]. L. Guo and S. Meyn, Adaptive control for time-varying systems: A combination of martingale and Markov Chain techniques, *Int. J. of Adaptive Control and Signal Processing*, **3**, No.1, 1989, 1-14.

[24]. L. Guo, On adaptive stabilization of time-varying stochastic systems, *SIAM J. Control and Optimization*, **28**(1990), 6, 1432-1451.

[25]. L. Guo, Existence and convergence of continuous-time AML, submitted for publication, 1991.

[26]. L. Guo, Estimating time-varying parameters by Kalman filter based algorithm: stability and convergence, *IEEE Trans. on Automatic Control*, **AC-35**, No.2, 141-147, 1990.

[27]. L. Guo and H.F. Chen, Convergence rate of ELS based adaptive trackers, *Syst. Sci. & Math. Scis.*, **1**(1988), 2, 131-138.

[28]. L. Guo and H.F. Chen, Åström-Wittenmark self-tuning regulator revisited and ELS-based tracker, *IEEE Trans. Autom. Control* (to appear).

[29]. L. Guo, H. F. Chen and J.F. Zhang, Consistent order estimation for linear stochastic feedback control systems (CARMA model), *Automatica*, 25(1989), 1, 147-151.

[30]. F. Giri, M. M'Saad, L. Dugard and J.M. Dion, Robust pole placement indirect adaptive control *Int. J. of Adaptive Control and Signal Processing*, 2(1988), 33-47.

[31]. F. Giri, M. M'Saad, L. Dugard and J.M. Dion, A cautious approach to robust adaptive regulation, *Int. J. of Adaptive Control and Signal Processing*, 2(1988), 273-290.

[32]. G.C. Goodwin and K. S. Sin, *Adaptive Filtering, Prediction and Control*, Prentice-Hall, Englewood Cliffs, 1984.

[33]. G.C. Goodwin, P.J. Ramadge and P.E. Caines, Discrete time stochastic adaptive control, *SIAM J. Control and Optimization*, 19(1981), 6, 829-853.

[34]. E. J. Hannan and M. Deistler, *The Statistical Theory of Linear Systems*, Wiley, New York, 1988.

[35]. E. M. Hermerly and M. H. Davis, Recursive order estimation of stochastic control systems, *Mathematical Systems Theory*, 22(1989), 323-346.

[36]. O. B. Hijab, The adaptive LQG problem, Part 1, *IEEE Trans. on Automatic Control*, AC-28, 1983, 171-178.

[37]. D.W. Huang and L. Guo, Estimation of nonstationary ARMAX models based on Hannan-Rissanen method, *The Annals of Statistics* (to appear).

[38]. P. R. Kumar, Optimal adaptive control of linear-quadratic-Gaussian systems, *SIAM J. Control & Optim.*, Vol.21, 1983, 163-178.

[39]. P.R. Kumar, Convergence of adaptive control schemes using least squares parameter estimates, *IEEE Trans. Autom. Control*, 35(1990), 4, 416-424.

[40]. P.R. Kumar and L. Praly, Self-tuning tracker, *SIAM J. Control & Optimization*, Vol.25, No.4, 1987, 1053-1071.

[41]. P. R. Kumar and P. P. Varayia, *Stochastic Systems: Estimation, Identification, Adaptive Control*, Prentice-Hall, Englewood Cliffs, 1986.

[42]. T. L. Lai and C. Z. Wei, Least squares estimates in stochastic regression models with application to identification and control of dynamic system, *Ann. Stat.*, 10(1983), No.1, 154-166.

[43]. T. L. Lai and C.Z. Wei, Asymptotic properties of general autoregressive models and strong consistency of least-squares estimates of their parameters, *Journal of Multivariate Analysis*, 13(1983), 1-23.

[44]. T.L. Lai and C.Z. Wei, Extended least squares and their applications to adaptive control and prediction in linear systems, *IEEE Trans. Autom. Control*, 31(1986), 898-906.

[45]. T. L. Lai and Z. Ying, Parallel recursive algorithms in asymptotically efficient adaptive control of linear stochastic systems, *Tech. Report*(1989), Dept. of Stochastics, Stanford University.

[46]. L. Ljung and S. Gunnarsson, Adaptive and tracking in system identification, *Tech. Report*(1988), Linköping University.

[47]. S.P. Meyn and P.E. Caines, A new approach to stochastic adaptive control, *IEEE Trans. Autom. Control*, 32(1987),220-226.

[48]. L. Praly, Robustness of indirect adaptive control based on pole placement design, *Proc. IFAC Workshop on Adaptive System in Control and Signal Processing*, San Francisco, 1983.

[49]. L. Praly, Towards a globally stable direct adaptive control scheme for not necessarily minimum phase systems, *IEEE Trans. on Automatic Control*, AC-29(1984), No.10, 946-949.

[50]. L. Praly, S. F. Lin and P. R. Kumar, A robust adaptive minimum variance controller, *SIAM J. Control & Optim.*, Vol.27, No.2, 1989, 235-266.

[51]. M. Radenkovic and A. N. Michel, Verification of the self-stabilization mechanism in robust stochastic adaptive control using Lyapunov function arguments, submitted for publication, 1991.

[52]. R. Scattlini, A multivariable self-tuning controller with integral action, *Automatica*, 22(1986), 619-627.

[53]. C. Samson, Stability analysis of adaptively controlled systems subject to bounded disturbances, *Automatica*, Vol.19(1983), 81-86.

[54]. J. F. Zhang, L. Guo and H.F. Chen, L_p-stability of estimation errors of Kalman filter for tracking time-varying parameters, *Int. J. of Adaptive Control and Signal Processing*, 5(1991), 155-174.

Adaptive Control of Admissions and Routing in an ATM Network [1]

C. Courcoubetis
  Institute of Computer Science, Research Center of Crete, Heraklion, Crete.
G. Kesidis, A. Ridder, J. Walrand
  EECS Dept., Cory Hall, University of California, Berkeley, CA94720.
R. Weber
  Queen's College, University of Cambridge, Cambridge, U.K.

Abstract

Asynchronous transfer mode (ATM) networks transport information as 53-byte cells that follow one another from the source to the destination. The different streams of cells (called virtual circuits) share buffers in the switching nodes of the network. Typically, cell streams are bursty. These fluctuations in the streams may lead switch buffers to overflow and, therefore, to lose cells. In this paper, we explain a procedure for estimating the spare capacity of switches in the network. The network can use these estimates to decide whether or not to accept new calls and how to route them. The difficulty that we address is obtaining fast and accurate estimates of the low loss probability. We propose quick estimators based on results from the theory of large deviations.

1 Introduction

Asynchronous transfer mode (ATM) is a form of packet switching that is proposed for broadband networks. In ATM, data is divided into 53 byte cells that are multiplexed on a time-slotted channel. When network traffic is bursty, ATM's use of statistical multiplexing results in an efficient use of bandwidth [1]. Every cell of a call (traffic stream) will use the same route. These routes are called virtual circuits (VCs).

Calls share buffers in switches. When the traffic offered to a buffer exceeds the buffer server's capacity, cells begin to accumulate. When a cell arrives at a full buffer, it is lost. Since cell losses are rare and delays are small, the statistics of a call do not significantly change along its virtual circuit. Therefore we assume that calls of the same type (e.g., video, speech, etc.) produce input traffic streams with identical statistics at every buffer they go through in the network. Traditional quality-of-service requirements for a call accepted through the network require that its cell loss rate be less than a predefined amount that depends on the type of service provided by the call.

Our goal is to design a routing algorithm that guarantees bounds on the cell loss probability (CLP) of the calls due of buffer overflows. The method we propose does not require models describing the statistics of the traffic. This contrasts with algorithms based on parametric models that attempt to estimate the parameters from the traffic. We choose the former approach because realistic models may be complex and slow to fit. We make an analogy with direct vs. indirect adaptive control. In indirect adaptive control, first a parametric model is fitted to the observed traffic. The optimal policy for the estimated parameters is then used. In the direct approach, the quantity to be

---

[1] Work supported by: NSERC of Canada, Pacific Bell, Micro Grant of the State of California, ARO Grant Number DAAL03-89-K-0128

optimized is measured. The control actions are selected to optimize future values of this quantity.

Thus, by monitoring its occupancy, each buffer constantly estimates its spare capacity to accept new calls. The algorithm then accepts and routes calls by using these estimates. To estimate its spare capacity, the buffer has to evaluate its CLP if more calls were added. Since the CLP is very small, estimators based on fractions of lost cells have a very large variance and are therefore very slow. To reduce the variance, we estimate the CLP if the buffer were smaller. That is, we keep track of a virtual buffer occupancy corresponding to a smaller buffer size. To relate the statistics of cell losses of this smaller buffer to those of the actual buffer, we use results on the shape of the loss probability as a function of the buffer size. These shape results are derived using the theory of large deviations[2].

2 Monitor to Infer Network Overflow Statistics

We describe an algorithm which can be used by a switch to predict its spare capacity. This algorithm is a monitor to infer network overflow statistics (MINOS).

As explained above, we want to estimate the loss probability in the switch buffers. Consider a given buffer of size B cells, with $N > 0$ virtual circuits sharing the buffer, and served by a fiber with transmission rate c cells/s. For simplicity, we assume that all the virtual circuits carry calls of the same type. The case of different traffic types is discussed in 4. Denote by $F_n = F(N_n, B_n, c_n)$ the current CLP at buffer n, for all buffers n in the network. Assuming a first-come-first-serve queuing discipline in each buffer, F_n is the CLP at buffer n by *each* call that uses buffer n. If call i uses buffers $1, 2, ..., m$, the CLP for that call is $1 - \prod_{n=1}^{m}(1 - F_n) \approx \sum_{n=1}^{m} F_n$.

Now say we are trying to route a new call. Using the above method, buffer n estimates $F_n' = F(N_n(1 + \epsilon), B_n, c_n)$, where $\epsilon = 1/N$. We attempt to find a path for the new call that satisfies

$$G^{\text{new}} \geq \sum_{n \in \text{path}} F_n'$$

where G^{new} is the CLP acceptable to the new call. Moreover, the router must ensure that, by choosing a particular path for the new call, the above constraint is not violated for any existing, previously routed call i (with guarantee G^i) which uses all or part of that path. If no path is found that satisfies these constraints, the new call is refused.

Consider now an arbitrary buffer and let $F(N, B, c)$ be the CLP at that buffer. We want to estimate the CLP, $F(N(1 + \epsilon), B, c)$, when a fraction ϵ more calls are added. In [2] we show that F has the following form for large B,

$$F(N, B, c) \approx \exp(-BI(\frac{c}{N}) + o(B)),$$

where I is a positive function. Thus, for large B,

$$F(N(1 + \epsilon), B, c) \approx F(N, B, \frac{c}{1 + \epsilon}).$$

Therefore we can estimate F when a fraction ϵ more calls are added by estimating F with the current number of calls, N, and the service rate reduced by the same fraction.

To estimate $F(N(1 + \epsilon), B, c)$, a device is added to the buffer that keeps track of the buffer occupancy, $X(t)$, when the service rate is $\frac{c}{1+\epsilon}$. Specifically, when a cell arrives at the buffer, $X(t)$ is incremented by one. Also, $X(t)$ is decremented by one every $\frac{1+\epsilon}{c}$ seconds when $X(t) > 0$. This function could be realized by a chip implementing the above algorithm. The problem now is to estimate $F(N, B, \frac{c}{1+\epsilon})$–which is very small (typically about 10^{-8}) –by monitoring the buffer. Consequently, a direct estimator based on the fraction of lost cells has a very large variance.

To improve the estimator, the device will estimate the losses for smaller buffers (called virtual buffers) so as to increase the frequency of buffer overflows, and therefore speed-up the collection of "important" samples. There is a tradeoff in choosing the size of the virtual buffers. If these virtual buffers are still too large, our estimates will be too slow. However, if these virtual buffers are too small, the original system system is over-distorted and we have a large error when we extrapolate back to B. Let B/k be the size of a virtual buffer for some $k > 1$.

Taking the $e^{o(B)}$ term in the expression for F above to be AB^ξ, we obtain estimates of

$$F(N, \frac{B}{k}, \frac{c}{1+\epsilon}) = A \left(\frac{B}{k}\right)^\xi \exp\left(-\frac{B}{k} I(N, \frac{c}{1+\epsilon})\right).$$

Because we have three unknowns (A, ξ, I), we will carry out this estimate for three values of k: $k_0 > k_1 > k_2 > 1$. These three equations can be solved for A, ξ and $I(N, \frac{c}{1+\epsilon})$. We can then plug in A, ξ and $I(N, \frac{c}{1+\epsilon})$ into the expression for $F(N, B, \frac{c}{1+\epsilon})$, and thus compute the desired quantity $F(N(1+\epsilon), B, c) \approx F(N, B, \frac{c}{1+\epsilon})$. This estimator's performance in finite time can be improved by using more complex expressions for the $o(B)$ term.

To summarize the above, the estimation algorithm in the device keeps track of three "virtual buffer" occupancy processes with buffers of size B/k_i, $i = 1, 2, 3$, and service rate $\frac{c}{1+\epsilon}$. Note that these computations can be done in parallel with the normal operation of the switch so that the estimates of $F(N, \frac{B}{k_i}, \frac{c}{1+\epsilon})$ are constantly available to the routing algorithm. We also describe a way to further reduce the variance of an estimate of F using the Kullback-Leibler distance.

3 Variance Reduction

3.1 Virtual Buffers

Recall that we proposed using virtual buffers to increase the frequency of "important" samples (buffer overflows) in order to reduce the variance of the estimate of F. The virtual buffers have sizes B/k_i, $i = 0, 1, 2$, with $k_0 > k_1 > k_2 > 1$. We will now explain a way to express the quantity we want to estimate

$$F \equiv F(N(1 + \epsilon), B, c) \approx F(N, B, \frac{c}{1+\epsilon}) \approx AB^{-\xi} e^{-BI}$$

in terms of the virtual buffer estimates

$$F_i \; \equiv \; F(N, \frac{B}{k_i}, \frac{c}{1+\epsilon}) \approx A \left(\frac{B}{k_i}\right)^{-\xi} \exp(-\frac{B}{k_i}I)$$

$i = 0, 1, 2$. Let $a_{i,j} = \frac{1}{k_i} - \frac{1}{k_j}$, $e_0 = \frac{-k_0(k_2-1)}{k_0-k_2}$, $e_2 = \frac{k_2(k_0-1)}{k_0-k_2}$, and $\gamma = \log(k_0^{e_0} k_2^{e_2})/\log(k_0^{a_{1,2}} k_1^{a_{2,0}} k_2^{a_{0,1}})$. Solving for the three unknowns, A, ξ, and I, in terms of the F_i, and substituting into the expression for F we get,

$$\log F \; = \; l_0 \log F_0 + l_1 \log F_1 + l_2 \log F_2$$

where $l_0 = e_0 + \gamma a_{2,1}$, $l_1 = \gamma a_{0,2}$, and $l_2 = e_2 + \gamma a_{1,0}$.

3.2   Analysis of Variance Reduction using Virtual Buffers

In this section, we estimate the variance reduction achieved by using the virtual buffers. For simplicity we take $\xi = 0$ and consider the variance reduction achieved by using two virtual buffers (instead of three) to estimate Φ (instead of F). Thus, we estimate

$$\Phi \equiv \; \Phi(N(1+\epsilon), B, c) \approx \Phi(N, B, \frac{c}{1+\epsilon}) \approx Ae^{-BI}$$

from the virtual buffer estimates

$$\Phi_i \; \equiv \; \Phi(N, \frac{B}{k_i}, \frac{c}{1+\epsilon}) \approx A \exp(-\frac{B}{k_i}I)$$

$i = 0, 2$. Substituting for A and I we get $\Phi = \Phi_0^{e_0} \Phi_2^{e_2}$.

Assume the time to estimate the $\Phi_{(i)}$s, n, measured in *busy cycles* is fixed and is the same for both virtual buffers. Let σ_i be the standard deviation of the estimate of Φ_i so that $\sigma_i = \sqrt{\Phi_i(1 - \Phi_i)/n} \approx \sqrt{\Phi_i/n}$, $i = 0, 2$. Thus, the relative error of the estimate of Φ, σ/Φ, satisfies

$$\frac{\sigma}{\Phi} \; \leq \; -\frac{\sigma_0}{\Phi_0}e_0 + \frac{\sigma_2}{\Phi_2}e_2 =: f(k_0, k_2)$$

for σ_i sufficiently small. Note that $e_0 < 0$ and f is an upper bound for σ/Φ because we have ignored the fact that the Φ_i are positively correlated.

Minimizing f over (k_0, k_2) we get that the optimal k_0 is very large and the optimal k_2 minimizes $g(k) \equiv (k - 1)\sqrt{1 - A} + k\sqrt{\exp(BI/k) - A}$. Let n_k and n_{TA} be the number of cycles required to achieve $\epsilon \times 100\%$ relative error with 95% confidence [3] using two virtual buffers and direct time averaging respectively. A simple computation yields: $n_k/n_{TA} = g^2(k_2)\exp(-BI)$. In our simulations, we found $A << 1$ (which implies the optimal $k_2 \approx 0.4BI$), and $BI \approx 8$, so that $n_k/n_{TA} \approx 1/17$. The speed up factor is actually larger than 17 because $\sigma/\Phi < f$; using sample standard deviations, we found a speed up was about 100. Unfortunately, fixing $\xi = 0$ results in estimates of Φ that are consistently one order of magnitude too small. These calculations give us a rule of thumb for choosing the k_i, $i = 0, 1, 2$: choose k_0 large and k_2 small (the tradeoff discussed in section 2).

3.3 Variance Reduction using the Kullback-Leibler Distance

We now describe a faster method for estimating the probability of buffer overflow in a cycle, Φ (it has been shown that $F = \Phi e^{o(B)}$ for some traffic models; see [4] and [2]). This method is useful when the estimation has to be performed very quickly, on the basis of few observations. The main point of this section is that estimators that improve upon those based on virtual buffers are possible. Instead of using three virtual buffers etc., we monitor the *peak buffer occupancy* in every cycle (call it Z_i for the ith cycle). Let n be a given number of cycles, and $B^* = B/k_0$. For integers $b \geq B^* - 1$, define the empirical tail distribution of Z_i:

$$p(b) = \frac{1}{n} \times \left\{ \begin{array}{ll} \sum_{i=1}^{n} 1\{Z_i < B^*\} & \text{if } b = B^* - 1 \\ \sum_{i=1}^{n} 1\{Z_i = b\} & \text{if } b \geq B^*. \end{array} \right.$$

Also define

$$\phi(A, I, b) = \left\{ \begin{array}{ll} 1 - A\exp(-B^* I) & \text{if } b = B^* - 1 \\ A\exp(-bI) - A\exp(-(b+1)I) & \text{if } b \geq B^* \end{array} \right.$$

where we have taken $\xi = 0$. The Kullback-Leibler distance [5] between ϕ and p is

$$K(A, I) = \sum_{b=B(*)-1}^{\infty} p(b) \log \left(\frac{p(b)}{\phi(A, I, b)} \right).$$

The values of A and I that minimize K are given by

$$I = \log \left(1 + \frac{1 - p(B^* - 1)}{\sum_{b=B^*}^{\infty} bp(b) - B^*(1 - p(B^* - 1))} \right) \text{ and}$$
$$A = (1 - p(B^* - 1))\exp(B^* I).$$

These expressions for A and I can be easily updated at the end of every cycle. Taking $\xi \neq 0$ so that K is a function of three parameters, we found no simple closed form solution to $\partial K/\partial \xi = \partial K/\partial I = \partial K/\partial A = 0$.

4 Discussion and Conclusions

The above method can be used to handle multiple types of calls sharing a buffer. Say there are six voice calls (same type) and two video calls currently using the buffer, and we wish to estimate the effect of adding a video call. Define a new type of call that is the sum of three voice calls and one video call. Therefore there are two calls of that type currently using the buffer. Instead of estimating the CLP when another video call is added, we estimate F when another call of that type is added. This, of course, may be a very conservative estimate of the affect of another video call on the buffer.

In order to estimate the number of Mips required by one virtual buffer to estimate F_i, we let the peak arrival rate into the buffer be $p \times c$ cells/s. The worst case occurs during cell loss when we have to handle the buffer occupancy and perform a comparison every $(pc + c)^{-1}$ seconds, and update the cells lost and cells arrived counters every $(pc)^{-1}$

seconds. Thus we require $2(pc + c) + 2pc = (4p + 2)c$ Mips. For $c = 3.5 \times 10^5$ cells/s (150 Mbps) and $p = 5$ we get 7.7 Mips required by one virtual buffer.

In summary, we have described an algorithm for estimating the spare capacity of buffers. This method monitors the traffic in a switch buffer and makes quick and direct estimates of the effect of routing more calls through that buffer on the CLP at that buffer. The method can be used by a call acceptance and routing algorithm. The method is robust: it has been shown, in principle, to work for batch Poisson and Markov fluid sources [2]. Finally, simulations were conducted which demonstrated the predictive property of the algorithm as well as the significant variance reduction with finite buffer size.

References

[1] J. Hui, "Network, transport, and switching integration for broadband communications," *IEEE Network*, vol. March, pp. 40–51, 1988.

[2] C. Courcoubetis, G. Kesidis, A. Ridder, J. Walrand, and R. Weber, "Admission control and routing in ATM networks using inferences from measured buffer occupancy," *submitted to IEEE Trans. Comm.*

[3] S. M. Ross, *A Course in Simulation.* New York, NY: Macmillan, 1990.

[4] M. Butto, E. Cavallero, and A. Tonietti, "Effectiveness of the leaky bucket policing mechanism in ATM networks," *IEEE JSAC*, vol. 9 No. 3, pp. 335–342, April 1991.

[5] J. Bucklew, *Large Deviation Techniques in Decision, Simulation and Estimation.* New York, NY: John Wiley and Sons, Inc., 1990.

Identification of Linear Systems

Manfred Deistler
Institute of Econometrics, Operations Research and Systems Theory
University of Technology, Vienna
Argentinierstraße 8, A-1040 Vienna, Austria

Abstract: We give a short survey on linear system identification, stressing the basic features of the problem. Emphasis is put on what we call the main stream approach but also a number of alternative approaches is discussed. The problem is decomposed into three "modules", namely realization and parametrization, estimation for given dynamic specification and finally determination of the dynamic specification from data (e.g. estimation of the order).

1. Introduction

Identification is concerned with the problem of finding a good model from data. This is a central issue in many branches of science. Problems of identification are often far from being trivial and in many cases share some common features. For these reasons systematic, formal approaches to identification have been developed. Here a special, however important case is considered, namely identification of linear systems from discrete (and equally spaced) time series data. Both, with respect to applications and with respect to the existing body of methods and theories, linear system identification is an extensive and rich subject now (see e.g. Caines 1988, Hannan and Deistler 1988 and Ljung 1987). The aim of this contribution is to give some insight into the basic features of the problem, rather than to give an account of existing methods and theories.

In system identification one has to specify:

(i)   The model class; i.e. the class of all a priori feasible candidate systems to be fitted to the data.

(ii)   The class of observations (y_t)

(iii)   The identification procedure which is a rule (a function in the automatic case) attaching to every finite part of the observations of the form $y_1,...,y_T$, $T \in N$ a system from the model class.

Identification of linear systems has a lot of different aspects and facets which depend - among others factors - on the amount of a priori information available about the phenomenon to be modeled and on the intended use of the model.

Identification is performed for instance for the following purposes:

-       to encode data (using a short, approximate description of the data by system parameters)
-       for spectral estimation
-       for prediction, filtering or interpolation
-       for control
-       for testing theories and for estimation of parameters (which have a "physical" meaning) in models obtained from theory.

Taking into account the wide range of applications, it is not surprising that approaches to identification have been developed in a number of different and partly quite separated areas such as system- and control engineering, statistics, in particular time series analysis, signal processing, econometrics or geophysics. In mathematics identification problems appear under the headings inverse problems and approximation theory.

Now let us make a few brief remarks concerning the history of the subject. Of course, the following list of names is not complete or even balanced. Evidently, system identification is closely related to time series analysis. Formal, systematic approaches to time series analysis date back to the late eighteenth and the early nineteenth century, where the search for hidden periodicities and trends, e.g. in the orbits of the planets, was one central question. The search for such unobserved components occupied great mathematicions such as Euler, Fourier, Lagrange and Laplace. Subsequently the periodogram was introduced by Stokes and used by Schuster as an instrument to search for hidden periodicities in the second part of the last century.

MA and AR systems for time series modeling have been introduced by Yule in the twenties of our century; in particular in order to model "non-exact periodicities" in time series as business cycles in economic series. In a certain sense this can be seen as the beginning of modern system identification.

In the thirties and fourties of our century the linear theory of weak-sense stationary processes was developed; in particular the problems of spectral representation, Wold representation, factorization of spectral densities, prediction, filtering and interpolation have been investigated by Cramer, Kolmogorov, Wiener, Wold and others.. All this work was based on the assumption of known (population) second moments; thus it was probability theory rather than statistics.

A rather modern approach to system identification was taken by the work of the Cowles Commission (in particular by T.W. Anderson, Haavelmo, Klein and Koopmans) in early econometrics. Here for the first time a theory of identifiability and of Maximum Likelihood (ML)-estimation for multi-input, multi-output (MIMO) systems with white noise errors was developed. Later, a number of numerically fast estimators, such as two stage least squares were developed.

In the more recent history (approximately starting with the end of the second world war), some of the most important achievements were the following:

The development of methods for estimation of spectra and of transfer functions, triggered in particular by Tukey.

The asymptotic theory for ML-type estimators for AR and ARMA parameters. Here the maximum lag lengths were assumed to be known. First the single-input single-output (SISO) case has been dealt with in particular by T.W. Anderson, Hannan and Walker and later, using the results of structure theory, the MIMO case was investigated, in particular by Hannan.

The introduction of state space models by Kalman, Kalman filtering and later the analysis of the structure of linear (in general MIMO) systems, also triggered by Kalman.

The book by Box and Jenkins had a great influence on applications; one reason being that it provided an "integrated approach" (including the treatment of certain nonstationarities, (non-auomatic) lag length determination from data and algorithms for optimizing the likelihood) which met some of the main needs of application. The Box-Jenkins approach has been developed mainly for the single-output case.

The development of (automatic) estimation procedures for dynamic specification (e.g. order estimation); in particular the estimators based on information criteria introduced by Akaike and Rissanen and the asymptotic properties of the corresponding procedures derived by Hannan should be mentioned here. These approaches also include the MIMO case.

As has been stated already, linear system identification has many different aspects and facets. Nevertheless, a "main stream theory" of linear system identification has evolved during the last decades (which includes the MIMO case). The basic framework for the main stream case may be described as follows:

(i)   The model class consists of linear, finite-dimensional, constant parameter, causal and stable systems only; the classification of the variables into inputs and outputs is given a priori.

(ii)   Noise is modeled by stochastic models; in particular by stationary, ergodic processes with rational spectral densities; these processes are represented by linear systems with unobserved, white noise inputs. In most cases the noise characteristics are estimated, too.

(iii)   The inputs are assumed to be noisefree; all noise is added to the outputs or to the equations. Furthermore, the noise is assumed to be orthogonal to the inputs (in econometrics this is called the "errors in equations" approach)

(iv)   The criteria for goodness of fit of a system to the data are of (Gaussian) maximum likelihood (ML) type. This includes criteria based on the prediction error variance.

(v)   In general, the model class will be so large, that estimates obtained from optimizing criteria of goodness of fit only, will not give reasonable results, basically due to problems of overfitting. In these cases the model class is decomposed into (finite-dimensional) subclasses. Each subclass is described by a dynamic specification (e.g. by maximum lag lengths) expressed by a number of integers (a so-called multiindex). The dynamic specification has to be determined from data too. Once the subclass is fixed, optimizing a criterion of goodness of fit gives a parametric estimation procedure. The overall approach (including a data driven determination of the dynamic specification) thus could be called semiparametric.

(vi)   As far as the desirable properties of the estimators are concerned, the main focus is on consistency and asymptotic efficiency.

Except for the last section, we will restrict ourselves to the main stream case; however, many ideas generalize to alternative approaches.

From the point of view of the mathematical analysis it turns out to be convenient to decompose the problem of identification into three steps or modules, where each module may be treated separately.

The first module is concerned with <u>structure theory,</u> in particular with questions of realization and parametrization. Since we commence from population second moments of the observations or from transfer functions and not from data in this module, we are not dealing with statistics in the narrow sense here. However, the results obtained turn out to be very important, e.g. in proving asymptotic properties of estimators.

In the second module we are concerned with the estimation of real-valued parameters for a given dynamic specification. Problems considered here are e.g. efficient (in a numerical sense) algorithms for optimizing the likelihood function or asymptotic properties (consistency and asymptotic efficiency) of estimators.

In the third module we are concerned with determining the dynamic specification from data, in particular with procedures for estimation of the integers characterizing the dynamic specification.

2. Representation of Linear Systems

There is a number of different representations for linear systems; the most common are the input-output representation, the ARMA(X) and the state space (SS) representation.

The <u>input-output representation</u> is of the form

$$y_t = \sum_{j=0}^{\infty} K_j u_{t-j} \tag{2.1}$$

where y_t are the s-dimensional outputs, u_t are the m-dimensional inputs and $K_j \in R^{s \times m}$ are the weighting matrices. In general $u_t' = (z_t', \varepsilon_t')$ consists of observed inputs z_t and of an s-dimensional white noise component ε_t (i.e. $E\varepsilon_t = 0$, $E\varepsilon_s \varepsilon_t' = \delta_{st} \Sigma$) which is not observed. The transfer function is of the form

$$k(z) = \sum_{j=0}^{\infty} K_j z^j \tag{2.2}$$

where z is used for the complex variable as well as for the backward shift on Z

$$\left\{ i.e. \ z(y_t | t \in Z) = (y_{t-1} | t \in Z) \right\} \tag{2.3}$$

The <u>ARMA(X) representation</u> is of the form

$$a(z)y_t = b(z)u_t = c(z)z_t + d(z)\varepsilon_t$$

where

$$a(z) = \sum_{j=0}^{p} A_j z^j \quad ; \quad A_j \in R^{sxs}$$

and

$$b(z) = \sum_{j=0}^{q} B_j z^j \quad ; \quad B_j \in R^{sxm}$$

are polynomial matrices. We have $b(z) = (c(z), d(z))$ In the main stream case we assume $a(z)$ to be stable and $d(z)$ to be miniphase

Then the transfer function is defined by

$$k(z) = a^{-1}(z) b(z) = \sum_{j=0}^{\infty} K_j z^j$$

For stationary inputs when z_t has spectral density f_z we have

$$f_y = (2\pi)^{-1} a^{-1}.d.\Sigma.d^* a^{-1*} + a^{-1}.c.f_z c^*.a^{-1*} \qquad (2.4)$$

and

$$f_{yz} = a^{-1}.c.f_z \qquad (2.5)$$

Here we have used f_y and f_{yz} for the spectral density of (y_t) and the cross spectrum between (y_t) and (z_t) respectively, a^* for instance is defined as

$$\sum_{j=0}^{p} A_j' z^{-j}$$

and the spectral densities are defined on C rather than on $[-\pi, \pi]$.

Equations (2.4) and (2.5) show the relation between the internal parameters (namely the realvalued parameters in the A_j and B_j and in Σ and the integer-valued parameters, which in the simplest case are p and q) and the external characteristics f_y, f_z and f_{yz} .For given external characteristics, the determination of the internal parameters may be decomposed into two steps, namely first determining the transfer function k and Σ, and then determining the internal parameters from the transfer function k. The first is relatively simple: Under the socalled persistent excitation conditions (e.g. if f_z is nonsingular except for a finite number of points) $a^{-1}.c$ is uniquely determined from (2.5) and determining $a^{-1}.d$

from (2.4) leads to a spectral factorization problem which is unique if we in addition assume $a(0) = d(0)$ and $\Sigma > 0$.

The third main representation is the <u>state space representation</u>

$$x_{t+1} = Fx_t + Gu_t = Fx_t + Lz_t + K\varepsilon_t \tag{2.6}$$

$$y_t = Hx_t + \varepsilon_t \tag{2.7}$$

where x_t is the (n-dimensional) state at time t and $F \in R^{n \times n}$, $G = (L,K) \in R^{n \times m}$ and $H \in R^{s \times n}$ are the parameter matrices (containing the realvalued parameters); the state dimension is an example for an integer-valued parameter.

Again, the stability and the miniphase assumptions are imposed.

In this case, the transfer function is defined by

$$k(z) = H(I - Fz)^{-1}zG + (0,I)$$

3. Structure Theory

Here we are concerned with the relation between the second (population) moments of the observations (or the transfer function k, which makes no difference for our analysis) and the internal (real and integer valued) parameters.

The main parts of the structure theory are concerned with <u>realization</u>, i.e. with finding an ARMA or SS representation from the second moments of the observations or from the transfer function and with <u>parametrization</u>, i.e. with the problem how to parametrize a given class (or subclass) of transfer functions.

Note that every ARMA system (satisfying our assumptions) as well as every SS system has a causal, rational transfer function which is analytic in a disc containing the closed unit disc. Conversely, for every causal, rational transfer function which is analytic in a disc containing the closed unit disc, there is an ARMA as well as a state space realization.

We have to distinguish between cases where additional a priori information is available, coming e.g. from physical theories, and cases where no such a priori information is used. We will only deal with the latter case here. Unless the contrary is stated explicitly, we restrict ourselves to ARMA realizations; for SS realizations most problems are analogous. Let T_A denote the set of all (a,b) (for fixed s,m but for variable p,q) satisfying our assumptions, let U_A denote the set of all corresponding transfer functions $k = a^{-1}b$ and let the mapping $\pi : T_A \to U_A$ be defined by $\pi(a,b) = a^{-1}b$. A set $T_\alpha \subset T_A$ is called <u>identifiable</u> if π restricted to T_α is injective.

In this case there exists a mapping

$$\varphi_\alpha = \pi(T_\alpha) = U_\alpha \to T_\alpha$$

such that $\varphi(\pi(a,b)) = (a\,,b)$ for all $(a,b) \in T_\alpha$ holds; such a mapping is called an ARMA parametrization of U_α. For $(a,b) \in T_\alpha$, in general not all entries in the parameter matrices A_j, B_j will be needed for a unique description of (a,b) due to constraints. A vector τ made of entries in the A_j, B_j such that (a,b) is uniquely determined from τ, such that τ has constant dimension, d_α say, over T_α, and such that d_α has minimal dimension, is called a vector of free parameters for T_α (or for U_α). We will assume in addition that the non-free parameters in A_j, B_j depend on τ in a continuously differentiable way. We will identify τ with $(a,b) \in T_\alpha$. Note that by $\tau \in R^{d_\alpha}$ we have implicitly assumed that the parametrization is finite-dimensional.

The set U_A is endowed with the topology which corresponds to the relative topology in the product space $(R^{s \times m})^{Z_+}$ for the power series coefficients $(K_j | j \in Z_+)$ of the transfer function $k(z)$. This topology for U_A is called the pointwise topology T_{pt}.

Now, T_A is not a "good" parameter space, since T_A is neither identifiable, nor finite dimensional nor does there exist a continuous parametrization for U_A (for the latter see Hazewinkel and Kalman 1976, Hazewinkel 1979). Thus T_A and U_A have to be broken into bits, T_α and U_α, say, in order to allow for a "convenient" parametrization $\varphi_\alpha : U_\alpha \to T_\alpha$. Here α is a multiindex (of integers) ranging over an indexset I. Thereby convenient means:

(i) T_α is identifiable

(ii) T_α (as a set of free parameters) is a subset of some Euclidian space R^{d_α} and T_α contains an open nonvoid set of R^{d_α}

(iii) $\varphi_\alpha : U_\alpha \to T_\alpha$ is T_{pt} - continuous. In addition, U_α should be open in its closure $\overline{U_\alpha}$

(iv) $\bigcup_{\alpha \in I} U_\alpha = U_A$ (covering property)

There exists a number of realization procedures, most of which commence from the block Hankel matrix

$$H_K = \begin{pmatrix} K_1, & K_2, & K_3 \ldots \\ K_2, & K_3, & K_4 \ldots \\ \ldots & \ldots & \ldots \end{pmatrix}$$

of the transfer function k. By selecting basis rows for the row space of H_K and expressing other rows of H_K as linear combinations of these basis rows, we obtain a linear equation for the parameters of $a(z)$; then $b(z)$ is directly determined from a(z) and k(z). The multiindex α marks the positions of the basis rows in H_K. According to different rules for selecting the basis rows in H_K, we obtain e.g. Echelon canonical form (where the corresponding multiindex consists of the Kronecker indices), Hermite canonical form or the overlapping description of the manifold M(n) of all systems of order n (this is described in detail e.g. in Hannan and Deistler, chap.2).

Using such procedures, state space realizations can be obtained in an analogous way and may even have the same free parameters. It should be noted, however, that, although there is a one-to-one relation between the equivalence classes of SS and of ARMA systems corresponding to the same k, the equivalence classes themselves are different. Thus there are state space realizations such as balanced realizations (see e.g. Moore 1981) which have no ARMA counterpart.

4. Estimation for Given Dynamic Specification

Here we consider the estimation of the realvalued parameters, i.e. of the entries in the vector τ of free parameters and of the free elements of Σ (which are its on and above digital entries) for given dynamic specification α, and thus for given T_α and U_α.

Most estimators are obtained from optimizing a criterion of goodness of fit over T_α. (Gaussian-) ML - estimators are of particular importance here. We restrict ourselves to the case $u_t = \varepsilon_t$, for simplicity of notation here. Then $-2T^{-1}$ times the Gaussian Likelihood Function is, up to a constant, of the form

$$L_T(\tau,\Sigma) = T^{-1}\log\det\Gamma_T(\tau,\Sigma) + T^{-1}.y(T)'.\Gamma_T(\tau,\Sigma).y(T) \qquad (4.1)$$

where T is sample size, y(T) is the stacked vector of observations $(y_1',......,y_T')'$, $u_t = \varepsilon_t$ and

$$\Gamma_T(\tau,\Sigma) = \left(\int e^{-i\lambda(r-t)}f_y(\lambda;\tau,\Sigma)d\lambda\right)_{r,t=1...T} \qquad (4.2)$$

denotes the sT x sT matrix of second moments of a vector $(y_1',......,y_T')'$ for an ARMA process with parameters τ and Σ; f_y denotes the corresponding spectral density. The maximum likelihood estimators (MLEs) of τ and Σ then are defined by

$$(\hat{\tau}_T,\hat{\Sigma}_T) = \arg\min_{(\tau,\Sigma)\in T_\alpha\times\underline{\Sigma}} L_T(\tau,\Sigma)$$

where $\underline{\Sigma} = \{\Sigma > 0\}$

It is easily seen from (4.1) and (4.2) that the likelihood depends on τ only via k(z). This allows for a "coordinate free" approach showing properties (e.g. consistency) of the MLE $\hat{k}_T$ for k irrespective of the special parametrization used.

It should be mentioned that in minimizing the likelihood certain boundary points of $U_\alpha \times \underline{\Sigma}$ cannot be excluded and have to be taken into account (see e.g. Hannan and Deistler 1988, chap. 4). Also the existence of a minimum of L_T is not a trivial problem (see e.g. Deistler and Pötscher 1984).

From a practical point of view serious problems are caused by the fact that in general there will be no explicit expression for the MLEs; thus numerical minimization procedures have to be applied, each of which in a strict sense defines an estimator. In this connection often problems connected with multiple relative minima of L_T and with the appropriate choice of initial estimates arise.

As far as asymptotic properties are concerned, general results concerning <u>consistency</u> of $\hat{k}_T$ and $\hat{\Sigma}_T$ which in particular do not require an unnatural compactness assumption for the parameter space, are available (Dunsmuir and Hannan 1976, see also Hannan and Deistler 1988, chap. 4).

The asymptotic normality results for $\hat{\tau}_T$ and $\hat{\Sigma}_T$ are rather straightforward to obtain.

Even if the "true" system, i.e., the system generating the observation is not contained in $U_\alpha \times \underline{\Sigma}$ (or in its boundary), a <u>generalized consistency</u> results can be shown: Let us define the set D of best approximants to the true transferfunction k_0 within $U_\alpha \times \underline{\Sigma}$ (where boundary points of U_α may be included) in the sense that D consists of the minimizers of the asymptotic likelihood function

$$L(k,\Sigma) = \log\det\Sigma + (2\pi)^{-1} \cdot \int_{-\pi}^{\pi} \mathrm{tr}\left\{(k(e^{-i\lambda})\Sigma k(e^{-i\lambda})^*)^{-1}(k_0(e^{-i\lambda})\Sigma_0 k_0(e^{-i\lambda})^*\right\}d\lambda$$

where Σ_0 corresponds to the true white noise process. Note that in general, D is not necessarily a singleton. Then the MLEs, $\hat{k}_T$ can be shown to converge to the set D.a.e. (see Ljung 1978, Hannan and Deistler, chap. 4). This is an important result, also from a conceptual point of view.

5. Dynamic specification

In many applications, the dynamic specification is not known a priori and has to be determined from the data, too. The development and evaluation of data based procedures for dynamic specification has constituted one of the most important contributions to the subject during the last three decades. These procedures may be classified into non-automatic and automatic ones. In the non-automatic case, subjective decisions have to be made at a certain stage and thus some experience is required. A procedure of this kind was developed in Box and Jenkins (1970).

Let us consider the problems of estimating the order first. Note that, if $\overline{M(n)}$ denotes the closure of the set of all transfer functions of order n, then $\overline{M(n_1)} \subset \overline{M(n_2)}$ holds for $n_1 < n_2$, and $M(n_1)$ has smaller dimension than $M(n_2)$. Thus by optimizing a criterion of goodness of fit, we will usually obtain a transfer function of the largest allowed order. One way to overcome this problem is to use criteria which define a tradeoff between goodness of fit and complexity measured by the dimension of the parameter space. An important class of these criteria is of the form

$$A(n) = \log \det \hat{\Sigma}_T'(n) + (2ns)\frac{c(T)}{T}; 0 \le n \le N_T \tag{5.1}$$

Here $\hat{\Sigma}_T(n)$ is the MLE of Σ_0 over $M(n) \times \Sigma$ with sample size T (which is a measure of misfit), $2ns$ is the dimension of $M(n)$ and $c(T)$ and N_T are constants depending on T) which have to be prescribed. An estimator of n is obtained by minimizing $A(n)$. N_T may be constant or increase with T (e.g. with the order $(\log T)^a$, $a<\infty$). The main question, however, is how to choose the constant $c(T)$, which defines the tradeoff. Important special cases are the <u>AIC criterion</u>

$$AIC(n) = \log \det \hat{\Sigma}_T(n) + (2ns)\frac{2}{T} \tag{5.2}$$

introduced by Akaike (1969, 1977) and the <u>BIC criterion</u>

$$BIC(n) = \log \det \hat{\Sigma}_T(n) + (2ns)\frac{\log T}{T} \tag{5.3}$$

introduced by Akaike, Rissanen (1983) and Schwarz(1978).

The idea behind the AIC can be described as follows (see Findley 1985): Let $-L_T(\tau, \Sigma)$ denote $(2T^{-1}$ times) the log likelihood (up to a constant) and let $\theta_i = (\hat{\tau}_i, \hat{\Sigma}_i), i = 1, 2$ denote the MLEs for $M(n_i)$, $i=1,2$ respectively. Finally, let $E_T(\theta) = E_T(L_T(\theta))$ denote the expectation with respect to the true (data generating) probability structure. Then it is suggested to prefer $\hat{\theta}_1$ over $\hat{\theta}_2$ if

$$E_T(\hat{\theta}_1) - E_T(\hat{\theta}_2) < 0$$

holds. Clearly, $E_T(\hat{\theta}_1) - E_T(\hat{\theta}_2)$ is unknown. It can be shown that $AIC(n_1) - AIC(n_2)$ provides an estimate of this quantity, which under some additional assumptions is asymptotically unbiased.

Rissanen derived BIC from the requirement of finding a code of minimal length for the data. Schwarz derived BIC from Bayesian arguments.

As far as the asymptotic properties are concerned, the following important results are available:

If the true transfer function k_0 is contained in $M(n_0)$, the BIC gives an a.e. consistent estimator for the true order n_0; more generally, this is true if

$$\liminf_{T \to \infty} \ (c(T)/\log T) > 0$$

and
$$c(T)/T \to 0$$

hold. Moreover, AIC is not consistent for the true order (Hannan 1980, 1981).

Nevertheless, AIC has virtues, in particular in the case where the true order is infinite. As has been shown by Shibata (1980, 1981) (for the autoregressive case), AIC has asymptotic optimality properties with respect to the estimated prediction error variance and the corresponding spectral estimators.

The Kronecker indices α can also be estimated by criteria of the form (5.1) (see e.g. Akaika 1966, Hannan and Kavalieris 1984, Hannan and Deistler 1988, Poskitt 1990); in particular, BIC is also consistent in this case.

Alternative approaches are based on the investigation of the linear dependence structure of the (estimated) Hankel matrix. Such an approach is appropriate in particular for estimating (for already determined order) the dynamic indices of the overlapping parametrization. As in this case the local neighborhoods are of the same dimension as $M(n)$ and are dense in $M(n)$, criteria of the type (5.1) are not appropriate.

6. Alternative approaches

There is a number of alternative approaches and extensions to the main stream approach. The most important ones probably are:

- Identification, in particular tracking of parameters, for time varying parameter systems. Closely related is the problem of detecting change points.

- Identification of unstable systems. In this case, the asymptotic theory is significantly different from the main stream case; the convergence rates for consistency of the estimators for the real valued parameters are faster than $\sqrt{T}$ here and the limiting distributions are no longer normal in general. In econometrics, special unstable systems, where the unstable roots are unity, have attracted considerable attention; one reason for this is that many economic time series have a "trend in variances".

- Feedback causes special problems in identification.

- Model reduction techniques such as Hankel norm approximation can be used instead of ML estimation.

- Identification in the context of control poses a number of important problems.

Now, we will describe a particular alternative approach in somewhat greater detail, namely errors-in-variables modeling. This is mainly because this topic is part of the author's current research interests.

In errors-in-variables (EV) modeling, also the observed inputs may be contaminated by noise. Such a generalization is important for instance:

- if we are interested in the true system (rather than e.g. in prediction) and if we cannot be sure a priori that the observed inputs are not corrupted by noise. This is related to identification of systems under feedback.

- if we want to approximate a high dimensional data vector by a small number of factors (factor analysis). A related issue is to reduce the dimension of the parameter space.

- if we have no sufficient a priori information about the number of independent equations in the system or about the classification of the variables into inputs and outputs; then we have to perform a more symmetric system modeling which in turn demands a more symmetric noise model.

Without distinguishing between inputs and outputs we write the system as

$$w(z)\hat{v}_t = 0 \qquad\qquad (6.1)$$

when $\hat{v}_t$ is the s+m dimensional vector of latent (i.e., in general unobserved) variables (i.e., the stacked vector of "true" inputs and outputs) and where

$$w(z) = \sum_{j=-\infty}^{\infty} W_j z^j; \, W_j \in R^{s \times (s+m)}$$

is the so-called relation function. The observed variables are of the form

$$v_t = \hat{v}_t + n_t \qquad\qquad (6.2)$$

where n_t is the noise vector. We assume that $E\hat{v}_t n_s' = 0$ holds for all s,t.

Clearly, (6.1) and (6.2) provide a symmetric way of system and noise modeling; also in general the number of equations is not assumed to be known a priori; thus this is a fairly general approach. However, unless some additional assumptions are imposed, the problem is completely undetermined.

Here, we additionally assume that the spectral density of (n_t) is diagonal. The identification problem in this case is far from being solved in the general case. However, recently a realization theory has been developed (Kalman 1982, Picci and Pinzoni 1986, van Schuppen 1989, Anderson and Deistler 1990, Deistler and Scherrer 1990).

From (6.2) we have

$$f_v = f_{\hat{v}} + f_n \tag{6.3}$$

where f_v, $f_{\hat{v}}$ and f_n are the spectral densities of $(v_t),(\hat{v}_t)$ and (n_t), respectively. For given $f_{\hat{v}}$, the relation function $w(z)$ is determined from

$$w.f_{\hat{v}} = 0 \tag{6.4}$$

The main features of this realization theory are as follows:

(i) In general, the underlying system (neither $w(z)$ nor $f_{\hat{v}}$) is not uniquely determined from the second moments of the observations f_v. Thus lack of knowledge concerning the noise mechanism may create uncertainty in addition to that coming from sampling variation. An important problem in this context is the description of the sets F and W of (w.r. to f_v) observationally equivalent $f_{\hat{v}}$ and $w(z)$, respectively.

(ii) Also the subsets F_s and W_s of F and W corresponding to the $f_{\hat{v}}$ of corank s and the systems with s outputs, respectively, are of interest. Of particular interest is the maximum corank of $f_{\hat{v}}$ among all $f_{\hat{v}}$ from F, s* say, and the corresponding sets F_{s^*} and W_{s^*}. The integer s* is called the Frisch corank. Up to now, there are no general results available concerning a complete description of the above-mentioned sets; however, a number of topological and geometric properties (e.g. regarding boundedness and manifold structures) (see e.g. Deistler and Scherrer 1990)has been shown for the general case and a rather complete analysis is available for the cases s*=1 and s*=n-1 (Deistler and Anderson 1991, Anderson and Deistler 1990)

(iii) Since the main goal is the identification of the above-mentioned sets (rather than the identification of single systems), the continuity of the mapping attaching e.g. the set F to f_v is an important property, e.g. for consistency of estimation procedures. For such results see e.g. Deistler and Scherrer (1990).

(iv) Another problem is to describe the set of all spectral densities corresponding to a given Frisch corank.

(v) As opposed to the main stream case, here higher order moments, (e.g. if the true variables are non-Gaussian distributed)may provide additional identifying information.

References

Akaike, H. (1969). Fitting autoregressive models for prediction.*Ann. Inst. Statist. Math* **21**, 243-247.

Akaike, H. (1976). canonical correlation analysis of time series and the use of an information criterion. In *System Identification: Advances and case studies* (eds. R.H. Mehra and D.G. Lainiotis), 27-96. New York, Academic Press.

Akaike, H. (1977). On entropy maximization principle. In *Applications of statistics* (ed. P.R. Krishnaiah), 27-41. Amsterdam, North Holland.

Anderson, B.D.O. and M. Deistler (1990). Identification of dynamic systems from noisy data: The case m*=n-1., *mimeo.*

Box, G.E.P. and G.M. Jenkins (1970). *Time Series Analysis, Forecasting and Control,* San Francisco, Holden Day.

Caines, E.P.(1988). *Linear Stochastic Systems,* New York, John Wiley and Sons.

Deistler, M. and B.D.O. Anderson (1991). Identification of dynamic systems from noisy data: The case m*=1, In *Mathematical System Theory: The Influence of R.E. Kalman* (ed. A.C. Antoulas), 423-435, Berlin, Springer Verlag.

Deistler, M. and B.M. Pötscher (1984). The behavior of the likelihood function for ARMA models, *Adv. Appl. Probab.* **16**, 843-865.

Deistler, M. and W. Scherrer (1990). Identification of linear systems from noisy data, *mimeo.*

Dunsmuir, W. and E.J. Hannan (1976). Vector linear time series models. *Adv. Appl. Probab.* **8**, 339-364.

Findley, D.F. (1985). On the unbiasedness property of AIC for exact or approximating linear stochastic time series models. *J. Time Series Anal.* **6**, 229-252.

Hannan, E.J. (1980). The estimation of the order of an ARMA process. *Ann. Statist.* **8**, 1071-1081.

Hannan, E.J. (1981). Estimating the dimension of a linear system. *J. Multivariate Anal.* **11**, 459-473.

Hannan, E.J. and M. Deistler (1988). *The Statistical Theory of Linear Systems.* New York, John Wiley.

Hannan, E.J. and L. Kavalieris (1984). Multivariate linear time series models. *Adv. Appl. Prob.* **16**, 492-561.

Hazewinkel, M. (1979). On identification and the geometry of the space of linear systems. *Lecture Notes in Control and Information Sciences.* vol. 16, 401-415. Berlin, Springer Verlag.

Hazewinkel, M. and R.E. Kalman (1976). Invariants, canonical forms and moduli for linear, constant, finite dimensional, dynamical systems. In *Lecture notes in Economics and Mathematical Systems*, vol. 131 (eds. G. Marchesini and S.K. Mitter), 48-60. Berlin, Springer Verlag.

Kalman, R.E. (1982). System identification from noisy data. In *Dynamical Systems II, A University of Florida international Symposium* (eds. A. Bednarek and L. Cesari), Academic Press, New York.

Ljung, L. (1978). Convergence analysis of parametric identification methods, *IEEE Trans. Autom. Control* **AC-23**, 770-783.

Ljung, L. (1987) *System Identification - Theory for the User*. Prentice Hall, Inc., Englewood Cliffs, New Jersey 07632.

Moore, B.C. (1981). Principal component analysis in linear systems: controllability, observability and model reduction. *IEEE Trans. Automatic Control*, **AC-26**, 17-27.

Picci, G. and S. Pinzoni (1986). Dynamic factor analysis models for stationary processes. *IMA Journal of Mathematical C ontrol and Information*, **3**, 185-210.

Rissanen, J. (1983). Universal prior for parameters and estimation by minimum description length. *Ann. Statist.* **11**, 416-431.

Schwarz, G. (1978). Estimating the dimension of a model. *Ann. Statist.* **6**, 461-464.

Shibata, R. (1980). Asymptotically efficient selection of the order of the model for estimating parameters of a linear process. *Ann. Statist.* **8**, 147-164.

Shibata, R. (1981). An optimal autoregressive spectral estimate. *Ann. Statist.* **9**, 300-306.

Van Schuppen, J. (1989). Stochastic realization problems. In *Three Decades of Mathematical System Theory*, Sproinger Lecture Notes in Control and Information Sciences (eds. H. Nijmeijer and J.M. Schumacher), **135**, 480-523. Berlin, Springer Verlag.

FINITE DIMENSIONAL FILTERS
RELATED TO MARKOV CHAINS

Robert J. Elliott
Department of Statistics and Applied Probability
University of Alberta
Edmonton, Alberta
Canada T6G 2G1

July, 1991

Abstract: New finite dimensional filters and smoothers are obtained which are related to the Wonham filter of a noisily observed Markov chain. In particular, finite dimensional, recursive filters and smoothers are given for the number of jumps from state i to state j, for the occupation time of state i, and for a stochastic integral related to the drift in the observations. These filters allow easy application of the EM algorithm for the estimation of the parameters of the Markov chain and observation process.

1. INTRODUCTION

By the term "finite dimensional filter" we shall mean a filter in which sufficient statistics of the signal process are given by a closed, finite dimensional system of equations. Usually the filtering equation for the conditional mean of a process involves the conditional expectation of the square, the equation for the conditional expectation of the square involves the conditional expectation of the cubic power, and so on, each equation introducing a higher power. For the Kalman filter, the conditional expectation of the cubic power turns out to be the conditional expectation of a centered Gaussian random variable, and so equals zero. The above process stops and the Kalman filter involves, therefore, just two equations, one for the conditional mean and a deterministic equation for the variance. The conditional estimate is a Gaussian random variable, and so these two statistics are sufficient.

Acknowledgements: The support of NSERC Grant A7964 is gratefully acknowledged. This research was completed while the author was a Visiting Fellow at the Systems Engineering Department, Research School of Physical Sciences, Australian National University, in May 1991.

As we indicate below, the Wonham filter for a finite state Markov chain is finite dimensional because the state process is, in effect, an indicator function and indicator functions are idempotent. Consequently, the square of the state process can be expressed in terms of the process itself and no higher order terms arise. Exploiting this idea we determine some finite dimensional filters related to the Wonham filter. Specializing our results we obtain finite dimensional filters and smoothers for the following processes:

1) the state of the Markov chain,

2) the number of jumps N_t^{ij} of the chain from state i to state j,

3) the occupancy time J_t^i of the Markov chain in state i,

4) a stochastic integral G_t^i related to the observation process.

The filtered estimate of the state is the Wonham filter [7]. The smoothed estimate of the state is given in Clements and Anderson [2]. A finite dimensional filter for the number of jumps N_t^{ij} was obtained by Zeitouni and Dembo [3], [9], and used to estimate the parameters of the Markov chain and the observation process. However, this estimation also involves J_t^i and G_t^i for which finite dimensional filters are not given in [9].

Our filters allow, therefore, the application of the EM algorithm, an extension of the discrete-time Baum–Welch algorithm. See [3], [9]. Unlike the Baum–Welch method our equations are recursive and can be implemented by the usual methods of discretization; no forward-backward estimates are required. Proofs are outlined in this paper; details appear in [6]. Applications of our methods will be given in subsequent papers.

The author is grateful to Professor John Moore for suggesting the author look at the Baum–Welch and EM algorithms, and to Vikram Krishnamurthy for interesting conversations. This work was completed while the author was a Visiting Fellow in the Systems Engineering Department, Research School of Physical Sciences, Australian National University, in May 1991. The hospitality of the department and the A.N.U. is gratefully acknowledged.

2. THE MARKOV CHAIN

For any finite set $\Sigma = \{s_1, s_2, \ldots, s_N\}$ consider the functions ϕ_i, $1 \le i \le N$, defined by $\phi_i(s_j) = \delta_{ij}$, and the vector function $\phi(s) := (\phi_1(s), \phi_2(s), \ldots, \phi_N(s))^*$. (We wish to consider column vectors, so * denotes transpose.) Then ϕ is a bijection of Σ and the set $S = \{e_1, e_2, \ldots, e_N\}$ of unit vectors $e_j = (0, 0, \ldots, 1, \ldots, 0)^*$ of R^N. Using such a bijection the state space of a finite state space Markov chain can, without loss of generality, be taken to be a set S.

Suppose, therefore, that X_t, $t \geq 0$, is a Markov chain defined on a probability space $(\Omega, \mathcal{F}, P)$ with state space $S = \{e_1, e_2, \ldots, e_N\}$. Writing $p_t^i = P(X_t = e_i)$, $1 \leq i \leq N$, we suppose for some Q-matrix $A = (a_{ij})$ the probability distribution $p_t = (p_t^1, p_t^2, \ldots, p_t^N)^*$ satisfies the forward equation

$$\frac{dp_t}{dt} = Ap_t. \tag{2.1}$$

Recall that in a Q-matrix

$$\sum_{i=1}^{N} a_{ij} = 0. \tag{2.2}$$

The process X_t is not observed directly; rather we suppose there is a (scalar) observation process given by

$$y_t = \int_0^t g(X_r)dr + B_t. \tag{2.3}$$

(The extension to vector processes y is straightforward.) Here, B_t is a standard Brownian motion on $(\Omega, \mathcal{F}, P)$ which is independent of X_t. Because X takes values in S the function g is given by a vector $g = (g_1, g_2, \ldots, g_N)^*$, so that $g(X) = \langle X, g \rangle$ where $\langle \cdot, \cdot \rangle$ denotes the scalar product in R^N.

Write

$$\mathcal{F}_t^0 = \sigma\{X_s, y_s : s \leq t\}$$

$$\mathcal{Y}_t^0 = \sigma\{y_s : s \leq t\}$$

and $\{\mathcal{F}_t\}$, $\{Y_t\}$, $t \geq 0$, for the corresponding right-continuous, complete filtrations.

Note $\mathcal{Y}_t \subset \mathcal{F}_t$ for all t.

NOTATION 2.1. If ϕ_t is an integrable, measurable process write $\hat{\phi}_t$ for its $\mathcal{Y}$-optional projection under measure P, so $\hat{\phi}_t = E[\phi_t \mid \mathcal{Y}_t]$ a.s. and $\hat{\phi}_t$ is the filtered estimate of ϕ_t.

For $s \leq t$, $\pi_t(\phi_s)$ will denote the $\mathcal{Y}$-optional projection of the constant process ϕ_s, so

$$\pi_t(\phi_s) = E[\phi_s \mid \mathcal{Y}_t] \quad \text{a.s.}$$

and $\pi_t(\phi_s)$ is the smoothed estimate of ϕ_s.

Of course, $\pi_t(\phi_t) = \hat{\phi}_t$. Write $\Phi(t,s) = \exp A(t-s)$ for the transition matrix associated with A, so that

$$\frac{d}{dt}\Phi(t,s) = A\Phi(t,s) \tag{2.4}$$

and for $s \leq t$

$$E[X_t \mid \mathcal{F}_s] = E[X_t \mid X_s]$$

$$= \Phi(t,s)X_s. \tag{2.5}$$

We first recall the following result:

LEMMA 2.2.

$$M_t := X_t - X_0 - \int_0^t AX_r dr$$

is a (vector) $\mathcal{F}_t$-martingale under P.

Proof. The proof follows from (2.4) and (2.5).

REMARKS 2.3. The semimartingale representation of the Markov chain X is, therefore,

$$X_t = X_0 + \int_0^t AX_r dr + M_t. \tag{2.6}$$

Note $\int_0^t AX_r dr = \int_0^t AX_{r-} dr$ because $X_r(\omega) = X_{r-}(\omega)$ a.s. except for a countable number of values of r. Similar identifications will be made below.

INNOVATIONS 2.4. The observation process y can be written either as

$$y_t = \int_0^t \langle X_r, g \rangle dr + B_t \tag{2.7}$$

or, in innovations form

$$y_t = \int_0^t \langle \hat{X}_r, g \rangle dr + \nu_t. \tag{2.8}$$

This equation defines the innovations Brownian motion process ν.

We shall also consider the Zakai equation. For this we introduce the probability measure P_0 by putting

$$\left.\frac{dP_0}{dP}\right|_{\mathcal{F}_t} = K_t = \exp\left(-\int_0^t \langle X_r, g \rangle dB_r - \frac{1}{2}\int_0^t \langle X_r, g \rangle^2 dt\right).$$

Now K_t is a martingale under P and

$$K_t = 1 + \int_0^t K_r \langle X_r, g \rangle dB_r.$$

By Girsanov's theorem, (see [4]), y is a standard Brownian motion under P_0. Define the process Λ_t by

$$\Lambda_t = 1 + \int_0^t \Lambda_r \langle X_r, g \rangle dy_r \tag{2.9}$$

so that $\Lambda_t = \exp\left(\int_0^t \langle X_r, g \rangle dy_r - \frac{1}{2}\int_0^t \langle X_r, g \rangle^2 dr\right)$ and $\Lambda_t K_t = 1$. Note Λ_t is an $\mathcal{F}$-martingale under P_0. However, it is under P that

$$y_t = \int_0^t \langle X_r, g \rangle dr + B_t$$

and so has the form of the observation process influenced by the Markov chain.

NOTATION 2.5. If ϕ_t is an $\mathcal{F}_t$-adapted, integrable process a version of Bayes' theorem states that

$$\hat{\phi}_t = E[\phi_t \mid \mathcal{Y}_t] = E_0[\Lambda_t \phi_t \mid \mathcal{Y}_t]/E_0[\Lambda_t \mid \mathcal{Y}_t] \tag{2.10}$$

$$= \sigma(\phi_t)/\sigma(1), \quad \text{say,}$$

where E_0 denotes expectation with respect to P_0 and $\sigma(\phi_t)$ is the $\mathcal{Y}$-optional projection of $\Lambda_t \phi_t$ under measure P_0. Consequently, $\sigma(1) = E_0[\Lambda_t \mid \mathcal{Y}_t] = \overline{\Lambda}_t$, say, is the $\mathcal{Y}$-optional projection of Λ_t under P_0.

Further, if $s \leq t$ we shall write $\sigma_t(\phi_s)$ for the $\mathcal{Y}$-optional projection of $\Lambda_t \phi_s$ under P_0, so that

$$\sigma_t(\phi_s) = E_0[\Lambda_t \phi_s \mid \mathcal{Y}_t] \quad \text{a.s.}$$

and $\sigma_t(\phi_t) = \sigma(\phi_t)$.

3. A GENERAL FINITE DIMENSIONAL FILTER

Consider a vector process H_t of the form

$$H_t = H_0 + \int_0^t \alpha_r dr + \int_0^t \beta_r \cdot dM_r + \int_0^t \delta_r dy_r \tag{3.1}$$

where α, β, δ are $\mathcal{F}$-predictable, square integrable processes of appropriate dimensions.

THEOREM 3.1. $\widehat{H}_t$ is given by the following filter

$$\widehat{H}_t = \widehat{H}_0 + \int_0^t \hat{\alpha}_r dr + \int_0^t \gamma_r^1 d\nu_r \tag{3.2}$$

where

$$\gamma_r^1 = \hat{\delta}_r + \widehat{\langle X_r, g \rangle H_r} - \langle \widehat{X}_r, g \rangle \widehat{H}_r. \tag{3.3}$$

Proof. The problem is the determination of the integrand γ^1. This can be effected following the methods of [5], where the product $y_t \cdot \widehat{H}_t$ is evaluated in two ways. Because $y_t \cdot \widehat{H}_t$ is a special semimartingale the predictable, bounded variation terms in each expression, (that is, the dt integrals) can be equated so giving γ_t^1.

The following steps are also used:

STEP 1: The optional projection of an integral minus the integral of projections is a martingale, (i.e., $\pi_t \left(\int_0^t \phi_r dr \right) - \int_0^t \pi_r(\phi_r) dr$ is a martingale). This is a simple consequence of Fubini's theorem.

STEP 2: If μ is an $\mathcal{F}$-martingale then $\hat{\mu}$ is a $\mathcal{Y}$-martingale; this is a consequence of repeated conditioning.

STEP 3: If μ is a square integrable $\mathcal{Y}$-martingale then there is a square-integrable $\mathcal{Y}$-predictable process γ such that

$$\mu_t = \mu_0 + \int_0^t \gamma_r d\nu_r,$$

where ν is the innovations process of (2.7). A proof of this result is given in Theorem 16.22 of [4].

REMARKS 3.2. A problem with the integrand γ^1 is that, although we were trying to filter H_t, we have ended up with a product term $\widehat{H_r \langle X_r, g \rangle}$. We, therefore, consider a filter for the process $H_t \cdot X_t$.

Write $B = \text{diag } g$ for the matrix with diagonal entries $g_1, g_2, \ldots, g_N$.

THEOREM 3.3.

$$\widehat{H_t \cdot X_t} = \widehat{H_0 X_0} + \int_0^t \widehat{\alpha_r X_r} dr + \int_0^t \widehat{H_r A X_r} dr$$

$$+ \sum_{i,j=1}^N \int_0^t \langle \widehat{\beta_r^j X_r} - \widehat{\beta_r^i X_r}, e_i \rangle a_{ji} dr (e_j - e_i) + \int_0^t \gamma_r^2 d\nu_r \qquad (3.4)$$

where

$$\gamma_r^2 = \widehat{\delta_r X_r} + B\widehat{H_r X_r} - \langle \widehat{X}_r, g \rangle \widehat{H_r X_r}.$$

Proof. Recall

$$X_t = X_0 + \int_0^t A X_r dr + M_t. \qquad (3.5)$$

From Remarks 2.3 this equals

$$X_t = X_0 + \int_0^t A X_{r-} dr + M_t. \qquad (3.6)$$

From (3.1) and (3.6), and using Remarks 2.3,

$$H_t \cdot X_t = H_0 X_0 + \int_0^t \alpha_r X_r dr + \int_0^t \beta_r X_{r-} dM_r$$

$$+ \int_0^t \delta_r X_{r-} dy_r + \int_0^t H_r A X_r dr + \int_0^t H_{r-} dM_r$$

$$+ \sum_{0 < r \le t} (\beta_r \cdot \Delta X_r) \Delta X_r. \qquad (3.7)$$

Now

$$(\beta_r \cdot \Delta X_r) \Delta X_r = \sum_{i,j=1}^N (\beta_r^j - \beta_r^i) \langle X_{r-}, e_i \rangle \langle X_r, e_j \rangle (e_j - e_i)$$

$$= \sum_{i,j=1}^N (\beta_r^j - \beta_r^i) \langle X_{r-}, e_i \rangle \langle X_r - X_{r-}, e_j \rangle (e_j - e_i),$$

so

$$\sum_{0 < r \le t} (\beta_r \cdot \Delta X_r) \Delta X_r = \sum_{i,j=1}^N \int_0^t (\beta_r^i - \beta_r^i) \langle X_{r-}, e_i \rangle \langle e_j, dX_r \rangle (e_j - e_i)$$

and, using the form (3.6),

$$dX_r = AX_{r-}dr + dM_r.$$

Therefore,

$$\sum_{0<r\leq t} (\beta_r \cdot \Delta X_r)\Delta X_r = \sum_{i,j=1}^{N} \int_0^t (\beta_r^j - \beta_r^i)\langle X_{r-}, e_i\rangle\langle e_j, dM_r\rangle(e_j - e_i)$$

$$+ \sum_{i,j=1}^{N} \int_0^t \langle \beta_r^j X_r - \beta_r^i X_r, e_i\rangle a_{ji}dr(e_j - e_i).$$

Again in the last integral we have replaced X_{r-} by X_r, as noted in Remarks 2.3. Substituting in (3.7) we have

$$H_t X_t = H_0 X_0 + \int_0^t \alpha_r X_r dr + \int_0^t \beta_r X_{r-}dM_r$$

$$+ \int_0^t \delta_r X_{r-}dy_r + \int_0^t H_r AX_r dr + \int_0^t H_{r-}dM_r$$

$$+ \sum_{i,j=1}^{N} \int_0^t (\beta_r^j - \beta_r^i)\langle X_{r-}, e_i\rangle\langle e_j, dM_r\rangle(e_j - e_i)$$

$$+ \sum_{i,j=1}^{N} \int_0^t \langle \beta_r^j X_r - \beta_r^i X_r, e_i\rangle a_{ji}dr(e_j - e_i).$$

We are, therefore, in a situation where Theorem 3.1 can be applied, with $H_t X_t$ replacing H_t,

$$\alpha_r X_r + H_r AX_r + \sum_{i,j=1}^{N} \langle \beta_r^j X_r - \beta_r^i X_r, e_i\rangle a_{ji}(e_j - e_i)$$

replacing α_r, and $\delta_r X_{r-}$ replacing δ_r. (Note β does not appear in γ^1.) Consequently, from Theorem 3.1

$$\widehat{H_t X_t} = \widehat{H_0 X_0} + \int_0^t \widehat{\alpha_r X_r}dr + \int_0^t \widehat{H_r AX_r}dr$$

$$+ \sum_{i,j=1}^{N} \int_0^t \langle \widehat{\beta_r^j X_r} - \widehat{\beta_r^i X_r}, e_i\rangle a_{ji}dr(e_j - e_i) + \int_0^t \gamma_r^2 d\nu_r$$

where

$$\gamma_r^2 = \widehat{\delta_r X_r} + \overbrace{\langle X_r, g \rangle H_r X_r} - \langle \widehat{X}_r, g \rangle \widehat{H_r X_r}.$$

However, $\langle X_r, g \rangle X_r = \sum_{k=1}^{N} \langle X_r, e_k \rangle g_k e_k$ so $\overbrace{\langle X_r, g \rangle H_r X_r} = \sum_{k=1}^{N} \langle \widehat{H_r X_r}, e_k \rangle g_k e^k$, and it is easily checked this equals $B \widehat{H_r X_r}$. Therefore, we see γ_r^2 has the stated form. That is,

$$\widehat{H_t X_t} = \widehat{H_0 X_0} + \int_0^t \widehat{\alpha_r X_r} dr + \int_0^t \widehat{H_r A X_r} dr$$

$$+ \sum_{i,j=1}^{N} \int_0^t \langle \widehat{\beta_r^j X_r} - \widehat{\beta_r^i X_r}, e_i \rangle a_{ji} dr(e_j - e_i)$$

$$+ \int_0^t (\widehat{X_r \delta_r} + B \widehat{H_r X_r} - \langle \widehat{X}_r, g \rangle \widehat{H_r X_r}) d\nu_r. \tag{3.8}$$

REMARKS 3.4. The advantage of the filter for $\widehat{H_t X_t}$, compared with that for $\widehat{H}_t$, is that the stochastic integrand γ_t^2 involves only $\widehat{H_t X_t}$ itself and $\widehat{X}_t$. Now $\widehat{X}_t$ is given by the Wonham filter, (which is a special case of Theorem 3.3).

Key steps in the above proof are the representations

$$\sum_{0 < r \le t} (\beta_r \cdot \Delta X_r) \Delta X_r = \sum_{i,j=1}^{N} \int_0^t (\beta_r^j - \beta_r^i) \langle X_{r-}, e_i \rangle \langle e_j, dX_r \rangle (e_j - e_i)$$

and

$$\langle X_r, g \rangle H_r X_r = \sum_{k=1}^{N} \langle H_r X_r, e_k \rangle g_k e_k = B H_r X_r,$$

which use the idempotent property of X_r. Because higher powers of X_r do not arise in γ^2 we are able to obtain finite dimensional filters. Indeed, if α_r, δ_r and β_r are each of the form

$$a + b \langle X_r, h \rangle + c H_r + d H_r \langle X_r, f \rangle,$$

where a, b, c, d are (different) real constants for α, β and δ then $\widehat{H_t X_t}$ is given by a finite dimensional filter.

If $1 = (1, 1, \ldots, 1)^*$, because $X_t \in S$ we have $\langle X_t, 1 \rangle = 1$ for all t. Therefore, if we have a filter estimate for $\widehat{H_t X_t}$ we know $\widehat{H}_t = \langle \widehat{H_t X_t}, 1 \rangle$.

To determine the Zakai equation for $\sigma(H_t X_t)$ we first require an equation for $\overline{\Lambda}_t = \sigma(1) = E_0[\Lambda_t | \mathcal{Y}_t]$.

THEOREM 3.5.

$$\overline{\Lambda}_t = 1 + \int_0^t \overline{\Lambda}_r \cdot \langle \widehat{X}_r, g \rangle dy_r. \tag{3.9}$$

Proof. This equation is a filtering result under measure P_0, for which the observation process y_t is a Brownian motion and $\Lambda_t = 1 + \int_0^t \Lambda_r \langle X_r, g \rangle dy_r$ is a P_0-martingale. Therefore, the result follows as a special case of Theorem 3.1 with $H_t = \Lambda_t$, $\alpha_r = 0$, $\beta_r = 0$ and $\delta_r = \Lambda_r \langle X_r, g \rangle$. Consequently, $\overline{\Lambda}_t = 1 + \int_0^t \sigma(\delta_r) dy_r$ where

$$\sigma(\delta_r) = E_0[\Lambda_r \langle X_r, g \rangle \mid \mathcal{Y}_r]$$

$$= \overline{\Lambda}_r \cdot \langle \widehat{X}_r, g \rangle,$$

by Bayes' formula (2.10).

The Zakai equation is a recursive equation for the unnormalized estimate $\sigma(H_t X_t)$.

THEOREM 3.6.

$$\sigma(H_t X_t) = \widehat{H_0 X_0} + \int_0^t \sigma(\alpha_r X_r) dr + \int_0^t \sigma(H_r A X_r) dr$$

$$+ \int_0^t \sum_{i,j=1}^N \langle \sigma(\beta_r^j X_r - \beta_r^i X_r), e_i \rangle a_{ji} dr(e_j - e_i)$$

$$+ \int_0^t (\sigma(\delta_r X_r) + B\sigma(H_r X_r)) dy_r. \tag{3.10}$$

Proof. From Bayes' formula (2.10)

$$\sigma(H_t X_t) = \overline{\Lambda}_t \cdot \widehat{H_t X_t}.$$

Computing this product using (3.8) and (3.9), and using Bayes' formula on the integrands, we obtain (3.10).

REMARKS 3.7. The advantages of the Zakai equation over (3.8) are

1) the product term $\langle \widehat{X}_r, g \rangle \widehat{H_r X_r}$ is no longer present, so the equation is linear in conditional expectations.

2) the Zakai equation (3.10) is driven by the observations process y, rather than the innovations ν.

The filter and Zakai equations are recursive, so for $s \leq t$ we have the following forms:

COROLLARY 3.8.

$$\widehat{H_t X_t} = \widehat{H_s X_s} + \int_s^t \widehat{\alpha_r X_r} dr + \int_s^t \widehat{H_r A X_r} dr$$

$$+ \sum_{i,j=1}^N \int_s^t \langle \widehat{\beta_r^j X_r} - \widehat{\beta_r^i X_r}, e_i \rangle a_{ji} dr (e_j - e_i)$$

$$+ \int_s^t (\widehat{\delta_r X_r} + B \widehat{H_r X_r} - \langle \widehat{X}_r, g \rangle \widehat{H_r X_r}) d\nu_r. \tag{3.11}$$

Here, the initial condition is $E[H_s X_s \mid \mathcal{Y}_s]$, and is $\mathcal{Y}_s$-measurable.

COROLLARY 3.9.

$$\sigma(H_t X_t) = \sigma(H_s X_s) + \int_s^t \sigma(\alpha_r X_r) dr + \int_s^t \sigma(H_r A X_r) dr$$

$$+ \sum_{i,j=1}^N \int_s^t \langle \sigma((\beta_r^j - \beta_r^i) X_r), e_i \rangle a_{ji} dr (e_j - e_i)$$

$$+ \int_s^t (\sigma(\delta_r X_r) + B \sigma(H_r X_r)) dy_r. \tag{3.12}$$

Here, the initial condition is $E_0[\Lambda_s H_s X_s \mid \mathcal{Y}_s]$, again a $\mathcal{Y}_s$-measurable random variable.

4. SPECIAL CASES

We now obtain particular finite dimensional filters and smoothers, in both their normalized and unnormalized (Zakai) form, by specializing the results of Section 3.

4.1. The Wonham Filter and Smoother

Take $H_t = H_0 = 1$, $\alpha_r = 0$, $\beta_r = 0 \in R^N$, $\delta_r = 0$. Applying Theorem 3.3 we obtain

$$\widehat{X}_t = \widehat{X}_0 + \int_0^t A \widehat{X}_r dr + \int_0^t (B \widehat{X}_r - \langle \widehat{X}_r, g \rangle \widehat{X}_r) d\nu_r. \tag{4.1}$$

This is the Wonham filter, [7], and is a single, finite dimensional equation for the conditional distribution $\widehat{X}_t$.

From Theorem 3.6 the Zakai form for

$$\sigma(X_t) = E_0[\Lambda_t X_t \mid \mathcal{Y}_t] = \overline{\Lambda}_t \cdot X_t$$

is

$$\sigma(X_t) = \widehat{X}_0 + \int_0^t A\sigma(X_r)dr + \int_0^t B\sigma(X_r)dy_r. \tag{4.2}$$

For the smoothed estimates of X_s given $\mathcal{Y}_t$, $s \leq t$, take

$$H_t = H_s = \langle X_s, e_i \rangle, \quad s \leq t, \quad \alpha_r = 0, \quad \beta_r = 0, \quad \delta_r = 0$$

and apply Corollaries 3.8 and 3.9. (Rather than taking $H_t = \langle X_s, e_i \rangle$ and estimating $P(X_s = e_i \mid \mathcal{Y}_t) = \pi_t(\langle X_s, e_i \rangle)$ we could consider all states of X_s simultaneously by taking $H_t = X_s$, $s \leq t$; however, the product $H_t X_t$ would then have to be interpreted as a tensor, or Kronecker, product $H_t \otimes X_t$.) This smoother was discussed by Clements and Anderson [2]. From (3.11)

$$\widehat{\langle X_s, e_i \rangle X_t} = \widehat{\langle X_s, e_i \rangle X_s} + \int_s^t \widehat{A \langle X_s, e_i \rangle X_r} dr$$

$$+ \int_s^t (B(\widehat{\langle X_s, e_i \rangle X_r}) - \langle \widehat{X}_r, g \rangle \widehat{\langle X_s, e_i \rangle X_r}) d\nu_r. \tag{4.3}$$

This is a finite dimensional filter for $\langle X_s, e_i \rangle X_r$. Taking the inner product with $\mathbf{1} = (1, 1, \ldots, 1)^*$ gives

$$\widehat{\langle X_s, e_i \rangle \langle X_t, \mathbf{1} \rangle} = \pi_t(\langle X_s, e_i \rangle) = P(X_s = e_i \mid \mathcal{Y}_t).$$

Also, note that because $\sum_{j=1}^N a_{ji} = 0$

$$\langle X_s, e_i \rangle \langle A X_r, \mathbf{1} \rangle = 0.$$

Substituting $H_t = \langle X_s, e_i \rangle$, $s \leq t$, in (3.12) gives the Zakai form of the smoother:

$$\sigma_t(\langle X_s, e_i \rangle X_t) = \sigma_s(\langle X_s, e_i \rangle X_s) + \int_s^t A\sigma(\langle X_s, e_i \rangle X_r)dr + \int_s^t B\sigma_r(\langle X_s, e_i \rangle X_r)dy_r. \tag{4.4}$$

This is a single equation finite dimensional filter, for $\sigma_t(\langle X_s, e_i \rangle X_t)$ $= E_0[\Lambda_t \langle X_s, e_i \rangle X_t \mid \mathcal{Y}_t]$, driven by y. Taking the inner product with $\mathbf{1}$ gives $\sigma_t(\langle X_s, e_i \rangle)$.

4.2. Finite Dimensional Filters and Smoothers for the Number of Jumps

For $e_i, e_j \in S$, $i \neq j$, consider the stochastic integral

$$M_t^{ij} = \int_0^t \langle X_{r-}, e_i \rangle \langle e_j, dM_r \rangle.$$

Note the integrand $\langle X_{r-}, e_i \rangle e_j$ is predictable, so M_t^{ij} is a martingale. Now

$$\langle X_{r-}, e_i \rangle \langle e_j, dX_r \rangle = \langle X_{r-}, e_i \rangle \langle e_j, X_r - X_{r-} \rangle$$

$$= \langle X_{r-}, e_i \rangle \langle X_r, e_j \rangle$$

$$= I(X_{r-} = e_i \text{ and } X_r = e_j).$$

Write N_t^{ij} for the number of jumps from e_i to e_j in the time interval $[0, t]$. Then using (3.6)

$$N_t^{ij} = \int_0^t \langle X_{r-}, e_i \rangle \langle e_j, dX_r \rangle$$

$$= \int_0^t \langle X_{r-}, e_i \rangle \langle e_j, AX_{r-} \rangle dr + M_t^{ij}. \tag{4.5}$$

Now $\langle X_{r-}, e_i \rangle \langle e_j, AX_{r-} \rangle = \langle X_{r-}, e_i \rangle a_{ji}$ so

$$N_t^{ij} = \int_0^t \langle X_{r-}, e_i \rangle a_{ji} dr + M_t^{ij}. \tag{4.6}$$

This is the semimartingale decomposition of N_t^{ij}. To obtain the filter equations, in normalized and Zakai form, take

$$H_t = N_t^{ij}, \qquad H_0 = 0, \qquad \alpha_r = \langle X_r, e_i \rangle a_{ji},$$

$$\beta_r = \langle X_r, e_i \rangle e_j, \qquad \delta_r = 0.$$

Substituting in Theorem 3.3 and noting $\langle X_r, e_i \rangle X_r = \langle X_r, e_i \rangle e_i$ and

$$\sum_{i,j=1}^N \langle (\beta_r^j - \beta_r^i) X_r, e_i \rangle = \langle X_r, e_i \rangle^2 = \langle X_r, e_i \rangle,$$

we have

$$\widehat{N_t^{ij}X_t} = \int_0^t \langle \hat{X}_r, e_i \rangle a_{ji} e_j dr + \int_0^t A\widehat{N_r^{ij}X_r}dr + \int_0^t (B\widehat{N_r^{ij}X_r} - \langle \hat{X}_r, g \rangle \widehat{N_r^{ij}X_r})d\nu_r, \quad (4.7)$$

because, again, $\sum_{k=1}^N \langle \widehat{N_r^{ij}X_r}, e_k \rangle g_k e_k = B\widehat{N_r^{ij}X_r}$ where $B = \text{diag } g$. Together with the Wonham filter for $\hat{X}_t$, (4.7) gives a finite dimensional filter for $N_t^{ij}X_t$. Taking the inner product with $\underline{1}$ we have an expression for $\langle \widehat{N_t^{ij}X_t}, \underline{1} \rangle = \hat{N}_t^{ij}$.

The Zakai equation for $\sigma(N_t^{ij}X_t)$ is obtained by substituting in Theorem 3.6:

$$\sigma(N_t^{ij}X_t) = \int_0^t \langle \sigma(X_r), e_i \rangle a_{ji} e_j dr + \int_0^t A\sigma(N_r^{ij}X_r)dr + \int_0^t B\sigma(N_r^{ij}X_r)dy_r. \quad (4.8)$$

The smoothed estimates of N_s^{ij} given $\mathcal{Y}_t$, $s \leq t$, are obtained from Corollaries 3.8 and 3.9 by taking $H_t = H_s = N_s^{ij}$, $s \leq t$, $\alpha_r = 0$, $\beta_r = 0$ and $\delta_r = 0$. Then, from (3.11) we have the finite dimensional smoother

$$\widehat{N_s^{ij}X_t} = \widehat{N_s^{ij}X_s} + \int_s^t A\widehat{N_s^{ij}X_r}dr + \int_s^t (B\widehat{N_s^{ij}X_r} - \langle \hat{X}_r, g \rangle \widehat{N_s^{ij}X_r})d\nu_r. \quad (4.9)$$

Here $\widehat{N_s^{ij}X_t} = E[N_s^{ij}X_t \mid \mathcal{Y}_t]$ and taking the inner product with $\underline{1}$ gives $E[N_s^{ij} \mid \mathcal{Y}_t] = \pi_t(N_s^{ij})$.

The Zakai form of the smoother is obtained by substituting in (3.12):

$$\sigma(N_s^{ij}X_t) = \sigma(N_s^{ij}X_s) + \int_s^t A\sigma(N_s^{ij}X_r)dr + \int_s^t B\sigma(N_s^{ij}X_r)dy_r. \quad (4.10)$$

4.3. Finite Dimensional Filters and Smoothers for the Occupation Time

The time spent by the process in state e_i is given by

$$J_t^i = \int_0^t \langle X_r, e_i \rangle dr, \quad 1 \leq i \leq N.$$

Take

$$H_t = J_t^i, \qquad H_0 = 0, \qquad \alpha_r = \langle X_r, e_i \rangle,$$

$$\beta_r = 0 \in R^N, \qquad \delta_r = 0.$$

Substituting in Theorem 3.3, and noting again that $\langle X_r, e_i \rangle X_r = \langle X_r, e_i \rangle e_i$, we have

$$\widehat{J_t^i X_t} = \int_0^t \langle \hat{X}_r, e_i \rangle e_i dr + \int_0^t A \widehat{J_r^i X_r} dr + \int_0^t (B \widehat{J_r^i X_r} - \langle \hat{X}_r, g \rangle \widehat{J_r^i X_r}) d\nu_r. \tag{4.11}$$

Together with the filter for $\hat{X}_t$ we have a finite dimensional filter for $\widehat{J_t^i X_t}$, $1 \leq i \leq N$. Taking the inner product with $\mathbf{1}$ gives $\hat{J}_t^i = \langle \widehat{J_t^i X_t}, \mathbf{1} \rangle$. The Zakai form of (4.11) is obtained by substituting in Theorem 3.6:

$$\sigma(J_t^i X_t) = \int_0^t \langle \sigma(X_r), e_i \rangle e_i dr + \int_0^t A \sigma(J_r^i X_r) dr + \int_0^t (B \sigma(J_r^i X_r)) dy_r. \tag{4.12}$$

Finite dimensional smoothers are obtained for J_s^i by taking $H_t = H_s = J_s^i$ for $s \leq t$ and $\alpha_r = 0$, $\beta_r = 0$, $\delta_r = 0$. Applying Corollaries 3.8 and 3.9 gives

$$\widehat{J_s^i X_t} = \widehat{J_s^i X_s} + \int_s^t A \widehat{J_s^i X_r} dr + \int_s^t (B \widehat{J_s^i X_r} - \langle \hat{X}_r, g \rangle \widehat{J_s^i X_r}) d\nu_r. \tag{4.13}$$

Here $\widehat{J_s^i X_t} = E[J_s^i X_t \mid \mathcal{Y}_t]$ and taking the inner product with $\mathbf{1}$ gives the smoothed estimate $\langle \widehat{J_s^i X_t}, \mathbf{1} \rangle = E[J_s^i \mid \mathcal{Y}_t] = \pi_t(J_s^i)$. The Zakai form here is

$$\sigma(J_s^i X_t) = \sigma(J_s^i X_s) + \int_s^t A \sigma(J_s^i X_r) dr + \int_s^t (B \sigma(J_s^i X_r)) dy_r. \tag{4.14}$$

4.4. Finite Dimensional Filters and Smoothers Related to the Drift Coefficient

In the next section we shall see that the estimation of the drift coefficient $g = (g_1, g_2, \ldots, g_N)^*$ of the observation process involves the filtered estimate of the processes

$$G_t^i = \int_0^t \langle X_r, e_i \rangle dy_r.$$

Taking $H_t = G_t^i$, $H_0 = 0$, $\alpha_r = 0$, $\beta_r = 0$ and $\delta_r = \langle X_r, e_i \rangle$ we shall apply Theorem 3.3, noting again that $X_r \delta_r = X_r \langle X_r, e_i \rangle = \langle X_r, e_i \rangle e_i$. Therefore, together with the Wonham filter we have the following finite dimensional filter for $\widehat{G_t^i X_t}$:

$$\widehat{G_t^i X_t} = \int_0^t A \widehat{G_r^i X_r} dr + \int_0^t (\langle \hat{X}_r, e_i \rangle e_i + B \widehat{G_r^i X_r} - \langle \hat{X}_r, g \rangle \widehat{G_r^i X_r}) d\nu_r. \tag{4.15}$$

The Zakai form here is

$$\sigma(G_t^i X_t) = \int_0^t A \sigma(G_r^i X_r) dr + \int_0^t (B \sigma(G_r^i X_r) + \langle \sigma(X_r), e_i \rangle e_i) dy_r. \tag{4.16}$$

Taking $H_t = G_s^i$ for $s \le t$, $\alpha_r = 0$, $\beta_r = 0$ and $\delta_r = 0$, we obtain from Corollaries 3.8 and 3.9 the following finite dimensional smoothers:

$$\widehat{G_s^i X_t} = \widehat{G_s^i X_s} + \int_s^t A\widehat{G_s^i X_r}\,dr + \int_s^t (B\widehat{G_s^i X_r} - \langle \hat{X}_r, g\rangle \widehat{G_s^i X_r})d\nu_r, \qquad (4.17)$$

with a Zakai form

$$\sigma(G_s^i X_t) = \sigma(G_s^i X_s) + \int_s^t A\sigma(G_s^i X_r)dr + \int_s^t B\sigma(G_s^i X_r)dy_r. \qquad (4.18)$$

REMARK 4.5. In all the above smoothing equations when we take an inner product with 1 the integral involving A will vanish because $\sum_{j=1}^{N} a_{ji} = 0$.

5. PARAMETER ESTIMATION OF A NOISILY OBSERVED MARKOV CHAIN

This problem is nicely discussed by Zeitouni and Dembo in [3] and [9]. We first review their formulation in our setting.

Suppose, as above, that X_t, $t \ge 0$, is a Markov chain with state space $S = \{e_1, e_2, \ldots, e_N\}$ and Q-matrix generator $A = (a_{ij})$. Then from Lemma 2.2

$$X_t = X_0 + \int_0^t AX_r dr + M_t. \qquad (5.1)$$

Again, suppose X_t is observed through the process

$$y_t = \int_0^t \langle X_r, g\rangle dr + B_t, \qquad (5.2)$$

where B is a Brownian motion independent of X. We have noted that the function g is determined by a vector $(g_1, g_2, \ldots, g_N)^*$.

The above model, therefore, is determined by the set of parameters

$$\theta := (a_{ij}, \ 1 \le i, j \le N, \ g_i, \ 1 \le i \le N).$$

Suppose the model is first determined by a set of parameters

$$\theta' = (a_{ij}, \ g_i, \ 1 \le i, j \le N)$$

and we wish to determine a new set

$$\theta = (a_{ij}, \; g_i, \; 1 \leq i,j \leq N)$$

which maximizes the log-likelihood defined below. Write $P_{\theta'}$ and P_θ for their respective probability measures.

From (4.6) we have, under $P_{\theta'}$, that

$$N_t^{ij} = \int_0^t \langle X_r, e_i \rangle a'_{ji} dr + M_t^{ij}. \tag{5.3}$$

From Theorem T2 of Chapter VI of Brémaud [1] we see that to modify the intensity of the counting process N_t^{ij}, that is, to change a'_{ji} to a_{ji}, we should introduce the Radon–Nikodym derivative L_t^{ij} given by:

$$L_t^{ij} = \left(\frac{a_{ji}}{a'_{ji}} \right)^{N_t^{ij}} \exp \Big\{ \sum_{i,j=1}^N \int_0^t (a'_{ji} - a_{ji}) \langle X_r, e_i \rangle dr \Big\}.$$

Clearly, if the chain jumps from state i to state j at time t it cannot jump from state i' to state j' at the same time, if $(i,j) \neq (i',j')$. Therefore,

$$[M^{ij}, M^{i'j'}]_t = \sum_{0 < r \leq t} \Delta M_r^{ij} \Delta M_r^{i'j'}$$

$$= 0$$

and the martingales M^{ij} and $M^{i'j'}$ are orthogonal if $(i,j) \neq (i',j')$. Consequently, to change all the a'_{ji} to a_{ji} and to change the g'_i to g_i, we should define

$$\frac{dP_\theta}{dP_{\theta'}}\Big|_{\mathcal{F}_t} = L_t := \prod_{i \neq j} L_t^{ij} \cdot \exp \Big\{ \int_0^t \langle X_r, g - g' \rangle dy_r - \frac{1}{2} \int_0^t (\langle X_r, g \rangle^2 - \langle X_r, g' \rangle^2) dr \Big\}$$

$$= \prod_{\substack{i,j=1 \\ i \neq j}}^N \left(\frac{a_{ji}}{a'_{ji}} \right)^{N_t^{ij}} \exp \Big\{ \sum_{i,j=1}^N \int_0^t (a'_{ji} - a_{ji}) \langle X_r, e_i \rangle dr \Big\}$$

$$\times \exp \Big\{ \sum_{i=1}^N (g_i - g'_i) \int_0^t \langle X_r, e_i \rangle dy_r - \frac{1}{2} \sum_{i=1}^N (g_i^2 - g_i'^2) \int_0^t \langle X_r, e_i \rangle dr \Big\}.$$

The log-likelihood is, therefore,

$$\log \frac{dP_\theta}{dP_{\theta'}}\Big|_{\mathcal{F}_t} = \log L_t$$

$$= \sum_{\substack{i,j=1 \\ i \neq j}}^{N} \left\{ N_t^{ij} \left(\log \frac{a_{ji}}{a'_{ji}} \right) + \int_0^t (a'_{ji} - a_{ji})\langle X_r, e_i \rangle dr \right\}$$

$$+ \sum_{i=1}^{N} \left\{ (g_i - g'_i) \int_0^t \langle X_r, e_i \rangle dy_r - \frac{1}{2}(g_i^2 - g_i'^2) \int_0^t \langle X_r, e_i \rangle dr \right\}$$

and, using the notation of Section 4, this is

$$= \sum_{\substack{i,j=1 \\ i \neq j}}^{N} (N_t^{ij} \log a_{ji} - a_{ji} J_t^i) + \sum_{i=1}^{N} (g_i G_t^i - \frac{1}{2} g_i^2 J_t^i) + R(\theta'),$$

where $R(\theta')$ does not involve any of the parameters of $\theta = (a_{ji}, g_i)$. Therefore,

$$E\left[\log \frac{dP_\theta}{dP_{\theta'}} \mid \mathcal{Y}_t \right] = \sum_{\substack{i,j=1 \\ i \neq j}}^{N} (\widehat{N}_t^{ij} \log a_{ji} - a_{ji} \widehat{J}_t^i) + \sum_{i=1}^{N} (g_i \widehat{G}_t^i - \frac{1}{2} g_i^2 \widehat{J}_t^i) + \widehat{R}(\theta'). \quad (5.4)$$

The unique maximum of (5.4) over θ, obtained by equating to zero the partial derivatives of (5.4) in a_{ji} and g_i, is therefore, given by

$$a_{ji} = \widehat{N}_t^{ij} / \widehat{J}_t^i = \sigma(N_t^{ij})/\sigma(J_t^i)$$

$$g_i = \widehat{G}^i / \widehat{J}_t^i = \sigma(G_t^i)/\sigma(J_t^i),$$

using Bayes' formula (2.8). This parameter set gives P_θ, the next probability measure in the sequence of steps in the EM procedure.

The sequence of log-likelihoods $\frac{dP_\theta}{dP_{\theta'}}$ constructed this way is increasing and so converges. The convergence of the sequence of θ is discussed in [3] and [9].

REMARKS 5.1. In [3] $\widehat{G}_t^i$ is written as $\int_0^t E[\langle X_r, g \rangle \mid \mathcal{Y}_t] dy_r$, a non-adapted stochastic integral which is not defined, at least in [3]. Also, it is not clear the reference to Yao [8] in [9] provides a finite dimensional filter for $\widehat{J}_t^i$ in the general case.

However, the results of Section 4 above give explicit finite dimensional filters, (and smoothers), for $\widehat{N}_t^{ij}$, $\widehat{J}_t^i$ and $\widehat{G}_t^i$, $1 \leq i,j \leq N$. Forward-backward algorithms of Baum–Welch type are not required.

References

1. P. Brémaud, *Point Processes and Queues.* Springer–Verlag, New York, Heidelberg, Berlin, 1981.

2. D. Clements and B.D.O. Anderson, "A non-linear fixed lag smoother for finite state Markov processes," *IEEE Trans. Inform. Theory*, vol. 1T-21, pp. 446–452, 1975.

3. A. Dembo and O. Zeitouni, "Parameter estimation of partially observed continuous time stochastic processes via the EM algorithm," *Stoch. Proc. and App.* 23, pp. 91–113, 1986.

4. R.J. Elliott, *Stochastic Calculus and Applications*, Applications of Mathematics, Vol. 18. Springer-Verlag, Berlin, Heidelberg, New York, 1982.

5. R.J. Elliott, "The nonlinear filtering equations." Lecture Notes in Control and Information Sciences 43, Springer–Verlag, Berlin, Heidelberg, New York, pp. 168–178, 1982.

6. R.J. Elliott, "New finite dimensional filters and smoothers related to the Wonham filter." Preprint, July 1991, Department of Statistics and Applied Probability, University of Alberta.

7. W.M. Wonham, "Some applications of stochastic differential equations to optimal non-linear filtering," *SIAM Jour. Control*, 2, pp. 347–369, 1965.

8. Y.C. Yao, "Estimation of noisy telegraph process: nonlinear filtering versus nonlinear smoothing," *IEEE Trans. Inform. Theory*, vol. 1T-31, pp. 444–446, 1985.

9. O. Zeitouni and A. Dembo, "Exact filters for the estimation of the number of transitions of finite-state continuous time Markov processes," *IEEE Trans. Inform. Theory*, vol. 1T-34, pp. 890–893, 1988.

ADAPTIVE CONTROL OF A PARTIALLY OBSERVED
CONTROLLED MARKOV CHAIN *

EMMANUEL FERNÁNDEZ-GAUCHERAND†, ARISTOTLE ARAPOSTATHIS‡,
AND STEVEN I. MARCUS§.

Abstract

We consider an adaptive finite state controlled Markov chain with partial state information, motivated by a class of replacement problems. We present parameter estimation techniques based on the information available after actions that reset the state to a known value are taken. We prove that the parameter estimates converge w.p.1 to the true (unknown) parameter, under the feedback structure induced by a certainty equivalent adaptive policy. We also show that the adaptive policy is self-optimizing, in a long-run average sense, for any (measurable) sequence of parameter estimates converging w.p.1 to the true parameter.

* This work was supported in part by the Texas Advanced Technology Program under Grant No. 003658-093, in part by the Air Force Office of Scientific Research under Grants AFOSR-91-0033, F49620-92-J-0045, and F49620-92-J-0083, and in part by the National Science Foundation under Grant CDR-8803012.

† Systems and Industrial Engineering Department, The University of Arizona, Tucson, Arizona 85721.

‡ Department of Electrical and Computer Engineering, The University of Texas at Austin, Austin, Texas 78712-1084.

§ Department of Electrical Engineering and Systems Research Center, The University of Maryland, College Park, Maryland 20742.

I. Introduction

In recent years, there has been a considerable amount of work in stochastic adaptive control [10]–[11]. However, aside from results for linear systems, little progress has been made on problems with incomplete or noisy state observations. An initial step in this direction was taken in [1], where the adaptive estimation of the state of a finite state Markov chain, with incomplete state information, and with the state transition probabilities depending on unknown parameters, is studied. This adaptive estimation problem is that of computing recursive estimates of the conditional probability vector of the state at time t, given all the past observations, when the transition matrix P is not completely known, i.e., it depends on a vector of unknown parameters θ — this dependence is expressed as $P(\theta)$. In [1] we use the previously derived recursive filter for the conditional probabilities, and simultaneously recursively estimate the parameters, using the most recent parameter estimates to update the filter. This adaptive estimation algorithm is then analyzed via the Ordinary Differential Equation (ODE) Method [12]–[13]. The convergence of the recursive parameter estimates is established, and optimality of the adaptive state estimator is proved, in a long-run average sense.

In [7]–[8], we began to investigate the application of similar techniques to the control of adaptive finite state Markov chains with incomplete observations. One interesting set of problems for which some results are available when the parameters are *known* are those involving quality control, replacement, and repair of a unit in a manufacturing system or communication network [9], [15], [18]. We formulated the adaptive version of a problem of this type in the above references; however, the presence of feedback makes this problem much more difficult than that of [1]. Discontinuities in the optimal control strategies lead to averaged ODE's with discontinuous right-hand sides that cannot be handled by currently available methods.

In this paper we present parameter estimation techniques based on the information available after actions that reset the state to a known value are taken. At these times, the (augmented) state process *regenerates*, its future evolution becoming independent of the past. We prove (by means of the ODE method) w.p.1 convergence of the parameter estimates to the true (unknown) parameter θ_0, for a parameter estimation scheme of this type. Then, given *any* sequence of parameter estimates which converges w.p.1 to θ_0, and which is measurable with respect to the filtration generated by the observations, we show that a certainty equivalent adaptive policy is *self-optimizing*. The latter is obtained by an analysis which uses the known (threshold) structure of optimal policies for problems with *known* parameters. Our analysis is of particular interest since the nice formalism recently presented in [17] cannot be directly applied in the present situation: here the state is only partially observed and the optimal policy is not a continuous function of θ.

The methodology exposed in the analysis relies largely on the w.p.1 convergence to θ_0 of the parameter estimates, and the continuity in the parameterization of quantities in the model, like $P(\theta)$ and the solutions to the corresponding optimality equations. Hence, this methodology is also applicable to a more general situation than the one presented here; see [6]. In addition, we note that the feedback structure induced by our adaptive policy obviates the need for, e.g. forced choice schemes, c.f. [11].

II. A Partially Observed Binary Replacement Problem

Consider a situation in which a system, such as a machine, production process, or computer communications network can fail. The (core) state X_t of the system can either be *good* (0), or *failed* (1); let $\mathbf{X} := \{0, 1\}$. The available control actions (or decisions) are to *operate* the system in its current condition (0), or to *reset/replace* the system to an *as new* condition (1); let $\mathbf{U} := \{0, 1\}$. Assume for the moment that there is an underlying probability space $(\Omega, \mathcal{B}, \mathcal{P})$. The process $\{X_t\}_{t \in \mathbb{N}_0}$ is modeled as a controlled finite state Markov chain, where we have that

$$\mathcal{P}\{X_{t+1} = j \mid X_t = i, X_{t-1}, \ldots, X_0; U_t = u, U_{t-1}, \ldots, U_0\}$$
$$= [P(u)]_{i,j}; \qquad t \in \mathbb{N}_0 := \{0, 1, 2, \ldots\}, \qquad (2.1)$$

and the state transition probability matrices are given as

$$P(0) = \begin{bmatrix} 1-\theta & \theta \\ 0 & 1 \end{bmatrix}; \qquad P(1) = \begin{bmatrix} 1 & 0 \\ 1 & 0 \end{bmatrix}. \qquad (2.2)$$

Here $\theta \in [0, 1]$ gives the *failure rate* of the system. Only imperfect observations of $\{X_t\}_{t \in \mathbb{N}_0}$ are available in the form of a random process $\{Y_t\}_{t \in \mathbb{N}}$; Y_t gives a correct observation of X_t with probability q, when $U_{t-1} = 0$, whereas if $U_{t-1} = 1$ then $Y_t = X_t = 0$. More precisely, $Y_t \in \mathbf{Y} = \{0, 1\}$ and

$$\mathcal{P}\{Y_{t+1} = i \mid Y_t, \ldots, Y_1; X_{t+1} = i, X_t, \ldots, X_0; U_t = 0, U_{t-1}, \ldots, U_0\}$$
$$= \mathcal{P}\{Y_{t+1} = i \mid X_{t+1} = i; U_t = 0\} =: q, \qquad t \in \mathbb{N}_0. \qquad (2.3)$$

It suffices to consider only $0.5 \leq q \leq 1$. The cases $q = 0.5$ and $q = 1$ correspond to the *completely unobserved* and *completely observed* situations, respectively; we restrict our analysis to the situation of strict partial observability, i.e., $q < 1$. The one-step cost $c(x, u)$ is defined as $c(0, 0) = 0$, $c(1, 0) = C$, $c(x, 1) = R$, where $0 < C < R$. Probability distribution vectors on $\mathbf{X}$ are elements of $\Delta := \{p \in \mathbb{R}^2 : p = [1-\rho, \rho], 0 \leq \rho \leq 1\}$. Thus, each $p \in \Delta$ can be uniquely identified with a scalar $\rho \in [0, 1]$, as indicated. Initially, there is a given probability $0 \leq \rho_0 \leq 1$ that the system is failed, an action is taken, and the state evolves according to (2.2); a first observation is received, another action is taken; and so on.

An (admissible) control *law, policy,* or *strategy* π is a rule for selecting the actions U_t, based on $h_t = (\rho_0, U_0, Y_1, \ldots, Y_{t-1}, U_{t-1}, U_t)$, where h_t is the available information at time t. The canonical sample path space is $\Omega = X \times U \times (X \times Y \times U)^\infty$, and B denotes the Borel σ-algebra obtained by endowing Ω with the discrete topology. Then to each admissible strategy π and $0 \le \rho_0 \le 1$, we associate the average cost

$$J(\pi, \rho_0) := \limsup_{n \to \infty} E^\pi_{\rho_0} \left[\frac{1}{n} \sum_{t=0}^{n-1} c(X_t, U_t) \right], \tag{AC}$$

where $E^\pi_{\rho_0}$ is the expectation with respect to an appropriate marginal of the (unique) probability measure $\mathcal{P}^\pi_{\rho_0}$ on B induced by ρ_0 and the strategy π; see [2], [10]. The *optimal (AC) control (or decision) problem* is that of selecting a strategy such that the average cost is minimized, over all admissible strategies. The optimal (AC) cost function is defined as $\Gamma(\rho_0) := \inf_\pi \{ J(\pi, \rho_0) : \pi \text{ is an admissible strategy} \}$, for $0 \le \rho_0 \le 1$.

A. Information States

It is well known that the conditional probability distribution process, whose i^{th} component is given by

$$p_t^{(i)} := \mathcal{P}^\pi_{\rho_0} \{ X_t = i \mid Y_t, \ldots, Y_1; U_{t-1}, \ldots, U_0 \}, \qquad t \in \mathbb{N}, \qquad p_0 := [1 - \rho_0, \rho_0],$$

constitutes an *information state* (or statistic sufficient for control) [2], [4], [5], [10], [11]; for this problem, it can be written as $p_t = [1 - \rho_t, \rho_t]$, where ρ_t is the conditional probability of the process being in the failed state.

A *separated* strategy is a sequence of maps $\pi = (\pi_0, \pi_1, \pi_2, \ldots)$, where $\pi_t : [0, 1] \to U$. When $\pi_t(\cdot) = \pi(\cdot)$ for all values of t, then the policy is said to be stationary. Then the partially observed, average cost problem is equivalent (i.e., equal minimum costs for each ρ_0) to the *completely observed* problem, with state ρ_t and state space $[0, 1]$, of finding a separated admissible strategy which minimizes

$$\overline{J}(\pi, \rho_0) := \limsup_{n \to \infty} E^\pi_{\rho_0} \left[\frac{1}{n} \sum_{t=0}^{n-1} \overline{c}(\rho_t, U_t) \right],$$

where $\overline{c}(\rho, u) = (1 - \rho)c(0, u) + \rho c(1, u)$. Note that $\overline{c}(\rho, 0) = \rho C$ and $\overline{c}(\rho, 1) = R$. Using Bayes' rule, it is easily shown that ρ_t can be computed recursively, as follows:

$$\rho_{t+1} = T(1, \rho_t, U_t) Y_{t+1} + T(0, \rho_t, U_t)(1 - Y_{t+1}), \tag{2.4}$$

where

$$V(1, \rho, 0) = (1 - q)(1 - \rho)(1 - \theta) + q[\rho(1 - \theta) + \theta] = 1 - V(0, \rho, 0), \tag{2.5}$$

$$V(1, \rho, 1) = 0, \qquad V(0, \rho, 1) = 1, \tag{2.6}$$

$$T(0,\rho,0) = \frac{(1-q)[\rho(1-\theta)+\theta]}{V(0,\rho,0)}, \qquad T(1,\rho,0) = \frac{q[\rho(1-\theta)+\theta]}{V(1,\rho,0)}, \qquad (2.7)$$

$$T(y,\rho,1) = 0; \qquad y = 0,1, \quad \rho \in [0,1]. \qquad (2.8)$$

Here $V(y,\rho,u)$ is interpreted as the (one-step ahead) conditional probability of the observation being y given the decision u and an a priori probability ρ of the state being failed. Likewise, $T(y,\rho,u)$ is interpreted as the a posteriori conditional probability of the unit being failed given that decision u was made, observation y obtained, and an a priori probability ρ. Let $I[A]$ denote the indicator function of the event A. A well known property of the process $\{\rho_t\}_{t=0}^\infty$ is the following [4].

Lemma 2.1. $\{\rho_t\}_{t=0}^\infty$ is a controlled Markov process, and its state transition probabilities are given by

$$\begin{aligned}
\mathcal{P}_{\rho_0}^\pi &\{\rho_{t+1} \in B \mid \rho_t = \rho; U_t = u\} \\
&= \sum_{y \in Y} V(y,\rho,u)\, I[T(y,\rho,u) \in B] =: \mathcal{K}(B \mid \rho, u), \qquad (2.9)
\end{aligned}$$

for all (Borel) subsets B of $[0,1]$.

III. The Structure of Optimal Policies.

Consider the optimal control problem corresponding to each parameter value $\theta \in [0,1]$. Then, the existence of solutions to the corresponding (average cost) optimality equation follows from the existence of a reset/repair action [6], [9], [15]–[16]. We summarize these results as follows; dependence on θ is made explicit.

Theorem 3.1. Assume $q \in [0.5,1)$, $\theta \in [0,1]$.

(i) There exist a constant $0 \le \Gamma_\theta^* \le R$ and a concave, nondecreasing map $h_\theta : [0,1] \to [0,R]$, with $h_\theta(0) = 0$, such that

$$\Gamma_\theta^* + h_\theta(\rho) = \min\left\{f_\theta(\rho)\,;\, R\right\}, \qquad (3.1)$$

where

$$f_\theta(\rho) := \rho C + \sum_{y=0}^{1} V(y,\rho,0;\theta)\, h_\theta(T(y,\rho,0;\theta)). \qquad (3.2)$$

(ii) Any stationary separated policy that achieves the minimum in (3.1) is average cost optimal; the minimum cost is Γ_θ^*, for any value of ρ_0.

The following will be used in the sequel; for the proof, see [19].

Corollary 3.1. *Assume $q \in [0.5, 1)$ and $\theta \in [0, 1]$.*

(i) *Any concave and nondecreasing solution $h_\theta(\cdot)$ of (3.1) is continuous on $[0, 1]$;*

(ii) *furthermore, there is only one such solution satisfying $h_\theta(0) = 0$.*

Henceforth, the dependence of $\mathcal{P}_{\rho_0}^\pi$ on ρ_0 will be omitted, in view of Theorem 3.1. Equation (3.1) can then be used to determine the structure of the optimal policies [6], [9], [15].

Theorem 3.2. *Assume $q \in [0.5, 1)$ and $\theta \in (0, 1]$.*

(i) *If*

$$\frac{C(1 + \theta)}{\theta} \leq R \iff \frac{C}{R - C} \leq \theta,$$

then the policy "operate $(U_t = 0)$ for all $\rho_t \in [0, 1]$" is average cost optimal.

(ii) *If*

$$R < \frac{C(1 + \theta)}{\theta} \iff \theta < \frac{C}{R - C},$$

then there exists a threshold policy which is average cost optimal; i.e., there exists $\alpha(\theta) \in (0, 1)$ such that it is optimal to operate $(U_t = 0)$ for $\rho_t \in [0, \alpha(\theta))$, and to repair $(U_t = 1)$ for $\rho_t \in [\alpha(\theta), 1]$.

IV. The Adaptive Binary Replacement Problem

If the parameter θ is unknown, we cannot compute ρ_t, nor can we directly solve the optimal control problem. The *enforced certainty equivalence* approach which we will adopt involves simultaneously computing recursive estimates $\hat{\theta}_t$ of the unknown parameter, and $\hat{\rho}_t$ of the information state, and using the latest available parameter estimate in the filtering equation (2.4) to compute the next estimate $\hat{\rho}_{t+1}$; the decision U_t is made taking $\hat{\theta}_t$ and $\hat{\rho}_t$ as if they were the true (correct) values. Let $\Theta_\delta := [\delta, \delta']$ be the parameter set in which $\hat{\theta}_t$ is allowed to take its values, where δ is an arbitrarily small positive number and $\delta' = \min\{1, \frac{C}{R-C} - \delta\}$. For decision-making, we define the set $\mathcal{OP} = \{\pi(\cdot; \theta)\}_{\theta \in \Theta_\delta}$ of optimal threshold policies described above, parameterized by θ. Thus, we conclude from Theorem 3.2 (ii) that $0 < \alpha(\theta) < 1$, for each $\theta \in \Theta_\delta$, where $\alpha(\theta)$ denotes the dependence of the threshold on θ. We also let θ_0 denote the (unknown) true value of the parameter, which we assume to be constant and an element of the interior of Θ_δ. The following result, on the continuity in θ of the optimal cost, the value function and the threshold, is proved in [3, Theorem A.1].

Theorem 4.1. *Assume $q \in [0.5, 1)$. Let $0 < \delta < 1$. Then for $\theta \in \Theta_\delta$, we have that:*

(i) *the pair $(\Gamma_\theta^*, h_\theta)$ is continuous in θ;*

(ii) *there exists a unique $\alpha(\theta) \in (0, 1)$ such that $f_\theta(\alpha(\theta)) = R$;*

(iii) *$\alpha(\cdot)$ is continuous on Θ_δ.*

Observe that by Theorem 4.1 (iii) and since Θ_δ is compact, there is a number $\alpha^* < 1$ such that, for all $\theta \in \Theta_\delta$, $0 < \alpha(\theta) \leq \alpha^*$.

A. Adaptive Policy.

Given a sequence of estimates $\{\hat{\theta}_t\}_{t=0}^\infty$ of θ_0, compute the control action at each time $t \in \mathbb{N}_0$ by

$$U_t = \pi(\hat{\rho}_t; \hat{\theta}_t), \qquad \pi(\cdot\,;\cdot) \in \mathcal{OP}, \tag{4.1}$$

where the conditional probability estimate is computed recursively via

$$\begin{aligned}
\hat{\rho}_{t+1} = &T(1, \hat{\rho}_t, \pi(\hat{\rho}_t; \hat{\theta}_t); \hat{\theta}_{t+1}) \cdot Y_{t+1} \\
&+ T(0, \hat{\rho}_t, \pi(\hat{\rho}_t; \hat{\theta}_t); \hat{\theta}_{t+1}) \cdot (1 - Y_{t+1}), \qquad \hat{\rho}_0 = \rho_0.
\end{aligned} \tag{4.2}$$

We will denote by π^a the policy given by (4.1) and (4.2).

B. Parameter Estimation.

There are a number of ways to compute the estimates $\hat{\theta}_t$; we consider here only recursive schemes. One method, discussed in [7]–[8], updates the parameter estimate $\hat{\theta}_t$ at each time step t, and is similar to that used for adaptive estimation in [1]. However, the analysis of convergence is very difficult, due to the complex feedback structure induced. We concentrate here on algorithms which update $\hat{\theta}_t$ after each repair. The advantage of this approach is that when a repair event occurs, the state of the system is reset to the "as new" state, and thus the processes of interest are identically distributed between these events. On the other hand, the convergence rate may be too slow, and thus some forcing may be needed to accelerate the convergence. Algorithms that take advantage of analogous *regenerative* behavior in some queueing problems, by updating after each busy period, have been presented in [13]. The next result is a direct consequence of [3, Theorem A.2].

Theorem 4.2. *Under the adaptive policy π^a, regeneration occurs infinitely often (i.o.), i.e.,*

$$\mathcal{P}^{\pi^a}\{U_t = 1, \text{ i.o.}\} = 1.$$

Let τ_k be the kth repair time under π^a (i.e., the kth time such that $U_t = 1$). Since $U_{\tau_k} = 1$, then $X_{\tau_k+1} = 0$, $U_{\tau_k+1} = 0$, and Y_{τ_k+2} is observed. Hence, the state is known perfectly at $\tau_k + 1$ and the observations $\{Y_{\tau_k+2} : k = 1, 2, \ldots\}$ form an independent identically distributed (i.i.d.) sequence of Bernoulli random variables, with $\mathcal{P}^{\pi^a}\{Y_{\tau_k+2} = j\} = \lambda_j(\theta_0)$, $j = 0, 1$, where

$$\lambda_1(\theta) := (1 - \theta)(1 - q) + \theta q = 1 - \lambda_0(\theta). \tag{4.3}$$

This sequence provides information about the transition from $X_{\tau_k+1} = 0$ to X_{τ_k+2}, and thus can be used to estimate θ_0. Define $\overline{Y}_k := Y_{\tau_k+2}$. The sequence $\{\overline{Y}_k\}_{k=0}^\infty$ is i.i.d., its distribution depending only on the true parameter θ_0 and the reliability of the measuring device (q).

Note that by Theorem 4.2 and the strong law of large numbers we have that

$$\frac{1}{n} \sum_{k=1}^n \overline{Y}_k \xrightarrow[n\to\infty]{} \lambda_1(\theta_0), \qquad \mathcal{P}^{\pi^*}\text{-a.s.}. \tag{4.4}$$

Let $\hat{\overline{\theta}}_n := \hat{\theta}_{\tau_n+2}$. Then, setting

$$\lambda_1(\hat{\overline{\theta}}_n) = \frac{1}{n} \sum_{k=1}^n \overline{Y}_k,$$

where $\lambda_1(\cdot)$ is defined in (4.3), we obtain a sequence of strongly consistent parameter estimates $\{\hat{\overline{\theta}}_n\}$. Also, a *prediction error-based* algorithm can be formulated. Since the observations take only the values $\{0, 1\}$, then the prediction error in this case is

$$\epsilon_n(\theta) = \overline{Y}_n - \lambda_1(\theta). \tag{4.5}$$

However, in order to have $\hat{\overline{\theta}}_n \in \Theta_\delta$, a projection mechanism is required. A stochastic approximation-type recursive algorithm which is designed to minimize $E^{\pi^*}[\frac{1}{2}\epsilon_n(\theta)^2]$ is then

$$\hat{\overline{\theta}}_{n+1} = \Pi_{\Theta_\delta}\left(\hat{\overline{\theta}}_n + \frac{1}{n+1} R_{n+1}^{-1} \psi_n \epsilon_n(\hat{\overline{\theta}}_n)\right), \qquad \hat{\overline{\theta}}_0 \in \Theta_\delta, \tag{4.6a}$$

where the map Π_{Θ_δ} is a projection into the interior of Θ_δ. Also, R_n can be computed in different ways, e.g. if $R_n = (2q-1)^2$, then we obtain a recursive (and projected) version of the scheme obtained from (4.4) above. We choose to use

$$R_{n+1} = R_n + \frac{1}{n+1}\left(\psi_n^2 - R_n\right), \qquad R_1 = 1,$$
$$\psi_n = -\frac{\partial}{\partial\theta}\,\epsilon_n(\theta)\Big|_{\theta=\hat{\overline{\theta}}_n} = \frac{\partial}{\partial\theta}\,\lambda_1(\theta) = 2q-1. \tag{4.6b}$$

The following can then be shown using the techniques in [12], [13].

Theorem 4.3. *Consider the algorithm (4.6). The sequence $\{\hat{\overline{\theta}}_n\}_{n=0}^\infty$ converges $\mathcal{P}^{\pi^*}$-a.s., as $n \to \infty$, to the set of limit points of the ODE*

$$\dot{\theta}(t) = -R^{-1}(t)(2q-1)^2(\theta(t) - \theta_0),$$
$$\dot{R}(t) = (2q-1)^2 - R(t). \tag{4.7}$$

Since θ_0 is assumed to lie in the interior of Θ_δ, all solutions of the ODE (4.7) leave the interior of Θ_δ invariant and thus the projection operator Π_{Θ_δ} need not be considered in the averaged equations. It is straightforward to show that (4.7) is globally asymptotically stable with unique limit point θ_0. In the natural way, we define $\hat{\theta}_t$ to be constant between updates: $\hat{\theta}_t := \hat{\overline{\theta}}_n, t \in \{\tau_n + 2, \tau_n + 3, \ldots, \tau_{n+1} + 1\}$. We thus have the following result, which is a direct consequence of Theorem 4.2.

Corollary 4.1. *Assume* $q \in (0.5, 1)$. *Then the sequence* $\{\hat{\theta}_t\}_{t=0}^{\infty}$ *converges to* θ_0, *as* $t \to \infty$, $\mathcal{P}^{\pi^a}$*-a.s.:*

Remark 4.1: Let π be any separated policy satisfying $\pi_t(0) = 0$, for all $t \in \mathbb{N}_0$, and $\mathcal{P}^{\pi}\{U_t = 1, \text{i.o.}\} = 1$. Then the results above will also hold if π is used instead of π^a.

V. Average Cost Optimality of the Adaptive Policy

We examine next the long-run average performance of the adaptive policy π^a given by (4.1) and (4.2). Let $\mathcal{F}_t$ be the σ-algebra generated by the observations up to time t, i.e., $\mathcal{F}_t = \sigma(Y_1, \ldots, Y_t)$. Note that $\{\hat{\theta}_t\}$ of Corollary 4.1 satisfies the following conditions:

(E1) $\hat{\theta}_t$ is $\mathcal{F}_t$-measurable, and $\hat{\theta}_t \in \Theta_\delta$, for all $t \in \mathbb{N}_0$;

(E2) $\hat{\theta}_t \to \theta_0$, $\mathcal{P}^{\pi^a}$-a.s..

Consider also the weaker condition:

(E2') $\hat{\theta}_t \to \theta_0$, in probability under $\mathcal{P}^{\pi^a}$.

Let $\{\hat{\theta}_t\}_{t=0}^{\infty}$ be *any* sequence of parameter estimates satisfying (E1) and (E2'); we will show that the corresponding adaptive policy π^a is *self-optimizing*, i.e., $\bar{J}(\pi^a, \rho_0) = \Gamma^*_{\theta_0}$, for all $0 \le \rho_0 \le 1$. In the case where $\{\hat{\theta}_t\}_{t=0}^{\infty}$ satisfies (E2), we will show the stronger sample path result

$$\lim_{n \to \infty} \frac{1}{n} \sum_{t=0}^{n-1} \bar{c}(\rho_t, U_t) = \Gamma^*_{\theta_0}, \quad \mathcal{P}^{\pi^a}\text{-a.s..} \tag{5.1}$$

The method we use to verify these self-optimizing properties of π^a is motivated by techniques in [14] and [17]. However, the verification here does not fit in the same framework, due to (a) discontinuity of $\pi(\cdot; \cdot) \in \mathcal{OP}$ in both its arguments and (b) the fact that the cost $\bar{c}(\rho, u)$ is an explicit function of u. We have that $T(y, \rho, u; \theta)$ is continuous in θ. Using this and the fact that regeneration occurs infinitely often, the following is shown in [19].

Lemma 5.1. *If* $\hat{\theta}_t \to \theta_0$, *as* $t \to \infty$, *in probability under* $\mathcal{P}^{\pi^a}$ *($\mathcal{P}^{\pi^a}$-a.s.), then* $|\hat{\rho}_t - \rho_t| \to 0$, *as* $t \to \infty$, *in probability under* $\mathcal{P}^{\pi^a}$ *($\mathcal{P}^{\pi^a}$-a.s.).*

Then, we have the following.

Theorem 5.1. *Assume* $q \in (0.5, 1)$.

(i) If $\{\hat{\theta}_t\}_{t=0}^{\infty}$ *satisfies (E1) and (E2'), then* π^a *is self-optimizing.*

(ii) If in addition $\{\hat{\theta}_t\}_{t=0}^{\infty}$ *satisfies (E2), then* π^a *is self-optimizing in a sample-path sense, i.e., (5.1) holds.*

Proof: (i) Let $\Phi_\theta(\cdot, \cdot)$ denote Mandl's discrepancy function, corresponding to the parameter value $\theta \in \Theta_\delta$, i.e., for $\rho \in [0, 1]$ and $u \in U$

$$\Phi_\theta(\rho, u) := \bar{c}(\rho, u) + \sum_{y \in Y} V(y, \rho, u; \theta) h_\theta(T(y, \rho, u; \theta)) - \Gamma^*_\theta - h_\theta(\rho).$$

Then by (2.5)–(2.8), Corollary 3.1 and Theorem 4.1, $\Phi_\theta(\rho, u)$ is continuous in both $\rho \in [0, 1]$ and $\theta \in \Theta_\delta$. Furthermore, since Θ_δ is compact, then $\Phi_\theta(\rho, u)$ is uniformly continuous and bounded in $(\rho, \theta) \in [0, 1] \times \Theta_\delta$; thus, $\Phi_{\hat{\theta}_t}(\hat{\rho}_t, u)$ is uniformly integrable, for each $u \in \mathrm{U}$. Therefore, for each $u \in \mathrm{U}$, we have

$$\left| \Phi_{\hat{\theta}_t}(\hat{\rho}_t, u) - \Phi_{\theta_0}(\rho_t, u) \right| \xrightarrow[t \to \infty]{} 0, \qquad L_1(\mathcal{P}^{\pi^*}),$$

and since U is finite,

$$E^{\pi^*} \left\{ \Phi_{\theta_0}(\rho_t, \pi(\hat{\rho}_t; \hat{\theta}_t)) \right\} \xrightarrow[t \to \infty]{} 0, \qquad (5.2)$$

where we used the fact that $\Phi_{\hat{\theta}_t}(\hat{\rho}_t, \pi(\hat{\rho}_t; \hat{\theta}_t)) = 0$, since $\pi(\cdot\,; \theta) \in \mathcal{OP}$ minimizes the optimality equation (3.1), for the parameter value $\theta \in \Theta_\delta$. The result then follows from (5.2); see [2], [10], [14], [17].

(ii) If the convergence is in the stronger $\mathcal{P}^{\pi^*}$–a.s. sense, then similarly as above, we obtain that

$$\Phi_{\theta_0}(\rho_t, \pi(\hat{\rho}_t; \hat{\theta}_t)) \xrightarrow[t \to \infty]{} 0, \qquad \mathcal{P}^{\pi^*}\text{–a.s.},$$

from which the result follows. $\qquad \qquad \square$

References

[1] A. Arapostathis and S. I. Marcus, "Analysis of an Identification Algorithm Arising in the Adaptive Estimation of Markov Chains," *Mathematics of Control, Signals and Systems,* vol. 3, 1990, pp. 1–29.

[2] A. Arapostathis, V. S. Borkar, E. Fernández-Gaucherand, M. K. Ghosh, and S. I. Marcus, "Discrete-Time Controlled Markov Processes with Average Cost Criterion: A Survey," submitted for publication.

[3] A. Arapostathis, E. Fernández-Gaucherand, and S. I. Marcus, "Analysis of an Adaptive Control Scheme for a Partially Observed Controlled Markov Chain," *Proc. 29th IEEE Conf. Decision and Control,* Honolulu, HI, 1990, pp. 1438–1444.

[4] K. J. Åström, "Optimal Control of Markov Processes with Incomplete State Information," *J. Math. Anal. Appl.,* vol. 10, 1965, pp. 174–205.

[5] D. P. Bertsekas, *Dynamic Programming: Deterministic and Stochastic Models,* Prentice-Hall, Englewood Cliffs, NJ, 1987.

[6] E. Fernández-Gaucherand, "Controlled Markov Processes on the Infinite Planning Horizon: Optimal & Adaptive Control," Ph.D. Dissertation, The University of Texas at Austin, August 1991.

[7] E. Fernández-Gaucherand, A. Arapostathis and S.I. Marcus, "On the Adaptive Control of Partially Observable Markov Decision Processes," *Proc. 27th IEEE Conf. Decision and Control,* Austin, TX, 1988, pp. 1204–1210.

[8] E. Fernández-Gaucherand, A. Arapostathis and S. I. Marcus, "On the Adaptive Control of a Partially Observable Binary Markov Decision Process," in *Advances in Computing*

and Control, W. A. Porter, S. C. Kak, J. L. Aravena, eds., Lecture Notes in Control and Information Sciences, vol. 130, Springer-Verlag, Berlin, 1989, pp. 217–228.

[9] E. Fernández-Gaucherand, A. Arapostathis and S. I. Marcus, "On the Average Cost Optimality Equation and the Structure of Optimal Policies for Partially Observable Markov Decision Processes," *Annals of Operations Research*, vol. 29, 1991, pp. 439–470.

[10] O. Hernández-Lerma, *Adaptive Markov Control Processes*, Springer Verlag, New York, 1989.

[11] P. R. Kumar and P. Varaiya, *Stochastic Systems: Estimation, Identification and Adaptive Control*, Prentice-Hall, Englewood Cliffs, NJ, 1986.

[12] H. J. Kushner, "An Averaging Method for Stochastic Approximations with Discontinuous Dynamics, Constraints, and State Dependent Noise," in *Recent Advances in Statistics*, Rizvi, Rustagi and Siegmund, Eds., Academic Press, New York, 1983, pp. 211–235.

[13] H. J. Kushner and D. S. Clark, *Stochastic Approximation Methods for Constrained and Unconstrained Systems*, Springer-Verlag, New York, 1978.

[14] P. Mandl, "Estimation and Control in Markov Chains," *Adv. Appl. Prob.*, vol. 6, 1974, pp. 40–60.

[15] M. Ohnishi, H. Mine and H. Kawai, "An Optimal Inspection and Replacement Policy Under Incomplete State Information: Average Cost Criterion," in *Stochastic Models in Reliability Theory*, S. Osaki and Y. Hatoyama, eds., Lecture Notes in Econ. and Math. Systems No. 235, Springer-Verlag, Berlin, 1984, pp. 187–197.

[16] L. K. Platzman, "Optimal Infinite-Horizon Undiscounted Control of Finite Probabilistic Systems," *SIAM J. Control Optim.*, Vol. 18, 1980, pp. 362–380.

[17] A. Shwartz and A. M. Makowski, "Comparing Policies in Markov Decision Processes: Mandl's Lemma Revisited," *Math. Oper. Res.*, vol. 15, 1990, pp. 155–174.

[18] C. C. White, "A Markov Quality Control Process Subject to Partial Observation," *Mang. Sci.*, Vol. 23, 1977, pp. 843–852.

[19] E. Fernández-Gaucherand, A. Arapostathis and S. I. Marcus, "Analysis of an Adaptive Control Scheme for a Partially Observed Controlled Markov Chain," Department of Systems and Industrial Eng. Working Paper #91-038, University of Arizona, Tucson, Arizona.

STRUCTURED SOLUTIONS FOR
STOCHASTIC CONTROL PROBLEMS *

Emmanuel Fernández–Gaucherand †, Steven I. Marcus §,
and Aristotle Arapostathis ‡.

Key Words: Stochastic Control, Controlled Markov Processes, Structured Solutions.

ABSTRACT

We consider the discrete-time stochastic control problem for controlled Markov processes (CMP), with an average cost criterion. We show how structural properties in the model can be used to obtain a functional characterization of optimal values and policies, in the form of an average cost optimality *equality* (ACOE). In particular, Convex CMP are defined as those models for which the (discounted) value functions are convex. This convexity is used to obtain the ACOE as a limit of the corresponding discounted optimality equations, as the discounting *vanishes*, i.e. as the discount factor tends to one. We further comment on the potential algorithmic impact of this and other structured solutions.

---

* This work was supported in part by the Texas Advanced Technology Program under Grant No. 003658-093, in part by the Air Force Office of Scientific Research under Grant AFOSR-91-0033, and in part by the National Science Foundation under Grant CDR-8803012.

† Systems and Industrial Engineering Department, The University of Arizona, Tucson, Arizona 85721. Email: emmanuel@sie.arizona.edu.

§ Department of Electrical Engineering and Systems Research Center, The University of Maryland, College Park, Maryland 20742. Email: marcus@src.umd.edu.

‡ Department of Electrical and Computer Engineering, The University of Texas at Austin, Austin, Texas 78712-1084. Email: ari@emx.utexas.edu.

I. Introduction

A Controlled Markov Process, is a discrete time stochastic dynamical system specified by the five-tuple $\langle \mathbf{X}, \mathbf{U}, \mathcal{U}, P, c \rangle$ where $\mathbf{X}$ is the *state space;* $\mathbf{U}$ is the *action,* or *control* space; $\mathcal{U}(x) \subseteq \mathbf{U}$ is the set of *feasible* actions (or control inputs) when the system is in state $x \in \mathbf{X}$; each pair (x, u) in $\mathbf{X} \times \mathbf{U}$ determines a *transition law* $P(\cdot \mid x, u)$; and $c : \mathbf{X} \times \mathbf{U} \to \mathbb{R}$ is the one-stage cost function. See [ABFGM], [BS], [HLM1] for more details. A control strategy, or policy, is a rule π for making decisions, based on the available information. At a given time t, the available information is the set h_t of observed states and actions taken up to that time, i.e. $h_t = (X_0, U_0, X_1, \ldots, U_{t-1}, X_t)$. Each policy π incurs a stream of costs $\{c(X_0, U_0), c(X_1, U_1), \ldots\}$. Depending upon the problem requirements, different criteria can be used to evaluate the performance of the system, under the policy used. The following criteria are frequently used in many diverse application areas. In the equations below, E_x^π denotes the expectation operator under the policy π, when $X_0 = x$.

Discounted Cost (DC): For $0 < \beta < 1$, the *discount factor,* and a policy π, the total discounted cost incurred by π over the infinite planning horizon is given by

$$J_\beta(x, \pi) := E_x^\pi \left[\sum_{t=0}^\infty \beta^t c(X_t, U_t) \right] ;$$

the optimal *value function,* i.e. the minimum $J_\beta(x, \pi)$ over all π, is denoted by $J_\beta^*(x)$.

Average Cost (AC): The expected long-run average cost incurred by the policy π is given by

$$J(x, \pi) := \limsup_{N \to \infty} E_x^\pi \left[\frac{1}{N} \sum_{t=0}^{N-1} c(X_t, U_t) \right] ;$$

the optimal average cost is denote by $J^*(x)$.

The (DC) and (AC) can be seen as two opposite extremes in the type of criteria that can be considered for infinite horizon problems, in the sense that the first captures primarily the performance of the system at the present and near future, due to the discounting, and the second only captures the performance at the distant future; see the comprehensive survey of this problem in [ABFGM]. To obtain a reasonable compromise, one can combine these criteria in a weighted sum. This approach was suggested by Feinberg [FEI], and recently it has been extensively studied by Krass et al. [KFS] for the case of finite state and control sets, and by the authors [FGM] for the situation of general (Borel) spaces.

II. The Stochastic Control Problem

As it is well known [ABFGM], [BS], [HLM1], the solution of the infinite horizon stochastic control problem under the above criteria, i.e. the functional characterization and computation of optimal values and policies, is related to the following dynamic programming-like functional equations.

The Discounted Cost Optimality Equation (DCOE):

$$J_\beta^*(x) = \inf_{u \in \mathcal{U}(x)} \left\{ c(x,u) + \beta \int_X J_\beta^*(y)P(dy \mid x,u) \right\} = T_\beta(J_\beta^*)(x), \qquad x \in X,$$

where the operator $T_\beta(\cdot)$ is defined in the obvious way.

The Average Cost Optimality Equation (ACOE):

$$\rho(x) + h(x) = \inf_{u \in \mathcal{U}(x)} \left\{ c(x,u) + \int_X h(y)P(dy|x,u) \right\} = T(h)(x), \quad x \in X,$$

where the operator $T(\cdot)$ is defined in the obvious way.

Among the most important consequences deriving from these optimality equations are that (under some conditions) they provide an optimality criterion for actions: minimizing actions are optimal; they allow iterative schemes to compute optimal values (value iteration); and they allow algorithmic methods to compute and improve decision rules (policy iteration).

When a (DC) criterion is used, a rather complete theory is available [BE], [BS], [DY], [HLM1], [KV]. For the average cost, one looks for conditions under which appropriate solutions to the ACOE exist. A solution is a pair $(\rho(\cdot), h(\cdot))$ of real-valued functions on X. There is a vast literature concerning the problem of existence and functional characterization of average cost optimal policies, when the state space X is *countable*, and/or the one stage cost function $c(\cdot, \cdot)$ is *bounded* [ABFGM], [BE], [HLM1]. However, this is not the case for the situation when the state space is a *general* (Borel) space, e.g. $X = \mathbb{R}^n$, and the one-stage cost function is unbounded. Necessary and sufficient conditions for a *bounded* solution to the ACOE have been recently given by the authors [FAM1]. However this type of solutions are not natural for problems involving an infinite number of states and an unbounded cost function. Recently, much research activity has been devoted to finding conditions for the functional characterization and existence results for average optimal values and policies, for the case of unbounded cost functions. Sennott [SEN1]-[SEN3] treated the situation of countable state space and finite action set, and Hernández-Lerma and Lasserre [HLL], [HLM2] extended these results to a general space setting. However, in these references the authors only show existence of solutions to an *average cost optimality inequality* (ACOI):

$$\rho(x) + h(x) \geq \inf_{u \in \mathcal{U}(x)} \left\{ c(x, u) + \int_X h(y) P(dy|x, u) \right\}.$$

Actually, it has been recently shown in [CC] that *strict inequality* is possible under the conditions in [HLL] and [SEN2]. The fact that equality is not shown prevents one from, e.g. quantifying the deviation from optimality for a policy π, via Mandl's discrepancy function [ABFGM, Theorem 6.3], [SM]; this discrepancy function is very important in order to, e.g. analyze the performance of *adaptive* schemes [FAM2], [HLM1], [SM]. Also, policy improvement in a policy iteration algorithm [TIJMS, Sect. 3.2] *cannot* be implemented if only an inequality result is available. In essence, the problem derives from the fact that the left-hand-side of the ACOI is not equal to the term on the right for optimal (minimizing) actions, and strictly smaller than it otherwise. Thus, although the ACOI gives a criterion for the *existence* of stationary average cost optimal policies, *it is not useful from an algorithmic standpoint*. Hence, the following question is very relevant:

> **What useful properties, shared by large and important problem classes, can be used to further show that an ACOE holds, and how can these properties be exploited to aid in the development of tractable algorithmic solutions?**

We address the above question, by concentrating on *structured* solutions to stochastic control models. By an structured solution we mean a model for which value functions and/or optimal policies have some special dependence on the (initial) state. For linear stochastic control problems, structural results have played a very important role, both theoretically and algorithmically, e.g. the celebrated LQG case. For more general (nonlinear) models like the ones we propose to study, structured solutions have been established and exploited only for specific models, but not for broad models classes that may share an unifying structure [HIN1]-[HIN2], [TIJMS]. Some useful structural properties for value functions and policies are: (i) monotonicity of the value functions (with respect to an appropriate order relation) on the state, this can be used to compute numerical approximations that monotonically interpolate the functions using a finite grid, and monotonicity with respect to the actions can lead to the search of optimal actions among the largest elements of $\mathcal{U}(x)$ only; (b) convexity of value functions on the state and actions leads likewise to convex interpolations, and also to search for optimal actions among the extreme points of $\mathcal{U}(x)$ (assumed to be a convex set). Although these ideas have been the object of some research [HIN1]-[HIN2], there is nevertheless still much to be done in the development and actual implementation of analytical and algorithmic solutions that can accommodate large problems classes, as well as the formulation of verifiable and unifying conditions that induce the required convex or monotone properties.

III. Convex Controlled Markov Processes

The use of, e.g. convexity of value functions in novel ways can further help in the analytical study of the infinite horizon stochastic control problem, under an average cost criterion. In [FG], it was shown how convexity properties of the discounted value function can be used to give a partial answer to the question previously posed. The main idea is to be able to extract an appropriately convergent subsequence, as $\beta \uparrow 1$, of the *differential* discounted value functions

$$h_\beta(x) := J_\beta^*(x) - J_\beta^*(\overline{x}), \qquad \forall x \in \mathbf{X},$$

where $\overline{x} \in \mathbf{X}$ (the reference state) is kept fixed, so that the ACOE can be obtained by taking limits in the DCOE (see [ABFGM, Sect.6]). If $\{h_\beta(\cdot)\}$ is bounded, uniformly in β, and *equicontinuous*, then the Arzela-Ascoli Theorem can be used to properly take the required limits. This idea was studied by Ross [RO]; however the required uniform boundedness and equicontinuity conditions are very difficult to show, in general, and most likely do not hold for problems with an uncountable state space and unbounded costs. It has been shown in [FG] that, under a convexity condition of the discounted value functions, *local* uniform boundedness and *local* equicontinuity properties can be shown for $\{h_\beta(\cdot)\}$, and thus the Arzela-Ascoli theorem can be successfully used to obtain the ACOE by taking limits in the DCOE. The framework is the following. Let $\mathbf{X}$ be an open convex subset of a separable Banach space, e.g. $\mathbb{R}^n$. In particular, $\mathbf{X}$ is a Borel space.

Definition: If the value functions $J_\beta^*(\cdot)$ are convex functions, then we say that $\langle \mathbf{X}, \mathbf{U}, \mathcal{U}, P, c \rangle$ is a *convex* CMP.

The next assumption is quite standard, c.f. [HLL], [SEN2].

Assumption A: For each $x \in \mathbf{X}$, $P(B \mid x, u)$ is continuous in $u \in \mathcal{U}(x)$, for each (Borel) sets $B \subseteq \mathbf{X}$; there exists a non-negative, upper semicontinuous function $b : \mathbf{X} \to \mathbb{R}$, a constant $M \geq 0$, and a sequence $\{\beta_n\} \subseteq (0,1)$ with $\beta_n \uparrow 1$, such that for all $x \in \mathbf{X}$

(i) $J^*_{\beta_n}(x) < \infty$;

(ii) $-M \leq h_{\beta_n}(x) \leq b(x)$;

(iii) $\int_{\mathbf{X}} b(y) P(dy \mid x, u) < \infty$; $\quad \forall u \in \mathcal{U}(x)$.

The next result is a simple consequence of Lemma 4.3 in [HLL]; see also [FG, Lemma 4.1].

Lemma: There exists a constant ρ^* and a subsequence $\beta_{n'} \uparrow 1$ of $\{\beta_n\}$, such that

$$\lim_{n \to \infty} (1 - \beta_{n'}) J^*_{\beta_{n'}}(x) = \rho^*, \qquad \forall x \in \mathbf{X}.$$

Finally, we assume the required convexity properties.

Assumption B: $J^*_{\beta_{n'}}(\cdot)$ is a convex function.

Then, the results in [HLL], [HLM2], [SEN2] can be strengthen as follows.

Theorem: Under Assumptions A-B, we have that

(i) there is a constant ρ^* and a convex function $h : \mathbf{X} \to \mathbb{R}$ with

$$-M \leq h(x) \leq b(x), \quad \forall x \in \mathbf{X},$$

such that the pair (ρ^*, h) is a solution to the ACOE, i.e.,

$$\rho^* + h(x) = \inf_{u \in \mathcal{U}(x)} \left\{ c(x, u) + \int_{\mathbf{X}} h(y) P(dy \mid x, u) \right\}; \tag{1}$$

(ii) there exists a stationary deterministic policy π^* which is average optimal, and every stationary deterministic policy π attaining the infimum in (1) is average optimal;

(iii) $J^*(x) = \rho^*$, for all $x \in \mathbf{X}$.

Proof: We make the usual *semicontinuity* assumptions on the transition kernel, and nonnegativity of the cost function [ABFGM], [BS], [DY]. In addition, we assume that the cost function has the *compact level sets* (CLS) property, i.e. for each $\lambda \in \mathbb{R}$, the set $\{(x, u)|c(x, u) \leq \lambda\}$ is compact. Then, under this conditions it can be shown that: (a) the DCOE holds, (b) a deterministic stationary policy is discount optimal if and only if it attains the infimum in the DCOE, and (c) one such policy exists [ABFGM], [BS], [FG], [HLM1].

(i) By Assumption B, $h_{\beta_n}(\cdot)$ is convex, and thus it exhibits a *local Lipschitz property*, as shown below. Let $x_0 \in \mathbf{X}$ be given, and let $\varepsilon > 0$ be small enough so that

$$\overline{B}(x_0, 2\varepsilon) := \left\{ y' \in \mathbf{X} \mid \|x_0 - y'\| \leq 2\varepsilon \right\},$$

is contained in $\mathbf{X}$.

Let $x, y \in B(x_0, \varepsilon) := \{y' \in \mathbf{X} \mid \|x_0 - y'\| < \varepsilon\}$, with $x \neq y$, and define

$$z := x + \frac{\varepsilon}{\|x - y\|}(x - y).$$

Thus $\|x_0 - z\| \leq \|x_0 - x\| + \varepsilon < 2\varepsilon$, and then

$$\|x - y\| z = \|x - y\| x + \varepsilon(x - y)$$

$$\implies x = \frac{\|x - y\|}{\|x - y\| + \varepsilon} z + \frac{\varepsilon}{\|x - y\| + \varepsilon} y,$$

hence x is a convex combination of $z, y \in B(x_0, 2\varepsilon)$. Since $b(\cdot)$ is upper semicontinuous, then it attains its maximum in $\overline{B}(x_0, 2\varepsilon)$. Thus there exists a constant $M(x_0, 2\varepsilon)$ such that

$$\left| h_{\beta_n}(y') \right| \leq M(x_0, 2\varepsilon), \qquad \forall y' \in \overline{B}(x_0, 2\varepsilon),$$

that is, $\{h_{\beta_n}(\cdot)\}$ is *locally uniformly bounded*. Now, by the convexity of $h_{\beta_n}(\cdot)$

$$h_{\beta_n}(x) \leq \frac{\|x - y\|}{\|x - y\| + \varepsilon} h_{\beta_n}(z) + \frac{\varepsilon}{\|x - y\| + \varepsilon} h_{\beta_n}(y),$$

and therefore

$$h_{\beta_{n'}}(x) - h_{\beta_{n'}}(y) \leq \frac{\|x - y\|}{\|x - y\| + \varepsilon}\left[h_{\beta_{n'}}(z) - h_{\beta_{n'}}(y)\right]$$

$$\leq \frac{\|x - y\|}{\|x - y\| + \varepsilon}[2M(x_0, 2\varepsilon)]$$

$$< \left[\frac{2M(x_0, 2\varepsilon)}{\varepsilon}\right]\|x - y\|.$$

Interchanging x and y, and since x_0 was arbitrary, we then conclude that $\{h_{\beta_{n'}}(\cdot)\}$ is *locally equi-Lipschitzian,* and in particular (locally) equicontinuous. Therefore, by the Arzela-Ascoli theorem [ROY], there exists a subsequence, which for simplicity we also denote as $\beta_{n'} \uparrow 1$, and a continuous function $h : X \to \mathbb{R}$, such that

$$h_{\beta_{n'}}(x) \xrightarrow[\beta_{n'}\uparrow 1]{} h(x), \qquad \forall x \in X,$$

the convergence being uniform on compact subsets of X. Furthermore, by the convexity of $h_{\beta_{n'}}(\cdot)$, $h(\cdot)$ is convex as well, and clearly $-M \leq h(x) \leq b(x)$, for all $x \in X$.

From the DCOE, we have that for all (x, u)

$$(1 - \beta_{n'})J^*_{\beta_{n'}}(\overline{x}) + h_{\beta_{n'}}(x) \leq c(x, u) + \beta_{n'} \int_X h_{\beta_{n'}}(y)P(dy \mid x, u). \tag{2}$$

Now we have that

$$0 \leq h_{\beta_{n'}}(\cdot) + M \leq b(\cdot) + M,$$

and then, taking limits in (2) as $\beta_{n'} \uparrow 1$, by the previous Lemma and by the dominated convergence theorem, we obtain

$$\rho^* + h(x) \leq c(x, u) + \int_X h(y)P(dy \mid x, u),$$

and since this holds for all (x, u), then

$$\rho^* + h(x) \leq \inf_{u \in \mathcal{U}(x)}\left\{c(x, u) + \int_X h(y)P(dy \mid x, u)\right\}. \tag{3}$$

Furthermore, let $\pi_{n'}^*$ be a $\beta_{n'}$-discount stationary deterministic optimal policy; thus for $\varepsilon > 0$ given, and for each $x \in \mathbf{X}$, there exists an integer $N(\varepsilon, x)$ such that for all $n' \geq N(\varepsilon, x)$

$$(1 - \beta_{n'})J_{\beta_{n'}}^*(\overline{x}) + h_{\beta_{n'}}(x) = c(x, \pi_{n'}^*) + \beta_{n'} \int_{\mathbf{X}} h_{\beta_{n'}}(y)P(dy \mid x, \pi_{n'}^*)$$

$$\leq \rho^* + h(x) + \varepsilon, \tag{4}$$

where, e.g., $c(x, \pi_{n'}^*) := c(x, \pi_{n'}^*(x))$. Recall that $\mathbf{X}$ is an open set, and that $h_{\beta_{n'}}(\cdot)$ is convex and bounded below; hence $h_{\beta_{n'}} \in \mathcal{L}(\mathbf{X})$, where $\mathcal{L}(\mathbf{X})$ denotes the set of lower semicontinuous and bounded below functions . Let

$$\mathbf{U}_{n'}(x, \varepsilon) := \left\{ u \in \mathcal{U}(x) \mid c(x, u) + \beta_{n'} \int_{\mathbf{X}} h_{\beta_{n'}}(y)P(dy \mid x, u) \leq \rho^* + h(x) + \varepsilon \right\}.$$

Then, by semicontinuity and condition (CLS), $\mathbf{U}_{n'}(x, \varepsilon)$ is closed, and $\pi_{n'}^*(x) \in \mathbf{U}_{n'}(x, \varepsilon)$, for all $n' \geq N(\varepsilon, x)$. Furthermore, we have that

$$\mathbf{U}_{n'}(x, \varepsilon) \subseteq \left\{ u \in \mathcal{U}(x) \mid c(x, u) \leq \rho^* + h(x) + \varepsilon + M \right\} =: \mathbf{U}_{\infty}(x) \subseteq \mathcal{U}(x)$$

and thus, by condition (CLS), $\{\pi_{n'}^*(x)\}$ is contained in a compact set, for each $x \in \mathbf{X}$. Therefore, there exists a subsequence $\{\pi_{n_k}^*(x)\}$ which converges to an action $u_x^* \in \mathbf{U}_{\infty}(x)$; denote by $\{\beta_{n_k}(x)\}$ the corresponding subsequence of $\{\beta_{n'}\}$. Hence, taking the limit inferior as $\beta_{n_k}(x) \uparrow 1$ in (4), we obtain

$$\rho^* + h(x) \geq c(x, u_x^*) + \int_{\mathbf{X}} h(y)P(dy \mid x, u_x^*), \tag{5}$$

by lower semicontinuity of $c(\cdot, \cdot)$ and $h(\cdot)$, the strong continuity of $P(B \mid x, \cdot)$, and the Generalized Fatou's Lemma [ROY]. Therefore, from (3) and (5), we obtain the ACOE, proving (i).

(ii)-(iii) By standard Tauberian arguments [FG], [HLL], we have that

$$\limsup_{\beta \uparrow 1}(1 - \beta)J_\beta^*(x) \leq J^*(x), \qquad \forall x \in \mathbf{X},$$

and hence $\rho^* \leq J^*(x)$, by the previous Lemma. Furthermore, since $h(\cdot)$ is convex and bounded below, and since $\mathbf{X}$ is an open set, then $h \in \mathcal{L}(\mathbf{X})$, and thus there exists a stationary deterministic policy π^* which attains the infimum in the ACOE, (see [ABFGM], [FG, Theorem A.2]). Therefore, we have that for such a policy, for all $x \in X$,

$$\rho^* + h(x) = E_x^{\pi^*} \left[c(X_0, U_0) + h(X_1) \right],$$

and by iterating the above

$$N\rho^* = \sum_{t=0}^{N-1} E_x^{\pi^*} \left[c(X_t, U_t) \right] + E_x^{\pi^*} \left[h(X_N) \right] - h(x)$$

$$\geq \sum_{t=0}^{N-1} E_x^{\pi^*} \left[c(X_t, U_t) \right] - \left[M + h(x) \right].$$

Dividing by N and taking the limit superior, we obtain

$$J(x, \pi^*) \leq \rho^* \leq J^*(x) \leq J(x, \pi^*),$$

and (ii) and (iii) follow.□

Remark: Related work to the developments above is that in linear quadratic control problems, where convexity of value functions is obtained [BE], see also [FG] and [HLM2]. Also, Dynkin [DYN] introduced the concept of stochastic concave dynamic programming, but his interest was on concavity properties of one-stage cost functions in the control actions, and its implications on the existence of minimizing actions.

The assumption that $J_{\beta_{n'}}^*(\cdot)$ is convex is the cornerstone of our developments. Therefore, in order to envision the type of problems for which our results may be applicable, it is necessary to identify conditions under which such convexity is obtained; see [HIN1] for some conditions of this nature. One natural approach to the above problem is by using induction via the value iteration scheme [ABFGM], [BS], [HLM1], to propagate the desired property [BTE], [HIN1], [STI]:

- First, for the null function $f(x) = 0$, for all $x \in X$, conditions on $c(\cdot, \cdot)$ are given such that

$$J_\beta^{(1)}(x) = \inf_{u \in \mathcal{U}(x)} \left\{ c(x, u) \right\}, \qquad \forall x \in X,$$

is convex (concave).

- Next, for the inductive step, assuming that $J_\beta^{(k)}(\cdot)$ is convex (concave), then give conditions for the dynamic programming map $T_\beta(\cdot)$ to preserve this property.

- Finally, convexity (concavity) of $J_\beta^*(\cdot)$ follows by the convergence of value iteration to the optimal value function.

REFERENCES

[ABFGM] A. Arapostathis, V. Borkar, E. Fernández-Gaucherand, M.K. Ghosh and S.I. Marcus, Controlled Markov Processes with an Average Cost Criterion: A Survey, SIE Working Paper #91-032 (submitted to *SIAM Journal on Control & Optimization*).

[BE] D.P. Bertsekas, *Dynamic Programming: Deterministic and Stochastic Models*, Prentice-Hall, Englewood Cliffs, 1987.

[BS] D.P. Bertsekas and S.E. Shreve, *Stochastic Optimal Control: The Discrete Time Case*, Academic Press, New York, 1978.

[BTE] F.J. Beutler and D. Teneketzis, Routing in Queueing Networks Under Imperfect Information: Stochastic Dominance and Thresholds, *Stochastics & Stochastics Reports*, 26 (1989) 81–100.

[CC] R. Cavazos-Cadena, A Counterexample on the Optimality Equation in Markov Decision Chains with the Average Cost Criterion, *Syst. Control Lett.* 16 (1991) 387-392.

[DY] E.B. Dynkin and A.A. Yushkevich, *Controlled Markov Processes*, Springer Verlag, New York, 1979.

[DYN] E.B. Dynkin, Stochastic Concave Dynamic Programming, *Math. USSR Sbornik* 16 (1972) 501–515.

[FAM1] E. Fernández-Gaucherand, A. Arapostathis and S.I. Marcus, Remarks on the Existence of Solutions to the Average Cost Optimality Equation in Markov Decision Processes, *Syst. Control Lett.* 15 (1990) 425–432.

[FAM2] E. Fernández-Gaucherand, A. Arapostathis and S.I. Marcus, Analysis of an Adaptive Control Scheme for a Partially Observed Controlled Markov Chain,

SIE Working paper #91-038 (under revision for IEEE Transactions in Automatic Control).

[FEI] E.A. Feinberg, Controlled Markov Processes with Arbitrary Numerical Criteria, *Theory Probab. Appl.* **27** (1982) 486-503.

[FG] E. Fernández-Gaucherand, *Controlled Markov Processes on the Infinite Planning Horizon: Optimal & Adaptive Control*, Ph.D. Thesis, The University of Texas at Austin, August 1991.

[FGM] E. Fernández-Gaucherand, M. K. Ghosh, and S.I. Marcus, Controlled Markov Processes in the Infinite Planning Horizon with a Weighted Cost Criterion, to appear in the *Proceedings of the IV Latin American Congress in Probability and Mathematical Statistics*, México City, México, 1990.

[HIN1] K.F. Hinderer, On the Structure of Solutions of Stochastic Dynamic Programs, in: *Proc. 7th Conf. on Probability Theory*, Brasov, Romania, (1984) 173-182.

[HIN2] K.F. Hinderer, Increasing Lipschitz Continuous Maximizers of Some Dynamic Programs, *Annals of Operations Research*, **29** (1991) 565-586.

[HLL] O. Hernández-Lerma and J.B. Lasserre, Average Cost Optimal Policies for Markov Control Processes with Borel State Space and Unbounded Costs, *Syst. Control Lett.* **15** (1990) 349–356.

[HLM1] O. Hernández-Lerma, *Adaptive Markov Control Processes*, Springer Verlag, New York, 1989.

[HLM2] O. Hernández-Lerma, Average Optimality in Dynamic Programming on Borel Spaces: Unbounded Costs and Controls, *Syst. Control Lett.* **17** (1991) 237-242.

[KFS] D. Krass, J.A. Filar and S. Sinha, A Weighted Markov Decision Process, to appear in *Operations Research*.

[KV] P.R. Kumar and P. Varaiya, *Stochastic Systems: Estimation, Identification and Adaptive Control*, Prentice-Hall, Englewood Cliffs, 1986.

[RO] S.M. Ross, Arbitrary State Markovian Decision Processes, *Ann. Math. Stat.* **39** (1968) 2118-2122.

[ROY] H.L. Royden, *Real Analysis*, 2nd. ed., Macmillan, New York, 1968.

[SEN1] L.I. Sennott, A New Condition for the Existence of Optimal Stationary Policies in Average Cost Markov Decision Processes, *Oper. Res. Lett.* **5** (1986) 17–23.

[SEN2] L.I. Sennott, Average Cost Optimal Stationary Policies in Infinite State Markov Decision Processes with Unbounded Costs, *Oper. Res.* **37** (1989) 626–633.

[SEN3] L.I. Sennott, Average Cost Semi-Markov Decision Processes and the Control of Queueing Systems, *Probab. in Eng. & Info. Sci.* **3** (1989) 247–272.

[SM] A. Shwartz and A.M. Makowski, Comparing Policies in Markov Decision Processes: Mandl's Lemma Revisited, *Math. Oper. Res.* **15** (1990) 155–174.

[STI] S. Stidham, Scheduling, Routing, and Flow Control in Stochastic Networks, in *Stochastic Differential Systems, Stochastic Control Theory and Applications*, W. Fleming and P.L. Lions, Eds., The IMA Volumes in Mathematics and Its Applications **10** 529–561, Springer Verlag, Berlin, 1988.

[TIJMS] H.C. Tijms, *Stochastic Modelling and Analysis: A Computational Approach*, John Wiley, Chichester, 1986.

Risk Sensitive Optimal Control and Differential Games

Wendell H. Fleming [1] and William M. McEneaney

Abstract

Risk-sensitive stochastic control problems for nonlinear systems described by stochastic differential equations are considered. A logarithmic transformation is applied to the optimal cost function. The value function for a zero-sum, two-controller differential game is obtained in the limit, as a small parameter which represents noise intensity tends to zero. Convergence to the value function is proved by viscosity solution methods for nonlinear partial differential equations.

1 Introduction.

There are several different approaches to dealing with disturbances in control systems. In stochastic control theory, disturbances are modelled as stochastic processes. A typical control objective is to minimize some expected cost criterion. In another approach, disturbances are modelled deterministically as functions of time. Optimization according to a minimax design criterion is often used to obtain good system performance in the presence of unfavorable disturbances.

The H - infinity, robust control method for the disturbance attenuation problem for linear control systems is a notable instance of the latter approach. When a continuous time, state space formulation of the disturbance attenuation problem is used, the analysis leads to the study of certain linear – quadratic differential games. See Basar - Bernhard [3].

In stochastic control, the well known linear – quadratic – gaussian (LQG) model leads to a linear feedback control policy which is insensitive to the intensity of additive gaussian white noise entering the system. In 1973, Jacobson [14] considered instead a linear – exponential – quadratic – gaussian

[1] Partially supported by NSF under grant DMS-900038, by ARO under grant DAAL03-86-K-0171 and by AFOSR under grant AFOSR-89-0015

(LEQG) problem, for which the linear feedback optimal control policy is sensitive to noise intensity. Jacobson also considered a linear – quadratic differential game closely related to the LEQG problem. Glover and Doyle [11] [12] found a connection between the LEQG problem and H - infinity control via a minimum entropy principle.

An interesting question is to find for nonlinear systems, or nonquadratic cost criteria, similar connections between stochastic control and deterministic (minimax) approaches to dealing with disturbances. Whittle [15] [16] introduced an interesting approach to this question, through the theory of "risk sensitive control" for Markov processes. In this approach, a small parameter (denoted here by ε) is introduced into the problem. Large deviations ideas are used to obtain in the limit as $\varepsilon \to 0$ a minimax control problem. In continuous time, this often becomes a differential game. Whittle's mathematical arguments are to a considerable extent formal. In the present paper, we show how the passage to the limit as $\varepsilon \to 0$ can be done in a mathematically precise way, for a particular class of models in which the disturbance appears simply as an additive forcing term in a differential equation which describes the state dynamics.

Large deviations ideas appear in a similar way in Hijab's thesis [13], which connects nonlinear filtering and Mortensen's least squares, deterministic approach to nonlinear estimation. There is also a similarity with work of Baras, Bensoussan and James [1] on nonlinear observers as small - noise limits of nonlinear filters.

We consider risk sensitive stochastic control problems only on a finite time interval. For problems on an infinite time horizon, a different kind of large deviation formulation may be relevant (see end of Section 6.)

After this paper was written, the authors learned of rather closely related work by James [17].

2 Problem formulation and main results.

We consider a $\mathbb{R}^n$ - valued, nearly deterministic Markov diffusion process x_t^ε on a finite time interval $s \leq t \leq T$, governed by the stochastic differential equation

$$(2.1) \qquad dx_t^\varepsilon = f(t, x_t^\varepsilon, u_t)dt + \sqrt{\frac{\varepsilon}{2\gamma^2}}db_t,$$

with initial data $x_s^\varepsilon = x(x \in \mathbb{R}^n)$. In (2.1) $\varepsilon > 0$ is a "small" parameter, $\gamma > 0$ is a constant and b_t is a standard n-dimensional brownian motion. The control chosen at time t is u_t, with $u_t \in U$ (U the control space.) Under the boundedness and compactness assumptions (2.3) - (2.5) below, γ is arbitrary. However, a restriction on γ is needed if these assumptions are relaxed, as in case of the LEQG problem (Section 6.)

The controller wishes to choose a control process u. which minimizes the expectation of an exponential cost criterion:

$$(2.2) \qquad J^\varepsilon = E \exp \frac{1}{\varepsilon} \int_s^T L(t, x_t^\varepsilon, u_t)dt.$$

Let $Q = [0, T] \times \mathbb{R}^n$. We assume that

$$(2.3) \qquad U \text{ is compact}, \ U \subset \mathbb{R}^m.$$

$$(2.4) \qquad f \text{ is of class } C^1(Q \times U) \text{ and } f, f_x \text{ are bounded.}$$

$$(2.5) \qquad L \text{ is of class } C^1(Q \times U), \ L \geq 0, \text{ and } L, L_x \text{ are bounded.}$$

Here, f_x, L_x denote the gradients in x of $f(s, x, v), L(s, x, v)$.

We consider only the case of "completely observed states" x_t^ε and use dynamic programming. (Problems with partially observed states, also considered by Whittle [16], involve additional mathematical difficulties.)

For the case of completely observed states, $u.$ is allowed to be any U - valued, progressively measurable control process. We write $J^{\mathcal{E}} = J^{\mathcal{E}}(s, x;\ u.)$ in (2.2) to indicate dependence on the initial data (s, x). Let $\phi^{\mathcal{E}}$ denote the optimal cost function:

$$(2.6) \qquad \phi^{\mathcal{E}}(s, x) = \inf_{u.} J^{\mathcal{E}}(s, x;\ u.).$$

(For a somewhat more precise formulation in terms of reference probability systems, see Fleming - Soner [7, Section 4.2].) From results about uniformly parabolic partial differential equations and a verification theorem in stochastic control theory, $\phi^{\mathcal{E}}$ is of class $C^{1,2}(Q)$ and is the unique bounded, positive solution of the dynamic programming partial differential equation (PDE)

$$(2.7) \qquad 0 = \phi^{\mathcal{E}}_s + \frac{\varepsilon}{4\gamma^2}\Delta_x\phi^{\mathcal{E}} + \min_{v\in U}[f(s, x, v) \cdot \nabla_x\phi^{\mathcal{E}}$$

$$+ \frac{1}{\varepsilon}L(s, x, v)\phi^{\mathcal{E}}]$$

with the terminal (Cauchy) data

$$(2.8) \qquad \phi^{\mathcal{E}}(T, x) = 1.$$

See Fleming - Soner [7, Secs. 4.3, 4.4], and for closely related results Fleming - Rishel [6, Secs. 6.4, 6.6].

Whittle's idea is to show that $\phi^{\mathcal{E}} \sim \exp(\varepsilon^{-1}V)$, where $V(s, x)$ turns out to be the value function for a differential game. More precisely, we wish to show that $V^{\mathcal{E}} \to V$, where

$$(2.9) \qquad V^{\mathcal{E}} = \varepsilon \log \phi^{\mathcal{E}}.$$

From (2.7), $V^{\mathcal{E}}$ satisfies the PDE

$$(2.10^{\mathcal{E}}) \qquad 0 = V^{\mathcal{E}}_s + \frac{\varepsilon}{4\gamma^2}\Delta_x V^{\mathcal{E}} + \frac{1}{4\gamma^2}|\nabla_x V^{\mathcal{E}}|^2$$

$$+ \min_{v\in U}[f(s, x, v) \cdot \nabla_x V^{\mathcal{E}} + L(s, x, v)]$$

with the terminal data

$$(2.11) \qquad\qquad V^\varepsilon(T, x) = 0.$$

For $\varepsilon = 0$, the corresponding PDE (2.10^0) is of first order, and one cannot expect smooth solutions to it. However, under our assumptions, (2.10^0) - (2.11) turns out to have a unique continuous solution in the viscosity sense [7, Chap. 2]. Moreover, V^ε tends to V^0 as $\varepsilon \to 0$ uniformly on compact subsets of Q. See Theorem 4.1.

In Section 5, we interpret $V^0(s, x)$ as the value of a differential game. For this interpretation, we write

$$(2.12) \qquad \frac{1}{4\gamma^2}|\nabla_x V^0|^2 = \max_{w \in I\!\!R^n}[w \cdot \nabla_x V^0 - \gamma^2|w|^2].$$

Then (2.10^0) becomes the Isaacs partial differential equation for a zero-sum differential game played on the time interval $s \le t \le T$. The game dynamics and payoff are described in Section 5. See (5.1), (5.2). Both players have "complete state information". A precise formulation of the game can be given by introducing strategies, for example in the Elliott - Kalton sense. The value function V^0 turns out to be the unique bounded continuous viscosity solution to (2.10^0) - (2.11). See Evans - Souganidis [5]. Since max over w and min over v appear separately in (2.10), the Isaacs minimax condition holds. Hence, the upper and lower values of the differential game are both equal to V^0.

Remark 1. Whittle also considered the case of "risk seeking" control, in which ε is replaced by $-\varepsilon$ in (2.2). In that case, instead of (2.9) one can consider $\tilde{V}^\varepsilon = -\varepsilon \log \phi^\varepsilon$. In (2.10^ε), the sign of the term $(4\gamma^2)^{-1}|\nabla_x V^\varepsilon|^2$ becomes negative. The limiting game is then "cooperative" since both controls v_t, w_t are chosen to minimize the payoff.

Remark 2. Risk sensitive control problems and differential games have been treated by Barron and Jensen [2], in the context of financial economics.

3   A priori estimates.

Let us next show that the functions V^ε in (2.9) are uniformly bounded and equicontinuous. Since $L \geq 0$ and L is bounded, by (2.2), (2.6)

$$1 \leq \phi^\varepsilon \leq \exp[\varepsilon^{-1}\|L\|(T-s)],$$

where $\|\ \|$ is the sup norm. Hence, by (2.9)

(3.1) $$0 \leq V^\varepsilon \leq \|L\|(T-s).$$

Lemma 3.1. *There exists a constant C such that*

(3.2) $$|V^\varepsilon(s,x) - V^\varepsilon(\tilde{s},\tilde{x})| \leq C[|x - \tilde{x}| + |s - \tilde{s}|^{\frac{1}{2}}]$$

for all $s, \tilde{s} \in [0,T], x, \tilde{x} \in \mathbb{R}^n$.

Proof. Let us first consider $\tilde{s} = s$. Let x_t^ε be as in (2.1), and $\tilde{x}_t^\varepsilon$ be the corresponding solution to (2.1) with $\tilde{x}_s^\varepsilon = \tilde{x}$. A standard argument, using Gronwall's inequality gives

$$|\tilde{x}_t^\varepsilon - x_t^\varepsilon| \leq C_1|\tilde{x} - x|$$

for some constant C_1 depending on $T - s$ and $\|f_x\|$. Hence, for every progressively measurable control process $u.$,

$$\int_s^T |L(t, \tilde{x}_t^\varepsilon, u_t) - L(t, x_t^\varepsilon, u_t)|dt \leq C_1\|L_x\|\ |\tilde{x} - x|(T-s),$$

$$J^\varepsilon(s,x;u.) \leq J^\varepsilon(s,\tilde{x};u.)\exp[\frac{C_1}{\varepsilon}\|L_x\|\ |\tilde{x} - x|(T-s)],$$

$$V^\varepsilon(s,x) \leq V^\varepsilon(s,\tilde{x}) + C_1\|L_x\|\ |\tilde{x} - x|(T-s).$$

The same inequality holds if x and $\tilde{x}$ are exchanged. Thus,

(3.3) $$|V^\varepsilon(s,x) - V^\varepsilon(s,\tilde{x})| \leq C_2|\tilde{x} - x|$$

if $C_2 \geq C_1\|L_x\|T$. Inequality (3.3) is equivalent to

$$(3.4) \qquad |\nabla_x V^\mathcal{E}(s,x)| \leq C_2.$$

We next take $x = \tilde{x}$, and without loss of generality consider $s < \tilde{s} \leq T$. We rewrite $(2.10^\mathcal{E})$ in the form

$$(3.5^\mathcal{E}) \qquad 0 = V_s^\mathcal{E} + \frac{\mathcal{E}}{4\gamma^2}\Delta_x V^\mathcal{E} + \min_{v \in U}[g^\mathcal{E}(s,x,v) \cdot \nabla_x V^\mathcal{E} + L(s,x,v)].$$

$$g^\mathcal{E}(s,x,v) = \frac{\mathcal{E}}{4\gamma^2}\nabla_x V^\mathcal{E}(s,x) + f(s,x,v).$$

By (2.4) and (3.4), $g^\mathcal{E}$ is bounded independent of $\mathcal{E}$. Then $(3.5^\mathcal{E})$ is the dynamic programming equation for the following stochastic control problem, with state dynamics $\eta_t^\mathcal{E}$ satisfying

$$d\eta_t^\mathcal{E} = g^\mathcal{E}(t,\eta_t^\mathcal{E},u_t)dt + \sqrt{\frac{\mathcal{E}}{2\gamma^2}}db_t,$$

with $\eta_s^\mathcal{E} = x$. Since $g^\mathcal{E}$ is bounded and $g^\mathcal{E}(t,\cdot,v)$ satisfies a local Lipschitz condition, the solution $\eta_t^\mathcal{E}$ exists (pathwise). The expected cost criterion to be minimized is

$$E\int_s^T L(t,\eta_t^\mathcal{E},u_t)dt.$$

By the dynamic programming principle

$$V^\mathcal{E}(s,x) = \inf_{u.} E[\int_s^{\tilde{s}} L(t,\eta_t^\mathcal{E},u_t)dt + V^\mathcal{E}(\tilde{s},\eta_{\tilde{s}}^\mathcal{E})].$$

Therefore,

$$|V^\mathcal{E}(s,x) - V^\mathcal{E}(\tilde{s},x)| \leq \|L\|(\tilde{s}-s)$$
$$+ \sup_{u.} E|V^\mathcal{E}(\tilde{s},\eta_{\tilde{s}}^\mathcal{E}) - V^\mathcal{E}(\tilde{s},x)|.$$

By (3.3)

$$|V^\mathcal{E}(\tilde{s},\eta_{\tilde{s}}^\mathcal{E}) - V^\mathcal{E}(\tilde{s},x)| \leq C_2 E|\eta_{\tilde{s}}^\mathcal{E} - x|.$$

Since $\|g^\mathcal{E}\|$ is bounded independent of $\mathcal{E}$, a standard estimate gives

$$E|\eta_{\tilde{s}}^\mathcal{E} - x| \leq C_3(\tilde{s}-s)^{\frac{1}{2}}.$$

From these estimates, for suitable C

$$|V^\varepsilon(s,x) - V^\varepsilon(\tilde{s},x)| \leq C(\tilde{s}-s)^{\frac{1}{2}}.$$

This implies (3.10) provided $C_2 \leq C$.

$\square$

4 Viscosity solution limit.

When $\varepsilon = 0$, consider the first - order partial differential equation

(4.1) $$0 = V_s^0 + \frac{1}{4\gamma^2}|\nabla_x V^0|^2 + \min_{v \in U}[f(s,x,v) \cdot \nabla_x V^0 + L(s,x,v)]$$

in $Q = [0,T] \times I\!R^n$, with the terminal data

(4.2) $$V^0(T,x) = 0.$$

(In Section 2, we referred to equation (4.1) as (2.10^0).) There need not be a smooth (class C^1) solution. However, (4.1) - (4.2) has a unique bounded uniformly continuous solution V^0 in the viscosity sense. See [7, Theorem 2.9.1]. Moreover, it follows easily from (3.1) and Lemma 3.1 that V^ε tends to V^0 as $\varepsilon \to 0$. See [7, Lemma 2.6.2]. More precisely:

Theorem 4.1 *Let V^ε be as in (2.9) for $\varepsilon > 0$. Then $V^\varepsilon \to V^0$ uniformly on compact subsets of Q as $\varepsilon \to 0$, where V^0 is the unique bounded uniformly continuous viscosity solution to (4.1) - (4.2).*

Remark. To obtain Theorem 4.1, Lemma 3.1 is not actually needed. Another proof using a comparison principle for semicontinuous viscosity sub and super solutions to (4.1) can be given. This proof uses the uniform bound (3.1), but not the equicontinuity of V^ε implied by (3.2). See [7, Chap. 7].

5 Differential game interpretation.

Let us next interpret the unique viscosity solution V^0 in Theorem 4.1 as the value function of the following zero-sum differential game. The minimizing player chooses a measurable function $v.: [s,T] \rightarrow U$ and the maximizing player chooses a function $w. \in L^2([s,T]; \mathbb{R}^n)$. The game dynamics are

(5.1)
$$\frac{d\xi_t}{dt} = f(t,\xi_t,v_t) + w_t, \ s \le t \le T,$$

with $\xi_s = x (x \in \mathbb{R}^n)$. The payoff is

(5.2)
$$P(s,x; \ v.,w.) = \int_s^T [L(t,\xi_t,v_t) - \gamma^2 |w_t|^2] dt.$$

The Isaacs partial differential equation for the value $V(s,x)$ of the differential game is

(5.3)
$$0 = V_s + \min_{v \in U}[f(s,x,v) \cdot \nabla_x V + L(s,x,v)]$$

$$+ \max_{w \in \mathbb{R}^n}[w \cdot \nabla_x V - \gamma^2 |w|^2].$$

A precise formulation of the differential game can be given using the Elliott-Kalton definition of strategy, as follows. Let

$$M(s) = L^2([s,T]; \ \mathbb{R}^n)$$

$$N(s) = L^\infty([s,T]; \ U).$$

A *strategy* for the minimizing player is a function $\beta : M(s) \rightarrow N(s)$ with the following property; for $s \le \tau \le T$, $w_t = \hat{w}_t$ for almost all $t \in [s,\tau]$ implies

$$\beta[w]_t = \beta[\hat{w}]_t \text{ for almost all } t \in [s,\tau].$$

The lower Elliott-Kalton value $V^-(s,x)$ is defined as

(5.4⁻)
$$V^-(s,x) = \inf_\beta \sup_{w. \in M(s)} P(s,x; \ w.,\beta[w].).$$

In a similar way, by introducing strategies $\alpha : N(s) \rightarrow M(s)$ for the maximizing player, the upper Elliott-Kalton value $V^+(s,x)$ is defined as

(5.4⁺)
$$V^+(s,x) = \sup_\alpha \inf_{v. \in N(s)} P(s,x; \alpha[v].,v.).$$

Theorem 5.1. $V^-(s,x) = V^+(s,x) = V^0(s,x)$, with V^0 as in Theorem 4.1.

Let us merely indicate the proof. By (2.12), the Isaacs PDE (5.3) is the same as (4.1). Since min and max appear separately in (5.3), the Isaacs minimax condition holds. By essentially the same method as in Evans - Souganidis [5], V^+ and V^- are both bounded, uniformly continuous viscosity solutions of (4.1) - (4.2). By uniqueness of such solutions, $V^+ = V^- = V^0$.

Since upper and lower values both equal $V^0(s,x)$, we call $V^0(s,x)$ the *Elliott - Kalton value* of the differential game.

Remark. In [5] the control spaces for both the maximizing player and the minimizing player are compact. Although $u_t \in U$ and U is compact, the maximizing player chooses $w_t \in \mathbb{R}^n$ and $\mathbb{R}^n$ is not compact. However, using the bound (3.4) the differential game is in fact equivalent to one with compact control spaces as follows. Let us introduce the artificial bound $|w| \leq K$, where $K \geq \gamma^{-2}C_2$ with C_2 as in (3.4). If $|p| \leq C_2$, then

$$(5.5) \qquad \max_{|w| \leq K}[w \cdot p - \gamma^2|w|^2] = \max_{w \in \mathbb{R}^n}[w \cdot p - \gamma^2|w|^2] = \frac{1}{4\gamma^2}|p|^2.$$

By (3.4) and (5.5) with $p = |\nabla_x V^\varepsilon|$, V^ε is a classical solution to

$$(5.6^a) \qquad 0 = V_s^\varepsilon + \frac{\varepsilon}{2}\Delta_x V^\varepsilon + \min_{v \in U}[f(s,x,v) \cdot \nabla_x V^\varepsilon + L(s,x,v)]$$

$$+ \max_{|w| \leq K}[w \cdot \nabla_x V^\varepsilon - \gamma^2|w|^2]$$

with terminal data (2.11). The same proof as for Theorem 4.1 shows that V^0 is a viscosity solution of (5.6^0) - (4.2). By uniqueness, $V^0 = V_K$, where V_K is the Elliott - Kalton value with the control constraint $|w| \leq K$ for the maximizing player.

6. Final Remarks. In the LEQG stochastic control problem, one has

$$f(s,x,v) = A(s)x + B(s)v$$

$$L(s,x,v) = x'Q(s)x + v'R(s)v,$$

with $v \in \mathbb{R}^m$ for some m, and $Q(s), R(s)$ symmetric matrices which are respectively nonnegative definite and positive definite. The differential game in Section 5 has in this case linear dynamics (5.1) and quadratic payoff (5.2) of indefinite sign. The minimizing player chooses $v_t \in \mathbb{R}^m$ and the maximizing player chooses $w_t \in \mathbb{R}^n$. Neither of the control spaces $\mathbb{R}^m, \mathbb{R}^n$ is compact. Let us require that $0 \leq s \leq T$. The value $V^0(s, x)$ is defined only under some restrictions on γ and T. There exists $\gamma_1 > 0$ such that T can be arbitrarily large if $\gamma \geq \gamma_1$. However, if $\gamma < \gamma_1$, then one must have $T < T_1 < \infty$. If $T < T_1$, then $V^0(s, x)$ is quadratic in x for $0 \leq s \leq T$ and V^0 can be found by solving a matrix Riccati differential equation. See [3] [14].

We have considered risk sensitive control problems on a finite time interval $[s, T]$. In order to connect the LEQG problem and H_∞ - control, Glover [11, p.390] let $T - s$ tend to infinity. His calculation corresponds to a result concerning large deviations of occupation measures from ergodicity, of so-called Donsker-Varadhan type [4]. On the other hand, the kind of small-noise large deviations results on a finite time interval considered by Whittle and in this paper are of Freidlin -Wentzell type [10]. It might be interesting to examine risk sensitive stochastic control for nonlinear systems in the context of Donsker - Varadhan large deviations theory.

References

1. J. S. Baras, A. Bensoussan and M. R. James, "Dynamic observers as asymptotic limits of recursive filters: special cases", SIAM J. Appl. Math 48 (1988) 1147–1158.

2. E. N. Barron and R. Jensen, "Total risk aversion, stochastic optimal control and differential games", Appl. Math. and Optimiz 19 (1989) 313–327.

3. T. Basar and P. Bernhard, "H^∞ - optimal Control and Related Minimax Design Problems" Birkhauser, Boston 1991.

4. M. D. Donsker and S. R. S. Varadhan, "Asymptotic evaluation of certain Markov process expectations for large time", I, II, III, Comm. Pure Appl. Math. 28 (1975) 1–45, 279–301; 29 (1976) 389–461.

5. L. C. Evans and P. E. Souganidis, "Differential games and representation formulas for solutions of Hamilton - Jacobi equations", Indiana Univ. Math. J. 33 (1984) 773–797.

6. W. H. Fleming and R. W. Rishel, "Deterministic and Stochastic Optimal Control", Springer Verlag, 1975.

7. W. H. Fleming and H. M. Soner, "Controlled Markov Processes and Viscosity Solutions", Springer Verlag, 1992.

8. W. H. Fleming and P. E. Souganidis, "PDE - viscosity solution approach to some problems of large deviations", Annali Scuola Normale Sup. Pisa, Ser. IV 23 (1986) 171–192.

9. W. H. Fleming and C-P Tsai, "Optimal exit probabilities and differential games", Applied Math. Optimiz. 7 (1981) 253–282.

10. M. I. Freidlin and A. D. Wentzell, "Random Perturbations of Dynamical Systems", Springer Verlag, 1984.

11. K. Glover, "Minimum entropy and risk-sensitive control: the continuous time case", Proc. 28th IEEE Conf. on Decision and Control, Dec. 1989, 388–391.

12. K. Glover and J. C. Doyle, "State-space formulae for all stabilizing controllers that satisfy an H^∞ - norm bound and relations to risk sensitivity", Systems and Control Letters 11, (1988) 167–172.

13. O. Hijab, "Minimum energy estimation," Ph.D. dissertation, Univ of Calif. Berkeley, 1980.

14. D. H. Jacobson, "Optimal stochastic linear systems with exponential criteria and their relation to deterministic differential games", IEEE Trans. Automat. Control AC - 18 (1973) 124–131.

15. P. Whittle, "Risk-sensitive Optimal Control", Wiley, 1990.

16. P. Whittle, "A risk - sensitive maximum principle", Systems and Control Lett. 15 (1990) 183–192.

17. M. R. James, "Asymptotic analysis of nonlinear stochastic risk-sensitive control and differential games", Preprint.

FIXED GAIN ESTIMATION AND TRACKING

László Gerencsér*
Systems and Control Theory Laboratory
Computer and Automation Institute of the
Hungarian Academy of Sciences

Abstract: We present a few basic strong approximation results for fixed gain estimators. Both off-line and on-line estimators are considered. The results are presented for time-invariant systems in Section 2. Once the appropriate techniques and results are developed the extension of the analysis for time-varying systems is quite easy. A basic result on tracking, the law of the cubic root is also explained.

Keywords: linear stochastic systems; time-varying systems; adaptive control; Ljung's scheme; L-mixing processes; receding horizon control.

1. INTRODUCTION

A basic problem of control theory is the adaptive control of slowly time-varying stochastic systems. While the importance of this problem has been recognized for a long time progress has been slow. The main reason seems to be that a fundamental technique of the theory of system identification proposed in [16] and in [4] had not been analyzed in depth until recently (c.f. [13]). The objective of this work is to present the basic elements of a rigorous mathematical foundation for a time-varying Ljung's scheme. Since many interesting adaptive control problems can be formulated in terms of Ljung's scheme as shown recently in [11], the results of this paper provide a foundation for the solution of some of the basic problems of the theory of stochastic adaptive control.

A time varying Ljung's scheme is obtained using exponential forgetting in the off-line case and fixed gain in the on-line case. In Section 2 we present results when these

---

* The first version of this paper had been completed and submitted for publication while the author was visiting the Department of Electrical Engineering, McGill Univesity Montreal, Quebec, Canada.

estimation schemes are applied for a time-invariant system. These results contain most of the essesntial ingredients of the results for time varying systems.

To have a feeling of what can be expected let us consider the very simple model

$$y_n = \theta^* + e_n$$

where (e_n) is an i.i.d. sequence of M-bounded random variables (c.f.below). Then the fixed gain-estimation of θ^* is given by

$$\widehat{\theta}_N = \sum_{n=1}^{N} (1-\lambda)^{N-n} \lambda y_n$$

with some $0 < \lambda < 1$. (I.e. small λ means poor forgetting.) We choose this notation for the forgetting factor in order to conform with the continuous time case. Then it is easy to see that

$$\widehat{\theta}_N - \theta^* = \sum_{n=1}^{N} (1-\lambda)^{N-n} \lambda e_n + o_M(1) = O_M(\lambda^{1/2}) + o_M(1).$$

by the Marczinkiewicz-Zygmund inequality. For the notation $O_M(.)$ c.f. Definition 1.1 below. Analogous results will be presented under much more general conditions in Section 2.

Now if the above model is time-varying, say

$$y_n = \theta_n + e_n$$

and

$$\dot{S} \stackrel{\triangle}{=} \sup_n |\theta_{n+1} - \theta_n| < \infty$$

then we get using the same argument that

$$\widehat{\theta}_N - \theta^* = O_M(\lambda^{1/2}) + o_M(1) + O(\dot{S}/\lambda).$$

where the last term is a deterministic tracking error.

From here the optimizing λ is found to be $\lambda = O(\dot{S}^{2/3})$ and with this we get

$$\widehat{\theta}_N - \theta^* = O_M(\dot{S}^{1/3}) + o_M(1).$$

I.e. the tracking error is proportional to the 1/3-rd power of the rate change.

To put the paper into perspective we mention some earlier works. An off-line estimation method for time-varying systems was presented in [7]. The first significant progress towards a recursive version of these algorithms was presented in [8]. The results presented here enabled us to study the asymptotic behaviour of a stochastic complexity associated with time-varying systems and this study has lead us to the design of a very effective change point detection method (c.f.[2] and[4]).

The continuous time extension of these results, or to put it in a different way, the fixed gain version of earlier continuous time results such as [5] or [15] is the subject of an ongoing research.

We start the technical discussions with definitions. Let $D \subset \mathbb{R}^p$ be an open domain and let the stochastic process $(u_n(\theta))$ be defined on $\mathbf{Z} \times D$, where $\mathbf{Z}$ denotes the set of natural numbers.

Definition 1.1 (c.f. [9])   We say that $u = (u_n(\theta))$ is M-bounded if for all $1 \leq q < \infty$

$$M_q(u) = \sup_{\substack{n \geq 0 \\ \theta \in D}} E^{1/q} |u_n(\theta)|^q < \infty.$$

Obviously the definition is applicable also if the process $u_n(\theta)$ does not depend on a parameter. If u is M-bounded we shall also write $u = O_M(1)$.

Let $(\mathcal{F}_n), n \geq 0$ be a family of monotone increasing σ-algebras, and $(\mathcal{F}_n^+), n \geq 0$ be a monotone decreasing family of σ-algebras. We assume that for all $n \geq 0, \mathcal{F}_n$ and $\mathcal{F}_n^+$ are independent. For $n \leq 0$ we set $\mathcal{F}_n^+ = \mathcal{F}_0^+$. A typical example is provided by the σ-algebras

$$\mathcal{F}_n = \sigma\{e_i : i \leq n\} \qquad \mathcal{F}_n^+ = \sigma\{e_i : i > n\}$$

where (e_i) is an i.i.d. sequence of r.v.'s.

Definition 1.2. A stochastic process $(u_n(\theta)), n \geq 0$ is L-mixing with respect to $(\mathcal{F}_n, \mathcal{F}_n^+)$ uniformly in θ if it is $\mathcal{F}_n$-progressively measurable, M-bounded and if we set

$$\gamma_q(\tau, u) = \gamma_q(\tau) = \sup_{\substack{n \geq \tau \\ \theta \in D}} E^{1/q} |u_n(\theta) - E(u_n(\theta)|\mathcal{F}_{n-\tau}^+)|^q$$

where τ is a positive integer then

$$\Gamma_q = \Gamma_q(u) = \sum_{\tau=1}^{\infty} \gamma_q(\tau) < \infty.$$

The definitions of $M_q(u)$ and $\Gamma_q(u)$ can be extended to $q = \infty$.

The concept of L-mixing processes was introduced in [9]. The slightly different concept of "exponentially stable processes" was used in [3].

The general estimation scheme proposed in [15] and [4] can be described as follows. Let us consider a parameter-dependent state space equation in which we incorporate a "true system-parameter" ψ^* end a test parameter θ:

$$\overline{\overline{x}}_{n+1}(\theta, \psi^*) = A(\theta, \psi^*)\overline{\overline{x}}_n(\theta, \psi^*) + B(\theta, \psi^*)e_n. \tag{1.1}$$

with $\psi \in D_\psi \subset \mathbb{R}^r, \theta \in D_{\theta 1} \subset \mathbb{R}^p, \overline{x}_n(\theta) \in \mathbb{R}^m$, where D_ψ and $D_{\theta 1}$ are bounded open domains. The dimensions and the interpretations of θ and ψ may be different. E.g. θ may denote some controller parameter-vector of a direct adaptive controller as in the Åström-Wittenmark regulator.

Condition 1.1 It is assumed that $A(\theta, \psi)$ is stable for $(\theta, \psi) \in D_{\theta 1} \times D_\psi$ and that the matrix-valued functions $(A(\theta, \psi))$ and $(B(\theta, \psi))$ are smooth for $(\theta, \psi) \in D_{\theta 1} \times D_\psi$.

Condition 1.2 Let (e_n) be a second order stationary process which is L-mixing with respect to a pair of families of σ-algebras $(\mathcal{F}_n, \mathcal{F}_n^+)$.

In practice the initial condition $\overline{\overline{x}}_0(\theta, \psi^*)$ is assumed to be M-bounded and $\mathcal{F}_0$-measurable, but otherwise arbitrary random variable. An important assumption is that the process $\overline{\overline{x}}_n(\theta, \psi)$ is computable in the sense that for each fixed θ, $\overline{\overline{x}}_n(\theta, \psi)$ is the output of a system which is obtained by coupling a physical system with a known system, performing identification and control.

Let Q be a quadratic function from $\mathbb{R}^r$ to $\mathbb{R}^p$, and define

$$G(\theta, \psi^*) = \lim_{n \to \infty} EQ(\overline{\overline{x}}_n(\theta, \psi^*)).$$

We will say that the function $G(\theta, \psi^*)$ is computable. $G(\theta, \psi^*)$ is well-defined and is smooth in $D_{\theta 1}$. The estimation problem can now be formulated as follows: solve the nonlinear algebraic equation

$$G(\theta, \psi^*) = 0. \tag{1.2}$$

Condition 1.3 We assume that (3.2) has a unique solution in $D_{\theta 1}$, say $\theta^*(\psi)$, and that $G_\theta(\theta^*, \psi^*)$ is nonsingular and thus the mapping $\psi^* \to \theta^*(\psi^*)$ is smooth from D_ψ into $D_{\theta 1}$,

Remark. The above scheme is the simplest possible and can be generalized in many different ways. E.g. Q may depend directly on θ, i.e. we may have functions of the form $Q(x, \theta)$. This is the case in multivariable identification. As an extreme case it may even happen that x does not depend on θ at all, as in the case of the adaptive filtering problems. Or there may be some equality constraints imposed on θ. However the analysis of these extensions does not differ in any significant degree from the analysis of the simplest case.

2. FIXED GAIN ESTIMATON

The off-line estimator of θ^* using exponential forgetting, say $\hat{\theta}_n^\lambda$ is obtained as "the solution" of the nonlinear algebraic equation

$$U_N(\theta, \psi^*) \triangleq \sum_{n=1}^{N} (1 - \lambda)^{N-n} \lambda Q(\overline{\overline{x}}_n(\theta, \psi^*)) = 0 \tag{2.1}$$

where $0 < \lambda < 1$ is the forgetting factor. (In the limiting case $\lambda \searrow 0$ the empirical "index of performance" $U_N(\theta, \psi^*)$ becomes the arithmetic mean of the terms $Q(\overline{\overline{x}}_n(\theta, \psi^*))$, i.e. we get the usual procedure of off-line estimation for time-invariant systems). By "the solution of (2.1)" we mean a $D_{\theta 1}$-valued random variable $\hat{\theta}_N$ which is equal to the unique solution of (2.1) in $D_{\theta 1}$ if a $D_{\theta 1}$-valued unique solution exists. Such a random variable exists by the measurable selection theorem.

Lemma 2.1. For any $d > 0$ and $s > 0$ the equation (2.1) has a unique solution in D with the property that it is also in the sphere $\{|\theta - \theta^*| < d\}$ with probability at least $1 - O(\lambda^s)$ where the constant in the error term $O(\lambda^s) = C\lambda^s$ depends only on d and s.

The extension of Theorem 2.1 of [10] is then the following theorem (c.f. Theorem 10.2 of [12]).

Theorem 2.2 ("The third representation theorem")   We have for small λ's

$$\theta_N^\lambda - \theta^* = -G_\theta^{-1}(\theta^*, \psi^*) \sum_{n=1}^{N} (1 - \lambda)^{N-n} \lambda Q(\overline{\overline{x}}_n(\theta^*, \psi^*)) + r_N$$

where $r_N = O_M(\lambda) + o_M(1)$. Here and in the sequel $o_M(1)$ is an exponentially delaying term.

It is easy to see that the first term on the right hand side is of order $O_M(\lambda^{1/2})$. Also it is easy to see that this dominant term is an L-mixing process, so we may expect that $\theta_n^\lambda - \theta^*$ itself will be an L-mixing process. However, since θ_n^λ is not uniquely defined on a set of nonvanishing probability, some precaution is needed. It can be shown that the solutions $\widehat{\theta}_N^\lambda$ are uniquely defined, except for a set

$$B_N = \{\omega : \delta_N \geq c > 0\} \tag{2.2}$$

where (δ_N) is an appropriately defined L-mixing process, and $\delta_N = O_M(\lambda^{1/2})$. However, $\chi_{B_N}(\omega) = \chi_{x>c}(\delta_n)$, where $\chi_{B_N}(\omega)$ is the characteristic function of B_N, is not necessarily L-mixing.

Definition 2.1   A stochastic process (x_n), $n \geq 0$ is L_0-mixing with respect to $(\mathcal{F}_n, \mathcal{F}_n^+)$, if it is L-mixing and for any $q \geq 1, m > 1, \gamma_q(\tau, x) = O(\tau^{-m})$.

Lemma 2.3   Let (x_n), $n \geq 0$ be an L_0-mixing process, and let $I \subset \mathbb{R}$ be a fixed nonempty open interval. Then there exist a sequence of real numbers $\delta_n \in I$ such that the process $y_n = \chi_{x>\delta_n}(x_n)$ is L_0-mixing.

Using Lemma 2.3 we can sharpen Theorem 2.2 as follows:

Theorem 2.4   If (e_n) is L_0-mixing then in the representation given in Theorem 2.2 we have $|r_N| \leq r_N^*$ where r_N^* is an L-mixing process such that $r_N^* = O_M(\lambda) + o_M(1)$.

The recursive estimator of θ^* is defined by the following fixed gain estimation scheme in which λ denotes the forgetting rate. Assume that an initial value $\widehat{\theta}_0^\lambda \in D_{\theta 1}$ is given .Then the estimator $\widehat{\theta}_n^\lambda$ of the parameter θ^* is generated recursively by the following equations:

$$x_{n+1} = A(\widehat{\theta}_n^\lambda, \psi^*)x_n + B(\widehat{\theta}_n^\lambda, \psi^*)e_n \tag{2.3}$$

$$\widehat{\theta}_{n+1}^\lambda = \widehat{\theta}_{n+1}^\lambda + \lambda Q(x_{n+1}) \tag{2.4}$$

with $x_0 = 0$.

We shall relate this estimator process to the solution of an associated ODE which is defined as

$$\dot{y}_t = \lambda G(y_t, \psi^*) \tag{2.5}$$

The initial value for (2.5) is assumed to be $y_s = \xi \in D_\xi$, where $D_\xi \subset D$ is a suitable compact domain. Let the general solution of (2.5) be denoted by $y(t, s, \xi)$. It is well-known that $y(t, s, \xi)$ is a continuously differentiable function of (t, s, ξ) and even $(\partial^2/\partial\xi^2)y(t, s, \xi)$ exists and is continuous in (t, s, ξ).

Let $D' \subset D$ be any subset of D and consider the set of points which are reachable from D' along the trajectories of (2.5), i.e. we consider the set $\{y : y = y(t, s, \xi)$ for some $t \geq s, \xi \in D'\}$. The closure of this set will be denoted by $y(D')$. The ε' neighborhood of the set D' will be denoted by $S(D', \varepsilon')$ i.e. $S(D', \varepsilon') = \{x : |x - z| < \varepsilon'$ for some $z \in D'\}$.

Condition 2.1 There exist compact domains $D_\xi \subset D_y \subset D_{\theta 2} \subset D_{\theta 1} \subset D_\theta$ such that we have for some $\varepsilon' > 0$ $y(D_\xi) \subset D_y$, $S(D_y, \varepsilon') \subset D_{\theta 2}$ and $y(D_{\theta 2}) \subset D_0$.

Condition 2.2 We have for $x \in D_{\theta 2}$, $t > s > 0$ and some $\alpha > 0$

$$\|\frac{\partial y}{\partial x} y(t, s, x)\| \leq C_0 e^{-\lambda \alpha(t-s)}.$$

Then we have the following theorem (c.f. Theorem 10.5 of [12]):

Theorem 2.5 ("The fourth representation theorem")   Let (e_n) be a bounded process, i.e. let $M_\infty(e) < \infty$ and such that $\Gamma_\infty(e) < \infty$. Then if $M_\infty(e)$ is sufficiently small then $\widehat{\theta}_n^\lambda$ does not leave $D_{\theta 1}$ and

$$\widehat{\theta}_N^\lambda - \theta^* = \sum_{n=1}^N \frac{\partial y^\lambda}{\partial \xi}(N, n, \theta^*) \lambda Q(\bar{\bar{x}}_n(\theta^*, \psi^*)) + r_N$$

where $|r_N|$ is majorized by an L-mixing process r_N^* such that $r_N^* = O_M(\lambda)$.

This theorem gives a very nice insight into the structure of the estimator and a new interpretation of the "ODE method". The proof of this theorem has some common elements with [6].

The above results are very useful in analyzing stochastic complexities associated with fixed gain estimators. A predictive form of stochastic complexity has been introduced in [17],[18] and by now has become one of the most important concept in the statistical analysis of linear stochastic systems. It was originally meant to be a device for solving model selection problems but has since outgrown of this framework. Thus e.g. in adaptive filtering the effect of parameter uncertainty on prediction is a central issue. Similarly in adaptive control the effect of parameter uncertainty on performance will likely to be a central issue in the years to come.

To put the problem into a general setting let us consider a real-valued quadratic function $S = S(x)$ and consider the index of ideal performance $S(\bar{x}_n(\theta^*))$ at time n. We assume that

$$W(\theta) = ES(\bar{x}_n(\theta^*))$$

is minimized at $\theta = \theta^*$ and that

$$T^* = \frac{\partial^2}{\partial \theta^2} W(\theta) \Big|_{\theta = \theta^*}$$

is positive definite.

For example we may consider a self-tuning regulator for an ARMAX system, where the function $G(\theta)$ consists of cross-correlations, and the function $S(x)$ is the square of the output (c.f [1]). We can now ask the classical question in the theory of stochastic complexity: how much do we lose in performance if we substitute θ^* by its estimate $\widehat{\theta}_{n-1}$?

In the general case let $S(x)$ be a performance index given abvove and assume that (e_n) is stationary. Assuming that the processes have been started at $n = -\infty$ let

$$\delta\theta = \sum_{n=-\infty}^{0} \frac{\partial y^\lambda}{\partial \xi}(0, n, \theta^*) \lambda Q(\bar{x}_n(\theta^*))$$

and let $(R^{**})^{-1} = E\delta\theta \cdot \delta\theta^T / \lambda$.

Theorem 2.6   Assume that the conditions of Theorem 2.4 are satisfied and (e_n) is stationary. Then

$$\overline{\lim}_{N\to\infty} |\frac{1}{N} \sum_{n=1}^{N}(S(x_n) - S(\bar{x}_n(\theta^*))) - \lambda \mathrm{Tr} T^*(R^{**})^{-1}| \le C\lambda^{1/2}$$

where C is a nonrandom constant.

3. TIME-VARYING SYSTEMS

A time-varying system is obtained if the systems parameter ψ^* in (1.1) is replaced by ψ_n^*. The system is then described by the following equation:

$$\bar{x}_{n+1}(\theta) = A(\theta, \psi)\bar{x}_n(\theta, \psi_n) + B(\theta, \psi_n)e_n. \tag{3.1}$$

Condition 3.1 We assume that $\psi_n \in D_\psi$. and that the time-varying system is slowly time-varying i.e.

$$\dot{S} \triangleq \sup_{n\ge 0} |\psi_{n+1} - \psi_n| < \infty,$$

and consequently $\sup_{n\ge 0} |\theta_{n+1} - \theta_n| \le C\dot{S}$ with some constant C. The quantity $\dot{S}$ is called the rate of change.

It is well known that if $\dot{S}$ is sufficiently small then (3.4) is "exponentially stable". The above description of slowly time-varying systems is a special case of a more general definition in [20]. An extension in another direction is given in [19] where the time-variation is not necessarily slow. The central issue in this work is the Lie-algebraic structure of the system.

The off-line estimator of θ_n say $\widehat{\theta}_n^\lambda$ is obtained as "the solution" of the nonlinear algebraic equation

$$U_N(\theta) \triangleq \sum_{n=1}^N (1-\lambda)^{N-n}\lambda Q(\bar{x}_n(\theta)) = 0 \tag{3.2}$$

where $0 < \lambda < 1$ is the forgetting factor (c.f. (2.1)). We define an approximation of θ^* as the solution of the nonrandom nonlinear equation

$$G_{vN}(\theta) \triangleq EU_N(\theta) = 0.$$

Let this solution be θ_{vN}^*. The extension of Theorems 2.1 and 2.3 is then the following (c.f. [12]):

Theorem 3.1 If (e_n) is L_0-mixing then

$$\widehat{\theta}_N^\lambda - \theta_{vN}^* = -G_{vN\theta}^{-1}(\theta_v^*) \sum_{n=1}^N (1-\lambda)^{N-n}\lambda Q(\bar{x}_n(\theta_v^*)) + r_N.$$

where we have $|r_N| \leq r_N^*$ where r_N^* is an L-mixing process such that $r_N^* = O_M(\lambda) + o_M(1)$.

Since it is easy to show that $|\theta_{vN}^* - \theta_N^*| = O(\dot{S}/\lambda)+o(1)$, we get a nice decomposition of the tracking error. This decomposition captures and separates the effects of noise and time variation. The optimizing value of λ is easily found to be $\lambda = O(\dot{S}^{2/3})$ which yields the tracking error $O_M(\dot{S}^{1/3})$. Thus we get a "law of the cubic root", which states that the tracking error can be made proportional to the cubic root of the rate of change.

The recursive estimator of $\theta^*(\psi^*) = \theta_n^*$ is defined by the following fixed gain estimation scheme in which λ denotes the forgetting rate. Assume that an initial value $\widehat{\overline{\theta}}_0^\lambda \in D_{\theta 1}$ is given .Then the estimator $\widehat{\overline{\theta}}_n^\lambda$ of the time-varying parameter θ_n is generated recursively by the following equations:

$$x_{n+1} = A(\widehat{\overline{\theta}}_n^\lambda, \psi_n)x_n + B(\widehat{\overline{\theta}}_n^\lambda, \psi_n)e_n \tag{3.3}$$

$$\widehat{\overline{\theta}}_{n+1}^\lambda = \widehat{\overline{\theta}}_n^\lambda + \lambda Q(x_{n+1}) \tag{3.4}$$

with $x_0 = 0$.

We shall relate this estimator process to be the solution of an associated ODE which is now defined as

$$\dot{y}_t = \lambda G(y_t, \psi_t) \tag{3.5}$$

where $\psi_t = \psi_n$ for $n \leq t < n+1$. The initial value for (3.5) is assumed to be $y_s = \xi \in D_\xi$, where $D_\xi \subset D$ is a suitable compact domain. Let the general solution of (3.5) be denoted by $y(t, s, \xi)$.

A nontrivial problem for time-varying systems is to guarantee the exponential stability of the associated time-varying ODE.

Theorem 3.2 Under Conditions 2.1 and 2.2 the solution of (3.5) is defined for all $t \geq s$ whenever $\dot{S}/\lambda$ is sufficiently small. Moreover the closure of the set

$$y(s, D_\xi) = \{y : y = y(t, s, \xi) \text{ for some } t \geq s \geq 0, \xi \in D_\xi\}$$

is a compact subset of D. If θ_t denotes the continuous-time extension of θ_n then

$$|y_t - \theta_t| \leq C\dot{S}/\lambda + o(1).$$

where $o(1)$ tends to zero exponentially fast when t tends to infinity. Finally (3.5) is exponentially stable if $\lambda \geq c\dot{S}$ where c is a system's constant and we then have

$$\|\frac{\partial}{\partial \xi} y(t, s, \xi)\| \leq c_0 e^{-c\lambda(t-s)}$$

where c_0 and c are systems constants.

Theorem 3.3 Under the conditions of Theorem 2.5 $\widehat{\widehat{\theta}}_N^\lambda$ will not leave D_0 and we have

$$\widehat{\widehat{\theta}}_N^\lambda - y_N = \sum_{n=1}^{N} \frac{\partial y^\lambda}{\partial \xi}(N, n, y_n) \lambda Q(\bar{x}_n(y_n)) + r_N$$

where $|r_N|$ is majorized by a L-mixing process r_N^* such that $r_N^* = O_M(\lambda)$.

Now by Theorem 3.2 $|y_n - \theta_n^*| \leq C\dot{S}/\lambda + o(1)$, and hence we get a decomposition of the tracking error similar to the one given after Theorem 3.1, and the corresponding law of the cubic root holds.

4. A LOWER BOUND

Let θ_t be a time-varying real-valued parameter which is continuously differentiable with respect to t and the absolute value of its derivative $\dot{\theta}_t$ is bounded by say $\dot{S}$. Assume that we have noisy observations in the form

$$dy_t = \theta_t dt + dw_t \qquad y_0 = 0$$

where w_t is a standard Wiener-process. A linear estimator of θ_T is defined as

$$\widehat{\theta}_T = \int_0^T f(t)dy_t$$

where $(f(t))$ is a nonnegative deterministic function of t belonging to $L_2[0,T]$. We call a linear estimator unbiased if it is unbiased for the case $\theta_t = \theta, 0 \leq t \leq T$. This means that we require

$$\int_0^T f(t)dt = 1.$$

Theorem 4.1 Let $\theta_t^* = \dot{S}t$ and let $\widehat{\theta}_T$ be an unbiased linear estimator of θ_T^* . Then

$$E^{1/2}|\widehat{\theta}_T - \theta_T^*|^2 \geq \frac{1}{2\sqrt{2}}\dot{S}^{1/3}.$$

Acknowledgements

This research was partially supported by the Natural Science and Engineering Research Council of Canada under Grant 01329 and by the Natural Scientific Research Foundation of Hungary under Grant 2042. The author wishes to thank to the NSF and the University of Kansas for their financial support and to Zsuzsanna Vágó for her help in preparing this paper.

REFERENCES

[1] K.J. Åström, and B. Wittenmark, *On self-tuning regulators*. Automatica, 9 (1973), pp.185-199.

[2] J. Baikovicius and L. Gerencsér, *Change point detection in a stochastic complexity framework*. To appear in the Proc of the 29-th IEEE CDC, Vol 6, (1990), 3554-3555.

[3] P.E. Caines, *Linear Stochastic Systems*, Wiley, (1988).

[4] D.P. Djereveckii and A.L. Fradko, *Applied theory of discrete adaptive control systems*, (In Russian), Nauka, Moscow (1981).

[5] T.E. Duncan, and B. Pasik-Duncan, *Adaptive Control of Continuous-Time Linear Stochastic Systems*, to appear in Mathematics of Control, Signals and Systems (1991).

[6] S. Geman, *Some averaging and stability results for random differential equations*, SIAM J. Appl. Math., 36 (1979) 87-105.

[7] L. Gerencsér, *Parameter Tracking of Time-Varying Continuous-Time Linear Stochastic Systems*, In Modelling, Identification and Robust Control (eds. Ch. I. Byrnes and A. Lindquist), North Holland, (1986), 581-595.

[8] L. Gerencsér, *Pathwise stability of random differential equation*. Preprint of the Department of Mathemetics, Chalmers University of Technology, 1986:19 Göteborg.

[9] L. Gerencsér, *On a class of mixing processes*. Stochastics, 26 (1989), pp.165-191.

[10] L. Gerencsér, *On the martingale approximation of the estimation error of ARMA parameters*. Systems and Control Letters, 15, (1990), 417-423.

[11] L. Gerencsér, *Closed loop parameter identifiability and adaptive control of a linear stochastic system*. Systems and Control Letters, 15, (1990), 411-416.

[12] L. Gerencsér, *Strong approximaton results in estimation and adaptive control*. In: Topics in Stochastic Systems: Modelling, Estimation and Adaptive Control (eds: L. Gerencsér, P.E. Caines), pp.268-299, Springer-Verlag Berlin, Heidelberg 1991.

[13] L. Gerencsér, *Rate of convergence of recursive estimators*. To appear in SIAM J. Control and Optimization, 1992.

[14] L. Gerencsér, and J. Baikovicius, *Change and point detection with stochastic complexity*. To appear in Proc. of 9-th IFAC/IFORS Symposium on Identification and System Parameter Estimation, Budapest, (1991), Pergamon Press, Oxford.

[15] L. Gerencsér, and Zs. Vágó, *From fine asymptotics to model selection*. in "Identification of Continuous-Time Systems" (eds. N.K. Sinha and G.P.Rao), Kluwer Academic Publishers, Dordrecht, The Netherlands, Chapter 19, 575-585. (1991).

[16] L. Ljung, *Analysis of recursive stochastic algorithms*, IEEE Trans. Aut. Cont., AC-22 (1977), pp.551-575.

[17] J. Rissanen, *Stochastic complexity and modeling*, Annals of Statistics 14 (1986), pp.1080-1100.

[18] J. Rissanen, *Stochastic complexity in statistical inquiry*. World Scientific Publisher (1989).

[19] F.Szigeti, *A differential-algebraic condition for controllabity and observability of time varying linear systems*. Submitted for publication.

[20] G. Zames, and L.Y. Wang, Local-Global Double Algebras for Slow H^∞ Adaptation: Part I–Inversion and Stability, *IEEE Trans. Aut. Cont.*, 36, 2, (1991), pp. 130-142.

OPTIMAL CONTROL OF SWITCHING DIFFUSIONS
MODELLING A FLEXIBLE MANUFACTURING SYSTEM*

MRINAL K. GHOSH†, ARISTOTLE ARAPOSTATHIS‡
AND STEVEN I. MARCUS†

ABSTRACT

A controlled switching diffusion model is developed to study the hierarchical control of flexible manufacturing systems. The existence of a homogeneous Markov nonrandomized optimal policy is established by a convex analytic method. Using the existence of such a policy, the existence of a unique solution in a certain class to the associated Hamilton-Jacobi-Bellman equations is established and the optimal policy is characterized as a minimizing selector of an appropriate Hamiltonian.

1. INTRODUCTION

We study a controlled switching diffusion process that arises in numerous applications of systems with multiple modes or failure modes, including the hierarchical control of flexible manufacturing systems. A flexible manufacturing system (FMS) consists of a set of workstations capable of performing a number of different operations and interconnected by a transportation mechanism. An FMS produces a family of parts related by similar operational requirements or by belonging to the same final assembly [23]. The rapidly growing range of applicability of FMS includes metal cutting, assembly of printed circuit boards, integrated circuit fabrication, automobile assembly lines, etc. Due to their tremendous flexibility, FMS are significantly more efficient in many ways than traditional manufacturing systems. However, the high capital cost of an FMS demands very efficient management of production and maintenance (repair/replacement) scheduling so that uncertain events such as random demand fluctuations, machine failures, inventory spoilages, sales returns, etc. can be taken care of. The large size of the system and its associated complexities make it imperative to divide the control or management into a hierarchy consisting of a number of

† Systems Research Center, 2167 A.V. Williams Bldg., University of Maryland, College Park, MD 20742.
‡ Dept. of Electrical and Computer Engineering, The University of Texas at Austin, Austin, TX 78712-1084.
*This work was supported in part by the Texas Advanced Technology Program under Grant No. 003658-093, in part by the Air Force Office of Scientific Research under Grants AFOSR-91-0033, F49620-92-J-0045, F49620-92-J-0083, and in part by the National Science Foundation under Grant CDR-8803012.

levels. Thus the overall complex problem is reduced to a number of managable subproblems at each level, and these levels are linked by means of a hierarchical integrative system. We refer to [1], [17], [23] for a detailed description of these hierarchical schemes. We will confine our attention to the top two levels, viz.

(i) Generation of decision tables, which is accomplished by developing a suitable mathematical model describing the dynamical evolution of the system. This is done off-line.

(ii) The flow control level: This plays the central role in the system. This determines on line the production and maintenance scheduling and continuously feeds the routing control level which calculates route splits, and which in turn governs the sequence controller which determines scheduling times at which to dispatch parts.

Since the top two levels directly govern the rest, it is of paramount importance to develop and study an appropriate mathematical model which will facilitate to find on line implementable optimal feedback policies.

We first present a heuristic description of our model, which is a modified version of the model in [1], [17], [23]. The FMS consists of L workstations, with each workstation having a number L_m of identical machines ($m = 1, 2, \ldots, L$). A family of N types of different parts is produced. Let $u(t) = [u_1(t), \ldots, u_N(t)]^T \in \mathbb{R}^N$ and $d(t) = [d_1(t), \ldots, d_N(t)]^T \in \mathbb{R}^N$ denote the production rate (a control variable) and the downstream demand rate vectors of this family of parts, respectively. Also, $X(t) = [X_1(t), \ldots, X_N(t)]^T \in \mathbb{R}^N$ denotes the downstream buffer stock. A negative value of $X_j(t)$, $j = 1, \ldots, N$, indicates a backlogged demand for part j, while a positive value is the size of the inventory stored in the buffers. The evolution of $X(t)$ is governed by the following stochastic differential equations

$$\frac{dX(t)}{dt} = u(t) - d(t) + \text{diag}(\sigma_1, \ldots, \sigma_N)\xi(t) \tag{1.1}$$

where $\sigma_i > 0$, $i = 1, \ldots, N$ and $\xi(t) = [\xi_1(t), \ldots, \xi_N(t)]^T$ is an N-dimensional white noise which can be interpreted as "sales returns," "inventory spoilage," "sudden demand fluctuations," etc. (see [5]).

If $S_m(t)$ denotes the number of operational machines in station m at time t, then the state of the workstations may be represented by the L-tuple $S(t) = (S_1(t), \ldots, S_L(t))$. The evolution of $S(t)$ is influenced by the inventory size and production scheduling, and can also be controlled by various decisions such as produce, repair, replace, etc. The dynamics of $S(t)$ can be described as follows:

$$P\{S_m(t + \delta t) = \ell + 1 \mid S_m(t) = \ell\} = \begin{cases} (L_m - \ell)v_m(t)\,\delta t & \text{for } 0 \le \ell < L_m \\ 0 & \text{otherwise} \end{cases} \tag{1.2}$$

where $v_m(t)$, $m = 1, \ldots, L$ are suitable control variables. In the uncontrolled case, $v_m(t) = \gamma_m$, which represents the infinitesimal repair rate at station m. These repair rates may implicitly depend on $X(t)$. This model also allows for a control variable reflecting the decision as to whether to repair or replace on the basis of the inventory size. Also,

$$P\{S_m(t+\delta t) = \ell - 1 \mid S_m(t) = \ell\} = \begin{cases} \ell\, p_m\big(X(t), u(t)\big)\delta t & \text{for } 0 \le \ell < L_m \\ 0 & \text{otherwise} \end{cases} \tag{1.3}$$

where p_m models the infinitesimal failure rate at the mth station. Equations (1.2) and (1.3) imply that

$$P\{S_m(t+\delta t) = \ell_1 \mid S_m(t) = \ell_2\} = 0, \qquad \text{for } |\ell_1 - \ell_2| > 1.$$

With i and j denoting two states of the system, we define

$$\lambda_{ij}(\cdot)\delta t = P\{S(t+\delta t) = j \mid S(t) = i\}, \qquad i \ne j$$

and

$$\lambda_{ii}(\cdot) = -\sum_{j \ne i} \lambda_{ij}(\cdot).$$

The machine state $S(t)$ can thus be modeled as a continuous time controlled jump process taking values in a finite state space. In the uncontrolled case, $S(t)$ becomes a continuous time homogeneous Markov chain with the infinitesimal generator given by the matrix $[\lambda_{ij}]$.

The choice of the production rate at each instant is constrained by the capacity of the currently operational machines. This translates into the requirement that at each time t the production rates must lie in some set $\Gamma(S(t))$ which depends on the machine state.

Let $y_{mn}^k(t)$ be the number of type n parts which undergo operation k at the mth station per unit interval of time and $\tau_{mn}^k(t)$ be the length of time required for the completion of this operation. The product $y_{mn}^k(t)\tau_{mn}^k(t)$ is the portion of each unit time interval that one or more operational machines at station m must dedicate to perform operation k on type n parts, as dictated by the flow rate $y_{mn}^k(t)$. Since the amount of work completed at each station per unit time interval cannot exceed the time available at the operational machines, the following constraint applies

$$\sum_n \sum_k y_{mn}^k(t)\tau_{mn}^k(t) \le S_m(t), \qquad \text{for all } m. \tag{1.4}$$

Also, assuming that no material is allowed to accumulate within the system, the throughput $u_n(t)$ of type n parts must satisfy

$$u_n(t) = \sum_m y_{mn}^k(t), \qquad \text{for all } k \text{ and } n. \tag{1.5}$$

Therefore, for each state i the set $\Gamma(i)$ is defined as the collection of all production rates $u = [u_1, \ldots, u_N]^T$ for which, with the machine state $S(t) = i$, there exist feasible flow rates $y_{mn}^k(t)$ satisfying (1.4) and (1.5).

The flow control problem can now be stated. Given an initial buffer state $X(0) = x$ and machine state $S(0) = i$, we wish to specify a production plan and maintenance (repair/replacement) policy that minimizes the performance index

$$J(x, i, u, v) = E\left[\int_0^\infty e^{-\alpha t} c\big(X(t), S(t), u(t), v(t)\big)\, dt \ \Big| \ X(0) = x, S(0) = i\right] \qquad (1.6)$$

where $c(\cdot)$ is a 'cost' function, $\alpha > 0$ is a discount factor, $u(\cdot)$ is the production rate, and $v(\cdot)$ is the maintenance rate. The objective is to find $u(\cdot)$, $v(\cdot)$ for which the minimum is achieved in (1.6). The ideal production and maintenance policy for a wide class of cost functions would minimize J by producing parts at exactly the demand rate, thereby keeping the buffer at zero. Such a policy is generally impossible because of the failures of the machines and various other uncertainties.

This FMS model motivates the study of a stochastic optimization problem in a more abstract setting which will subsume the flow control problem in the FMS as a special case. This abstract problem is manifested in numerous other situations. In [14] it is encountered in a hybrid model proposed for the study of dynamic phenomena in large scale interconnected power networks. Sworder [31], [32] describes possible applications to macroeconomic models and dynamic renewal problems in general. In addition, it should be useful at other levels of the hierarchy described in [17]. We refer to [1], [2], [4–6], [11–17], [23–25], [27–32], [34–35] for related work.

Our paper is structured as follows. A rigorous description of the mathematical model of the FMS is given in Section 2. The optimization problem is formulated and subsequently reduced to an equivalent convex optimization problem, via the study of associated occupation measures. The proof of existence of optimal policies is established in Section 3. Section 4 deals with the characterization of optimal policies via dynamic programming equations. In Section 5, we apply our theory to a simplified model and derive some interesting results. Finally, Section 6 contains some concluding remarks. We have omitted proofs and other details at several points; [18] is a more detailed version of this paper.

2. MATHEMATICAL MODEL AND PRELIMINARIES

Let U be a compact metric space and $S = \{1, \ldots, M\}$. Let $\overline{b} = [\overline{b}_1, \ldots, \overline{b}_N]^T : I\!R^N \times S \times U \to I\!R^N$. For each $i \in S$, $\overline{b}(\cdot, i, \cdot)$ is assumed to be bounded, continuous and Lipschitz in its first argument uniformly with respect to the third. For $i, j \in S$, let $\overline{\lambda}_{ij} : I\!R^N \times U \to I\!R$ be bounded, continuous and Lipschitz in its first argument uniformly with respect to the second. Also, assume that for $i, j \in S$, $i \neq j$, $\overline{\lambda}_{ij} \geq 0$, and $\sum_{j=1}^M \lambda_{ij} = 0$ for any $i \in S$. Let $\sigma_i > 0$, $i = 1, 2, \ldots, N$ be prescribed numbers. For a Polish space Y, $\mathfrak{B}(Y)$ will denote its Borel σ-field and $\mathcal{P}(Y)$ the space of probability measures on $\mathfrak{B}(Y)$ endowed with the Prohorov topology, i.e. the topology of weak convergence [7]. Let $\mathfrak{M}(Y)$ be the set of all non-negative integer-valued, σ-finite measures on $\mathfrak{B}(Y)$. Let $\mathfrak{M}_\sigma(Y)$ be the smallest σ-field on $\mathfrak{M}(Y)$ with respect to which all maps from $\mathfrak{M}(Y)$ into $I\!N \cup \{\infty\}$ of the form $\mu \longmapsto \mu(B)$ with $B \in \mathfrak{B}(Y)$, are measurable. $\mathfrak{M}(Y)$ will always be assumed to be endowed with this measurability structure. Let $V = \mathcal{P}(U)$ and $b = [b_1, \ldots, b_N]^T : I\!R^N \times S \times V \to I\!R^N$ be defined by

$$b_i(\cdot, \cdot, v) := \int_U \overline{b}_i(\cdot, \cdot, u) \, v(du), \qquad v \in V, \quad i = 1, \ldots, N. \tag{2.1}$$

Similarly for $i, j \in S$, $\lambda_{ij} : I\!R^N \times V \to I\!R$ is defined as

$$\lambda_{ij}(\cdot, v) := \int_U \overline{\lambda}_{ij}(\cdot, u) \, v(du), \qquad v \in V, \quad i, j \in S. \tag{2.2}$$

For $i, j \in S$, $x \in I\!R^N$, $v \in V$, we construct the intervals $\Delta_{ij}(x, v)$ of the real line in the following manner (see also [10], [14]):

$$\Delta_{12}(x, v) = \big[0, \lambda_{12}(x, v)\big)$$
$$\Delta_{13}(x, v) = \big[\lambda_{12}(x, v), \lambda_{12}(x, v) + \lambda_{13}(x, v)\big)$$
$$\vdots$$
$$\Delta_{1M}(x, v) = \left[\sum_{j=2}^{M-1} \lambda_{1j}(x, v), \sum_{j=2}^{M} \lambda_{1j}(x, v)\right)$$
$$\Delta_{21}(x, v) = \left[\sum_{j=2}^{M} \lambda_{1j}(x, v), \sum_{j=2}^{M} \lambda_{1j}(x, v) + \lambda_{21}(x, v)\right)$$
$$\vdots$$
$$\Delta_{2M}(x, v) = \left[\sum_{j=2}^{M} \lambda_{1j}(x, v) + \sum_{\substack{j=1 \\ j \neq 2}}^{M-1} \lambda_{2j}(x, v), \sum_{j=2}^{M} \lambda_{1j}(x, v) + \sum_{\substack{j=1 \\ j \neq 2}}^{M} \lambda_{2j}(x, v)\right)$$

and so on. For fixed x and v, these are disjoint intervals, and the length of $\Delta_{ij}(x, v)$ is $\lambda_{ij}(x, v)$. Now define a function

$$h : I\!R^N \times S \times V \times I\!R \longrightarrow I\!R$$

by

$$h(x, i, v, z) = \begin{cases} j - i & \text{if } z \in \Delta_{ij}(x, v) \\ 0 & \text{otherwise.} \end{cases} \tag{2.3}$$

Let $\big(X(t), S(t)\big)$ be the $(I\!R^N \times S)$-valued *controlled switching diffusion* process given by the following stochastic differential equations

$$dX(t) = b\big(X(t), S(t), v(t)\big) \, dt + \text{diag}(\sigma_1, \ldots, \sigma_N) \, dW(t)$$
$$dS(t) = h\big(X(t), S(t-), v(t), z\big) \, \mathfrak{p}(dt, dz) \tag{2.4}$$

for $t \geq 0$ with $X(0) = X_0$ and $S(0) = S_0$ where

(i) X_0 is a prescribed $I\!\!R^N$-valued random variable.

(ii) S_0 is a prescribed S-valued random variable.

(iii) $W(\cdot) = [W_1(\cdot), \ldots, W_N(\cdot)]^T$ is an N-dimensional standard Wiener process independent of X_0, S_0.

(iv) $\mathfrak{p}(dt, dz)$ is an $\mathfrak{M}(I\!\!R_+ \times I\!\!R)$-valued Poisson random measure with intensity $dt \times m(dz)$, where m is the Lebesgue measure on $I\!\!R$.

(v) $\mathfrak{p}(\cdot, \cdot)$ and $W(\cdot)$ are independent.

(vi) $v(\cdot)$ is a $\mathcal{V}$-valued process with measurable sample paths satisfying the following non-anticipativity property. Let $\mathfrak{F}_t^v = \sigma\{v(s) : s \leq t\}$,

$$\mathfrak{F}_{[t,\infty)}^{W,\mathfrak{p}} = \sigma\{W(s) - W(t), \mathfrak{p}(A, B) : A \in \mathfrak{B}([s, \infty)), B \in \mathfrak{B}(I\!\!R), s \geq t\}.$$

Then $\mathfrak{F}_t^v$ and $\mathfrak{F}_{[t,\infty)}^{W,\mathfrak{p}}$ are independent.

Such a process $v(\cdot)$ will be called an *admissible* (control) *policy*. If $v(\cdot)$ is a Dirac measure, i.e. $v(\cdot) = \delta_{u(\cdot)}$, where $u(\cdot)$ is a U-valued process, then it is called an admissible *nonrandomized* policy. An admissible policy $v(\cdot)$ is called *feedback* if $v(\cdot)$ is progressively measurable with respect to the natural filtration of $(X(\cdot), S(\cdot))$. A particular subclass of feedback policies is of special interest. A feedback policy $v(\cdot)$ is called a (non-homogeneous) *Markov* policy if $v(\cdot) = \tilde{v}(\cdot, X(\cdot), S(\cdot))$ for a measurable map $\tilde{v} : I\!\!R_+ \times I\!\!R^N \times S \to \mathcal{V}$. With an abuse of notation the map $\tilde{v}$ itself is called a Markov policy. If $\tilde{v}$ has no explicit time dependence, it is called a *homogeneous* Markov policy. Thus a homogeneous Markov nonrandomized policy can be identified with a measurable map $v : I\!\!R^N \times S \to U$.

If $(W(\cdot), \mathfrak{p}(\cdot, \cdot), X_0, S_0, v(\cdot))$ satisfying the above are given on a prescribed probability space $(\Omega, \mathfrak{F}, P)$, then under our assumptions on b and λ, equation (2.4) will admit an a.s. unique strong solution [20, Chap. 3], [22, Chap. 3, Sect. 2c] and $X(\cdot) \in C(I\!\!R_+; I\!\!R^N)$, $S(\cdot) \in D(I\!\!R_+; S)$. However, if $v(\cdot)$ is a feedback policy, then there exists a measurable map

$$f : I\!\!R_+ \times C(I\!\!R_+; I\!\!R^N) \times D(I\!\!R_+; S) \longrightarrow \mathcal{V}$$

such that for each $t \geq 0$, $v(t) = f(t, X(\cdot), S(\cdot))$ and is measurable with respect to the σ-field generated by $\{X(s), S(s) : s \leq t\}$. Thus $v(\cdot)$ cannot be specified a priori in (2.4). Instead, one has to replace $v(t)$ in (2.4) by $f(t, X(\cdot), S(\cdot))$ and (2.4) takes the form

$$dX(t) = b(X(t), S(t), f(t, X(\cdot), S(\cdot))) \, dt + \operatorname{diag}(\sigma_1, \ldots, \sigma_N) \, dW(t) \qquad (2.5a)$$

$$dS(t) = h(X(t), S(t-), f(t, X(\cdot), S(\cdot)), z) \, \mathfrak{p}(dt, dz) \qquad (2.5b)$$

for $t \geq 0$ with $X(0) = X_0$ and $S(0) = S_0$. In general, (2.5) will not even admit a weak solution. However, if the feedback policy is a Markov policy, then one can establish the existence of a unique strong solution [34]. We now introduce some notation which will be used throughout. Define

$$L^1(\mathbb{R}^N \times S) = \{f : \mathbb{R}^N \times S \longrightarrow \mathbb{R} : \text{ for each } i \in S, f(\cdot, i) \in L^1(\mathbb{R}^N)\}.$$

$L^1(\mathbb{R}^N \times S)$ is endowed with the product topology of $(L^1(\mathbb{R}^N))^M$. Similarly, we define $C_0^\infty(\mathbb{R}^N \times S)$, $W_{loc}^{2,p}(\mathbb{R}^N \times S)$, etc. For $f \in W_{loc}^{2,p}(\mathbb{R}^N \times S)$, $u \in U$ we write

$$L^u f(x, i) = L_i^u f(x, i) + \sum_{j=1}^M \overline{\lambda}_{ij}(x, u)[f(x, j) - f(x, i)] \tag{2.6}$$

where

$$L_i^u f(x, i) = \frac{1}{2} \sum_{j=1}^N \sigma_j^2 \frac{\partial^2 f(x, i)}{\partial x_j^2} + \sum_{j=1}^N \overline{m}_j(x, i, u) \frac{\partial f(x, i)}{\partial x_j} \tag{2.7}$$

and more generally, for $v \in \mathcal{V}$,

$$L^v f(x, i) = \int_U L^u f(x, i) \, v(du). \tag{2.8}$$

Theorem 2.1. *Under a Markov policy v, (2.4) admits an a.s. unique strong solution such that $(X(\cdot), S(\cdot))$ is a Feller process with extended generator L^v.*

The Optimization Problem.

Let $\overline{c} : \mathbb{R}^N \times S \times U \to \mathbb{R}_+$ be a bounded, continuous cost function, and let $c : \mathbb{R}^N \times S \times \mathcal{V} \to \mathbb{R}_+$ be defined as

$$c(\cdot, \cdot, v) = \int_U \overline{c}(\cdot, \cdot, u) \, v(du).$$

Let $\alpha > 0$ be a prescribed discount factor. Let $v(\cdot)$ be an admissible policy and $(X(\cdot), S(\cdot))$ the corresponding process. Then the total α-discounted cost under $v(\cdot)$ is defined as

$$J_v(x, i) := E\left[\int_0^\infty e^{-\alpha t} c(X(t), S(t), v(t)) \, dt \,\Big|\, X(0) = x, S(0) = i\right]. \tag{2.9}$$

If the laws of X_0, S_0 are $\pi \in \mathcal{P}(\mathbb{R}^N)$, $\xi \in \mathcal{P}(S)$ respectively, then

$$J_v(\pi, \xi) = \sum_i \int_{R^N} J_v(x, i) \, \pi(dx) \, \xi(i). \tag{2.10}$$

Let

$$V(x, i) := \inf_{v(\cdot)} \{J_v(x, i)\}, \tag{2.11}$$

$$V(\pi, \xi) := \inf_{v(\cdot)} \{J_v(\pi, \xi)\}. \tag{2.12}$$

The function $V(x,i)$ is called the (α-discounted) value function. An admissible policy $v(\cdot)$ satisfying

$$J_v(\pi,\xi) = V(\pi,\xi)$$

is called an optimal policy for the initial law (π,ξ). An admissible policy is called optimal if it is optimal for any initial law. Our aim is to find an admissible optimal policy which is homogeneous Markov and nonrandomized.

We now introduce (discounted) occupation measures [9]. Let $v(\cdot)$ be an admissible policy and $(X(\cdot), S(\cdot))$ the corresponding process with initial law (π,ξ). Define the occupation measure $\nu[\pi,\xi;v] \in \mathcal{P}(\mathbb{R}^N \times S \times U)$ by

$$\int f \, d\nu[\pi,\xi;v] = \alpha E\left[\int_0^\infty e^{-\alpha t} \int_U f(X(t), S(t), u)\, v(t)(du)\, dt\right] \qquad (2.13)$$

for $f \in C_b(\mathbb{R}^N \times S \times U)$. Let

$$M_1[\pi,\xi] = \{\nu[\pi,\xi;v] : v(\cdot) \text{ is admissible}\} \qquad (2.14)$$

$$M_2[\pi,\xi] = \{\nu[\pi,\xi;v] : v(\cdot) \text{ is homogeneous Markov}\} \qquad (2.15)$$

$$M_3[\pi,\xi] = \{\nu[\pi,\xi;v] : v(\cdot) \text{ is homogeneous nonrandomized Markov}\} \qquad (2.16)$$

In terms of these occupation measures

$$J_v(\pi,\xi) = \alpha^{-1} \int \bar{c}\, d\nu[\pi,\xi;v]. \qquad (2.17)$$

We now state the following important result.

Theorem 2.2. $M_1[\pi,\xi] = M_2[\pi,\xi]$ and $M_2[\pi,\xi]$ is compact, convex and $M_2^e[\pi,\xi] \subset M_3[\pi,\xi]$, where $M_2^e[\pi,\xi]$ is the set of extreme points of $M_2[\pi,\xi]$.

Remark 2.1. For a fixed initial law, the optimization problem (2.9) will reduce to a convex optimization problem in view of (2.17).

3. EXISTENCE OF AN OPTIMAL POLICY

Theorem 3.1. *There exists a homogeneous Markov optimal policy.*

Proof. Let $(\pi,\xi) \in \mathcal{P}(\mathbb{R}^N) \times \mathcal{P}(S)$ such that $\text{supp}(\pi) = \mathbb{R}^N$ and $\text{supp}(\xi) = S$. Since $\bar{c}$ is bounded and continuous the map $M_2[\pi,\xi] \ni \nu \longmapsto \int \bar{c}\, d\nu$ is continuous. Thus there exists a homogeneous Markov policy v^* such that

$$J_{v^*}(\pi,\xi) = \min_v \{J_v(\pi,\xi) : v \text{ is homogeneous Markov}\}.$$

By Theorem 2.2, it follows that

$$J_{v^*}(\pi, \xi) = V(\pi, \xi).$$

Therefore v^* is optimal for the initial law (π, ξ). We will show that v^* is optimal for any initial law. It suffices to show that v^* is optimal for any initial condition $(x, i) \in \mathbb{R}^N \times S$. Suppose there exist $(x_0, i_0) \in \mathbb{R}^N \times S$ and a homogeneous Markov policy v such that

$$J_v(x_0, i_0) < J_{v^*}(x_0, i_0). \tag{3.1}$$

Using the fact that the solution of (2.4) under a Markov policy is a Feller process, it can be easily shown that the function $J_v(x, i)$ is continuous in x for each v. Thus (3.1) holds in a neighborhood B of x_0. Define a policy v' as follows

$$v'(t) = v^*(X(t), S(t)) I\{X_0 \notin B\} + v'(X(t), S(t)) I\{X_0 \in B\}$$

where $(X(\cdot), S(\cdot))$ is governed by $v'(\cdot)$. Then it is easily shown that

$$J_{v'}(\pi, \xi) < J_{v^*}(\pi, \xi)$$

which is a contradiction. Thus v^* is optimal. $\square$

Theorem 3.2. *There exists a homogeneous Markov nonrandomized optimal policy.*

Proof. Let v^* be as in Theorem 3.1. Let $M_2^e[\pi, \xi]$ be the set of extreme points of $M_2[\pi, \xi]$. Since $M_2[\pi, \xi]$ is compact, by Choquet's theorem [26], $\nu[\pi, \xi; v^*]$ is the barycenter of a probability measure m supported on $M_2^e[\pi, \xi]$. Therefore,

$$\int \bar{c} \, d\nu[\pi, \xi; v^*] = \int_{M_2^e[\pi, \xi]} \left(\int \bar{c} \, d\mu \right) m(d\mu). \tag{3.2}$$

Since v^* is optimal, it follows from (3.2) that there exists a $\nu[\pi, \xi; v] \in M_2^e[\pi, \xi]$ such that

$$\int \bar{c} \, d\nu[\pi, \xi; v^*] = \int \bar{c} \, d\nu[\pi, \xi; v].$$

Thus v is also optimal. By Theorem 2.2 it is nonrandomized. $\square$

4. DYNAMIC PROGRAMMING EQUATIONS

Using the existence results of the previous section, we will now derive the dynamic programming or Hamilton-Jacobi-Bellman equations (HJB) which in our case will be a weakly coupled system of quasilinear elliptic equations, and then characterize the optimal policy as a minimizing selector of an appropriate "Hamiltonian". The HJB equations for our problem are

$$\alpha\psi(x, i) = \inf_{u \in U} \{ L^u \psi(x, i) + \bar{c}(x, i, u) \}. \tag{4.1}$$

Theorem 4.1. *The value function $V(x,i)$ is the unique solution of (4.1) in the space* $W_{loc}^{2,p}(I\!\!R^N \times S) \cap C_b(I\!\!R^N \times S)$ *for any $2 \leq p < \infty$.*

Proof. We have already seen in the proof of Theorem 3.1 that $V(x,i) \in C_b(I\!\!R^N \times S)$. Let v^* be a homogeneous Markov nonrandomized optimal policy and $(X(\cdot), S(\cdot))$ the corresponding solution of (2.4). Then for $(x,i) \in I\!\!R^N \times S$

$$V(x,i) = E\left[\int_0^\infty e^{-\alpha t}\overline{c}(X(t), S(t), v^*(X(t), S(t)))\, dt \,\Big|\, X(0) = x, S(0) = i\right]. \qquad (4.2)$$

By standard arguments [8], $V(x,i)$ is the unique solution in $W_{loc}^{2,p}(I\!\!R^N \times S) \cap C_b(I\!\!R^N \times S)$ for any $2 \leq p < \infty$ of

$$\alpha V(x,i) = L^{v^*(x,i)}V(x,i) + \overline{c}(x,i,v^*(x,i)). \qquad (4.3)$$

Suppose there exist $x_0 \in I\!\!R^N$, $i_0 \in S$, $u \in U$ and $\delta > 0$ such that

$$\alpha V(x_0, i_0) > L^u V(x_0, i_0) + \overline{c}(x_0, i_0, u) + \delta.$$

Then by the continuity of $V(\cdot, i_0)$ the above will hold in a neighborhood $N(x_0)$ of x_0. Define a homogeneous Markov nonrandomized policy $\bar{v}$ as follows:

$$\bar{v}(x,i) = \begin{cases} v^*(x,i) & \text{if } (x,i) \notin N(x_0) \times S \\ u & \text{if } (x,i) \in N(x_0) \times S. \end{cases}$$

Then

$$\alpha V(x, i_0) > L^{\bar{v}(x,i_0)}V(x, i_0) + \overline{c}(x, i_0, \bar{v}(x, i_0)) + \delta I\{x \in N(x_0)\}.$$

Now it is easily seen that

$$V(x, i_0) \geq J_{\bar{v}}(x, i_0) + \delta'$$

for some $\delta' > 0$, which is a contradiction. Hence $V(x,i)$ satisfies (4.1). Let V' be another solution of (4.1) in the desired class. Then it can be shown using standard arguments (cf. [8, Thm. III.2.4, pp. 69–70]) that

$$|V(x,i) - V'(x,i)| \leq 2Ke^{-\alpha t}$$

where $K > 0$ is a constant. Letting $t \to \infty$, $V \equiv V'$. $\square$

Corollary 4.1. *Assume that for each $i \in S$, $\overline{c}(\cdot, i, \cdot)$ is Lipschitz in its first argument uniformly with respect to the third. Then $V(x,i)$ is the unique solution of (4.1) in $C^2(I\!\!R^N \times S) \cap C_b(I\!\!R^N \times S)$.*

Proof. It suffices to show that V is C^2. Since $V(x,i) \in W_{loc}^{2,p}(I\!\!R^N \times S)$ for any $2 \leq p < \infty$, by Sobolev's imbedding theorem $V(x,i) \in C^{1,\gamma}(I\!\!R^N \times S)$ for $0 < \gamma < 1$, γ arbitrarily close to 1, and hence by our assumptions on $\overline{b}$, $\overline{\lambda}$, $\overline{c}$, it is easy to see that

$$\alpha V(x,i) - \inf_{u \in U}\left\{\sum_{j=1}^N \overline{m}_j(x,i,u)\frac{\partial V(x,i)}{\partial x_j} + \sum_{j=1}^M \overline{\lambda}_{ij}(x,u)(V(x,j) - V(x,i)) + \overline{c}(x,i,u)\right\}$$

is in $C^{0,\gamma}$. By elliptic regularity [21, p. 287] applied to (4.1) (V replacing ψ), we conclude that $V \in C^{2,\gamma}$. $\square$

Theorem 4.2. *A homogeneous Markov nonrandomized policy v is optimal if and only if*

$$\sum_{j=1}^{N} \overline{m}_j\big(x, i, v(x,i)\big) \frac{\partial V(x,i)}{\partial x_j} + \sum_{k=1}^{M} \overline{\lambda}_{ik}\big(x, v(x,i)\big)\big(V(x,k) - V(x,i)\big) + \overline{c}\big(x, i, v(x,i)\big)$$

$$= \inf_{u \in U} \Bigg\{ \sum_{j=1}^{N} \overline{m}_j(x, i, u) \frac{\partial V(x,i)}{\partial x_j} + \sum_{k=1}^{M} \overline{\lambda}_{ik}(x, u)\big(V(x,k) - V(x,i)\big)$$

$$+ \overline{c}(x, i, u) \Bigg\} \quad \text{a.e. } x \in \mathbb{R}^N, i \in \mathcal{S}. \quad (4.4)$$

Proof. The 'necessity' part is contained in the proof of Theorem 4.1. We establish the sufficiency. Let $v(\cdot, \cdot)$ satisfy (4.4). The existence of such a v is guaranteed by a standard measurable selection theorem [3, Lemma 1]. Let v' be any other homogeneous Markov nonrandomized policy. Then using standard arguments involving Ito's formula and the strong Markov property, it can be shown that

$$J_v(x, i) \leq J_{v'}(x, i)$$

a.e. $x \in \mathbb{R}^N$, $i \in \mathcal{S}$. Hence by Theorem 2.2,

$$J_v(x, i) \leq J_{\overline{v}}(x, i)$$

for any admissible policy $\overline{v}$. Thus v is optimal. $\square$

Remark 4.1. Thus far, we have assumed that the cost function $\overline{c}$ is bounded. However, this condition can be relaxed.

5. AN APPLICATION TO A SIMPLIFIED MODEL

We consider a modified version of the model studied in [2]. Suppose there is one machine producing a single commodity. Suppose that the demand rate is a constant $d > 0$. Let the machine state $S(t)$ take values in $\{0, 1\}$, $S(t) = 0$ or 1 according as the machine is down or functional. Let $S(t)$ be a continuous time Markov chain with generator

$$\begin{bmatrix} -\lambda_0 & \lambda_0 \\ \lambda_1 & -\lambda_1 \end{bmatrix}.$$

The inventory $X(t)$ is governed by the Ito equation

$$dX(t) = (u(t) - d)\, dt + \sigma\, dW(t) \quad (5.1)$$

where $\sigma > 0$. The production rate $u(t)$ is constrained by

$$\begin{cases} u(t) = 0 & \text{if } S(t) = 0 \\ u(t) \in [0, R] & \text{if } S(t) = 1. \end{cases}$$

Let $c : \mathbb{R} \to \mathbb{R}_+$ be the cost function which is assumed to be convex and Lipschitz continuous. Let $\alpha > 0$ be the discount factor and let the value function be denoted by $V(x, i)$. In this case $V(x, i)$ is the minimal non-negative C^2 solution of the HJB equation

$$
\begin{pmatrix} \frac{\sigma^2}{2} V''(x, 0) - dV'(x, 0) \\ \frac{\sigma^2}{2} V''(x, 1) - \min_{u \in [0, R]} \{(u - d)V'(x, 1)\} \end{pmatrix}
$$
$$
= \begin{bmatrix} \lambda_0 + \alpha & -\lambda_0 \\ \lambda_1 & \alpha - \lambda_1 \end{bmatrix} \begin{pmatrix} V(x, 0) \\ V(x, 1) \end{pmatrix} - \begin{pmatrix} 1 \\ 1 \end{pmatrix} c(x). \quad (5.2)
$$

Using the convexity of $c(\cdot)$ it can be shown as in [2] that $V(\cdot, i)$ is convex for each i. Hence there exists an x^* such that

$$
\begin{aligned} V'(x, 1) &\leq 0 \quad \text{for } x \leq x^* \\ &\geq 0 \quad \text{for } x \geq x^*. \end{aligned} \quad (5.3)
$$

From (5.2), it follows that the value of u which minimizes $(u - d)V'(x, 1)$ is

$$
u = \begin{cases} R & \text{if } x \leq x^* \\ 0 & \text{if } x \geq x^*. \end{cases}
$$

At $x = x^*$, $V'(x^*, 1) = 0$ and therefore any $u \in [0, R]$ minimizes $(u - d)V'(x, 1)$. Thus, in view of Theorem 4.2, we can choose any $u \in [0, R]$ at $x = x^*$. To be specific, we choose $u = d$ at $x = x^*$. It follows that the following homogeneous Markov nonrandomized policy is optimal

$$
v(x, 0) \equiv 0
$$
$$
v(x, 1) = \begin{cases} R & \text{if } x < x^* \\ d & \text{if } x = x^* \\ 0 & \text{if } x > x^*. \end{cases} \quad (5.4)
$$

We note at this point that the piecewise deterministic model, in general, would lead to a singular control problem when $V'(x, 1) = 0$ [2], [23]. In [2] Akella and Kumar have obtained the solution of the HJB equation (this would be (5.2) without the second order term) in closed form and computed an explicit expression for x^*. They have shown that a policy of the type (5.4) is optimal among all homogeneous Markov nonrandomized policies. In our case the additive noise in (5.1) induces a smoothing effect to remove the singular situation; in addition, our results imply that the policy (5.4) is optimal among *all admissible policies*. The only limitation of our model is that it would, in general, be very difficult to solve (5.2) analytically. Therefore, one must rely on numerical methods to compute our optimal policy of the type (5.4).

We now discuss the manufacturing model studied in [23] as described in the introduction. The machine state $S(t)$ is again a prescribed continuous time Markov chain taking values in $S = \{1, \ldots, M\}$. For each $i \in S$, the production rate $u = (u_1, \ldots, u_N)$ takes values in U_i which is a convex polyhedron in $I\!\!R^N$. The demand rate is $d = [d_1, \ldots, d_N]^T$. In this case, if the cost function $c : I\!\!R^N \to I\!\!R_+$ is Lipschitz continuous and convex, then it can be shown that for each $i \in S$, the value function $V(\cdot, i)$ is convex. But from this fact alone optimal policies of the type (5.4) cannot be obtained. However, since an optimal homogeneous Markov nonrandomized policy $v(x, i)$ is determined by minimizing

$$\sum_{j=1}^{N} (u_j - d_j) \frac{\partial V(x, i)}{\partial x_j}$$

over U_i, $v(x, i)$ takes values at extreme points of U_i. Thus for each machine state i, an optimal policy divides the buffer state space into a set of regions in which the production rate is constant. If the gradient $\nabla V(x, i)$ is zero or orthogonal to a face of U_i, a unique minimizing value does not exist. But again in view of Theorem 4.2 we may prescribe arbitrary production rates at those points where $\nabla V(x, i) = 0$, and if $\nabla V(x, i)$ is orthogonal to a face of U_i, we can choose any corner of that face. Hence once again we can circumvent the singular situation.

6. CONCLUDING REMARKS

We have analyzed the optimal control of switching diffusions with a discounted criterion on the infinite horizon. The model allows a very general form of coupling between the continuous and the discrete components of the process. We have shown that there exists a homogeneous, nonrandomized Markov policy which is optimal in the class of all admissible policies. Also, the existence of a unique solution in a certain class to the associated Hamilton-Jacobi-Bellman equations is established and the optimal policy is characterized as a minimizing selector of an appropriate Hamiltonian.

The primary motivation for this study is a class of control problems encountered in flexible manufacturing systems. By explicitly taking into account the noise present in the dynamics, we are able to remove singularities arising in the noiseless situation. In addition, we show that hedging type policies are optimal in a much wider class of non-anticipative policies than previously considered. We have confined our attention to the flow control level only. However, our results can be used to study control problems at other levels in hierarchical manufacturing systems [17], as well as control problems in other hybrid systems (see, e.g., [14], [30], [31]).

Here we have studied only the discounted criterion. Following [9], we can obtain similar results for the finite horizon and exit time criteria. However, the long-run average cost problem is more difficult and is currently under preparation [19]. To treat this problem, it is clear that one needs to study the ergodic behavior of the hybrid process $(X(t), S(t))$, which turns out to be quite involved. Even if each generator L_i gives rise to a positive recurrent diffusion and the parametrized Markov chain is ergodic, the switching diffusion can be anything from transient to positive recurrent. Under certain conditions, we have obtained some characterizations of ergodic switching diffusions. Using these results, we have established the existence of a homogeneous Markov nonrandomized policy which is almost surely optimal for the pathwise long-run average cost problem. This work will be reported elsewhere [19].

REFERENCES

[1] R. Akella, Y. Choong and S.B. Gershwin, *Performance of hierarchical production scheduling policy*, IEEE Trans. on Components, Hybrids and Manufacturing Technology **CHMT-7** (1984), 225–240.

[2] R. Akella and P.R. Kumar, *Optimal control of production rate in a failure prone manufacturing system*, IEEE Trans. Autom. Contr. **AC-31** (1986), 116–126.

[3] V.E. Beneš, *Existence of optimal strategies based on specified information for a class of stochastic decision problems*, SIAM J. Control **8** (1970), 179–188.

[4] A. Bensoussan and J.L. Lions, *Impulse Control and Quasi-Variational Inequalities*, Gauthier-Villars, 1984.

[5] A. Bensoussan, S.P. Sethi, R. Vickson and N. Derzko, *Stochastic production planning with production constraints*, SIAM J. Control Opt. **22** (1984), 920–935.

[6] T. Bielecki and P.R. Kumar, *Optimality of zero-inventory policies for unreliable manufacturing systems*, Oper. Res. **36** (1988), 532–546.

[7] P. Billingsley, *Convergence of Probability Measures*, Wiley, New York, 1968.

[8] V.S. Borkar, *Optimal Control of Diffusion Processes*, Pitman Research Notes in Math. Series 203, Longman, Harlow, 1989.

[9] V.S. Borkar and M.K. Ghosh, *Controlled diffusions with constraints*, J. Math. Anal. Appl. **152** (1990), 88–108.

[10] R.W. Brockett and G.L. Blankenship, *A representation theorem for linear differential equations with Markovian coefficients*, Proc. of 1977 Allerton Conference on Circuits and Systems Theory, Urbana, Illinois, 1977.

[11] E.K. Boukas and A. Haurie, *Optimality conditions for continuous time systems with controlled jump Markov disturbances: application to an FMS planning problem*, Analysis and Optimization of Systems (A. Bensoussan and J.L. Lions, eds.), Lecture Notes on Control and Information Sciences, vol. III, Springer-Verlag, 1988, pp. 633–676.

[12] ———, *Manufacturing flow control and preventive maintenance: a stochastic control approach*, IEEE Trans. Autom. Control **AC-35** (1990), 1024–1031.

[13] E.-K. Boukas, A. Haurie and P. Michel, *An optimal control problem with a random stopping time*, JOTA **64** (1990), 471–480.

[14] D.A. Castanon, M. Coderch, B.C. Levy, and A.S. Willsky, *Asymptotic analysis, approximation, and aggregation methods for stochastic hybrid systems*, Proc. 1980 Joint Automatic Control Conference, San Francisco, CA, 1980.

[15] M.H.A. Davis, *Stochastic control and nonlinear filtering*, Tata Institute of Fundamental Research, Bombay, 1984.

[16] W.H. Fleming, S.P. Sethi and H.M. Soner, *An optimal stochastic production planning with randomly fluctuating demand*, SIAM J. Control Opt. **25** (1987), 1494–1502.

[17] S.B. Gershwin, *Hierarchical flow control: a framework for scheduling and planning discrete events in manufacturing systems*, Proc. IEEE **77** (1989), 195–209.

[18] M.K. Ghosh, A. Arapostathis and S.I. Marcus, *Optimal control of switching diffusions with application to flexible manufacturing systems*, preprint.

[19] M.K. Ghosh, A. Arapostathis and S.I. Marcus, *Ergodic control of switching diffusions*, under preparation.

[20] I.I. Gihman and A.V. Skorohod, *Controlled Stochastic Processes*, Springer-Verlag, New York, 1979.

[21] P. Grisvard, *Elliptic Problems in Non-Smooth Domains*, Pitman, Boston, 1965.

[22] J. Jacod and A.N. Shiryayev, *Limit Theorems for Stochastic Processes*, Springer-Verlag, New York, 1980.

[23] J. Kimenia and S.B. Gershwin, *An algorithm for the computer control of a flexible manufacturing system*, IIE Trans. 15 (1983), 353–362.

[24] J. Lehoczky, S. Sethi, H.M. Soner and M. Taksar, *An asymptotic analysis of hierarchical control of manufacturing systems under uncertainty*, Math. O.R. 16 (1991), 596–608.

[25] G.J. Oldser and R. Suri, *Time optimal control of parts-routing in a manufacturing system with failure prone machines*, Proc. 19th IEEE Conference on Decision and Control, Albuquerque, New Mexico, 1980, pp. 722–727.

[26] R. Phelps, *Lectures on Choquet's Theorem*, Van Nostrand, New York, 1966.

[27] R. Rishel, *Dynamic programming and minimum principles for systems with jump Markov disturbances*, SIAM J. Control 13 (1975), 338–371.

[28] S. Sethi and M.I. Taskar, *Deterministic equivalent for a continuous time linear-convex stochastic control problem*, JOTA 64 (1990), 169–181.

[29] A. Sharifnia, *Production control of a manufacturing system with multiple machine states*, IEEE Trans. Autom. Control AC-33 (1988), 620–625.

[30] D.D. Sworder, *Feedback control of a class of linear systems with jump parameters*, IEEE Trans. Autom. Control AC-14 (1969), 9–14.

[31] _____, *Control of systems subject to sudden change in character*, Proc. IEEE 64 (1976), 1219–1225.

[32] _____, *Control of a linear system with non-Markovian modal changes*, J. Economic Dynamics and Control 2 (1980), 233–240.

[33] A. Ju. Veretennikov, *On strong solutions of Ito stochastic equations with jumps*, Theory of Probability and Its Applications 32 (1988), 148–152.

[34] D. Vermes, *Optimal control of piecewise deterministic Markov processes*, Stochastics 14 (1985), 165–208.

[35] W.M. Wonham, *Random differential equations in control theory*, Probabilistic Methods in Applied Mathematics (A.T. Bharucha-Reid, ed.), vol. 2, Academic Press, New York, 1970, pp. 131–212.

Stochastic Stability Analysis
of Nonlinear Gated Radar Range Trackers

Robert E. Gover,[†] Eyad H. Abed[‡] and Allen J. Goldberg[†]

[†]*Advanced Techniques Branch*
Tactical Electronic Warfare Division, Code 5753
Naval Research Laboratory
Washington, D.C. 20375 USA

[‡]*Department of Electrical Engineering*
and the Systems Research Center
University of Maryland
College Park, MD 20742 USA

ABSTRACT

This study addresses a nonlinear system of differential equations with colored noise parameters which models the behavior of a gated radar range tracking system in the presence of two competing targets. Recently [1], the authors introduced nonlinear dynamic models for a class of gated radar range trackers with automatic gain control (AGC). The set-up of [1] is general in that the number and nature of the targets, as well as the weighting patterns of the electronic tracking system gates, are not prespecified. These models were employed to study track point stability in a multi-target environment. The ability of a gated range tracker to resolve closely spaced targets was also studied using these models. In this paper, the basic model developed in [1] is applied to the study of target resolution assuming randomly fluctuating targets. Here we consider the case in which two randomly fluctuating targets are present in the range gate, one fast and one slow (relative to the tracker bandwidth). We obtain a Markovian model by augmenting the state space to include the colored noise. Numerical computation of the evolution of the probability density is employed to study sensitivity to certain model parameters, especially noise correlation time, intensity of the random fluctuations, target separation and separation rate and the AGC bandwidth. These numerical results are used to motivate several challenging analytical problems for this nonlinear stochastic system, for which we present some preliminary results.

1   Introduction

This paper describes research being conducted at the Naval Research Laboratory (NRL) toward the solution of a set of nonlinear differential equations exhibiting bifurcations and driven by colored stochastic parameters. The problem is drawn from a practical application in Electronic Warfare (EW), but also bears similarities to equations being studied by physicists in the fields of optical bi-stability, nematic crystals, dye lasers, superfluid helium and quantum mechanics. NRL is pursuing deeper insights into these problems that are achieved by modern analytic methods beyond those provided by traditional digital simulation approaches. The authors believe the nonlinear stochastic control community will find challenges in one or more aspects of these problems.

The EW problem motivation is as follows. At some point in its attack, an anti-ship cruise missile (ASCM) limits its view of the world environment to the confines of a three dimensional cell. (See Figure 1.) The cell is used to guide the missile toward a particular

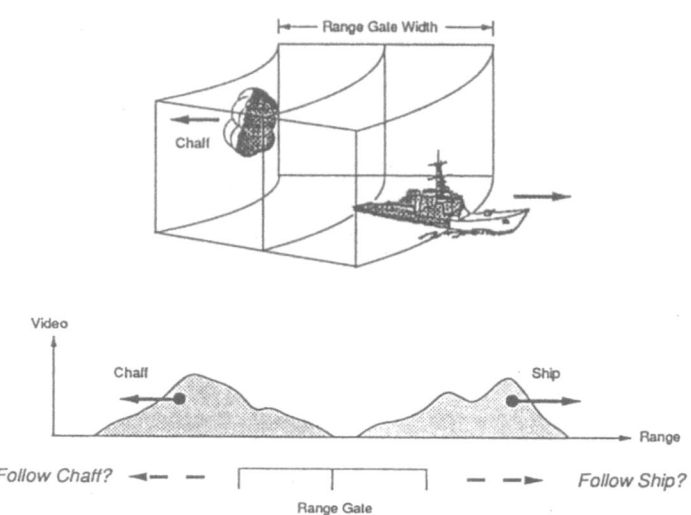

Figure 1: Range / Angle Cell

quarry and reject other unwanted objects. Feedback tracker loops are used to continuously re-position this cell in order to follow a moving target of interest. In an EW situation, alternative or false targets are deliberately introduced to compromise the tracker's intended purpose. An example of this is depicted in Figure 1, where a chaff cloud is in competition with a ship for the attention of the tracker's cell. As the chaff cloud separates from the ship beyond the dimensions of the cell, the feedback-positioned cell no longer can subtend both targets, and nonlinear dynamics forces the tracker to commit to following just one of them. The likelihood of committing to chaff rather than ship is an important measure of the value of the former as an EW countermeasure. The pathwise response of the feedback loop, though of some interest, is not as important as the notion of a region of attraction associated with a target and the time evolution of the probabilities of the tracker remaining in or leaving these regions.

In what follows, the cell trackers are formulated in one dimension only, namely relative range between threat missile and target(s). In [1], rigorous equations were developed for modeling the range tracker, which can be summarized via a block diagram. See Figure 2. The range tracker consists of two coupled nonlinear feedback loops, one for automatic gain control (AGC), and one for positioning the cell in range. The loop is driven by information $E(t)$ supplied by the radar about the external world. The signal $E(t)$, the so-called complex video envelope, can be meaningfully organized as a raster line-by-line scan representation $E(\sigma, k)$, where the line counter k indicates radar pulse repetition intervals (PRI's), and the continuous range variable σ keeps track of the relative range in units of time delay. (See Figure 3. Reference [1] may be consulted for the details of the raster representation and the relationship between $E(t)$ and $E(\sigma, k)$.)

2 Nonlinear Range Tracker Model

Referring again to Figure 2, the AGC gain $g[k]$ is first applied to the input signal $E(\sigma, k)$ and the combined signal is then presented to both the AGC and the range tracker. Tracing through the loops, we begin with the AGC. The AGC-adjusted signal $g[k] E(\sigma, k)$ is applied to the sum operator W_S and subsequently sampled at the end of each PRI. In Figure 2, we denote the sampling operation as ZOH to indicate a zero-order hold. The sampled output $S[k]$ is compared to a threshold to create an error which is then integrated with a time constant of T_{AGC}. The integrator output is sampled at the end of each PRI resulting in the state variable $\eta[k]$, which is the (sampled) AGC voltage signal. This is applied to the exponential function to produce the AGC gain g for the next PRI. The input to the range tracker loop is the AGC-adjusted signal $g[k] E(\sigma, k)$ which is applied to the difference operator W_D and sampled. The sampled output $D[k]$ is integrated with a gain K. The integrator output $\rho(\sigma, k)$ is then compared to a reference ramp function $\psi(\sigma) = \sigma$ that begins at zero on each transmit pulse to convert $\rho(\sigma, k)$ to the time delay relative to the instant of pulse transmission. The resulting time delay or range estimate $\rho[k]$ is used to position the weighting patterns in both of the operators W_D and W_S. The sampled outputs of the operators W_D and W_S, $D[k]$ and $S[k]$ are given by

$$D[k] = \int_{R_D} g[k]^2 |E(\sigma, k)|^2 w_D(\sigma - \rho) d\sigma, \text{ and} \tag{1}$$

$$S[k] = \int_{R_S} g[k]^2 |E(\sigma, k)|^2 w_S(\sigma - \rho) d\sigma, \tag{2}$$

respectively. In the above $|E(\sigma, k)|^2$ is the square envelope of the radar return video for PRI k; w_D is the weighting pattern of the range error comparator or range gate and is typically an odd function of compact support R_D; w_S is the weighting pattern of the AGC comparator and is typically an even function of compact support R_S; and for each k the integrations are performed over the compact support of the respective weighting patterns.

The dynamics of the tracker of Figure 2 are cast in state space form as

$$\rho[k+1] = \rho[k] + KT_{\text{PRI}} e^{-2\eta_d[k]} \int_{R_D} |E(\sigma, k-1)|^2 w_D(\sigma - \rho_d[k]) d\sigma, \tag{3}$$

$$\rho_d[k+1] = \rho[k], \tag{4}$$

$$\eta[k+1] = \eta[k] + \frac{T_{\text{PRI}}}{T_{\text{AGC}}} (e^{-2\eta_d[k]} \int_{R_S} |E(\sigma, k-1)|^2 w_S(\sigma - \rho_d[k]) d\sigma - 1), \tag{5}$$

$$\eta_d[k+1] = \eta[k]. \tag{6}$$

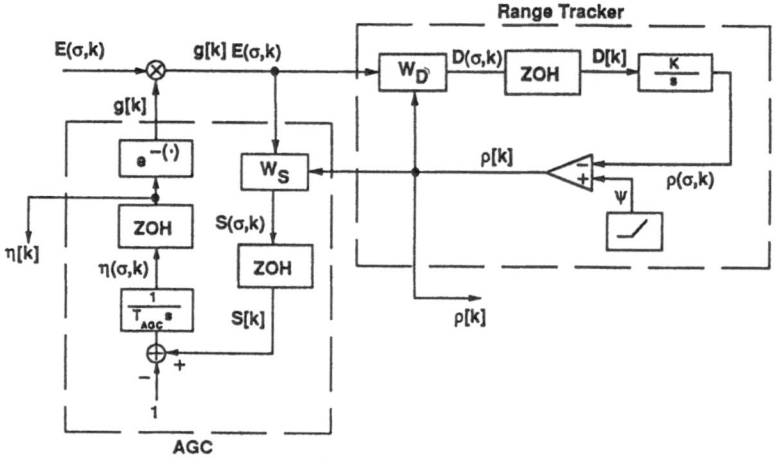

Figure 2: Tracker Block Diagram

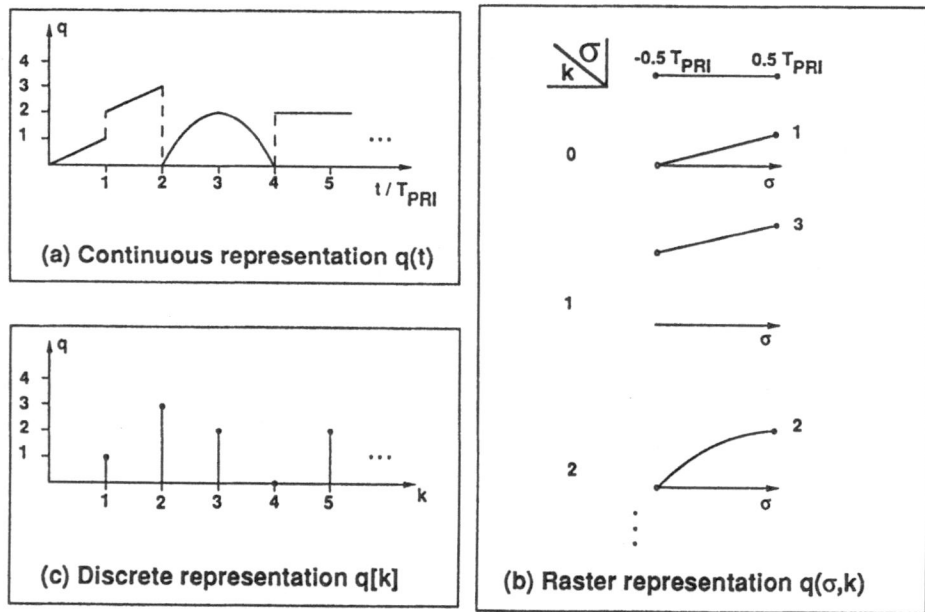

(a) Continuous representation q(t)

(c) Discrete representation q[k]

(b) Raster representation q(σ,k)

Figure 3. Signal Representations

In the above, the two additional states ρ_d and η_d are, respectively, the range estimate and the AGC voltage delayed by one PRI.

If T_{PRI} is very small relative to the system time constants, the discrete-time tracker equations above are well approximated by the differential equations [1]

$$\frac{d\rho}{dt} = Ke^{-2\eta} \int_{R_D} |E(\sigma, t)|^2 w_D(\sigma - \rho) d\sigma, \tag{7}$$

$$T_{\text{AGC}} \frac{d\eta}{dt} = e^{-2\eta} \int_{R_S} |E(\sigma, t)|^2 w_S(\sigma - \rho) d\sigma - 1, \tag{8}$$

where PRI epoch k has been replaced by elapsed time t. The integrations of the video signal envelope $|E|^2$ times the weighting patterns w_D and w_S are performed with respect to range σ for each t, providing an instantaneous error to drive the now continuous tracker equations.

The nonlinearities appearing in Eqs. (1)-(8) are essential for properly characterizing real tracker dynamics. For example, the nonlinearity of the difference error comparator function D with respect to relative range gate position $\sigma - \rho$, created by the range cut-off in the weighting pattern w_D, induces a bifurcation of the solutions corresponding to the phenomenon of resolution of individual targets as they move apart. Without this nonlinearity, the tracker always follows the target group centroid, behavior not representative of real tracking systems. In [1] we were successful in using these nonlinear equations to establish measures of target resolution in terms of track point existence and stability, and in comparing these results in important new ways with predictions of resolvability by classical information-theoretic methods, for the case of deterministic nonfluctuating signals.

2.1 A Two-Target Example

An example will now be used to illustrate several important characteristics of the problem before us. Analysis of this example will continue in the next section. Consider the case of two complex-valued target signals $x_r(t) + jx_i(t)$ and $y_r(t) + jy_i(t)$, located symmetrically about the σ (range) origin at $-d(t)$ and $d(t)$, respectively. The complex video E for this case becomes

$$E(\sigma, t) = (x_r(t) + jx_i(t))h(\sigma + d) + (y_r(t) + jy_i(t))h(\sigma - d). \tag{9}$$

If we further assume a rectangular radar transmit envelope of duration T and a matched filter IF response in the receiver, then h, the convolution of the transmit envelope with the receiver complex envelope impulse response, is given analytically and graphically as follows:

$$h(\sigma) = \begin{cases} 1 - \frac{|\sigma|}{T} & \text{for } -T \le \sigma \le T, \\ 0 & \text{otherwise.} \end{cases}$$

Let us also concentrate on a centroiding split-gate range tracker for which the weighting patterns w_D and w_S are given analytically and graphically as follows:

$$w_D(\sigma) = \begin{cases} -1 & \text{for } -T \le \sigma < 0, \\ 1 & \text{for } 0 < \sigma \le T, \\ 0 & \text{otherwise,} \end{cases}$$

and,

$$w_S(\sigma) = \begin{cases} 1 & \text{for } -T \le \sigma \le T, \\ 0 & \text{otherwise.} \end{cases}$$

For this example, the tracker equations (7), (8) become

$$\frac{d\rho}{dt} = Ke^{-2\eta}\left[a(t)f(\rho+d) + b(t)f(\rho-d) + 2(x_r(t)y_r(t) + x_i(t)y_i(t))\tilde{f}(\rho,d)\right], \quad (10)$$

$$T_{\text{AGC}}\frac{d\eta}{dt} = e^{-2\eta}\left[a(t)g(\rho+d) + b(t)g(\rho-d) + 2(x_r(t)y_r(t) + x_i(t)y_i(t))\tilde{g}(\rho,d)\right] - 1,$$

$$(11)$$

where

$$a(t) = x_r^2(t) + x_i^2(t),$$
$$b(t) = y_r^2(t) + y_i^2(t),$$
$$f(s) = \int_0^{s+T} h^2(\sigma)d\sigma - \int_{s-T}^0 h^2(\sigma)d\sigma,$$
$$\tilde{f}(s,d) = \int_0^{s+T} h(\sigma+d)h(\sigma-d)d\sigma - \int_{s-T}^0 h(\sigma+d)h(\sigma-d)d\sigma,$$
$$g(s) = \int_{s-T}^{s+T} h^2(\sigma)d\sigma, \text{ and}$$
$$\tilde{g}(s,d) = \int_{s-T}^{s+T} h(\sigma+d)h(\sigma-d)d\sigma.$$

In the above, f and $\tilde{f}$ are odd functions of ρ, and g and $\tilde{g}$ are even functions of ρ. The separation d is in general a function of time. Functions $\tilde{f}$ and $\tilde{g}$ account for the interactions between the two target signals. Each of these functions vanishes for $d > T$, since the finite support of h is such that no target interactions occur for separations greater than T.

With the separation d taken to be constant, and with constant nonfluctuating targets, the equilibrium points of Eq. (10) exhibit a bifurcation as d is quasistatically varied between 0 and T. This is perhaps most easily seen using a graphical representation of an effective reduced order model corresponding to the dynamics of ρ for small T_{AGC}. (This is precisely the case of a fast AGC, which will be returned to subsequently.) A reduced order model results from formally setting $T_{\text{AGC}} = 0$ (i.e., instantaneous AGC) in Eqs. (10), (11), and is given by

$$\frac{d\rho}{dt} = K\frac{a(t)f(\rho+d) + b(t)f(\rho-d) + 2(x_r(t)y_r(t) + x_i(t)y_i(t))\tilde{f}(\rho,d)}{a(t)g(\rho+d) + b(t)g(\rho-d) + 2(x_r(t)y_r(t) + x_i(t)y_i(t))\tilde{g}(\rho,d)}. \quad (12)$$

Understanding the relationship between the models (12) and (10)-(11) involves a singular perturbation argument, which is not pursued here. Figure 4 depicts the right side of Eq. (12) (denoted error(ρ) in the figure) vs. ρ for various values of d. Figure 4 corresponds to two constant deterministic targets of equal strength. The parameter values used to obtain Figure 4 are $x_r = y_r = 1, x_i = y_i = 0$, and $K = 1$. It is clear from the figure that a pitchfork bifurcation occurs for some value of d between 0 and T. Using a stability result in [1], we find that this critical value of d is given by $d = \frac{2}{3}T$.

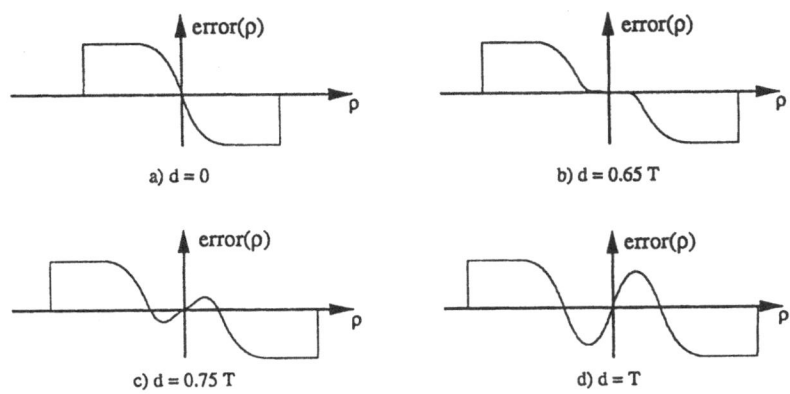

Figure 4: Tracker error law (Instantaneous AGC)

3   Stochastic Simulation

For a fluctuating video signal E, Eqs. (7), (8) are differential equations with random coefficients whose analysis may require new sophisticated mathematical techniques. The random coefficients are non-Gaussian, with large variance, and have bandwidths that are comparable to the tracker system bandwidth. Many standard stochastic analysis techniques are of limited effectiveness when applied to this problem. In this section, we first specialize the two target model of the previous section to a case involving two separating targets, one fluctuating and one constant. Next, we give representative sample paths showing the types of tracker behavior that are to be expected. Monte Carlo simulations are then given for various choices of relative target strengths and correlation time of the fluctuating target.

3.1   Separating Targets

Throughout this section, we consider a specialization of the two target model Eqs. (10), (11) in which one target is of constant signal strength, the other fluctuates randomly, and the targets separate at a constant velocity. Recall from Section 2.1 the targets are at range positions $\sigma = -d(t)$ and $\sigma = d(t)$. The target at $\sigma = -d(t)$ is constant with complex video envelope signal $c + jc$, and the fluctuating target is at $\sigma = d$ with complex video envelope signal $y_r(t) + jy_i(t)$. The separating target scenario begins with both targets at the origin. With this set-up, Eqs. (10), (11) specialize to

$$\frac{d\rho}{dt} = Ke^{-2\eta}\left[2c^2f(\rho + d) + b(t)f(\rho - d) + 2(cy_r(t) + cy_i(t))\tilde{f}(\rho, d)\right], \quad (13)$$

$$T_{AGC}\frac{d\eta}{dt} = e^{-2\eta}\left[2c^2g(\rho + d) + b(t)g(\rho - d) + 2(cy_r(t) + cy_i(t))\tilde{g}(\rho, d)\right] - 1, \quad (14)$$

$$d(t) = vt, \quad (15)$$

$$d(0) = 0. \quad (16)$$

Here, $b, f, \tilde{f}, g$ and $\tilde{g}$ are as specified in Section 2.1, and v is the speed of each target.

In the above, $y_r(t)$ and $y_i(t)$ are independent, identically distributed, stationary zero-mean Gauss-Markov processes, with autocorrelation function

$$E\{y_r(t + \tau)y_r(\tau)\} = E\{y_i(t + \tau)y_i(\tau)\} = e^{-|\tau|/\tau_c}.$$

It is straightforward to show that $b(t)$ is a stationary, exponentially distributed process with autocovariance function

$$E\{b(t + \tau)b(\tau)\} - E\{b(t)\}^2 = 4e^{-2|\tau|/\tau_c}. \quad (17)$$

In the simulations, the AGC is taken to be much faster than the range tracker and faster than the target fluctuation rate. That is $1/T_{AGC} >> K$, and the correlation time τ_c satisfies $\tau_c < T_{AGC}$, respectively. Specifically, in the simulations below, $T_{AGC} = 0.02$ and $K = 1$. The differential equations are solved numerically using Euler integration with a time step of 0.01. Our primary interest is in assessing the dependence of tracker behavior on the relative magnitudes of τ_c and $1/K$. However, we must also take into account the relative strengths of the fluctuating target and the constant target.

The power of the complex video signal $x_r(t) + jx_i(t)$ is $E\{x_r^2 + x_i^2\}$. In particular, the power of the constant target, denoted by P_C, is $P_C = 2c^2$. The power of the fluctuating target, denoted by P_F, is $P_F = E\{y_r^2 + y_i^2\} = E\{b\} = 2$. Each of the simulations in the next two subsections corresponds to either the case $P_C = 1.44670$ or the case $P_C = 0.72335$. The choice of these values for P_C deserves a brief explanation. For the purposes of this paper, it was desired to establish a baseline case for the simulations. The power of the fluctuating target, as noted above, is fixed at $P_F = 2$ in all simulations. The question arises, for a given value of the correlation time τ_c, as to the power level of the constant target for which the constant target and the fluctuating target is each chosen by the tracker 50% of the time on average. The baseline case for our purposes is associated with a situation in which the fluctuating target is fast (sometimes we say it exhibits *fast fades*), and, more specifically, with the case $\tau_c = 2.5T_{AGC} = 0.05$. Monte Carlo simulation resulted in the conclusion that this baseline case corresponds to the constant target power of $P_C = 1.44670$. The other constant target power used in the simulations, namely the value $P_C = 0.72335$ noted above, is simply half that of the baseline case.

3.2   Sample Path Simulations

Typical sample paths for the baseline case (recall that this involves fast fades for the fluctuating target) are shown in Figure 5. Figure 6 shows sample paths for a case in which $P_C = 1.44670$ as in the baseline case, but with the fluctuating signal exhibiting slow fades. Specifically, the simulation of Figure 6 was obtained using a correlation time $\tau_c = 4$. Figures 5a and 6a are plots of the instantaneous power ratio of the fluctuating target to the constant target. Figures 5b and 6b are plots of that same power ratio smoothed by a first order low-pass filter. Figures 5c and 6c show the positions of the constant and fluctuating targets as they separate (denoted by d_C and d_F, respectively), along with the tracker's range estimate ρ.

An important effect of the AGC, readily discernable by comparing Figure 5b with Figure 5c or Figure 6b with Figure 6c, is that it causes the tracker to be driven to a large extent by the power ratios of the targets. Since the tracker is the slowest element of the system, we choose the maximum tracker bandwidth as the bandwidth of the smoothing filter. In these examples, since the AGC is much faster than the tracker, the tracker will closely follow the smoothed power ratio until the targets separate beyond the limits of the range gate, forcing the tracker to commit to one of the targets. In Figures 5c and 6c, the point at which the decision to commit occurs is apparent. Notice that for the fast fade case (Figure 5c), the tracker position estimate ρ follows the centroid of the targets until it commits to a target. From Figure 6c, it would appear that in addition to relative target strength, another mechanism, the presence and duration of fades close to the bifurcation point may influence the final outcome of target selection by the range tracker. This will be explored in the next section.

3.3   Monte Carlo Simulations

The time evolution of the probability of the range estimate ρ is of far greater value to understanding the tracker than is the sample path solutions. Monte Carlo simulations were made using equations (13) to (16) under the same conditions as the sample path examples. Figure 7 shows the results of Monte Carlo simulations for two cases, the baseline case in Figure 7a, and in Figure 7b the only modification from the baseline case is that the correlation time is larger, namely $\tau_c = 4$ (slow fades). Figures 7a and 7b show estimates of the time evolutions of the probability density of tracker position ρ. Each of these figures is obtained using an ensemble of 1000 sample paths. In Figures 7a and 7b, at each time t along the abscissa the ordinate is a histogram of ρ. The gray scale indicates the probability, with level of darkness a monotonically increasing function of probability. In Figure 7a, it is observed that the tracker follows the average power centroid fairly closely until target commitment. However, it is apparent from Figure 7b that the slowly varying fluctuating target signal allows the tracker to jump repeatedly between the fluctuating target and the constant target. In the case represented in Figure 7b, the tracker commits to the fluctuating target only 35% of the time even though the constant signal power is the same as that in the simulations used to generate Figure 7a. So in addition to the relative target strengths, the final range tracker decision is also influenced by the occasional occurrence of long target fades near the

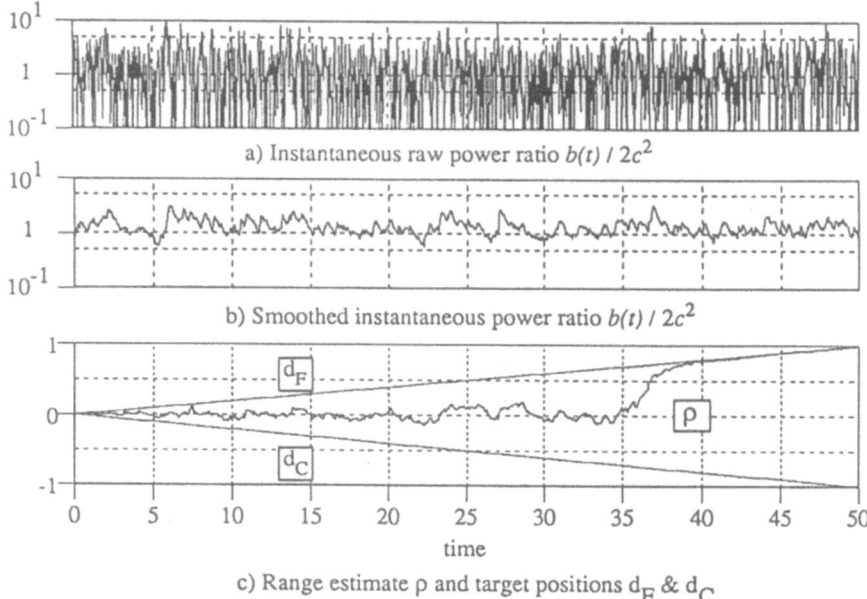

c) Range estimate ρ and target positions d_F & d_C

Figure 5: Sample paths for separating targets and fast fades

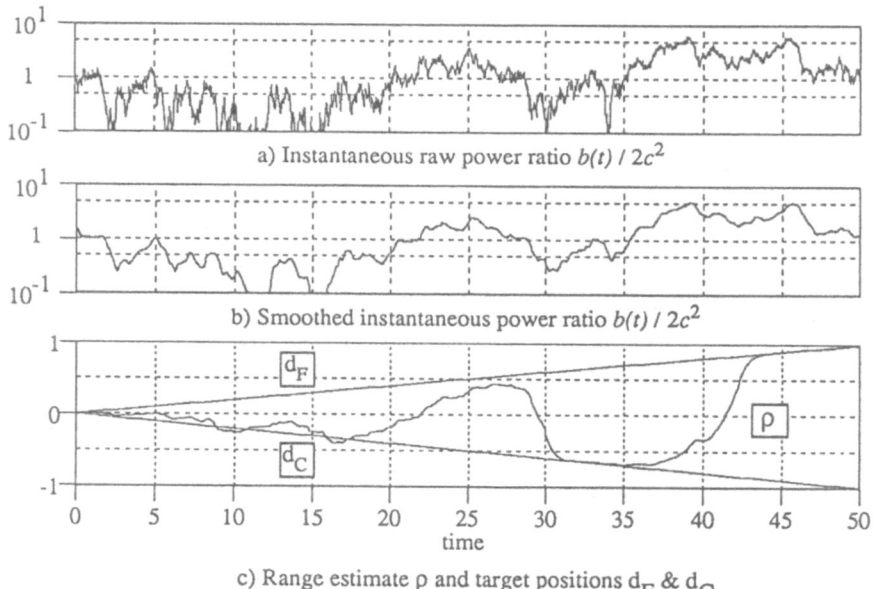

c) Range estimate ρ and target positions d_F & d_C

Figure 6: Sample paths for separating targets and slow fades

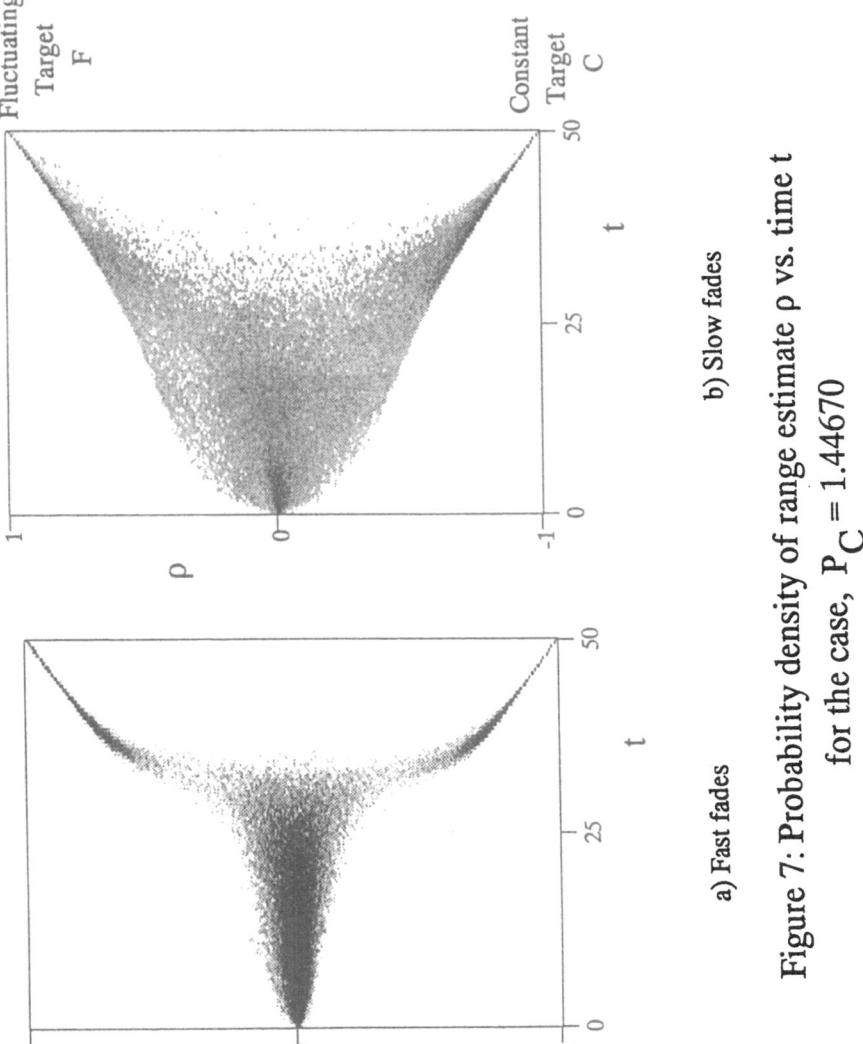

a) Fast fades

b) Slow fades

Figure 7: Probability density of range estimate ρ vs. time t

for the case, $P_C = 1.44670$

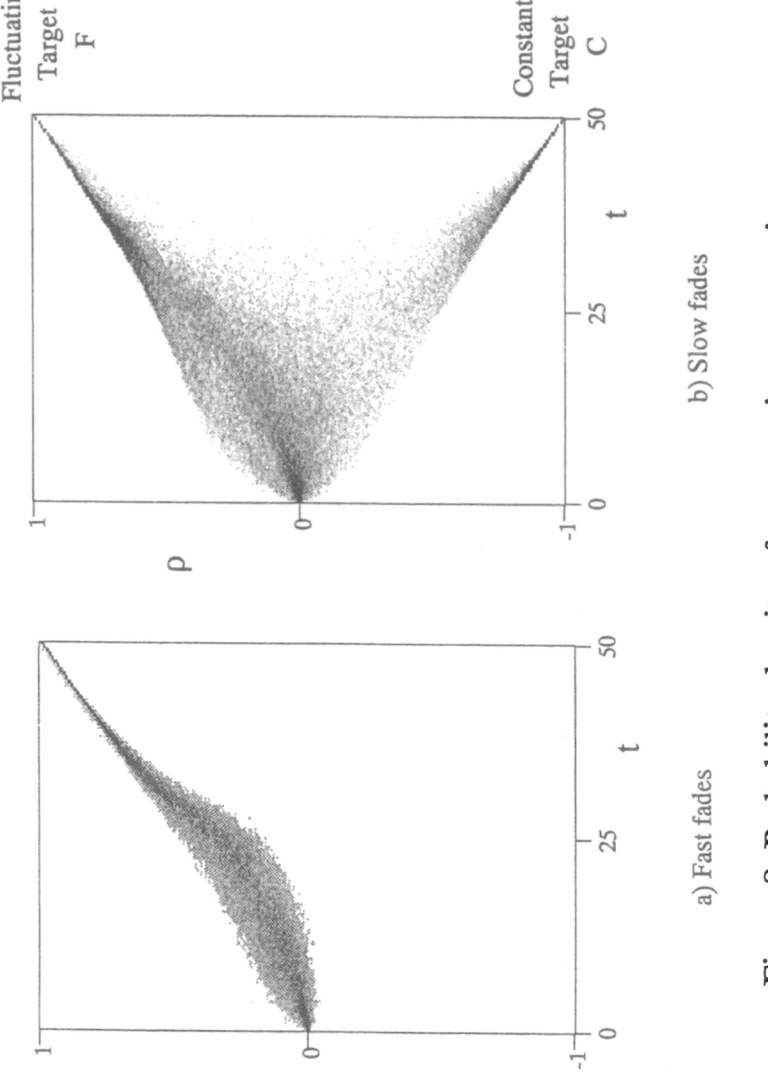

a) Fast fades

b) Slow fades

Figure 8: Probability density of range estimate ρ vs. time t
for the case, $P_C = 0.72335$

bifurcation. As a final observation, it appears from the figure, that the region where the tracker commits to a target is not as well defined as in the case of fast fades.

To better demonstrate the effect of slow fades on the final range estimate, the Monte Carlo simulation was repeated, but with the power of constant target cut by half ($P_C = 0.72335$). The resulting time evolutions of the range estimate histograms are shown in Figure 8. Notice that in Figure 8a, the range tracker follows the fast fluctuating target 100% of the time. However, in the case of slow target fluctuations, illustrated in Figure 8b, the occurrence of long sustained fades in the fluctuating target at commitment allows the weaker target to attract the range gate 32% of the time.

4 Liapunov Functions

In [1], local *deterministic* stability of the range tracker models was studied assuming nonfluctuating targets at spatially fixed positions. There are several directions in which these results may be generalized. In this section, we consider stability results which are not local, and which allow for random target fluctuation. First, Liapunov functions are constructed which can be used to assess the *domains of attraction* of track points assuming deterministic target video signals. Next, representative results on stochastic stability are given. These results employ stochastic Liapunov functions, and apply under the assumption of wide-band noise modulating the target video signals. We focus on the continuous-time tracker model.

4.1 Fast AGC

First we obtain Liapunov functions useful in studying the stability of Eqs. (7), (8) in the case of a fast AGC, i.e., for T_{AGC} small. Since T_{AGC} is small, the system (7), (8) is singularly perturbed. Saberi and Khalil [4] gave a technique for constructing Liapunov functions for nonlinear singularly perturbed systems, which we proceed to apply. In this approach, one associates with the singularly perturbed system of interest a slow subsystem and a fast subsystem, determines Liapunov functions for these subsystems, and then combines these Liapunov functions to result in a Liapunov function for the original system. For details, see [4]. Consider Eqs. (7), (8) and assume a nonfluctuating return signal, i.e., $E(\sigma, k) \equiv E(\sigma)$ is independent of k. Formally setting $T_{AGC} = 0$ in (7), (8) results in the slow subsystem

$$\frac{d\rho}{dt} = K \frac{\Delta(\rho)}{\Sigma(\rho)},$$

where

$$\Delta(\rho) := \int_{R_D} |E(\sigma)|^2 w_D(\sigma - \rho) d\sigma$$

and

$$\Sigma(\rho) := \int_{R_S} |E(\sigma)|^2 w_S(\sigma - \rho) d\sigma.$$

It is straightforward to check that, with the sufficient conditions for local stability found in [1] in effect, a Liapunov function for the slow subsystem is given by

$$v_S(\rho) := \int_0^\rho \frac{\Delta(u)}{\Sigma(u)} du.$$

The fast subsystem is obtained by rescaling time to $\tilde{t} := t/T_{\text{AGC}}$ and taking the limit as $T_{\text{AGC}} \to 0$, yielding

$$\frac{d\eta}{d\tilde{t}} = e^{-2\eta}\Sigma(\rho) - 1.$$

This system has a Liapunov function

$$v_F(\rho, \eta) = (e^{-2\eta}\Sigma(\rho) - 1)^2.$$

Using results of [4], we combine the Liapunov functions for the slow and fast subsystems to obtain a Liapunov function candidate for the overall system, given by

$$v(\rho, \eta) = \alpha v_S(\rho) + (1 - \alpha)v_F(\rho, \eta)$$

where α lies in the interval $(0, 1)$. Following the detailed analysis of [4], one can show that $v(\rho, \eta)$ is indeed a Liapunov function for some values of $\alpha \in (0, 1)$. We omit the details of this calculation, which will appear elsewhere.

4.2 Stochastic Liapunov Functions

Next, we present stochastic Liapunov functions for a more detailed tracker model. This can be applied as well to the simpler model of the previous analysis. The AGC-aided tracker model appearing in Section 2 applies in the case of a "first-order" tracker. However, in reference [1] the case of a second-order tracker is also considered. A continuous-time model given in [1] for a second-order tracker with AGC has the form

$$\frac{d\rho}{dt} = \nu + (g_2 + 0.5g_1 T_{\text{PRI}})e^{-2\eta} \int_{R_D} |E(\sigma, t)|^2 w_D(\sigma - \rho)d\sigma$$

$$\frac{d\nu}{dt} = g_1 e^{-2\eta} \int_{R_D} |E(\sigma, t)|^2 w_D(\sigma - \rho)d\sigma$$

$$T_{\text{AGC}}\frac{d\eta}{dt} = e^{-2\eta} \int_{R_S} |E(\sigma, t)|^2 w_S(\sigma - \rho)d\sigma - 1,$$

where ν is the velocity estimate along the slant range and T_{PRI} is the length of the pulse repetition interval.

Assume the AGC to be slow (i.e., that T_{AGC} is large), and suppose the target video signal is stochastic and of the form

$$E(\sigma, t) = n(t)E_0(\sigma).$$

Under these assumptions, the tracker system has the form

$$\frac{d\rho}{dt} = \nu + \alpha(1 + \epsilon\Psi(t)) \int_{R_D} |E_0(\sigma)|^2 w_D(\sigma - \rho)d\sigma$$

$$\frac{d\nu}{dt} = g(1 + \epsilon\Psi(t)) \int_{R_D} |E_0(\sigma)|^2 w_D(\sigma - \rho)d\sigma$$

where $\Psi(t)$ is a noise signal. If $\Psi(t)$ is white Gaussian noise, then the function

$$V(\rho, \nu) \equiv \frac{\nu^2}{2} - g \int_0^\rho \int_{R_D} |E_0(\sigma)|^2 w_D(\sigma - \mu) d\sigma d\mu$$

qualifies as a stochastic Liapunov function for this system. After accounting for the Wong-Zakai correction term, we find

$$E(\frac{dV}{dt}) = -\alpha g (\int_{R_D} |E_0(\sigma)|^2 w_D(\sigma - \rho) d\sigma)^2$$
$$\times [1 - \frac{\epsilon^2}{2} - \frac{\epsilon^2}{2}\eta + \frac{\alpha \epsilon^2}{2} \int_{R_D} |E_0(\sigma)|^2 w_D(\sigma - \rho) d\sigma].$$

5 Conclusions

In conclusion, a general nonlinear gated range tracker model was employed to study range resolution for competing targets in cases involving random fluctuations in the target video signals. These fluctuations are not necessarily small, and enter the dynamics nonlinearly. Preliminary results on this subject were given, using both Monte Carlo simulation and stochastic Liapunov functions, with either white or colored noise excitations. One important conclusion from this work is that the noise correlation times strongly influence tracker behavior. The approaches employed here are only two of the possible tools in the analysis of this problem. Indeed, it is expected that by drawing analogies to similar stochastic bistability problems in other application areas, results on target tracking probabilities, residence times, first passage times, and other stochastic qualitative results may be attainable.

References

[1] E.H. Abed, A.J. Goldberg and R.E. Gover, "Nonlinear modeling of gated range tracker dynamics with application to radar range resolution," IEEE Trans. Aerospace and Electronic Systems, Vol. 27, pp. 68-82, 1991; also Rept. NRL TR 9264, Naval Research Laboratory, to appear.

[2] H.J. Kushner, Stochastic Stability and Control, Academic Press, New York, 1967.

[3] H.J. Kushner, Weak Covergence Methods and Singularly Perturbed Stochastic Control and Filtering Problems, Birkhäuser, Boston, 1990.

[4] A. Saberi and H.K. Khalil, "Quadratic-type Lyapunov functions for singularly perturbed systems," IEEE Trans. Automatic Control, Vol. AC-29, pp. 542-550, 1984.

[5] R.J. Schlesinger, Principles of Electronic Warfare, Prentice-Hall, Englewood-Cliffs, NJ, 1961.

OPTIMAL CONTROL AND REPLACEMENT
WITH STATE–DEPENDENT FAILURE RATE*

ARTHUR C. HEINRICHER† AND RICHARD H. STOCKBRIDGE‡

Abstract. A class of stochastic control problems where the payoff depends on the running maximum of a diffusion process is described. The controller must make two kinds of decision: first, he must choose a work rate (this decision determines the rate of profit as well as the proximity of failure), and second, he must decide when to replace a deteriorated system with a new one. Preventive replacement is a realistic option if the cost for replacement after failure is larger than the cost of a preventive replacement.

We focus on the revenue and replacement cost for a single work cycle and solve the problem in two stages. First, the optimal feedback control (work rate) is determined by maximizing the payoff during a single excursion of a controlled diffusion away from the running maximum. This step involves the solution of the Hamilton-Jacobi-Bellman partial differential equation. The second step is to determine the optimal replacement set. The assumption that failure occurs only when the state is increasing restricts the optimal replacement set. This leads to a simple formula for the optimal replacement level in terms of the value function.

Key words. controlled diffusion, running maximum, dynamic programming

AMS(MOS) subject classifications. 93E20

1. Introduction. This paper is devoted to stochastic control problems motivated by optimal control and replacement problems for deteriorating systems. The *state* records the level of deterioration and the actions of the controller influence the *rate* of deterioration as well as the reliability of the system. For example, a high work rate may be attractive because it implies a high revenue rate, but the system may become less efficient and less profitable as it deteriorates. A high level of wear is also dangerous because it entails an increase in the probability of an expensive failure. The controller has the option to replace a worn, but still working, system with a new one and this may be optimal when replacement at or after failure is more expensive than a planned replacement.

Throughout the paper, the *state* is given by the pair (x_t, y_t) satisfying

$$
(1) \quad
\begin{cases}
dx_t = f(x_t, y_t, u_t)dt + \sigma(x_t, y_t)dw_t, & x_0 = x, \\
y_t = \max\{x_s : 0 \leq s \leq t\} \vee y, & y_0 = y \geq x.
\end{cases}
$$

Here $w = (w_t, \ 0 \leq t < \infty)$ denotes a standard, one-dimensional, Brownian motion and $u = (u_t, \ 0 \leq t < \infty)$ is a control process. Stochastic control problems involving the running max $y = (y_t, \ 0 \leq t < \infty)$ are studied in [10] (see also [3, 11, 12]).

We focus on a single working cycle and maximize the revenue collected minus the replacement cost. Revenue is accumulated at rate $h = h(x_t, y_t, u_t)$ for a control policy $u = (u_t, \ 0 \leq t < \infty)$ as long as the machine is working. A cost is incurred when the machine fails and/or is replaced. The controller specifies a replacement level Δ, with an associated replacement time $\tau(\Delta)$, but the system may fail at a random time $\zeta = \zeta(u)$ less than $\tau(\Delta)$.

For an initial state (x, y) with $x \leq y$, the total *profit* for the cycle is

$$
(2) \quad J(x, y; u, \Delta) - \overline{R}(\Delta) = E_{xy}\left[\int_0^{\zeta(u) \wedge \tau(\Delta)} h(x_t, y_t, u_t)dt - R(\Delta)\right].
$$

The objective is to determine a pair (u^*, Δ^*) in order to maximize this profit.

---

* The authors' research was supported by the National Science Foundation grant DMS-9006674.

† Department of Mathematics, University of Kentucky, Lexington, Kentucky 40506-0027.

‡ Department of Statistics, University of Kentucky, Lexington, Kentucky 40506-0027.

The replacement cost $R = R(\Delta)$ has two components. There is a cost R_2 for a preventive replacement and larger cost $R_1 \geq R_2$ for replacement at or after failure and so the cost associated with a replacement level Δ is

$$(3) \qquad R(\Delta) = R_1 1_{\{y(\zeta) \leq \Delta\}} + R_2 1_{\{y(\zeta) > \Delta\}} \ .$$

We assume that *failure can occur only while the running maximum is increasing*. In particular, if the system is working at level y, it will fail as the state increases from y to $y + \delta y$ with probability $k(y)\delta y + o(\delta y)$. This implies that

$$(4) \qquad P_{xy}(y(\zeta) > \Delta) = \exp\left(-\int_y^\Delta k(z)dz\right) \ .$$

It is important to notice that this is not a standard state–dependent failure mechanism; the failure rate is *per unit wear* and not *per unit time*.

We consider only replacement times which are first–passage times. That is, for a replacement level $\Delta \geq y$, the replacement time τ is

$$(5) \qquad \tau = \tau(\Delta) = \inf\{t \geq 0 : \ y_t \geq \Delta\},$$

where $\tau = +\infty$ if the set is empty (which will happen if and only if $\Delta = +\infty$). Thus the expected replacement cost becomes

$$(6) \qquad \overline{R}(\Delta) = E_{xy}[R(\Delta)] = R_1 - (R_1 - R_2)\exp\left(-\int_y^\Delta k(z)dz\right) \ .$$

The fact that y_t increases only when $x_t = y_t$ and the special structure of the failure rate implies that it is always optimal to work if $x < y$ (assuming that $h(x, y, u) > 0$ for some admissible control); the running maximum is constant in this region and failure cannot occur. This simplifies and separates the optimal replacement decision from the optimal control decision:
- The optimal replacement set is restricted to the main diagonal $\{(x, y) : \ x = y\}$.
- The optimal control policy maximizes the profit accrued up to the first time that $x_t = y_t$.

One of the fundamental works on optimal replacement for deteriorating systems (without control between replacements), is Taylor [19]. Anderson [1, 2] analyzes the variational and quasi-variational inequalities that arise in optimal replacement problems for general (monotone) Markov processes (again without control between replacements). The recent survey article by Valdez-Flores and Feldman [20] contains a wealth of additional references.

Conrad and McClamroch [7] consider a control and replacement problem with a (non-monotone) diffusion model for wear and an application in automated manufacturing (the machine is a drill press). But a reasonable model for wear should be monotone; tires and drill bits do not unwear. Taking the running maximum is a simple way to construct a continuous, monotone process from a (necessarily non-monotone) diffusion. The pair (x_t, y_t) defined in this way is an example of a *Markov additive* process [4]; one coordinate of a multi-dimensional Markov process is monotone and not a Markov process when considered alone. Erhan Çinlar [5, 6] has argued that this is perhaps the most general class of stochastic processes suitable for modeling wear.

1.1. Summary. We use dynamic programming methods to obtain sufficient conditions for optimality and to determine optimal control policies as well as optimal replacement levels. The optimal feedback control is characterized for all (x, y) with $x < y$ by solving a simpler "auxiliary" control problem: maximize the revenue collected up to the first time that $x_t = y_t$. This involves the solution of a (simpler) Hamilton–Jacobi–Bellman (HJB) equation on each excursion of the controlled diffusion below the running maximum. Our assumptions concerning the failure-rate restrict the optimal replacement set to the main diagonal ($x = y$) and so, with reasonable assumptions on the profit and wear rates, the optimal replacement level is determined by a simple one-dimensional maximization problem.

The problem discussed here is studied in greater detail in the papers [13, 14]. The dynamic programming approach is described in detail in the first paper [13]. The second paper [14] (see also [12]) focuses on the long-term average optimization problem and uses a different approach; the state is defined as the solution to a (controlled) constrained martingale problem [16, 17] and the control problem is formulated as an optimization problem on a space of invariant measures. The dynamic programming approach discussed here gives an explicit formula for an optimal feedback control, assuming that the HJB equation has a sufficiently smooth solution. The invariant measure approach does not require any regularity of the value function, but it does not lead directly to the optimal control policies.

2. Formulation of the problem. The technical formulation of our control problem follows the standard approach to controlled diffusions described, for instance, in the text by Fleming and Rishel [8] (see also [9]). For the admissible controls, take the collection of *nonanticipative controls* as defined in Chapter VI of [8, p. 162] and let $\mathcal{A}$ denote the collection of admissible controls. We assume that control processes $u = (u_t, 0 \leq t < \infty)$ take values in a *compact* subset U of the real numbers and that the coefficients of the problem satisfy conditions sufficient to provide polynomial growth for the value function and guarantee existence of solutions to (1).

We assume that

(A1) h is continuous and bounded on $\{(x, y, u) : x \leq y, y \geq 0, u \in U\}$;

(A2) f is a bounded C^1 function on $\{(x, y, u) : x \leq y, y \geq 0, u \in U\}$ and σ is a bounded C^1 function on $\{(x, y) : x \leq y, y \geq 0\}$ with

$$|f_x(x, y, u)| + |f_y(x, y, u)| + |f_u(x, y, u)| \leq K, \quad (x \leq y, \ y \geq 0, \ u \in U),$$
$$|\sigma_x(x, y)| + |\sigma_y(x, y)| \leq K, \quad (x \leq y, \ y \geq 0),$$

for a suitable constant K.

(A3)   there is a constant α such that

$$0 < \alpha \leq f(x, y, u), \quad (x \leq y, \ y \geq 0, u \in U).$$

(A4) $k(y) \geq 0$ for $y \geq 0$, $k(\cdot)$ is nondecreasing and right-continuous (possibly infinite) with

$$\exp\left(-\int_0^\infty k(z)dz\right) = 0.$$

Remark 2.1. These conditions are sufficient to guarantee the existence and pathwise uniqueness of solutions. Condition (A3) guarantees that $E_{xy}[\tau(\Delta)] < \infty$ if $\Delta < \infty$. Conditions (A3) and (A4) combine to guarantee that the expected failure time is finite.

3. Solve the control problem: Determine u^*. Assume that the replacement level Δ is fixed and the that controller seeks only to maximize the profit accrued up to this replacement level. Failure may occur, and this is accounted for by a discount factor, but the cost for replacement is not considered.

The objective is to choose an admissible control process $u = (u_t, \ 0 \leq t < \infty)$ to maximize

(7) $$J(x, y; u, \Delta) = E_{xy} \int_0^{\tau(\Delta)} e^{-\int_0^t k(y_s)dy_s} \ h(x_t, y_t, u_t)dt$$

where the state is the pair (x_t, y_t) defined in (1) and $\tau = \tau(\Delta)$ is the replacement time (5).

The following theorem is an extension of the standard sufficient conditions for optimality as presented in Chapter VI of [8] (see also [10]). The proof is an application of the generalized Itô formula and is included in [13].

THEOREM 3.1. *Let $V(x, y)$ be a solution of the dynamic programming equation*

(8) $$\max_{u \in U} \left\{ \frac{1}{2}\sigma(x, y)^2 \ V_{xx}(x, y) + f(x, y, u)V_x(x, y) + h(x, y, u) \right\} = 0,$$

in the region $x < y$, $0 < y < \Delta$, satisfying the boundary condition

(9) $$V_y(y, y) - k(y)V(y, y) = 0 \qquad (0 < y < \Delta),$$

as well as the terminal condition

(10) $$V(\Delta, \Delta) = 0 .$$

(If $\Delta = +\infty$, the terminal condition is not enforced.) In addition, suppose $V(x, y)$ is continuous, twice continuously differentiable with respect to x, and satisfies a polynomial growth condition

(11) $$|V(x, y)| \leq C(1 + |x|^p + |y|^p) \qquad (x \leq y, \ 0 \leq y \leq \Delta),$$

for appropriate constants C and p. Then:

(a) $V(x, y) \geq J(x, y; u, \Delta)$ *for any admissible control u and any $x \leq y$;*
(b) *If u^* is an admissible control which attains the maximum in (8), then u^* is optimal and $V(x, y) = J(x, y; u^*, \Delta)$ is the value function.*

Remark 3.2. Two properties of the state process introduce the boundary condition (9). The fact that y_t increases only when $x_t = y_t$ would require that $V_y(x, y) = 0$ when $x = y$; this condition is introduced in [10]. The assumption that failure can occur only when y_t is increasing has added the "killing term" $-k(y)V(y, y)$ to the boundary condition. This is in contrast to the usual sort of killing which would surface as a zeroeth order term in the HJB equation (8).

The key to the actual solution of the control problem is the realization that the value function can be constructed from a family of *auxiliary problems* where y is a fixed parameter. This decomposition is introduced in [10] for a simpler problem (without failure or replacement).

Fix $0 \leq y \leq \Delta$, let $x \leq y$, and define

(12) $$\theta = \theta(x, y; u) := \inf\{t \geq 0 : \ x_t = y\} .$$

We seek an admissible control u to maximize

(13) $$I(x, y; u) = E_{xy} \int_0^\theta h(x_t, y, u_t)dt.$$

Observe that the failure term does not enter this objective function because $x_t < y_t \equiv y$ for $0 \leq t < \theta$. Let $W(x, y)$ denote the value function for this auxiliary problem:

$$W(x, y) = \sup_{u \in \mathcal{A}} I(x, y; u).$$

The HJB equation satisfied by $W(x, y)$ is of the standard form:

$$(14) \qquad \max_{u \in U} \left\{ \frac{1}{2} \sigma(x,y)^2 \, W_{xx}(x,y) + f(x,y,u) W_x(x,y) + h(x,y,u) \right\} = 0$$

on the half-line $x < y$ with the terminal condition

$$(15) \qquad W(y,y) = 0.$$

There is a simple relationship between the value functions for the auxiliary problem and the original, single-cycle control problem. The proof is based on the principle of optimality (see [18]).

PROPOSITION 3.3. *The value functions for the single-cycle control problem and the auxiliary problem satisfy*

$$(16) \qquad V(x,y) = W(x,y) + V(y,y) \qquad (x \le y).$$

We can go one step further and represent the optimal value $V(y,y)$ on the diagonal in terms of the auxiliary value $W(x,y)$. In this way, the value function $V(x,y)$ is determined entirely in terms of the auxiliary value $W(x,y)$.

THEOREM 3.4. *For each $0 \le y \le \Delta$, let $W(x,y)$ be a solution of the dynamic programming equation* (14) *on the halfline $x < y$ satisfying the terminal condition* (15). *Then the single-cycle value function is given by*

$$(17) \qquad V(x,y) = W(x,y) + \int_y^\Delta e^{-\int_y^z k(s)ds} W_y(z,z) dz \qquad (x \le y, \ 0 \le y \le \Delta).$$

This is valid as long as $W(x,y)$ is continuous with respect to (x,y), differentiable along $x = y$, and twice continuously differentiable with respect to x and satisfies the polynomial growth condition

$$|W(x,y)| \le C(1 + |x|^p + |y|^p) \qquad (x \le y),$$

for appropriate constants C and p.

In addition, if $u^*(x,y)$ is an admissible control which attains the maximum in (14), *then $u^*(x,y)$ is an optimal control for the auxiliary problem as well as for the objective* (7).

Proof. Defining $V(x,y)$ as in (17), $V(x,y)$ inherits exactly the smoothness of $W(x,y)$; in particular, we have

$$V_x(x,y) = W_x(x,y), \quad V_{xx}(x,y) = W_{xx}(x,y).$$

Since $W(x,y)$ satisfies (14), $V(x,y)$ satisfies (8) as well as the boundary condition (9); differentiating (17) with respect to y and evaluating along the diagonal provides

$$V_y(y,y) - k(y) V(y,y) = W_y(y,y) - W_y(y,y) = 0.$$

Theorem 3.1 identifies $V(x,y)$ as the value function and $u^*(x,y)$ as the optimal control policy for the running max problem. $\square$

One application of the previous theorem provides a simple formula for the expected failure/replacement time when the system is replaced at level Δ and the control policy is constant.

LEMMA 3.5. *Assume that the drift is given by* $f(x,y,u) = f(u)$ *and that the diffusion coefficient* $\sigma(x,y) = \sigma$ *is constant. If the system starts in the initial state* (x,y) *with* $x \leq y$, *a constant control policy* $u \in U$ *is used, and the preventive replacement threshold is* $\Delta \geq y$, *then*

$$(18) \qquad E_{xy}\left[\zeta(u) \wedge \tau(\Delta)\right] = \frac{(y-x)}{f(u)} + \frac{1}{f(u)} \int_y^\Delta \exp\left(-\int_y^z k(s)ds\right) dz \,.$$

Proof. The representation (18) is obtained by solving a boundary value problem. The problem of interest is the dynamic programming partial differential equation (8) with the boundary condition (9) when the control set is the singleton $U = \{u\}$ and the coefficients f and σ are defined as in the lemma.

Begin by solving the auxiliary problem:

$$\frac{1}{2}\sigma^2 \psi_{xx}(x,y) + f(u)\psi_x(x,y) + 1 = 0, \quad (x < y),$$

with the terminal condition

$$\psi(y,y) = 0 \,.$$

(The solution is the expected time to return to the diagonal.) The unique solution for this problem (satisfying a polynomial growth condition as $x \to -\infty$) is

$$\psi(x,y) = \frac{(y-x)}{f(u)}, \quad (x \leq y).$$

The representation (17) in Theorem 3.4 provides the formula we seek

$$
\begin{aligned}
\phi(x,y) &= \psi(x,y) + \int_y^\Delta \exp\left(-\int_y^z k(s)ds\right) \psi_y(z,z)dz \\
&= \frac{(y-x)}{f(u)} + \frac{1}{f(u)} \int_y^\Delta \exp\left(-\int_y^z k(s)ds\right) dz \,.
\end{aligned}
$$

It is an application of the generalized Itô formula to verify that

$$E_{xy}\left[\zeta(u) \wedge \tau(\Delta)\right] = \phi(x,y) \,,$$

and the proof is complete. □

4. **Solve the replacement problem: Determine** Δ^*. Theorem 3.4 provides a representation (17) for the optimal revenue with a fixed replacement level Δ. We now bring in the replacement cost and optimize over $\Delta \geq 0$. For this section, assume that the system starts in the "new" state $(x,y) = (0,0)$.

Combining the representation (17) with the formula (6) for $\overline{R}(\Delta)$ and integrating, our goal is to determine $\Delta \geq 0$ to maximize

$$
\begin{aligned}
\Delta &\mapsto J(0,0; u^*, \Delta) - \overline{R}(\Delta) \\
&= \int_0^\Delta e^{-\int_0^z k(s)ds} W_y(z,z)dz - \overline{R}(\Delta) \\
&= \int_0^\Delta e^{-\int_0^z k(s)ds} \left[W_y(z,z) - (R_1 - R_2)k(z)\right] dz - R_2 \,.
\end{aligned}
$$

The failure rate $k(z)$ is nondecreasing, so the solution to this optimization problem is simple if we know something about the monotonicity of W_y on the diagonal.

THEOREM 4.1. *Let $W = W(x, y)$ satisfying the hypotheses of Theorem 3.4 be the auxiliary value function and assume that*

(19) $$z \mapsto W_y(z, z) \text{ is nonincreasing for } z \geq 0 .$$

Then the optimal replacement level is given by

(20) $$\Delta^* = \inf \{ z \geq 0 : W_y(z, z) - (R_1 - R_2)k(z) \leq 0 \} ,$$

with $\Delta^* = +\infty$ if the set is empty.

If we are modeling a deteriorating system, then it is natural to assume that (i) the revenue rate decreases as the machine ages and that (ii) an older machine has a higher wear rate. With these assumptions, (19) will hold.

PROPOSITION 4.2. *Assume that the state equation (1) admits a unique, strong solution for any admissible control policy u. Assume also that, in addition to (A1) and (A2), h and f satisfy*
(i) *$h = h(x, y, u)$ is nonincreasing in x and y for fixed u;*
(ii) *$f = f(x, y, u)$ is nonincreasing in y for fixed x and u.*
Then

$$z \mapsto W_y(z, z) \text{ is nonincreasing for } z \geq 0 .$$

Proof. It is sufficient to show that

(21) $$W(z_1, z_1 + \delta) - W(z_2, z_2 + \delta) \geq 0$$

for all $0 \leq z_1 < z_2$ and $\delta > 0$. Let $u^2 = (u_t^2, \ 0 \leq t < \infty)$ be an admissible control and let (x^2, y^2) denote the corresponding state if the initial conditions are $(x_0, y_0) = (z_2, z_2 + \delta)$. Use the same control process when the initial conditions are $(x_0, y_0) = (z_1, z_1 + \delta)$ and denote the state by (x^1, y^1).

The monotonicity assumption (ii) implies that

$$f(x, z_1 + \delta, u) \leq f(x, z_2 + \delta, u) \text{ for all } x \text{ and } u .$$

A comparison theorem for stochastic differential equations (see, for example [15, Section 5.2C]) implies that

(22) $$P(x_t^1 - x_0^1 \leq x_t^2 - x_0^2 \text{ for all } t \geq 0) = 1 ,$$

and hence

$$\theta_1 = \inf \left\{ t \geq 0; \ x_t^1 - x_0^1 = \delta \right\} \geq \theta_2 = \inf \left\{ t \geq 0; \ x_t^2 - x_0^2 = \delta \right\} .$$

The monotonicity of h then implies that

$$I(z_1, z_1 + \delta; u^2) - I(z_2, z_2 + \delta; u^2) \geq 0 .$$

The control process u^2 was arbitrary, so we have shown that the optimal auxiliary value starting at $(z_1, z_1 + \delta)$ is at least as large as the auxiliary value starting at $(z_2, z_2 + \delta)$. This implies (21) and completes the proof. □

REFERENCES

[1] R.S. ANDERSON, *Replacement with nonconstant operating cost*, SIAM J. Control Optim., 26 (1988), pp. 1076–1098.

[2] R.S. ANDERSON, *Long-run average maintenance problems*, preprint.

[3] E.M. BARRON, *The Bellman equation for control of the running max of a diffusion and applications to look-back options*, preprint.

[4] E. ÇINLAR, *Markov additive processes*, I & II, Z. Wahrsch. Verw. Gebiete., 24 (1972), pp. 85–121.

[5] _____, *Shock and wear models and Markov additive processes*, in Theory and Applications of Reliability, Vol. 1, Academic Press, New York, 1977.

[6] _____, *Markov and Semi-Markov models for deterioration*, Reliability Theory and Models, Academic Press, New York, 1984.

[7] C. CONRAD AND N.H. McCLAMROCH, *The drilling problem: A stochastic modeling and control example in manufacturing*, IEEE Trans. Automat. Control, 32 (1987), pp. 947–958.

[8] W.H. FLEMING AND R.W. RISHEL, *Deterministic and Stochastic Optimal Control*, Springer–Verlag, New York, Heidelberg, Berlin, 1975.

[9] I.I. GIHMAN AND A.V. SKOROHOD, *Controlled Stochastic Processes*, Springer–Verlag, New York, Heidelberg, Berlin, 1979.

[10] A.C. HEINRICHER AND R.H. STOCKBRIDGE, *Optimal control of the running max*, SIAM J. Control Optim., 29 (1991).

[11] _____, *Long-term average control for a continuous, monotone process*, submitted for publication.

[12] _____, *An infinite-dimensional LP solution to control of a continuous, monotone process*, to appear in Proceedings of the Joint U.S.–French Workshop on Applied Stochastic Processes, Springer Lecture Notes in Mathematics, 1991.

[13] _____, *Optimal control and replacement with state-dependent failure rate: Dynamic programming*, submitted for publication.

[14] _____, *Optimal control and replacement with state-dependent failure rate: An invarient measure approach*, submitted for publication.

[15] I. KARATZAS AND S.E. SHREVE, *Brownian Motion and Stochastic Calculus*, Springer–Verlag, New York, Heidelberg, Berlin, 1987.

[16] T.G. KURTZ, *Martingale problems for constrained Markov processes*, Recent Advances in Stochastic Calculus, J.S. Baras and V. Mirelli, eds., Springer–Verlag, New York, 1990. University of Maryland (to appear).

[17] _____, *A control formulation for constrained Markov processes*, Proceedings of the AMS-SIAM Summer Seminar on Mathematics of Random Media (to appear).

[18] P.L. LIONS, *Optimal control of diffusions and Hamilton-Jacobi equations. Part I: the dynamic programming principle and applications*, Comm. Partial Differential Equations, 8 (1983), pp. 1101–1174.

[19] H.M. TAYLOR, *Optimal replacement under additive damage and other failure models*, Naval Research Logistics Quarterly, 22 (1975), pp. 1–18.

[20] C. VALDEZ-FLORES AND R.M. FELDMAN, *A survey of preventive maintenance models for stochastically deteriorating single-unit systems*, Naval Research Logistics Quarterly, 36 (1989), pp. 419–446.

PARTIALLY OBSERVED CONTROL OF MARKOV PROCESSES

OMAR HIJAB

Temple University

The aim of this paper is to outline the proof of the existence of optimal controls in the partially observed setting in the simplest case: controlling the drift of a Brownian motion under partial observations. The general case is worked out in [3].

The point of view we adopt is a combination of results of P. L. Lions on the control of the Zakai equation [4] and results of the author on the control of diffusions in finite dimensions [2].

§1. THE THEOREM

The state equation is

$$(1.1) \qquad dx = u(t)dt + d\xi, x(0) \in \mathbf{R},$$

and the observation equation is

$$(1.2) \qquad dy = c(x)dt + d\eta, y(0) = 0 \in \mathbf{R},$$

where $u = u(t, y)$ depends only on the observations y and $(\xi, \eta) \in \mathbf{R}^2$ is a Brownian motion.

Fix a compact convex $U \subset \mathbf{R}$ and let $M = M(\mathbf{R})$ denote the set of probability measures on $\mathbf{R}$. Let $\|\varphi\|_k$ denote the norm in $C_b^k(\mathbf{R})$, let $\mu(\varphi)$ denote the integral of φ against μ over $\mathbf{R}$, and let $\|\mu\|_{-k} = \sup\{|\mu(\varphi)| : \|\varphi\|_k \le 1\}$ denote the dual norm, where $k \ge 0$ and $\mu \in M$.

A *control* is a progressively measurable map $u : [0, \infty) \times C([0, \infty); \mathbf{R}) \to U$, $u = u(t, y)$. Assume the signal c is bounded. It follows then by Girsanov's theorem that, for each fixed $m \in M$, there is a one-to-one correspondence between the set of controls and the set of *systems starting from m* i.e. the set of laws Q of processes (x, y) such that $(\xi, \eta) \in \mathbf{R}^2$ is a Brownian motion and the law of $x(0)$ is m.

Given u and m the corresponding cost is

$$(1.3) \qquad v^u(m) = E^Q \left(\int_0^\infty e^{-\lambda t} f(x(t), u(t))dt \right);$$

1980 *Mathematics Subject Classification* (1985 *Revision*). 60G35,49A10,93E11,93E20.

Key words and phrases. Markov Process, Partially Observed Control, Measure-Valued Diffusions, Infinite-Dimensional Bellman Equation.

Supported in part by a National Science Foundation Grant .

the *value function* is

(1.4) $$v(m) = \inf_u v^u(m).$$

Here the infimum is over all controls, $m \in M$ is arbitrary, and $\lambda > 0$ is the discount factor. The problem is to characterize, for each m, the control u *optimal at m*, i.e. the control satisfying $v^u(m) = v(m)$.

It turns out [3], [4] that $v : M \to \mathbf{R}$ is weakly continuous and bounded. To obtain further regularity of v we make some definitions.

Fix $\mu \in M$ and let ν be a signed measure on $\mathbf{R}$. We say ν is *tangent to M at μ* if $\mu + t\nu \in M$ for $|t|$ small. A functional $\Phi : M \to \mathbf{R}$ is *differentiable at μ* if there is a bounded function $\varphi : \mathbf{R} \to \mathbf{R}$ such that the limit

$$\left.\frac{d}{dt}\right|_{t=0} \Phi(\mu + t\nu)$$

exists and equals $\nu(\varphi)$ for all ν tangent to M at μ. In this case φ is denoted $D\Phi(\mu)$ and is unique up to an additive constant. We say Φ is *differentiable* if Φ is differentiable at μ for all $\mu \in M$.

In general Dv does not exist. Nevertheless [4] v solves, in the viscosity sense, an infinite-dimensional Bellman equation

(1.5) $$-\hat{A}v + H(\mu, Dv) + \lambda v = 0, \mu \in M,$$

where $\hat{A}$ is a diffusion generator on M and

(1.6) $$H(\mu, p) = \sup_{u \in U}(\mu(-A^u p) - L(\mu, u));$$

here $\mu \in M, p \in C_b^2(\mathbf{R})$, $A^u = u\partial_x + \frac{1}{2}\partial_x^2$ is the state generator, and $L(\mu, u) = \mu(f(\cdot, u))$.

Assume $f \in C_b^2(\mathbf{R}^2)$ and $f_{uu}(x, u) > 0$ on $\mathbf{R} \times U$. Then $L_{uu} > 0$ on $M \times U$ and standard convexity reasoning shows (see for example the Appendix in [2]) that the supremum in (1.6) is attained at a unique point $\mathbf{u}(\mu, p) \in U$ where the map $(\mu, p) \mapsto \mathbf{u}(\mu, p)$ satisfies

(1.7) $$|\mathbf{u}(\mu, p) - \mathbf{u}(\mu', p')| \leq C_r\|\mu - \mu'\|_{-1} + C_r\|p - p'\|_1,$$

$\mu, \mu' \in M, p, p' \in C_b^2(\mathbf{R}), \|p\|_2 \leq r, \|p'\|_2 \leq r$, for all $r > 0$.

For each $t \geq 0$ let $\mu_m^u(t)$ denote the (normalized) conditional probability distribution of $x(t)$ given $y(s), 0 \leq s \leq t$,

$$\mu_m^u(t)(\varphi) = E^Q(\varphi(x(t))|y(s), 0 \leq s \leq t).$$

Theorem. *Assume $c \in C_b^2(\mathbf{R})$, $f \in C_b^2(\mathbf{R}^2)$, and $f_{uu} > 0$ on $\mathbf{R} \times U$; then there exists $\lambda_1 \geq 0$ and $C > 0$ such that, for $\lambda > \lambda_1$, v is differentiable and*

  (1) $\|Dv(\mu)\|_2 \leq C, \mu \in M,$
  (2) $|v(\mu) - v(\mu')| \leq C\|\mu - \mu'\|_{-2}, \mu, \mu' \in M,$
  (3) $\|Dv(\mu) - Dv(\mu')\|_1 \leq C\|\mu - \mu'\|_{-1}, \mu, \mu' \in M.$

Moreover for each m and $\lambda > \lambda_1$ there exists exactly one control u_m^* optimal at m and a control u satisfies

$$u(t) = \mathbf{u}(\mu_m^u(t), Dv(\mu_m^u(t))), t \geq 0,$$

iff $u = u_m^*$.

The theorem remains valid for a wide class of systems: The state space may be taken to be $\mathbf{R}^d$, the convexity assumptions can be dropped, the drift coefficients in the state generator A^u can be controlled in a nonlinear fashion, and U can be an arbitrary complete separable metric space. What is necessary is the estimate (1.7). Moreover $L(\mu, u)$ may be any nonlinear functional on $M \times U$ satisfying estimates similar to (1), (2), (3). If the diffusion coefficients in A^u are controlled, then C_b^4 smoothness is required of c, f and the appropriate modification of the Theorem is valid [3].

§2. OUTLINE OF PROOF

We remark that the estimates below are stated for the general case [3]. In the special case under consideration, it is easy to improve some of them substantially.

Let $\Phi : M \to \mathbf{R}$ be weakly continuous and differentiable and let μ, ν be in M. Then $\mu(t) = (1-t)\mu + t\nu \in M$ and $\mu(t) = \mu(s) + (t-s)(\nu - \mu)$ for $0 \leq s < t \leq 1$; this implies $\nu - \mu$ is tangent to M at $\mu(t)$ for $0 < t < 1$ and so $f(t) = \Phi(\mu(t))$ is differentiable on $0 < t < 1$ with $f'(t) = (\nu - \mu)(D\Phi(\mu(t)))$. Thus

$$(2.1) \qquad \Phi(\nu) - \Phi(\mu) = \int_0^1 (\nu - \mu)(D\Phi(\mu(t)))dt.$$

Since $|\mu(\varphi)| \leq \|\varphi\|_2 \|\mu\|_{-2}$, it follows that Φ is Lipschitz on M relative to $\|\cdot\|_{-2}$ whenever $\|D\Phi(m)\|_2 \leq C$. This shows that (2) is implied by (1) in the Theorem. Note also that if $\mu_n \to \mu$ weakly then by Ascoli's theorem $\|\mu_n - \mu\|_{-2} \to 0$. This implies that $\Phi : M \to \mathbf{R}$ is weakly continuous whenever Φ is Lipschitz relative to $\|\cdot\|_{-2}$. This shows (2) in the Theorem implies v is weakly continuous. Note also that $\|\cdot\|_{-2} \leq \|\cdot\|_{-1} \leq \|\cdot\|_0 \leq 1$ on M.

We will need to work with weak sense controls following Fleming and Pardoux [1]. A *generalized control* is a filtered probability space $\alpha = (\Omega, \mathcal{Y}, \mathcal{Y}_t, P)$ equipped with $\mathcal{Y}_t$ progressively measurable processes (y, u) such that u is U-valued and y is a $\mathcal{Y}_t$ Brownian motion. A generalized control α is a (strict-sense) *control* if in addition u is a progressively measurable function of y.

Let x, $\mathcal{X}$, $\mathcal{X}_t$, $t \geq 0$, denote the canonical process, Borel σ-algebra, and canonical filtration respectively on $C([0, \infty); \mathbf{R})$. For each measurable $\beta : [0, \infty) \to U$ and $m \in M$ let E_m^β denote the expectation (on $\mathcal{X}$)

$$E_m^\beta(\Phi(x)) = \int_{\mathbf{R}} E\left(\Phi\left(a + \int_0^\cdot \beta(t)dt + \xi(\cdot)\right)\right) dm(a).$$

Given a generalized control α and $m \in M$ let $Q^0 = Q^0(\alpha, m)$ be the unique law on $\mathcal{X} \times \mathcal{Y}$ such that

(1) the marginal of Q^0 on $\mathcal{Y}$ is P and
(2) the conditional expectation of x given $\mathcal{Y}$, under Q^0, is E_m^u.

It follows then that y is an $\mathcal{X}_t \times \mathcal{Y}_t$ Brownian motion under Q^0 and hence

$$R(t) = \exp\left(\int_0^t c(x(s))dy(s) - \frac{1}{2}\int_0^t c(x(s))^2 ds\right), t \geq 0,$$

is well-defined.

The *generalized system* starting at m and corresponding to α is the law $Q = Q(\alpha, m)$ on $\mathcal{X} \times \mathcal{Y}$ satisfying $dQ/dQ^0 = R(t)$ on $\mathcal{X}_t \times \mathcal{Y}_t$ for all $t \geq 0$. Clearly the law of (x, y, u) under $Q = Q(\alpha, m)$ depends only on m and the law of (y, u) under P.

Let $v^\alpha(m)$ denote the right side of (1.3) where the expectation is against $Q = Q(\alpha, m)$ and let $v'(m) = \inf_\alpha v^\alpha(m)$. We say a generalized control is *optimal at* m if $v^\alpha(m) = v'(m)$.

For each measurable $\beta : [0, \infty) \to U$ and $(s, x) \in [0, \infty) \times \mathbf{R}$ let $E^\beta_{s,x}$ denote the expectation (on $\mathcal{X}$)

$$E^\beta_{s,x}(\Phi(x)) = E\left(\Phi\left(x + \int_s^{\cdot \vee s} \beta(t)dt + \xi(\cdot \vee s) - \xi(s)\right)\right).$$

Given a generalized control α and $(s, x) \in [0, \infty) \times \mathbf{R}$ let $Q^0 = Q^0(\alpha, s, x)$ be the unique law on $\mathcal{X} \times \mathcal{Y}$ such that

(1) the marginal of Q^0 on $\mathcal{Y}$ is P and
(2) the conditional expectation of x given $\mathcal{Y}$, under Q^0, is $E^u_{s,x}$.

It follows then that y is an $\mathcal{X}_t \times \mathcal{Y}_t$ Brownian motion under Q^0 and hence

$$R(t|s) = \exp\left(\int_s^t c(x(r))dy(r) - \frac{1}{2}\int_s^t c(x(r))^2 dr\right), t \geq s,$$

is well-defined. Let

$$T_{t,s}\varphi(x) = E^{Q^0}(\varphi(x(t))R(t|s)|\mathcal{Y}_t) = E^u_{s,x}(\varphi(x(t))R(t|s)), t \geq s.$$

Then $T_{t,s}$ is an operator-valued $\mathcal{Y}_t$-measurable random variable.

Differentiation under the expectation sign yields the following.

Lemma 1. *There exists $C > 0$ such that for all $t \geq s \geq 0$ there is a $\mathcal{Y}_t$-measurable random variable $M(t, s, x) \geq 0$ satisfying*

(1) $E^P(M(t, s, x)|\mathcal{Y}_s) \leq C(1 + (t - s))$ *for all $x \in \mathbf{R}$,*
(2) *for $\varphi \in C^2_b(\mathbf{R})$, $x \in \mathbf{R}$,*

$$|\partial^k_x T_{t,s}\varphi(x)| \leq M(t, s, x)\|\varphi\|_2, k = 0, 1, 2,$$

almost surely. □

Set

(2.2)
$$p(s) = p(s, \cdot) = E^P\left(\int_s^\infty e^{-\lambda(t-s)}T_{t,s}f(\cdot, u(t))dt\bigg|\mathcal{Y}_s\right).$$

Then $p(s)$ is a bounded function on $\mathbf{R}$ for each $s \geq 0$, almost surely. An immediate consequence of the above Lemma is

Lemma 2. *There exists $C > 0$ such that $p(t)$ is in $C_b^2(\mathbf{R})$ and*

$$\|p(t)\|_2 \leq C \frac{\lambda + 1}{\lambda^2}$$

for $t \geq 0$ and $\lambda > 0$, almost surely. $\square$

Let $\mu(t) \in M$ denote the conditional distribution $\mu(t)(\varphi) = E^Q(\varphi(x(t))|\mathcal{Y}_t)$ of $x(t)$ given $\mathcal{Y}_t$; here $Q = Q(\alpha, m)$. This is defined for any α and agrees with $\mu_m^u(t)$ as defined in §1 when α is strict-sense.

Lemma 3. *Let $\lambda > 0$ and suppose α is optimal at m. Then*

$$(2.3) \qquad u(t) = \mathbf{u}(\mu(t), p(t)), t \geq 0. \quad \square$$

This Lemma is obtained for $t = 0$ by an Euler-Lagrange argument and then extended to all $t > 0$ by the Markov property. This Lemma shows that $p(t) \in C_b^2(\mathbf{R})$ plays the role of the co-state (adjoint) process dual to $\mu(t) \in M$.

More explicitly, given a constant control $a \in U$ and $\epsilon > 0$ and an optimal (generalized) control $u(\cdot)$ define $u^\epsilon(t)$ by setting it equal to a for $t < \epsilon$ and equal to $u(t)$ elsewhere; let $v^\epsilon(m)$ denote the corresponding cost. Since $v^\epsilon(m) \geq v^0(m) = v^u(m)$, it follows that $(d/d\epsilon)v^\epsilon(m) \geq 0$ at $\epsilon = 0$. Performing the differentiation explicitly we obtain an inequality valid for all $a \in U$ which yields (2.3) at $t = 0$.

We now have a formula (2.2) yielding the dependence of $p(t)$ on $u(t)$ and a formula (2.3) yielding the dependence of $u(t)$ on $\mu(t)$, $p(t)$; to obtain a closed system we need to know the dependence of $\mu(t)$ on u. This is given by the Bayes formula as follows.

Lemma 4. *For $t \geq 0$, $\varphi \in C_b(\mathbf{R})$, $m \in M$, $\lambda > 0$, and any α,*

$$(2.4) \qquad \mu(t)(\varphi) = \frac{E_m^u(\varphi(x(t))R(t))}{E_m^u(R(t))},$$

almost surely $Q(\alpha, m)$. $\square$

Following Fleming-Pardoux [1] one can impose a notion of convergence on the set of generalized controls α such that

(1) the set of generalized controls is compact,
(2) the function $(m, \alpha) \mapsto v^\alpha(m)$ is continuous,

where the weak topology is on M. This leads to

Lemma 5. *For each m and $\lambda > 0$ there is a generalized control optimal at m.* $\square$

Since this aspect of the subject has been known since the early 1980's, we omit the definition of the appropriate notion of convergence but we note that to handle the nonlinear dependence of f on u one must (temporarily) work with a slightly larger class of generalized controls valued in $M(U)$ instead of U and modify Lemma 3 accordingly in order to obtain Lemmas 3 and 5 *as stated above*.

The crucial estimate is to compare two generalized controls, one optimal at m_1 and one optimal at m_2. However two generalized controls can be compared only if they are defined on the same probability space. Because of this, we use the Watanabe-Yamada technique to establish

Lemma 6. *Given two generalized controls* $(\Omega_i, \mathcal{Y}_t^i, P_i, y_i, u_i)$, $i = 1, 2$, *there exists a probability space* $(\Omega^*, \mathcal{Y}_t^*, P^*)$ *supporting processes* (y^*, u_1^*, u_2^*) *such that the law of* (y_i, u_i) *under* P_i *equals the law of* (y^*, u_i^*) *under* P^*. $\square$

The Watanabe-Yamada technique was invented to establish pathwise or strong uniqueness for initial value problems. Here we use it to establish Lipschitz dependence of $(\mu(t), p(t))$ on the *boundary* conditions at $t = 0$ and $t = \infty$, $\mu(0) = m$ and $e^{-\lambda t} p(t)|_{t=\infty} = 0$.

Now consider two generalized controls optimal at m_1 and m_2 respectively. By Lemma 6 we can assume they are defined over the same filtration $\mathcal{Y}_t$ with the same process y and probability measure P. Let $\mu_i(t)$, $p_i(t)$, denote the corresponding objects, $i = 1, 2$.

For $\rho : [0, \infty) \to [0, \infty)$ set

$$K\rho(t) = K_{C,\lambda}\rho(t) = C \int_0^t e^{C(t-s)} \rho(s)ds + C \int_t^\infty e^{(C-\lambda)(s-t)} \rho(s)ds.$$

Set

$$\rho(t) = E^P \left(\|\mu_1(t) - \mu_2(t)\|_{-1}^2 + \|p_1(t) - p_2(t)\|_1^2 \right).$$

Here E^P denotes expectation against the background probability common to both generalized controls. Then we have two closed systems (2.2), (2.3), (2.4), one for each generalized control. Straightforward but tedious estimates yield the crucial estimate:

Lemma 7. *There exists* $C > 0$ *such that for* $\lambda > 0$

$$\rho(t) \le Ce^{Ct} \|m_1 - m_2\|_{-1}^2 + K_{C,\lambda}\rho(t), t \ge 0. \quad \square$$

Now set $\lambda_1 = \max(6C, 1)$. Then for $\lambda > \lambda_1$ we have $\rho(t) \le C' \le C'e^{3Ct}$ for some C' and $K(e^{3Ct}) \le \delta e^{3Ct}$ with $\delta < 1$. Because of this iterating the inequality in Lemma 7 yields

Lemma 8. *For* $\lambda > \lambda_1$,

$$\rho(t) \le Ce^{3Ct} \|m_1 - m_2\|_{-1}^2, t \ge 0.$$

In particular

$$\|p_1(0) - p_2(0)\|_1 \le C\|m_1 - m_2\|_{-1}. \quad \square$$

Recalling the definition of $Q^0(\alpha, m)$, $Q(\alpha, m)$ it follows that

$$v^\alpha(m) = E^P \left(\int_0^\infty e^{-\lambda t} E_m^u(f(x(t), u(t))R(t))dt \right)$$

$$= \int_{\mathbf{R}} E^P \left(\int_0^\infty e^{-\lambda t} E_{0,x}^u(f(x(t), u(t))R(t))dt \right) dm(x) = m(p(0))$$

is affine in m. From this we have

Lemma 9. *For each* α, m, *and* $\lambda > 0$, $Dv^\alpha(m) = p(0)$ *and the map* $(\alpha, m) \mapsto Dv^\alpha(m)$ *is continuous in an appropriate sense.* $\square$

If we set $F(m)$ equal to $Dv^\alpha(m)$ for any α optimal at m, Lemmas 8 and 9 show that $F(m)$ is well-defined and Lipschitz relative to $\|\cdot\|_{-1}$, when $\lambda > \lambda_1$.

Lemma 10. *For $\lambda > \lambda_1$, v' is differentiable on M and $Dv' = F$. Moreover if α is optimal at m, then*

$$(2.5) \qquad u(t) = \mathbf{u}(\mu(t), Dv'(\mu(t))), t \geq 0. \quad \square$$

To establish this, fix $m \in M$ and ν tangent to M at m and let α_t be an optimal generalized control at $m + t\nu$. Then for $t > 0$

$$\frac{v'(m + t\nu) - v'(m)}{t} \leq \frac{v^{\alpha_0}(m + t\nu) - v^{\alpha_0}(m)}{t}.$$

Thus

$$\limsup_{t \downarrow 0} \frac{v'(m + t\nu) - v'(m)}{t} \leq \nu(Dv^{\alpha_0}(m)) = \nu(F(m)).$$

Also

$$\frac{v'(m + t\nu) - v'(m)}{t} \geq \frac{v^{\alpha_t}(m + t\nu) - v^{\alpha_t}(m)}{t} = \int_0^1 \nu(Dv^{\alpha_t}(m + st\nu))ds.$$

Passing to a subsequence $\alpha_t \to \alpha_0'$ we obtain, since $(\alpha, m) \mapsto v^\alpha(m)$ is continuous, α_0' is optimal at m and hence

$$\liminf_{t \downarrow 0} \frac{v'(m + t\nu) - v'(m)}{t} \geq \nu(Dv^{\alpha_0'}(m)) = \nu(F(m))$$

since $(\alpha, m) \mapsto \nu(Dv^\alpha(m))$ is continuous (Lemma 9). Thus

$$\lim_{t \downarrow 0} \frac{v'(m + t\nu) - v'(m)}{t} = \nu(F(m)).$$

Replacing ν by $-\nu$ shows that $Dv'(m) = F(m)$ for all m.

Since $Dv'(m) = p(0) = Dv^\alpha(m)$ we obtain (2.5) at $t = 0$ from Lemma 3. For $t > 0$ we use the Markov property to reduce to $t = 0$. This establishes Lemma 10.

Lemma 11. *Suppose $\lambda > \lambda_1$. Then every α optimal at m is necessarily induced by a strict-sense control. Moreover for each m there is exactly one strict-sense control u solving the feedback*

$$(2.6) \qquad u(t) = \mathbf{u}(\mu_m^u(t), Dv(\mu_m^u(t))), t \geq 0,$$

and $v(m) = v'(m)$.

To establish this one first notes the feedback $F(m) = \mathbf{u}(m, Dv'(m))$ is Lipschitz relative to $\|\cdot\|_{-1}$ for $\lambda > \lambda_1$ by Lemma 10. One then solves the fixed point formula (2.5) by iteration to produce a strong solution. This can be done for any feedback F Lipschitz relative to $\|\cdot\|_{-1}$. Moreover the same iteration procedure shows that any weak solution of the fixed point formula (2.5) equals the strong solution. This then implies $v = v'$ and hence (2.5) implies (2.6). The proof of Lemma 11 is entirely pathwise i.e. no moments are involved. The details are in [3].

Combining Lemmas 2, 9, 10, 11 yields the differentiability of v together with the estimates $\|Dv(m)\|_2 \le C$, $\|Dv(m) - Dv(m')\|_1 \le C\|m - m'\|_{-1}$, for $\lambda > \lambda_1$. This completes the outline of the proof.

We conclude with some remarks concerning the roles of the Kushner-Stratonovitch and Zakai equations for the normalized and unnormalized conditional distributions respectively.

A natural technique to estimate quantities such as $\|\mu(t)\|_{-k}$ or $\|\mu_1(t) - \mu_2(t)\|_{-k}$ is by time differentiation, using either of these equations, and solving the resulting differential inequalities; however this method is not available here as $\|\cdot\|_{-k} : M \to \mathbf{R}$ is not differentiable, at least not in a convenient sense. Because of this difficulty one must, if one follows this approach, use a differentiable norm by interpreting $\|\cdot\|_{-k}$ as an L^2 Sobolev norm and reformulating the Theorem accordingly. This is done in [4] for a different though related result. Since we instead deal directly with the Bayes formula (2.4) and do not use the above technique, we avoid this problem entirely. Nevertheless the idea that in these types of problems one *must* use the $(\|\varphi\|_1, \|\mu\|_{-1})$ duality instead of the simpler $(\|\varphi\|_0, \|\mu\|_0)$ duality first appears in [4].

Another remark concerns the fundamental estimate in Lemma 7. Throughout we work with the *normalized* conditional distribution. If instead in defining $\rho(t)$ we work with the *unnormalized* conditional distribution, the estimate in Lemma 7 fails. We explain this phenomenom in terms of the differential equations of filtering; although ultimately we avoid them, the difficulty persists. This is because the Zakai equation is *bilinear* in μ, u and so the moments of the difference $\|\mu_1(t) - \mu_2(t)\|_{-1}$ cannot be estimated in terms of the difference of two controls $|u_1(t) - u_2(t)|$. In the normalized case the Kushner-Stratonovitch equation is also bilinear (in fact a little worse) but we have now the *a priori* estimate $\|\mu(t)\|_0 = 1$ *almost surely* whose effect is to make the coefficients bounded, resolving the difficulty. This difficulty persists even for bilinear equations in finite dimensions (e.g. when the state space is finitely many points instead of $\mathbf{R}$).

References

1. W. H. Fleming & E. Pardoux, *Existence of Optimal Control for Partially Observed Diffusions*, SIAM J. Control 20 (1982), 261-283.
2. O. Hijab, *Control of Degenerate Diffusions in $\mathbf{R}^d$*, Trans. Amer. Math. Soc. (October 1991).
3. O. Hijab, *Partially Observed Control of Markov Processes, IV*, to appear, J. Funct. Anal. (1992).
4. P. L. Lions, *Viscosity Solutions of Fully Nonlinear Second-Order Equations and Optimal Stochastic Control in Infinite Dimensions, Part II: Optimal Control of Zakai's Equation*, CEREMADE preprint #8825 (1989).

DEPARTMENT OF MATHEMATICS, TEMPLE UNIVERSITY, BROAD & MONTGOMERY, PHILADELPHIA, PA 19122

E-mail: hijab@euclid.math.temple.edu

ASYMPTOTICALLY OPTIMAL POLICIES FOR CONTROLLED QUEUES IN HEAVY TRAFFIC

E. V. Krichagina
University of Toronto

M. I. Taksar
SUNY at Stony Brook

Abstract

We consider $GI/G/1$ queueing system operating in heavy traffic. There is a proportional holding cost with coefficient h. A controller can increase the service rate (thus moving the system into light traffic) incurring the cost proportional to the increased rate with coefficient l. Given the discount rate ρ the objective is to minimize the total expected discounted cost. We show that when the parameters δ associated with heavy traffic and $\bar{\delta} = h/l$ are small and ρ has an appropriate order this control problem can be approximated by a singular control problem for Brownian motion, namely, reflected follower problem. The optimal policy for the limiting system is characterized by a single number z^*, so that the optimal process is a reflected diffusion on $[0, z^*]$. We find a transcendental equation whose root yields z^*. Using the solution of the limiting problem, we find an approximately optimal policy for the original system. This policy is characterized by a critical level z_0. The service rate should be increased to maximum whenever the queue length process exceeds z_0.

Key words and phrases. Singular stochastic control, controlled queues, diffusion approximation, reflected Brownian motion, free boundary problem.

1 Introduction

We consider a controlled $GI/G/1$ queue with linear holding cost. A controller has an option to increase the service rate incurring the cost proportional to the service rate increase. The objective is to minimize the total discounted expected cost.

The queue length process in $GI/G/1$ model is not Markov, therefore the optimal control rate at each moment t should depend on the size of the queue, time since the arrival of the last customer and the integral of the service rate from the time of beginning of the last service prior to t. We are not aware of any works in which the problem considered in our paper is solved in a closed form. We therefore turn to the aid of an asymptotic approach based on the idea of diffusion approximation. The essential features of this approach can be described as follows. Based on the heavy-traffic limit theorems, the control problem for the original system is approximated by a limiting optimal control problem involving Brownian motion. This control problem is easier to analyze and in many cases it is explicitly solvable. The solution is then interpreted in the original terms and a certain control policy is derived for the original system.

In a number of papers [6], [7], [18] controlled diffusions have been used as models for controlled queueing networks. In these papers the original processes are replaced by Brownian motions. In thus modified models diffusion control theory can be used to derive the optimal policy. The optimal policies obtained are then interpreted in terms of the original system. Justification of the procedure employed is based usually on mere intuition or simulation.

Another kind of papers such as [12]-[16] deal with a sequence of systems in heavy traffic whose limit is a controlled diffusion. The convergence of the optimal costs is proved. If there exists an optimal control for the limiting system then one can construct a sequence of *asymptotically optimal* policies defined to be those for which the difference between the associated cost and the value function converges to zero as traffic intensity approaches its critical value.

The main purpose of studying diffusion approximations is to develop a reasonable policy for a given real system with traffic intensity close to one. To this end the real system is imbedded into a sequence of systems in heavy traffic, being identified with one of the elements of this sequence. The sequence has an asymptotically optimal policy and thus one has a reasonable policy for the original system. If the traffic intensity of the original system is close to one than one can expect that the policy suggested is close to optimal or is *nearly optimal* in a certain sense.

However, there are several issues which have to be considered in this context. First, the sequence of systems and their parameters are not uniquely determined by the original system and therefore one can construct infinitely many sequences which the original system can be imbedded into. To eliminate this ambiguity we show that the policy recommended for the original system can be expressed only in terms of its parameters and thus does not depend on what sequence was used to derive the answer. Second, asymptotic analysis of the systems in heavy traffic needs rescaling of the state space. The rescaling procedure introduces the difference of several orders of magnitude between the cost functional of the original system and that of its representation in the sequence. For this reason the relative error rather than the absolute error should be used to evaluate the quality of control, since the relative errors for the original system and for its representation in the sequence coincide. These two particularities of the use of diffusion approximation for real systems were first mentioned in [10], where asymptotic analysis via diffusion approximation was carried out for stochastic manufacturing system with unreliable machine.

The paper has the following structure. We start with a formulation of the optimal control problem for the original system S (Section 2). The system is assumed to operate in heavy traffic, i.e., the ratio of arrival and service rates is close to one. In Section 3 we construct a sequence $\{S_\epsilon\}$ of systems similar to the original one with parameters being functions of ϵ. This sequence is later used for asymptotic analysis. In Section 4 we introduce the system S_W which is the limiting system for the sequence S_ϵ as $\epsilon \to 0$. The corresponding optimal control problem is a singular control problem for a diffusion process (reflected follower problem). The nature of the solution to the reflected follower problem is similar to those of other papers on singular control (see for example [1], [4],[8]). There exists a level z^* such that the optimal process X^* is a Brownian motion on $[0, z^*]$ with reflection at the boundaries of this interval. The optimal control functional L^* is the one which induces the reflection at z^*, i.e. it coincides with the local time of X^* at z^*. Although the works by Karatzas and Shreve [8] and Baldurson [1] have a comprehensive analysis of the reflected follower problem an analytical expression for the optimal level z^* is not given there. We derive a transcendental equation, whose root yields the value of z^*.

The main theoretical results of the paper are formulated in Section 5. The optimal control problem for S_ϵ is related to that of for the limiting system S_W. We prove that the lower limit of the optimal costs for the sequence $\{S_\epsilon\}$ is bounded below by the optimal cost for the limiting system S_W. Using an explicit expressions for the optimal level z^*, we construct a threshold control policy for S_ϵ characterized by the parameter $z^\epsilon \epsilon^{-1}$ such that the corresponding costs converge to this lower bound as ϵ approaches zero. These policies are proved to be asymptotically optimal.

In Section 6 we apply these results to the original system S. The system S is identified with S_{ϵ_0} for some ϵ_0. We use the constructed control policy for S and show that the corresponding threshold value can be expressed in terms of the original system parameters.

2 Description of the Original System S

We start with two independent sequences of nonnegative i.i.d. random variables $\{\xi_i\}$ and $\{\eta_i\}$, $i = 1, 2, \ldots$ on a probability space $(\mathcal{A}, \mathcal{G}, P)$ representing the interarrival and service times. Denote

$$E\{\xi_1\} = \lambda^{-1} \ , \ \ E\{\eta_1\} = \mu^{-1} \ , \ \ \mathrm{Var}\{\xi_1\} = \sigma_1^2 \ \ , \ \ \mathrm{Var}\{\eta_1\} = \sigma_2^2 \ .$$

Put

$$A(t) = \max\{k \geq 1 : \xi_1 + \ldots + \xi_k \leq t\} \ , \ S(t) = \max\{k \geq 1 : \eta_1 + \ldots + \eta_k \leq t\} \ ,$$

assuming $\max\{\emptyset\} = 0$. In the absence of control the length of the queue at time t is given by

$$Q(t) = x + A(t) - S(B(t)) . \tag{2.1}$$

Here x is the initial number of customers present at $t = 0$ and

$$B(t) = \int_0^t I(Q(s) > 0) \, ds ,$$

where $I(A)$ is the indicator function of the set A (see [5]). At each time t the controller chooses a service rate $1 + p(t)$, $0 \le p(t) \le r$. Thus, $(r + 1)\mu$ is the maximal service rate. In the presence of a control the queue length is also given by formula (2.1) with

$$B(t) = \int_0^t (1 + p(s)) \, I(Q(s) > 0) \, ds . \tag{2.2}$$

Intuitively, control consists of increasing the speed of service process, without changing the probabilistic structure of the underlying renewal process, a sort of "making time run faster" for the server. The next definition specifies restrictions we impose on control functionals.

Definition 1. A real-valued process $(p(\cdot), t \ge 0)$ on $(\mathcal{A}, \mathcal{G}, P)$ is called an admissible policy, if the following three conditions are satisfied

(2.A) $0 \le p(t) \le r$,

(2.B) equations (2.1), (2.2) have a unique solution with probability one,

(2.C) the process $p(t)$ is adapted to the filtration $\mathcal{G}_t \equiv \sigma(A(s), S(B(s)), s \le t)$ and the sequence of random vectors $\{\xi_{A(t)+1+i}, \eta_{S(B(t))+1+i}\}$, $i = 1, 2\dots$ is independent of $\mathcal{G}_t$ and has the same distribution as $\{\xi_i, \eta_i\}, i = 1, 2 \dots$

We use $\mathcal{P}$ to denote the set of admissible policies. Note that in the definition of an admissible policy we do not fix the filtration $\mathcal{G}_t$ a priori for all $p(\cdot)$, rather it depends on the process $p(\cdot)$ itself. Condition (2.C) is an analog of a usual requirement for a control to be "nonanticipative" with respect to the future. A question arises whether admissible policies exist. In fact the class of admissible policies is large enough. One can verify explicitly that any feedback policy (i.e., $p(\cdot) = f(Q(\cdot))$) is admissible.

The objective is to minimize

$$J(x, p(\cdot)) = E\{h \int_0^\infty \exp(-\rho t)Q(t)dt + \int_0^\infty l \exp(-\rho t)p(t)dt\} , \tag{2.3}$$

over all admissible policies $p(\cdot) \in \mathcal{P}$. Here $\rho > 0$ is a given discount rate, h is holding cost coefficient and l is the cost for service rate increase. We assume that

$$h > \rho l \mu^{-1} . \tag{2.4}$$

We denote the corresponding value function as

$$v(x) = \inf_{\mathcal{P}} J(x, p(\cdot)) \ . \tag{2.5}$$

In the above formula x is integer. It stands for the initial position of the queueing process (see (2.1)).

A closed form solution for the optimal control problem (2.1)- (2.5) is very difficult of not impossible to obtain. In view of this, we assume that the system S is operating in heavy traffic and find an approximate solution by means of asymptotic analysis. The heavy traffic assumption means that

$$\delta = 1 - \frac{\lambda}{\mu} \tag{2.6}$$

is close to one.

3   Sequence of Systems S_ϵ in Heavy Traffic

Consider a family of systems $\{S_\epsilon\}$ indexed by parameter $\epsilon \to 0$ belonging to a countable set $\mathcal{E}$. Each system S_ϵ is a clone of the original system S. It is characterized by its maximal service rate r^ϵ, holding and service cost coefficients h^ϵ and l^ϵ, discount rate ρ^ϵ and families of i.i.d. random variables $\{\xi_i^\epsilon\}$, $\{\eta_i^\epsilon\}$ with corresponding mean and variance parameters indexed by ϵ as well. Our main assumptions on the family $\{S_\epsilon\}$ are the following. There exist constants $\hat{\lambda}, \hat{\mu}, \hat{r}, \hat{l}, \hat{h}, \hat{\sigma}_i^2, i = 1, 2$; $c > 0; \gamma > 0$ and $q > 0$ such that:

$$a) \quad \begin{aligned} &\lambda^\epsilon \to \hat{\lambda} \ , \mu^\epsilon \to \hat{\mu} \ , r^\epsilon \to \hat{r} \ , h^\epsilon \to \hat{h} \ , \\ &\sigma_{1\epsilon}^2 \to \hat{\sigma}_1^2 \ , \sigma_{2\epsilon}^2 \to \hat{\sigma}_2^2 \ , \epsilon^2 l^\epsilon \to \hat{l} \ , \epsilon \to 0 \ ; \end{aligned} \tag{3.1}$$

$$b) \quad c^\epsilon \equiv \epsilon^{-1}(\lambda^\epsilon - \mu^\epsilon) \to c, \ \epsilon \to 0 \ ; \tag{3.2}$$

$$c) \quad \gamma^\epsilon \equiv \rho^\epsilon \epsilon^{-2} \to \gamma \ , \ \epsilon \to 0 \ , \tag{3.3}$$

$$d) \quad \sup_\epsilon E(\xi_1^\epsilon)^{2+q} \le \infty \ , \ \sup_\epsilon E(\eta_1^\epsilon)^{2+q} \le \infty \ . \tag{3.4}$$

Note that according to (3.2) $\hat{\lambda} = \hat{\mu}$ and the traffic intensity of the system S_ϵ, i. e. $(1 - \lambda^\epsilon (\mu^\epsilon)^{-1})$ converges to one at the order of ϵ.

Let $p^\epsilon(\cdot)$ be an admissible policy for the system S_ϵ and $Q^\epsilon(\cdot)$ be defined similar to (2.1), namely

$$Q^\epsilon(t) = x + A^\epsilon(t) - S^\epsilon(B^\epsilon(t)) \ , \ \ B^\epsilon(t) = \int_0^t I(Q^\epsilon(s) > 0)(1 + p^\epsilon(s))ds \ , \tag{3.5}$$

where $A^\epsilon(t) = \max\{k : \xi_1^\epsilon + \ldots + \xi_k^\epsilon \leq t\}$ and $S^\epsilon(t) = \max\{k : \eta_1^\epsilon + \ldots + \eta_k^\epsilon \leq t\}$. For each initial position x in (3.5) and each $\epsilon > 0$, let J^ϵ be the following cost functional associated with $p^\epsilon(\cdot)$:

$$J^\epsilon(x, p^\epsilon(\cdot)) = \epsilon^3 E\{h^\epsilon \int_0^\infty \exp(-\rho^\epsilon t) Q^\epsilon(t) dt + l^\epsilon \int_0^\infty \exp(-\rho^\epsilon t) p^\epsilon(t) dt\} . \tag{3.6}$$

Put

$$v^\epsilon(x) = \inf_{p^\epsilon} J^\epsilon(x, p(\cdot)) . \tag{3.7}$$

Definition 2. A sequence of controls $\tilde{p}^\epsilon(\cdot)$, $\epsilon \in \mathcal{E}$ is said to be *asymptotically optimal* for the sequence S_ϵ in the sense of *absolute error* if

$$\lim_{\epsilon \to 0} |v^\epsilon(x\epsilon^{-1}) - J^\epsilon(x\epsilon^{-1}, \tilde{p}^\epsilon(\cdot))| = 0 ,$$

and *asymptotically optimal* in the sense of *relative error* if

$$\lim_{\epsilon \to 0} |v^\epsilon(x\epsilon^{-1}) - J^\epsilon(x\epsilon^{-1}, \tilde{p}^\epsilon(\cdot))|(v^\epsilon(x\epsilon^{-1}))^{-1} = 0 . \tag{3.8}$$

In the next section we construct a sequence of asymptotically optimal controls both in sence of absolute and relative errors using the solution to the limiting control problem.

4 Limiting System S_W

As a limiting model for the family of systems S_ϵ we consider the so-called reflected follower problem, i.e., singularly controlled Brownian motion with drift c equal to the right-hand side of (3.2) and variance

$$\sigma^2 = \hat{\lambda}^3 \hat{\sigma}_1^2 + \hat{\mu}^3 \hat{\sigma}_2^2 , \tag{4.1}$$

where $\hat{\lambda}, \hat{\sigma}_1^2, \hat{\mu}, \hat{\sigma}_2^2$ are defined by (3.1). In this problem we start with (c, σ^2) Brownian motion $W(t)$ on $(\Omega, \mathcal{F}, \mathcal{F}_t, P)$. Control is described by a nondecreasing process $L(t)$ adapted to $\mathcal{F}_t$. The dynamics of the system under control is

$$X(t) = x + W(t) - L(t) + R(t) , \tag{4.2}$$

where $x > 0$ is the initial position and

$$R(t) = -\min[\, 0, \, \min_{s \leq t}(x + W(s) - L(s)) \,] . \tag{4.3}$$

The functional $R(\cdot)$ in (4.2) is not a part of the control. It maintains reflection of $X(t)$ at 0, so that the controlled process always stays nonnegative. With each control functional L we associate the cost

$$J_x(L) = E\{\hat{h} \int_0^\infty e^{-\gamma t} X(t) dt + \hat{l}\hat{\mu}^{-1} \int_0^\infty e^{-\gamma t} dL(t)\} , \tag{4.4}$$

where $x = X(0)$.

Objective is to find

$$V(x) = \inf_L J_x(L) \tag{4.5}$$

and the functional $L^*(\cdot)$ such that $V(x) = J_x(L^*)$. We call $L^*(\cdot)$ the optimal policy or optimal control for the limiting system.

Put $\alpha = \sigma^{-2}(-c + \sqrt{c^2 + 2\gamma\sigma^2})$ and $\beta = \sigma^{-2}(c + \sqrt{c^2 + 2\gamma\sigma^2})$. Consider the following equation

$$\alpha(\frac{\hat{l}}{\hat{\mu}} - \frac{\hat{h}}{\gamma})e^{\beta z} + \beta(\frac{\hat{l}}{\hat{\mu}} - \frac{\hat{h}}{\gamma})e^{-\alpha z} + (\alpha + \beta)\frac{\hat{h}}{\gamma} = 0 \tag{4.6}$$

One can easily see that (4.6) has a unique solution. Indeed, let $m(z)$ be the left-hand side of (4.6). Due to (2.4) we may assume $\hat{h}\gamma^{-1} > \hat{l}\hat{\mu}^{-1}$. Therefore we have $m(0) = \hat{l}\hat{\mu}^{-1}(\alpha + \beta) > 0$ and $m(+\infty) = -\infty$. It is easy to verify that $m(\cdot)$ is decreasing function. Thus, (4.6) has unique positive solution z^*.

Let L^* be the functional such that (4.2), (4.3) and (4.7) below are satisfied

$$L^*(t) = \sup_{s \le t}[X(s) + R(s) - z^*]^+ . \tag{4.7}$$

Theorem 1 describes the solution to the reflected follower problem.

Theorem 1. *The function $V(x)$ given by (4.5) is twice continuously differentiable and satisfies the following conditions*

$$1/2\sigma^2 V''(x) + cV'(x) - \gamma V(x) + \hat{h}x = 0 , \qquad 0 \le x \le z^* ,$$
$$V'(0) = 0 , \quad V'(x) = \hat{l}\hat{\mu}^{-1} , \qquad x \ge z^* .$$

The functional L^* defined by (4.7) is optimal.

The proof is similar to that of Theorem 1 of Krichagina and Taksar [9].

5 Main Results

In this section we specify the relation between the structure of the optimal control for the limiting system and asymptotically optimal controls for the sequence $\{S_\epsilon\}$.

Theorem 2. *Let $V(x)$ be the optimal cost of the limiting problem defined by (4.5). Then for any x and any admissible control $p^\epsilon(\cdot)$ we have $V(x) \le \liminf_{\epsilon \to 0} J^\epsilon(x\epsilon^{-1}, p^\epsilon(\cdot))$.*

Define

$$\sigma_\epsilon^2 = (\lambda^\epsilon)^3 \sigma_{1\epsilon}^2 + (\mu^\epsilon)^3 \sigma_{2\epsilon}^2 , \tag{5.1}$$

$$\alpha^\epsilon = \sigma_\epsilon^{-2}(-c^\epsilon + \sqrt{(c^\epsilon)^2 + 2\gamma^\epsilon\sigma_\epsilon^2}) \ , \beta^\epsilon = \sigma_\epsilon^{-2}(c^\epsilon + \sqrt{(c^\epsilon)^2 + 2\gamma^\epsilon\sigma_\epsilon^2}) \ . \tag{5.2}$$

Let z^ϵ be the solution to the equation

$$\alpha^\epsilon(\frac{l^\epsilon\epsilon^2}{\mu^\epsilon} - \frac{h^\epsilon}{\gamma^\epsilon})e^{\beta^\epsilon z} + \beta^\epsilon(\frac{l^\epsilon\epsilon^2}{\mu^\epsilon} - \frac{h^\epsilon}{\gamma^\epsilon})e^{-\alpha^\epsilon z} + (\alpha^\epsilon + \beta^\epsilon)\frac{h^\epsilon}{\gamma^\epsilon} = 0 \ . \tag{5.3}$$

Note. One can verify directly that (5.3) as well as (6.5) below have unique solution.

Consider the policy

$$\bar{p}^\epsilon(t) = r^\epsilon I(Q^\epsilon(t) > z^\epsilon\epsilon^{-1}) \ . \tag{5.4}$$

Theorem 3. Let $\bar{p}^\epsilon(t)$ be defined by (5.4). Then $V(x) = \lim_{\epsilon \to 0} J^\epsilon(x\epsilon^{-1}, \bar{p}^\epsilon(\cdot))$.

Corollary 1. Let v^ϵ be defined by (3.7). Theorems 2 and 3 imply $\lim_{\epsilon \to 0} v^\epsilon(x\epsilon^{-1}) = V(x)$.

Corollary 2. The control policy $\bar{p}^\epsilon(\cdot)$ is asymptotically optimal both in the sense of absolute and relative errors.

Proof. Similarly to other problems with a heavy traffic assumption the proofs of the Theorems 2 and 3 are based on the rescaling procedure. We define the following rescaled processes

$$X^\epsilon(t) = \epsilon Q^\epsilon(\epsilon^{-2}t) \ , \tag{5.5}$$

$$L^\epsilon(t) = \mu^\epsilon\epsilon^{-1}\int_0^t I(X^\epsilon(s) > 0)p^\epsilon(\epsilon^{-2}s)ds \ , \tag{5.6}$$

$$R^\epsilon(t) = \mu^\epsilon\epsilon^{-1}\int_0^t I(X^\epsilon(s) = 0)ds \ , \tag{5.7}$$

$$W^\epsilon(t) = \epsilon[A^\epsilon(\epsilon^{-2}t) - \lambda^\epsilon\epsilon^{-2}t - S^\epsilon(B^\epsilon(\epsilon^{-2}t)) + \mu^\epsilon B^\epsilon(\epsilon^{-2}t)] \ . \tag{5.8}$$

Then the dynamic equation for the rescaled queue length process with the initial state $x\epsilon^{-1}$ can be written as

$$X^\epsilon(t) = x + W^\epsilon(t) + R^\epsilon(t) - L^\epsilon(t) \ . \tag{5.9}$$

The corresponding cost functional can be expressed in terms of the rescaled processes as follows

$$J_x^\epsilon(L^\epsilon(\cdot)) \equiv J^\epsilon(x\epsilon^{-1}, p^\epsilon(\cdot)) = E\{h^\epsilon\int_0^\infty e^{-\gamma^\epsilon t}X^\epsilon(t)dt + \frac{l^\epsilon\epsilon^2}{\mu^\epsilon}\int_0^\infty e^{-\gamma^\epsilon t}dL^\epsilon(t)\} \ . \tag{5.10}$$

The statement of Theorem 2 is equivalent to the following inequality

$$J_x(L^*) \leq \liminf_{\epsilon \to 0} J_x^\epsilon(L^\epsilon(\cdot)) \ . \tag{5.11}$$

One may assume that the right-hand side of (5.11) does not exceed a constant K, otherwise (5.11) is valid. Using this assumption, one can derive

$$E\{L^\epsilon(T)\} \leq K(T) \ , \quad E\{R^\epsilon(T)\} \leq K(T) \ , \tag{5.12}$$

where $K(T)$ is a constant depending on T. Consider the processes L^ϵ and R^ϵ as random elements with values in the space V^+ of all nondecreasing functions endowed with Skorohod M_1 topology [17]. Then (5.12) implies that the distributions of the processes $L^\epsilon(\cdot)$ and $R^\epsilon(\cdot)$ are tight in the space (V^+, M_1) (see [2]). Denote by $\overset{d}{\to}$ weak convergence of the processes with values in the space $D[0, \infty)$ endowed with Skorohod J_1 topology [17]. One can prove the convergence

$$W^\epsilon(\cdot) \overset{d}{\to} W(\cdot) , \tag{5.13}$$

where $W(\cdot)$ is (c, σ^2) Brownian motion. Consequently, the distributions of W^ϵ are tight in $(D[0, \infty), J_1)$. Applying the Skorohod representation theorem (see [3]) we may assume that

$$(W^\epsilon(\cdot), L^\epsilon(\cdot), R^\epsilon(\cdot)) \to (W(\cdot), L(\cdot), \hat{R}(\cdot)) , \quad \epsilon \to 0 \quad \text{a.s.} \tag{5.14}$$

Put

$$\hat{X}(t) = x + W(t) + \hat{R}(t) - L(t) .$$

In view of (5.14) we can use (5.9) to get

$$X^\epsilon(t) \to \hat{X}(t) , \quad L^\epsilon(t) \to L(t) \quad \text{as } \epsilon \to 0 \quad \text{a.s.} \tag{5.15}$$

for each t wich is a point of continuity of L and $\hat{R}$. Fatou's lemma and (5.15) yield

$$E\{\hat{h} \int_0^\infty e^{-\gamma t}\hat{X}(t)dt + \frac{\hat{l}}{\hat{\mu}} \int_0^\infty e^{-\gamma t}dL(t)\} \le$$
$$\liminf_{\epsilon \to 0} E\{h^\epsilon \int_0^\infty e^{-\gamma^\epsilon t}X^\epsilon(t)dt + \frac{l^\epsilon \epsilon^2}{\mu^\epsilon} \int_0^\infty e^{\gamma^\epsilon t}dL^\epsilon(t)\} \equiv \liminf_{\epsilon \to 0} \mathcal{J}^\epsilon_x(L^\epsilon(\cdot)) . \tag{5.16}$$

Since $X^\epsilon(t) \ge 0$ for all t and $\hat{X}(\cdot)$ has $cadlag$ trajectories we have $\hat{X}(t) \ge 0$ for all t a.s. Let $R(t)$ be given by (4.3) and $X(t)$ by (4.2). It is known that for any $\hat{R}$ such that $\hat{X}(t) = x + W(t) + \hat{R}(t) - L(t) \ge 0$ for all t the following inequality is valid $\hat{R}(t) \ge R(t)$. Therefore $\hat{X}(t) \ge X(t)$ for all t a.s. Using this fact, we derive from (5.16)

$$J_x(L(\cdot)) \le \liminf_{\epsilon \to 0} \mathcal{J}^\epsilon_x(L^\epsilon(\cdot)) . \tag{5.17}$$

The natural next step is to apply the results of Section 4 and conclude

$$J_x(L^*(\cdot)) \le J_x(L(\cdot)) . \tag{5.18}$$

To this end we show that $W(\cdot)$ is a Brownian motion with respect to the family of σ-fields $(\mathcal{F}_t, t \ge 0)$, where $\mathcal{F}_t = \sigma(W(s), L(s), s \le t)$. Then $L(\cdot)$ is admissible control for the limiting problem and (5.18) is valid. The statement of Theorem 2 follows from (5.17) and (5.18).

The idea of the proof of Theorem 3 is the following. We substitute (5.4) into (5.6) and get

$$L^\epsilon(t) = r^\epsilon \mu^\epsilon \epsilon^{-1} \int_0^t I(X^\epsilon(s) > z^\epsilon) ds \ . \tag{5.19}$$

Using (5.5), (5.7), (5.8) and (5.19) one can write the following representation for the rescaled queue length process

$$X^\epsilon(\cdot) = \Phi^\epsilon(x + W^\epsilon(\cdot))(\cdot) + \Delta^\epsilon(\cdot) \ ,$$

where Φ^ϵ is *reflection mapping* for the region $[0, z^\epsilon]$ and $\Delta^\epsilon(\cdot) \xrightarrow{d} 0$ as $\epsilon \to 0$. The operator Φ^ϵ maps a function $y(\cdot)$ into a function $\hat{y}(\cdot)$, such that $0 \le \hat{y}(t) \le z^\epsilon$ for all t. The difference between $y(\cdot)$ and $\hat{y}(\cdot)$ increases (decreases) only at those points t for which $\hat{y}(t) = 0$ (for which $\hat{y}(t) = z^\epsilon$). This mapping is similar to the one used to construct a reflected process on $[a, b]$ starting with a process on the whole real line (cf. [5]). One can prove that $z^\epsilon \to z^*$ as $\epsilon \to 0$. Let Φ be the reflection mapping corresponding to the region $[0, z^*]$. Then $\Phi^\epsilon(y^\epsilon)$ converges uniformly to $\Phi(y)$ whenever the function y^ϵ converges uniformly to y. Thus, one can apply the generalized continuous mapping theorem (see Theorem 4.2 [2]) and (5.13) to prove that

$$X^\epsilon(\cdot) \xrightarrow{d} X^*(\cdot) \ , \quad \epsilon \to 0 \ , \tag{5.20}$$

where X^* is (c, σ^2) Brownian motion in the region $[0, z^*]$ with reflection on the boundaries. Similarly one can prove that

$$L^\epsilon(\cdot) \xrightarrow{d} L^*(\cdot) \ , \quad \epsilon \to 0 \ , \tag{5.21}$$

where L^* is the local time of X^* at z^*, which is the optimal control functional for the limiting problem.

Next we use (5.20) and (5.21) to show the convergence of the cost functionals defined by (5.10) to the optimal cost of the limiting problem. To this end one needs to prove the uniform integrability of the family of random variables $X^\epsilon(t)$ and $L^\epsilon(t)$ for any fixed $t > 0$. This fact follows from the estimates

$$\sup_\epsilon E\{X^\epsilon(t)^{(1+q/2)}\} < \infty \ , \quad \sup_\epsilon E\{L^\epsilon(t)^{(1+q/2)}\} < \infty \ , \quad q > 0 \ , \tag{5.22}$$

where q is the same as in (3.2) d). The proof of (5.22) is based on the following general result.

Lemma. Let $\{\xi_k\}, k = 1, 2, \dots$ be nonnegative i.i.d. random variables with $E\xi_1 = \lambda$ and such that for some $p \ge 2$

$$E\xi_1^p = \mu(p) < \infty \ .$$

Let $A(t) = \max\{k \geq 1 : \xi_1 + ... + \xi_k \leq t\}$ $(\max\{\emptyset\} = 0)$. *Then there exists a constant c such that*

$$\{E(A(nt)/n)^p\}^{1/p} \leq c(t+1), \quad \{E \sup_{s \leq t} |n^{-1/2}(A(ns) - \lambda ns)|^p\}^{1/p} \leq c(t^{1/2} + 1).$$

The proof of this Lemma as well as detailed proofs of Theorems 2 and 3 can be found in [9].

6 Interpretation of the Results in Terms of the Original System

Our goal now is to draw conclusions for the original system S, using the sequence of asymptotically optimal policies constructed above. To this end we identify the original system dynamics and cost coefficients with those of S_{ϵ_0} for some fixed $\epsilon_0 \in \mathcal{E}$, that is,

$$h^{\epsilon_0} = h, \quad l^{\epsilon_0} = l, \quad \rho^{\epsilon_0} = \rho, \quad \lambda^{\epsilon_0} = \lambda, \quad \mu^{\epsilon_0} = \mu, \quad r^{\epsilon_0} = r, \quad \sigma^2_{i\epsilon_0} = \sigma^2_i, \quad i = 1, 2; \tag{6.1}$$

Let $p^{\epsilon_0}(\cdot)$ be any admissible control for the system S_{ϵ_0}. Put $p(\cdot) \equiv p^{\epsilon_0}(\cdot)$. Then in accordance with (2.3), (2.5) and (3.6), (3.7)

$$J(x, p(\cdot)) \equiv \epsilon_0^{-3} J^{\epsilon_0}(x, p^{\epsilon_0}(\cdot)), \quad v(x) \equiv \epsilon_0^{-3} v^{\epsilon_0}(x). \tag{6.2}$$

Thus, the only difference between the systems S_{ϵ_0} and S is in presence of a large factor ϵ_0^{-3} in the cost functional. Consider an element of the sequence of asymptotically optimal policies

$$\bar{p}^{\epsilon_0}(t) = r^{\epsilon_0} I(Q^{\epsilon_0}(t) > z^{\epsilon_0} \epsilon_0^{-1}),$$

where z^{ϵ_0} is determined via (5.1),(5.2)and (5.3) with $\epsilon = \epsilon_0$. Set $\bar{p}(\cdot) \equiv \bar{p}^{\epsilon_0}(\cdot)$. Obviously (6.2) implies

$$\left|\frac{J(x, \bar{p}(\cdot)) - v(x)}{v(x)}\right| \equiv \left|\frac{v^{\epsilon_0}(x) - J^{\epsilon_0}(x, \bar{p}^{\epsilon_0}(\cdot))}{v^{\epsilon_0}(x)}\right|. \tag{6.3}$$

Corollary 2 of the previous section ensure that the right hand side of (6.3) is small whenever ϵ_0 is small and therefore the policy $\bar{p}(\cdot)$ is close to optimal. However there is one important issue related to (6.3). Note that for a given system S a sequence $\{S_\epsilon\}$ can be constructed in many different ways. In fact, there exist infinitely many sequences $\{S_\epsilon\}$ which S can be imbedded into. The index ϵ_0 indicates its position in the sequence and therefore the value of ϵ_0 is not determined by the parameters of the original system S. Thus it is important to show that the policy derived for the system S is the same no matter which sequence $\{S_\epsilon\}$ is used. In other words we have to show that the value $z^{\epsilon_0} \epsilon_0^{-1}$ can be expressed only in terms of the parameters of the original system.

From (2.6) and (3.2) we have

$$\delta \equiv \delta^{\epsilon_0} = c^{\epsilon_0} \epsilon_0 (\mu^{\epsilon_0})^{-1} .$$ (6.4)

Consequently, $c^{\epsilon_0} = \delta \epsilon_0^{-1} \mu^{\epsilon_0}$ and due to (3.3) $\gamma^{\epsilon_0} = \rho^{\epsilon_0} \epsilon_0^{-2}$. Substituting these expressions into (5.1)-(5.3) with $\epsilon = \epsilon_0$ and using (6.1), one can easily verify that $z^0 = z^{\epsilon_0} \epsilon_0^{-1}$ satisfy the following equation

$$\alpha_0 (\frac{l}{\mu} - \frac{h}{\rho}) e^{\beta_0 z} + \beta_0 (\frac{l}{\mu} - \frac{h}{\rho}) e^{-\alpha_0 z} + (\alpha_0 + \beta_0) \frac{h}{\rho} = 0 ,$$ (6.5)

where

$$\sigma_0^2 = \lambda^3 \sigma_1^2 + \mu^3 \sigma_2^2 , \quad \alpha_0 = \sigma_0^{-2}(\delta\mu + \sqrt{\delta^2\mu^2 + 2\sigma_0^2\rho}) , \quad \beta_0 = \sigma_0^{-2}(-\delta\mu + \sqrt{\delta^2\mu^2 + 2\sigma_0^2\rho}) .$$

It means that z^0 is expressed entirely in terms of the original system parameters.

Thus, the relative error given by (6.3) depends only on the parameters of the system S and not on the value of ϵ. When δ is small the relative error is small as well. Following the usual practice in the literature (e.g. [12]-[15]), the corresponding threshold control policy is loosely referred to as *nearly* or *approximately optimal policy* for S. It should be emphasized once more that the relative rather than the absolute error should be used to evaluate the quality of approximation.

Another important practical question is to characterize the class of real systems for which the quality of the nearly optimal policy obtained via diffusion approximation procedure is good. Theoretically this issue is related to the rate of convergence in (3.8). Since there are no general results on this topic we try to provide some reasonable arguments clarifying this issue. Consider the original system defined by (2.1)-(2.3). Along with the traffic intensity parameter δ which is assumed to be small, we introduce $\bar{\delta} = \sqrt{h/l}$. Note that these parameters are independent, since the first characterizes the dynamics of the system while the second specify the structure of the cost functional.

Suppose that $\bar{\delta}$ is not a small parameter. Then rescaling the cost functional by δ and calculating the corresponding holding and control parameters, we see that the cost for control actions is much lower than the the holding cost. In this case the threshold value will be close to zero and the rescaled process, which is close to Brownian motion, will be very often affected by the threshold policy. In such cases reflected diffusion does not provide an accurate approximation for the original system.

On the other hand suppose both δ and $\bar{\delta}$ are small but $\bar{\delta}$ is much smaller than δ. In this case the holding cost in the limiting problem becomes negligible compared to the control cost and the policy with $p(t) \equiv 0$ is nearly optimal.

Thus, we suggest that diffusion approximation provides a reasonable answer for the original problem if

1) Both δ and $\bar{\delta}$ are small and δ is of the same or smaller order than $\bar{\delta} = \sqrt{h/l}$.

Another requirement is due to the fact that ρ should be of the order of $\bar{\delta}^2$. Otherwise after rescaling procedure the discount factor becomes very large. It remains to mention that due to (2.4) we have to ensure that $\rho < \mu\bar{\delta}^2$. Thus, the second requirement can be formulated as follows:

2) ρ is of the order of $\bar{\delta}^2 = h/l$ and $\rho < \mu\bar{\delta}^2$.

Asknowledgements

The authors asknowledge the financial support from NSERC Grant A4619, AFOSR Grant 88-0813 and NATO International Scientific Exchange Program under Grant CRG-900147. The U.S. government is authorized to reproduce and distribute reprints for Governmental purposes not withstanding any copy right notation thereon.

References

[1] Baldurson, F.M. (1987). Singular stochastic control and optimal stopping. *Stochastics* 21 1-40.

[2] Billingsley, P. (1968). *Convergence of Probability Measures.* Wiley, New York.

[3] Either, S.N. and Kurtz, T.G. (1986). *Markov processes. Characterization and Convergence.* Wiley, New York.

[4] Harrison, J.M. and Taksar, M.I. (1983). Instantaneous control of Brownian Motion. *Math. Oper. Res.* 8 439-453.

[5] Harrison, J.M. and Reiman, M.I. (1981). Reflected Brownian Motion on an Orthant. *Ann. Probab.* 9 302-309.

[6] Harrison, J.M. and Wein, L.M. (1989). Scheduling networks of queues: heavy traffic analysis of a simple open network. *Queueing Systems: Theory and Applications* 5 265-280.

[7] Harrison, J.M. and Wein, L.M. (1990). Scheduling networks of queues: heavy traffic analysis of a two-station closed network. *Oper. Res.* 38 1051-1064.

[8] Karatzas, I. and Shreve, S.E. (1985) Connection between optimal stopping and singular stochastic control II. Reflected follower problem. *SIAM J. Control Optim.* 23 433-451.

[9] Krichagina, E.V. and Taksar, M.I. (1991). Diffusion Approximation for GI/G/1 Queues. Submitted to *Queueing Systems: Theory and Applications*.

[10] Krichagina, E.V. , Lou S.X., Sethi, S.P. and Taksar, M.I. (1991). Production control in a Failure-Prone Manufacturing System: Diffusion Approximation and Asymptotic Optimality. Submitted to *Ann. Appl. Probab.*.

[11] Kushner H.J. and Rungaldier, W. (1987). Nearly optimal state feedback controls for stochastic systems with wideband noise disturbances. *SIAM J. Control Optim.* 25 289-315.

[12] Kushner, H.J. and Ramachandran, K.M. (1988). Nearly optimal singular controls for wideband noise driven systems. *SIAM J. Control Optim.* 26 569-591.

[13] Kushner H.J. and Ramachandran, K.M. (1989). Optimal and Approximately Optimal Control Policies for Queues in Heavy Traffic. *SIAM J. Control Optim.* 27 1293-1318.

[14] Kushner, H.J. and Martins, L.F. (1990). Routing and Singular Control for Queueing Network in Heavy Traffic. *SIAM J. Control Optim.* 28 1209-1233.

[15] Kushner, H.J. and Martins, L.F. (1991). Limit Theorems for Pathwise Average Cost per Unit Time Problems for Controlled Queues in Heavy Traffic. Preprint LCDS/CCS #91-2 Division of Applied Mathematics, Brown University.

[16] Kushner, H.J. and Martins, L.F. (1990). Numerical Methods for Stochastic Singularly Controlled Problems, Preprint LCDS # 89-21 Division of Applied Mathematics, Brown University.

[17] Skorohod, A.V. (1956). Limit theorems for stochastic processes. *Theor. Prob. Appl.* 1 261-284.

[18] Wein, L.M. (1990). Scheduling Networks of Queues: Heavy Traffic Analysis of a Two-Station Network With Controllable Inputs. *Oper. Res.* 38 1065-1078.

CERTAINTY EQUIVALENCE WITH UNCERTAINTY ADJUSTMENTS IN STOCHASTIC ADAPTIVE CONTROL

Tze Leung Lai

Department of Statistics, Stanford University, Stanford, CA 94305

Abstract

A useful technique for finding relatively simple and yet asymptotically optimal solutions to stochastic adaptive control problems is to incorporate into certainty-equivalence rules suitable adjustments for parameter uncertainty. We review some results in sequential testing and estimation theories in statistics and discuss their applications to the assessment of parameter uncertainty and to efficient adjustments of certainty-equivalence rules in stochastic adaptive control.

1. INTRODUCTION

To design controllers for a stochastic system whose dynamics depends not only on the control variables but also on certain unknown parameters, the "certainty-equivalence" approach first finds the optimal (or asymptotically optimal) control scheme when the system parameters are known and then replaces the parameter values in this control scheme by their sample estimates at every stage. It tries to mimic the optimal rule (assuming known system parameters) at every stage by updating the parameter estimates based on all the available data. It is particularly attractive when the optimal (or asymptotically optimal) control scheme assuming known system parameters has a simple recursive form that can be implemented in real time and when there are real-time recursive algorithms for updating the parameter estimates.

A serious pitfall of the certainty-equivalence approach to stochastic adaptive control is that in closed-loop identification the parameter estimates often differ substantially from the true parameter values even for very large sample sizes. To avoid this pitfall, one can in principle use a Bayesian approach, putting a prior distribution on the unknown system parameters and formulating the stochastic adaptive control problem as a dynamic programming problem in which the "state" is the conditional distribution of the original system state and parameter vector given the past observations. However, the dynamic programming equations are usually prohibitively difficult to handle, both computationally and analytically.

Instead of the complicated Bayesian approach, a much more practical and intuitively appealing alternative is to modify the simple certainty-equivalence approach by making suitable adjustments for parameter uncertainty. We have recently shown in [1] how such uncertainty adjustments to certainty-equivalence rules can be made to yield control schemes that are asymptotically as efficient as the Bayes procedures in three classical problems of stochastic adaptive control. They are (i) the multi-armed bandit problem, (ii) the multi-period control problem in econometrics, and (iii) adaptive control of an ARMAX system.

Some recent developments in sequential estimation and testing theories that are useful for the enhancement of certainty-equivalence control rules by making suitable uncertainty adjustments are discussed herein. In Section 2 we show how several results and analytical techniques in sequential testing theory can be applied to adjust certainty-equivalence rules in stochastic adaptive control problems with finite control sets. These ideas will be illustrated in the context of problem (i) and variants thereof. In Section 3 we review some recent work in the theory of recursive estimation and adaptive prediction in stochastic regression models and point out its usefulness in the development of adjusted certainty-equivalence solutions to problems (ii) and (iii). Section 4 discusses the general heuristic principle of certainty equivalence with uncertainty adjustments and gives several concluding remarks.

2. FINITE CONTROL SET AND SEQUENTIAL TESTING THEORY

Sequential testing theory provides important clues to the construction of asymptotically optimal adaptive control schemes by suitable modifications of certainty-equivalence rules when the control set is finite. To illustrate the usefulness of these clues, we shall apply them to the following adaptive allocation problem, which is a more complex variant of the "multi-armed bandit problem" discussed in [1]-[6].

Example. Consider three bivariate normal populations Π_1, Π_2, Π_3, with respective mean vectors $(\mu_1, \lambda), (\mu_2, \mu_3)$ and $(\mu_3, \mu_2 + \lambda)$, where $\mu_1, \mu_2, \mu_3, \lambda$ are unknown parameters. Suppose that each population is known to have the identity matrix as its covariance matrix. How should we sample $x_1, \cdots, x_N$ sequentially to maximize, in some sense, the expected value of the first component of the vector $S_N = x_1 + \cdots + x_N$?

If the parameter values μ_1, μ_2, μ_3 were known, the optimal rule would be to sample from Π_j if μ_j is the largest of μ_1, μ_2, μ_3, noting that we can choose any of the populations that tie for the largest μ_j value. In ignorance of the μ_j, the certainty-equivalence approach is to estimate them by maximum likelihood and to choose the Π_j that has the largest maximum likelihood estimate $\hat{\mu}_{j,n}$ at stage n. The difficulty with this rule is that we may have sampled too little from apparently inferior populations to get reliable estimates of the parameters and may thereby miss the actually superior population.

To see how uncertainty adjustments can be made to remedy this drawback of the certainty-equivalence rule, first note that the relevant information about the parameters we need for optimal control can be represented by the three hypotheses $H_j : \mu_j = \max(\mu_1, \mu_2, \mu_3)$, $j = 1, 2, 3$. In other words, we do not need to know the actual values of the parameters $\mu_1, \mu_2, \mu_3, \lambda$ but only need to determine which of μ_1, μ_2, μ_3 is the largest. Moreover, while information about μ_1 can only be obtained by sampling from Π_1, information about μ_2 and μ_3 can be obtained by sampling from Π_2 alone or from Π_3 and Π_1.

If we put a joint prior distribution Ψ on the unknown parameters $\lambda, \mu_1, \mu_2, \mu_3$, then the certainty-equivalence approach is essentially tantamount to choosing the hypothesis $H_{j(n)}$ with the largest posterior probability at every stage n and sampling from $\Pi_{j(n)}$. However, the Bayes rule that maximizes $\int E_{\lambda,\mu_1\mu_2,\mu_3}(S_N)d\Psi(\lambda, \mu_1, \mu_2, \mu_3)$ takes into consideration not only which H_j has the largest posterior probability but also how uncertain one is about the competing hypotheses. Specifically, the sampling scheme of the Bayes rule has the "dual" function of (i) sampling from Π_j to learn about μ_j so that one can have enough evidence against the false hypotheses and (ii) not sampling from an inferior Π_j once one is reasonably sure of its inferiority. The precise form of the Bayes rule, however, is very complicated although in principle it can be evaluated by numerical integration and backward induction; moreover, it involves the specification of a reasonable and tractable prior distribution Ψ.

For the much simpler two-armed bandit problem of sampling from two univariate standard normal populations, one with unknown mean μ having a normal prior distribution and the other with known mean 0, the precise Bayes rule has been evaluated and used in [2] to develop approximate Bayes solutions to the multi-armed bandit problem involving k univariate populations with densities from an exponential family and independent prior distributions for the parameters of the k populations. The reason why it is possible to obtain such simple approximations has been explained later in [3] by using (i) the connection between the asymptotic solutions of finite-horizon and discounted stochastic control problems, and (ii) Gittins' [4] characterization of the optimal Bayes procedure for the discounted multi-armed bandit problem in the case of independent prior distributions for the k parameters. In particular, in the case of k independent normal populations, Gittins' work suggests why it suffices to compare each individual poulation with a normal population with known mean. A basic idea of [2] is to start by considering the case of k normal populations and then using normal approximations to the signed log-likelihood ratio statistics from an exponential family of distributions for the k populations.

In the present example, we cannot put independent prior distributions on the unknown mean vectors of the three bivariate normal distributions since the mean vector of Π_3 can be completely determined from those of Π_1 and Π_2. Thus, Gittins' theory is not applicable to the discounted version of the present problem. Instead of the Bayesian approach, we shall use certainty equivalence with uncertainty adjustments that are suggested by sequential testing theory.

Let $(Y_{j,1}, Z_{j,1}), (Y_{j,2}, Z_{j,2}), \cdots$ denote successive observations from the bivariate population Π_j, and let $\bar{Y}_{j,t} = t^{-1} \sum_{i=1}^{t} Y_{j,i}$, $\bar{Z}_{j,t} = t^{-1} \sum_{i=1}^{t} Z_{j,i}$. An "adaptive allocation rule" is a sequence of random variables $\phi_1, \phi_2, \cdots$ such that the event $\{\phi_n = j\}$ ("sample from Π_j at stage n"), $j = 1, 2, 3$, belongs to the σ-field $\mathcal{F}_{n-1}$ generated by the previous observations. For an adaptive allocation rule ϕ, let $T_n(j) = \sum_{i=1}^{n} I_{\{\phi_i = j\}}$ denote the number of observations that ϕ samples from Π_j up to stage n. Let $\theta = (\lambda, \mu_1, \mu_2, \mu_3)$ and $\mu^* = \max(\mu_1, \mu_2, \mu_3)$. As shown in [5], the problem of maximizing $E_\theta S_n$ (over the class of all adaptive allocation rules) is equivalent to that of minimizing the regret

$$R_n(\theta) = n\mu^* - E_\theta S_n = \sum_{j: \mu_j < \mu^*} (\mu^* - \mu_j) E_\theta T_n(j). \tag{1}$$

The maximum likelihood (or equivalently, least squares) estimate $\widehat{\theta}_n = (\widehat{\lambda}_n, \widehat{\mu}_{1,n}, \widehat{\mu}_{2,n}, \widehat{\mu}_{3,n})$ of θ at stage n is given by

$$\widehat{\mu}_{1,n} = \bar{Y}_{1,T_n(1)}, \widehat{\mu}_{3,n} = (\sum_{t=1}^{T_n(2)} Z_{2,t} + \sum_{t=1}^{T_n(3)} Y_{3,t})/(T_n(2) + T_n(3)),$$

$$(T_n(2) + T_n(3))\widehat{\mu}_{2,n} + T_n(3)\widehat{\lambda}_n = \sum_{t=1}^{T_n(2)} Y_{2,t} + \sum_{t=1}^{T_n(3)} Z_{3,t}, \tag{2}$$

$$T_n(3)\widehat{\mu}_{2,n} + (T_n(1) + T_n(3))\widehat{\lambda}_n = \sum_{t=1}^{T_n(3)} Z_{3,t} + \sum_{t=1}^{T_n(1)} Z_{1,t}.$$

Let $\Theta_j = \{\theta : \mu_j \geq \max_{i \neq j} \mu_i\}$ and consider the three hypotheses $H_j : \theta \in \Theta_j (j = 1, 2, 3)$. Since with probability 1, $\widehat{\mu}_{1,n}, \widehat{\mu}_{2,n}, \widehat{\mu}_{3,n}$ are distinct by continuity of the underlying (normal) distributions, there is a unique population $\Pi_{j(n)}$ with the largest estimated mean $\max_{i \leq j \leq 3} \hat{\mu}_{j,n}$, or equivalently, $\hat{\theta}_n \in \Theta_{j(n)}$ for a unique $j(n)$. Although this suggests $H_{j(n)}$ as the most plausible hypothesis, it does not mean that the other hypotheses are incompatible with the data. An important step in modifying the certainty-equivalence rule is to assess if there is enough evidence in the data against these other hypotheses. Useful insights into how this assessment can be carried out are provided by asymptotic solutions to the following closely related sequential testing problems: Fix $j \in \{1,2,3\}$ and $\bar{\theta} \in \Theta_j$ with $\bar{\mu}_j > \max_{i \neq j} \bar{\mu}_i$. Find sequential tests of the hypotheses $H_i^{(j)} : \mu_i \geq \bar{\mu}_j$ $(i \neq j)$ to minimize the expected weighted sum $r_{\alpha,j}(\bar{\theta}) = \sum_{i \neq j}(\bar{\mu}_j - \bar{\mu}_i) E_{\bar{\theta}} \tau_i$ of the number τ_i of observations from Π_i, subject to the error constraints

$$P(\text{Reject } H_i^{(j)}) \leq \alpha \text{ when } H_i^{(j)} \text{ is true } (i \neq j). \tag{3}$$

First consider the testing problem in the case $j = 1$ and try to minimize $r_{\alpha,1}(\bar{\theta}) = (\bar{\mu}_1 - \bar{\mu}_2)E_{\bar{\theta}}\tau_2 + (\bar{\mu}_1 - \bar{\mu}_3)E_{\bar{\theta}}\tau_3$ at a given $\bar{\theta}$ with $\bar{\mu}_1 > \max(\bar{\mu}_2, \bar{\mu}_3)$ under the error constraints (3) for the corresponding sequential tests of $H_2^{(1)} : \mu_2 \geq \bar{\mu}_1$ and $H_3^{(1)} : \mu_3 \geq \bar{\mu}_1$. Since the cost criterion $r_{\alpha,1}(\bar{\theta})$ does not involve any sampling cost from Π_1, we can start by taking infinitely many observations from Π_1 and can therefore assume that (μ_1, λ) is known to have the value $(\bar{\mu}_1, \bar{\lambda})$. Noting that the normal random variables $Y_{2,i}$ and $Z_{3,i} - \bar{\lambda}$ have the same mean μ_2, and that the normal random variables $Z_{2,i}$ and $Y_{3,i}$ have the same mean μ_3, we can sample either from Π_2 or from Π_3 to learn about (μ_2, μ_3). Since the cost criterion $r_{\alpha,1}(\bar{\theta})$ says that each observation from Π_2 costs $\bar{\mu}_1 - \bar{\mu}_2$ while each observation from Π_3 costs $\bar{\mu}_1 - \bar{\mu}_3$, it then follows that we should sample from Π_2 if $\bar{\mu}_2 \geq \bar{\mu}_3$ and from Π_3 if $\bar{\mu}_2 < \bar{\mu}_3$.

Suppose that $\bar{\mu}_2 \geq \bar{\mu}_3$. A classical result in optimal stopping and sequential analysis says that to test $H_2' : \mu_2 = \bar{\mu}_1$ sequentially on the basis of successive observations $Y_{2,t}$ ($t = 1, 2, \cdots$) and under the error constraint $P_{\mu_2 = \bar{\mu}_1}(\text{Reject } H_2') \leq \alpha$, the one-sided sequential probability ratio test which stops sampling at stage

$$\tau_2^* = \inf\{n \geq 1 : \sum_{i=1}^{n} \log(f_{\bar{\mu}_2}(Y_{2,i})/f_{\bar{\mu}_1}(Y_{2,i})) \geq A_2^*\} \tag{4a}$$

and which rejects $H_2' : \mu_2 = \bar{\mu}_1$ upon stopping (where f_μ denotes the normal density with mean μ and variance 1 and A_2^* is so chosen that the test has probability α of wrongly rejecting H_2') is optimal in the sense of minimizing the expected number of observations from Π_2 under $\mu_2 = \bar{\mu}_2(< \bar{\mu}_1)$, cf. [7]. Moreover, $P_{\mu_2 \geq \bar{\mu}_1}(\tau_2^* < \infty) \leq P_{\mu_2 = \bar{\mu}_1}(\tau_2^* < \infty) = \alpha$ and $E_{\bar{\theta}} \tau_2^* \sim |\log \alpha|/I(\bar{\mu}_2, \bar{\mu}_1)$ as $\alpha \to 0$, where

$$I(\lambda, \mu) = E_\lambda \log(f_\lambda(Y_{2,i})/f_\mu(Y_{2,i})) = (\lambda - \theta)^2/2$$

is the Kullback-Leibler information number. Likewise, to test $H_3^{(1)} : \mu_3 \geq \bar{\mu}_1$ subject to the error constraint (3), we use the stopping rule

$$\tau_2^{**} = \inf\{n \geq 1 : \sum_{i=1}^{n} \log(f_{\bar{\mu}_3}(Z_{2,i})/f_{\bar{\mu}_1}(Z_{2,i})) \geq A_2^{**}\}, \tag{4b}$$

rejecting $H_3^{(1)}$ upon stopping, to minimize the expected number of observations from Π_2 under $\mu_3 = \bar{\mu}_3$. As $\alpha \to 0$, $E_{\bar{\theta}} \tau_2^{**} \sim |\log \alpha|/I(\bar{\mu}_3, \bar{\mu}_1) \leq |\log \alpha|/I(\bar{\mu}_2, \bar{\mu}_1)$. For the overall rule that samples from Π_2 until stage $\max(\tau_2^*, \tau_2^{**})$, we have $r_{\alpha,1}(\bar{\theta}) \sim (\bar{\mu}_1 - \bar{\mu}_2)|\log \alpha|/I(\bar{\mu}_2, \bar{\mu}_1)$.

In the case $\bar{\mu}_2 < \bar{\mu}_3 < \bar{\mu}_1$, we should sample from Π_3 instead of Π_2 to minimize $r_{\alpha,1}(\bar{\theta})$. Sampling from Π_3 is stopped at stage $\max(\tau_3^*, \tau_3^{**})$, where

$$\tau_3^* = \inf\{n \geq 1 : \sum_{i=1}^{n} \log(f_{\bar{\mu}_3}(Y_{3,i})/f_{\bar{\mu}_1}(Y_{3,i})) \geq A_3^*\},$$
$$\tau_3^{**} = \inf\{n \geq 1 : \sum_{i=1}^{n} \log(f_{\bar{\mu}_2}(Z_{3,i} - \bar{\lambda})/f_{\bar{\mu}_1}(Z_{3,i} - \bar{\lambda}) \geq A_3^{**}\}, \tag{5}$$

and $A_3^* \sim A_3^{**} \sim |\log \alpha|$. For this rule, $r_{\alpha,1}(\bar{\theta}) \sim (\bar{\mu}_1 - \bar{\mu}_3)|\log \alpha|/I(\bar{\mu}_3, \bar{\mu}_1)$ as $\alpha \to 0$.

We next consider the case $j = 2$ and try to minimize $r_{\alpha,2}(\bar{\theta}) = (\bar{\mu}_2 - \bar{\mu}_1)E_{\bar{\theta}} \tau_1 + (\bar{\mu}_2 - \bar{\mu}_3)E_{\bar{\theta}} \tau_3$ at a given $\bar{\theta}$ with $\bar{\mu}_2 > \max(\bar{\mu}_1, \bar{\mu}_3)$. Since the cost criterion $r_{\alpha,2}(\bar{\theta})$ does not involve any sampling cost from Π_2, we can take infinitely many observations from Π_2 and can therefore assume that (μ_2, μ_3) is known to assume the value $(\bar{\mu}_2, \bar{\mu}_3)$. Hence we do not need to sample from Π_3 to learn about μ_3 but need only sample from Π_1 to learn about μ_1. To test $H_1^{(2)} : \mu_1 \geq \bar{\mu}_2$, we sample from Π_1 until stage τ_1^* to minimize the expected number of observations from Π_1 under $\mu_1 = \bar{\mu}_1(< \bar{\mu}_2)$, where

$$\tau_1^* = \inf\{n \geq 1 : \sum_{i=1}^{n} \log(f_{\bar{\mu}_1}(Y_{1,i})/f_{\bar{\mu}_2}(Y_{1,i})) \geq A_1^*\} \tag{6}$$

and $A_1^* \sim |\log \alpha|$. For this rule, $r_{\alpha,2}(\bar{\theta}) \sim (\bar{\mu}_2 - \bar{\mu}_1)|\log \alpha|/I(\bar{\mu}_1, \bar{\mu}_2)$ as $\alpha \to 0$.

Finally consider the case $j = 3$ and try to minimize $r_{\alpha,3}(\bar{\theta}) = (\bar{\mu}_3 - \bar{\mu}_1)E_{\bar{\theta}} \tau_1 + (\bar{\mu}_3 - \bar{\mu}_2)E_{\bar{\theta}} \tau_2$ at a given $\bar{\theta}$ with $\bar{\mu}_3 > \max(\bar{\mu}_1, \bar{\mu}_2)$. We can take infinitely many observations from Π_3 without incurring any cost in the criterion $r_{\alpha,3}(\bar{\theta})$, and can therefore assume that $(\mu_3, \mu_2 + \lambda)$ is known to assume the value $(\bar{\mu}_3, \bar{\gamma})$. To learn about μ_1, we must sample from Π_1. In particular, to test $H_1^{(3)} : \mu_1 \geq \bar{\mu}_3$, we sample from Π_1 until stage

$$\tau_3' = \inf\{n \geq 1 : \sum_{i=1}^n \log(f_{\bar{\mu}_1}(Y_{1,i})/f_{\bar{\mu}_3}(Y_{1,i})) \geq A_1'\}, \tag{7a}$$

with $A_1' \sim |\log\alpha|$, to minimize the expected number of observations from Π_1 under $\mu_1 = \bar{\mu}_1$. To learn about $\mu_2 = \bar{\gamma} - \lambda$, which is the mean of the normal random variables $Y_{2,i}$ and $\bar{\gamma} - Z_{1,i}$, we can sample either from Π_2, or from Π_1, or from both. Since we already need τ_1' observations from Π_1 and since the cost criterion $r_{\alpha,3}(\bar{\theta})$ says that each observation from Π_1 (resp. Π_2) costs $\bar{\mu}_3 - \bar{\mu}_1$ (resp. $\bar{\mu}_3 - \bar{\mu}_2$), it is clear that to test $H_2^{(3)} : \mu_2 \geq \bar{\mu}_3$ we should take additional observations (if needed) from Π_1 if $\bar{\mu}_1 \geq \bar{\mu}_2$ and from Π_2 if $\bar{\mu}_1 < \bar{\mu}_2$. In particular, we sample from Π_1 until stage $\max(\tau_1', \tau_1'')$ in the case $\bar{\mu}_1 \geq \bar{\mu}_2$, where

$$\tau_1'' = \inf\{n \geq 1 : \sum_{i=1}^n \log(f_{\bar{\mu}_2}(\bar{\gamma} - Z_{1,i})/f_{\bar{\mu}_3}(\bar{\gamma} - Z_{1,i})) \geq A_1''\}, \tag{7b}$$

and sample from Π_2 until stage τ_2' in the case $\bar{\mu}_1 < \bar{\mu}_2$, where

$$\tau_2' = \inf\{n \geq 0 : \sum_{j=1}^{\tau_1'} \log \frac{f_{\bar{\mu}_2}(\bar{\gamma} - Z_{1,i})}{f_{\bar{\mu}_3}(\bar{\gamma} - Z_{1,i})} + \sum_{i=1}^n \log \frac{f_{\bar{\mu}_2}(Y_{2,i})}{f_{\bar{\mu}_3}(Y_{2,i})} \geq A_2'\}, \tag{7c}$$

with $A_1'' \sim A_2' \sim |\log\alpha|$. For this sampling scheme, it can be shown that as $\alpha \to 0$,

$$r_{\alpha,3}(\bar{\theta}) \sim (\bar{\mu}_3 - \bar{\mu}_1)|\log\alpha|/I(\bar{\mu}_1, \bar{\mu}_3), \quad \text{if } \bar{\mu}_2 \leq \bar{\mu}_1 < \bar{\mu}_3,$$

$$\sim \{\frac{\bar{\mu}_3 - \bar{\mu}_1}{I(\bar{\mu}_1, \bar{\mu}_3)} + \frac{I(\bar{\mu}_1, \bar{\mu}_3) - I(\bar{\mu}_2, \bar{\mu}_3)}{I(\bar{\mu}_1, \bar{\mu}_3)I(\bar{\mu}_2, \bar{\mu}_3)}(\bar{\mu}_3 - \bar{\mu}_2)\}|\log\alpha|, \quad \text{if } \bar{\mu}_1 < \bar{\mu}_2 < \bar{\mu}_3. \tag{8}$$

Summarizing, the preceding testing problem has an explicit solution in the form of the optimal stopping rules (4)-(7), which yield the asymptotically optimal cost given by (8) and

$$r_{\alpha,2}(\bar{\theta}) \sim (\bar{\mu}_2 - \bar{\mu}_1)|\log\alpha|/I(\bar{\mu}_1, \bar{\mu}_2) \quad \text{if } \bar{\mu}_2 > \max(\bar{\mu}_1, \bar{\mu}_3), \tag{9}$$

$$r_{\alpha,1}(\bar{\theta}) \sim (\bar{\mu}_1 - \bar{\mu}_2)|\log\alpha|/I(\bar{\mu}_2, \bar{\mu}_1) \quad \text{if } \bar{\mu}_1 > \bar{\mu}_2 \geq \bar{\mu}_3,$$

$$\sim (\bar{\mu}_1 - \bar{\mu}_3)|\log\alpha|/I(\bar{\mu}_3, \bar{\mu}_1) \quad \text{if } \bar{\mu}_1 > \bar{\mu}_3 > \mu_2. \tag{10}$$

These results suggest the following modification of the certainty-equivalence rule that samples at stage n from the population $\Pi_{j(n)}$ with $\hat{\mu}_{j(n),n} = \max_{1 \leq i \leq 3} \hat{\mu}_{j,n}$, where the $\hat{\mu}_{j,n}$ are the maximum likelihood estimates given by (2).

The basic idea of this modification is to ensure that the sampling scheme continues to learn about all the μ_j $(j \neq j(n))$ for which there is not enough evidence in the data to conclude that Π_j is indeed inferior to $\Pi_{j(n)}$. In this connection, the definition of the "leading population" $\Pi_{j(n)}$ will also be modified slightly, following an idea introduced in [5]. Specifically, we require the leader at stage n to be one from which we have sampled at least $n/3$ times so that the estimates of its parameters are relatively trustworthy. We therefore redefine $j(n)$ by

$$\widehat{\mu}_{j(n),n} = \max\{\widehat{\mu}_{j,n} : T_n(j) \geq n/3\}. \tag{11}$$

For $m > m_0$, at stage $3m + j$ with $j \in \{1, 2, 3\}$, we sample to learn about μ_j unless there is enough evidence that Π_j is inferior to $\Pi_{j(3m)}$, in which case we forgo learning about Π_j and sample from $\Pi_{j(3m)}$ instead. The details of how we learn about μ_j and how we test if Π_j should still be considered as a contender are given below and are suggested by the optimal stopping rules (4)-(7) of the preceding sequential testing problem in which we put $\alpha = (3m)^{-1}$.

Let $n = 3m$ and suppose $j(n) = 1$. At stage $n + 1$ sample from Π_1. If $\widehat{\mu}_{2,n} \geq \widehat{\mu}_{3,n}$, sample either from Π_2 to learn about μ_2 and μ_3 at stages $n + 2$ and $n + 3$, or from the "leading" population Π_1 at stage $n + 2$ if

$$\widehat{\mu}_{1,n} > \widehat{\mu}_{2,n} \text{ and } \sum_{i=1}^{T_n(2)} \log\{f_{\widehat{\mu}_{2,n}}(Y_{2,i})/f_{\widehat{\mu}_{1,n}}(Y_{2,i})\} \geq \log n,$$

as suggested by (4a), or from Π_1 at stage $n + 3$ if

$$\widehat{\mu}_{1,n} > \widehat{\mu}_{3,n} \text{ and } \sum_{i=1}^{T_n(2)} \log\{f_{\widehat{\mu}_{3,n}}(Z_{2,i})/f_{\widehat{\mu}_{1,n}}(Z_{2,i})\} \geq \log n$$

as suggested by (4b). If $\widehat{\mu}_{2,n} < \widehat{\mu}_{3,n}$, sample either from Π_3 to learn about μ_2 and μ_3 at stages $n + 2$ and $n + 3$, or from Π_1 at stage $n + 3$ if

$$\widehat{\mu}_{1,n} > \widehat{\mu}_{3,n} \text{ and } \sum_{i=1}^{T_n(3)} \log\{f_{\widehat{\mu}_{3,n}}(Y_{3,i})/f_{\widehat{\mu}_{1,n}}(Y_{3,i})\} \geq \log n,$$

(cf. (5)), or from Π_1 at stage $n + 2$ if

$$\widehat{\mu}_{1,n} > \widehat{\mu}_{2,n} \text{ and } \sum_{i=1}^{T_n(3)} \log\{f_{\widehat{\mu}_{2,n}}(Z_{3,i} - \bar{Z}_{1,T_n(1)})/f_{\widehat{\mu}_{1,n}}(Z_{3,i} - \bar{Z}_{1,T_n(1)})\} \geq \log n.$$

Suppose $j(n) = 2$. At stages $n+2$ and $n+3$, sample from Π_2. At stage $n+1$, sample from Π_2 if

$$\widehat{\mu}_{2,n} > \widehat{\mu}_{1,n} \text{ and } \sum_{i=1}^{T_n(1)} \log\{f_{\widehat{\mu}_{1,n}}(Y_{1,i})/f_{\widehat{\mu}_{2,n}}(Y_{1,i})\} \geq \log n,$$

(cf. (6)), otherwise sample from Π_1.

Finally suppose $j(n) = 3$. At stage $n+3$ sample from Π_3. At stage $n+1$, sample from Π_3 if

$$\widehat{\mu}_{3,n} > \widehat{\mu}_{1,n} \text{ and } \sum_{i=1}^{T_n(1)} \log\{f_{\widehat{\mu}_{1,n}}(Y_{1,i})/f_{\widehat{\mu}_{3,n}}(Y_{1,i})\} \geq \log n,$$

(cf. (7a)), otherwise sample from Π_1. At stage $n+2$, sample from Π_3 if

$$\widehat{\mu}_{3,n} > \widehat{\mu}_{2,n} \text{ and } \sum_{i=1}^{T_n(1)} \log \frac{f_{\widehat{\mu}_{2,n}}(\bar{Z}_{3,T_n(3)} - Z_{1,i})}{f_{\widehat{\mu}_{3,n}}(\bar{Z}_{3,T_n(3)} - Z_{1,i})} + \sum_{i=1}^{T_n(2)} \log \frac{f_{\widehat{\mu}_{2,n}}(Y_{2,i})}{f_{\widehat{\mu}_{3,n}}(Y_{2,i})} \geq \log n,$$

(cf. (7c)), otherwise sample from Π_2 in the case $\widehat{\mu}_{2,n} > \widehat{\mu}_{1,n}$ and from Π_1 in the case $\widehat{\mu}_{2,n} \leq \widehat{\mu}_{1,n}$.

We initialize by taking m_0 observations from each poulation during the first $3m_0$ stages. This completes the specification of the adaptive allocation rule, which we denote by ϕ^*. It can be shown that the regret (1) of the rule ϕ^* has the order of magnitude

$$
\begin{aligned}
R_n(\theta) &= O(1) \text{ if } \mu_1 = \max(\mu_2, \mu_3), \\
&\sim (\mu_1 - \mu_2)(\log n)/I(\mu_2, \mu_1) \text{ if } \mu_1 > \mu_2 \geq \mu_3, \\
&\sim (\mu_1 - \mu_3)(\log n)/I(\mu_3, \mu_1) \text{ if } \mu_1 > \mu_3 > \mu_2, \\
&\sim (\mu_2 - \mu_1)(\log n)/I(\mu_1, \mu_2) \text{ if } \mu_2 > \max(\mu_1, \mu_3), \\
&\sim (\mu_3 - \mu_1)(\log n)/I(\mu_1, \mu_3) \text{ if } \mu_2 = \mu_3 > \mu_1 \text{ or } \mu_3 > \mu_1 \geq \mu_2, \\
&\sim \{\frac{\mu_3 - \mu_1}{I(\mu_1, \mu_3)} + \frac{I(\mu_1, \mu_3) - I(\mu_2, \mu_3)}{I(\mu_1, \mu_3)I(\mu_2, \mu_3)}\}\log n \text{ if } \mu_3 > \mu_2 > \mu_1.
\end{aligned}
$$
(12)

The first relation in (12) can be proved by arguments similar to the proof of Theorem 2 of [6]. The other relations in (12) are analogous to (8)-(10) for the associated sequential testing problem, and can be proved by arguments similar to those used in the proof of Theorem 3 of [5]. Moreover, by developing an asymptotic lower bound analogous to that of Theorem 2 of [5] for the regret $R_n(\theta)$ of adaptive allocation rules satisfying $R_n(\theta) = o(n^a)$ for every $a > 0$ and every θ, it can be shown that the rule ϕ^* is asymptotically optimal in the sense that (12) is the same as the asymptotic lower bound.

Instead of describing uncertainty adjustments in terms of hypothesis testing, [5] and [2] modify the certainty-equivalence rule in the classical multi-armed bandit problem by using certain upper confidence bounds to replace the maximum likelihood estimates. This has the advantage of being readily connected to Gittins' theory [4] for the discounted problem, and the upper confidence bounds have been shown (cf. [3]) to be asymptotically equivalent to Gittins' dynamic allocation indices. In view of the well known duality between confidence interval estimation and hypothesis testing, instead of using the sequence of upper confidence bounds as in [2] and [5] to adjust the certainty-equivalence rule, we can use the hypothesis testing approach (similar to that in the above example) to obtain a modification of the certainty-equivalence rule having the same asymptotic properties as that using the confidence sequence approach. The hypothesis testing approach has the advantage of being directly linkable to the results and analytical techniques of sequential testing theory. Indeed, the probability calculations in [2] and [5] for the sequence of upper confidence bounds based on generalized likelihood ratio statistics are adaptations of those originally developed in [8] and [9] for sequential generalized likelihood ratio tests of composite hypotheses in a general exponential family.

For adaptive control problems in which the set of possible control actions is finite, one can extend the ideas in the above example to adjust the certainty-equivalence rule. The idea is to test sequentially each possible action against the apparent leader and to try to learn about the parameters related to that action if there is not enough evidence in the data to confirm that the action is indeed inferior. In [10] we have used this approach for adaptive control of finite-state Markov chains, improving the seminal work of Borkar and Varaiya [11], [12], and simplifying and generalizing the recent advances due to Agrawal, Teneketzis and Anantharam [13].

3. RECURSIVE ESTIMATION AND ADAPTIVE CONTROL IN STOCHASTIC REGRESSION MODELS

As in the previous section, we begin this section with an example to illustrate the main issues and ideas. Consider the ARX model defined by the linear stochastic difference equation

$$y_n + \alpha_1 y_{n-1} + \cdots + \alpha_p y_{n-p} = \beta_1 u_{n-1} + \cdots + \beta_q u_{n-q} + \epsilon_n, \qquad (13)$$

where $\{y_n\}, \{u_n\}$ and $\{\epsilon_n\}$ denote the output, input and disturbance sequences, respectively, and $\alpha_1, \cdots, \alpha_p, \beta_1, \cdots, \beta_q$ are unknown parameters with $\beta_1 \neq 0$. How should the inputs u_t be chosen, on the basis of current and past observations $y_t, y_{t-1}, u_{t-1}, \cdots$, to regulate the outputs, say, such that $\sum_{i=1}^{N} y_i^2$ is minimized, in some sense, at least asymptotically as $N \to \infty$? In the case $p = q = 1$, assuming $\alpha_1 = -1$ to be known, β_1 to have a normal prior distribution and the ϵ_i to be independent standard normal random variables, Åström [14] computed the Bayes rule in the case $N = 30$ by dynamic programming, which took 180 CPU hours on a VAX 11/780 computer, and found that the Bayes rule takes relatively large and irregular control actions to probe the system when the Bayes estimate $\widehat{\beta}_t$ of β has poor precision, and that the Bayes rule is well approximated by the certainty-equivalence rule $\widehat{\beta}_t u_t + y_t = 0$ (recalling that $\alpha_1 = -1$) when $\widehat{\beta}_t$ has high precision. Åström's numerical example again points to the need of assessing parameter uncertainties in the certainty-equivalence rule and of using probing inputs in the case of substantial uncertainties.

Let $\theta = (-\alpha_1, \cdots, -\alpha_p, \beta_1, \cdots, \beta_q)^T, \psi_n = (y_n, \cdots, y_{n-p+1}, u_n, \cdots, u_{n-q+1})^T$. Then we can rewrite (13) as a linear stochastic regression model

$$y_{n+1} = \theta^T \psi_n + \epsilon_{n+1}. \tag{14}$$

The least squares estimate of θ at stage n is $\widehat{\theta}_n = (\sum_{i=1}^{n} \psi_i \psi_i^T)^{-1} \sum_{i=1}^{n} \psi_i y_i$, which can be written in the recursive form

$$\widehat{\theta}_n = \widehat{\theta}_{n-1} + P_{n-1} \psi_{n-1}(y_n - \widehat{\theta}_{n-1}^T \psi_{n-1}), \tag{15a}$$

$$P_n = P_{n-1} - P_{n-1} \psi_n \psi_n^T P_{n-1}/(1 + \psi_n^T P_{n-1} \psi_n). \tag{15b}$$

Suppose that $\{\epsilon_n\}$ is a martingale difference sequence with respect to an increasing sequence of σ-fields such that $E(\epsilon_n^2|\mathcal{F}_{n-1}) = \sigma^2$ and $\sup_n E(|\epsilon_n|^\alpha|\mathcal{F}_{n-1}) < \infty$ a.s. (almost surely) for some $\alpha > 2$. Statistical properties of the least squares estimate $\widehat{\theta}_n$ are closely related to the matrix $P_n = (\sum_{i=1}^{n} \psi_i \psi_i^T)^{-1}$ given by (15b) and to σ^2 which can be estimated by $\widehat{\sigma}_n^2 = n^{-1} \sum_{i=1}^{n}(y_i - \widehat{\theta}_n^T \psi_{i-1})^2$. In particular, it is shown in [15] that

$$\|\widehat{\theta}_n - \theta\|^2 = O(\{\log \lambda_{\max}(P_n^{-1})\}/\lambda_{\min}(P_n^{-1})) \text{ a.s.}, \tag{16}$$

and that under certain conditions $\widehat{\theta}_n$ is asymptotically normal with mean θ and covariance matrix $\sigma^2 P_n$, i.e., $(\sigma^2 P_n)^{-1/2}(\widehat{\theta}_n - \theta)$ has a limiting normal distribution as $n \to \infty$. Hence the matrix $\widehat{\sigma}_n^2 P_n$ provides useful information about the accuracy of $\widehat{\theta}_n$ as an estimate of θ.

If the parameter vector θ were known, the optimal controller would choose the input at stage t so that $\theta^T \psi_t = 0$, and its output at stage $t+1$ would be ϵ_{t+1}. In ignorance of θ, the (least squares) certainty-equivalence rule defines the input at stage t by the equation $\hat{\theta}_t^T \psi_t = 0$, and its output at stage $t+1$ is $(\theta - \hat{\theta}_t)^T \psi_t + \epsilon_{t+1}$. This shows that uncertainty assessments and adjustments in the certainty-equivalence rule should be focussed on the accuracy of $\hat{\theta}_t^T \psi_t$ as an estimate of $\theta^T \psi_t$, instead of the accuracy of $\hat{\theta}_t$ as an estimate of θ. Under efficient control, the matrix P_n^{-1} is ill-conditioned for large n (cf. [16]) and some linear functionals of θ may be much better estimated than other linear functionals. What we need for efficient control is that the linear functionals $\theta^T \psi_t$ are well estimated by $\hat{\theta}_t^T \psi_t$. The following result of [17] gives a sharp bound on the cumulative squared difference between $\hat{\theta}_t^T \psi_t$ and $\theta^T \psi_t$:

$$\sum_{i=1}^{n}(\hat{\theta}_t^T \psi_t - \theta^T \psi_t)^2 \leq (\sigma^2 + o(1)) \log \, \det(P_n^{-1}) \text{ a.s. on } \{ \lim_{n \to \infty} \psi_n^T P_n \psi_n = 0\}. \qquad (17)$$

Important insights into what uncertainty adjustments should be incorporated into the certainty-equivalence rule are provided by (17) and the relatively tractable Bayes problem that assumes β_1 to be known and the other parameters to have some prior truncated normal distribution π. Let $R_N = \sum_{i=1}^{N}(y_i - \epsilon_i)^2$ be the cumulative squared difference in outputs between a given input sequence and the optimal controller assuming known system parameters (with output ϵ_i at stage i). This is called the "regret" of the input sequence (cf. [17]), and the problem of minimizing $E(\sum_{i=1}^{N} y_i^2) = E(R_N) + \sigma^2 N$ is clearly equivalent to that of minimizing the expected regret. Suppose that $\beta_1 \neq 0$ is known and let π be a standard multivariate normal distribution restricted to the following region of $\lambda \triangleq \beta_1^{-1}(-\alpha_1, \cdots, -\alpha_p, \beta_2, \cdots, \beta_q)^T$: $A(z) \triangleq 1 + \sum_{i=1}^{p} \alpha_i z^i \neq 0$ and $B(z) \triangleq \sum_{i=1}^{q} \beta_i z^{i-1} \neq 0$ for $|z| \leq 1$ and the polynomials $\alpha_1 z^{p-1} + \cdots + \alpha_p$ and $\beta_1 z^{q-1} + \cdots + \beta_q$ are relatively prime. Then as shown in [17],

$$\int E_\lambda(R_N) d\pi(\lambda) \geq (\sigma^2 + o(1))(p + q - 1) \log N, \qquad (18)$$

for all input sequences $\{u_n\}$ satisfying the conditions $u_n^2 = O(n^\delta)$ for some $0 < \delta < 1$ and

$$R_n/n \to 0 \text{ a.s.} \qquad (19)$$

In view of the asymptotic lower bound (18), we would like to construct an input sequence $\{u_n\}$ so that its regret R_N is no more than $(\sigma^2 + o(1)) (p + q - 1) \log N$, even when β_1 is unknown and λ lies outside the stability and identifiability region defined above as the support of π. Comparing this goal with (17) suggests that this goal may be achieved by incorporating uncertainty adjustments into the least squares certainty-equivalence rule to ensure that (i) $\log \det P_n^{-1} \le (p+q-1+o(1)) \log n$ a.s., (ii) $\psi_n^T P_n \psi_n \to 0$ a.s., and (iii) the equation $\widehat{\theta}_n^T \psi_n = 0$ has a unique solution so that the least squares certainty-equivalence rule is well defined. Note that the requirement (iii) is the same as that the component of $\widehat{\theta}_n$ estimating β_1 be nonzero. In [18], it is shown how such uncertainty adjustments can be constructed under stability assumptions on $A(z)$ and $B(z)$, giving an adjusted certainty-equivalence rule whose regret satisfies

$$R_n \le (\sigma^2 + o(1))(p+q-1) \log n \quad \text{a.s.} \tag{20}$$

Note that (20) is a much stronger conclusion than (19), which is called the "self-optimizing" property (cf. [16]).

The ideas of [18] for incorporating uncertainty adjustments into certainty-equivalence rules have been extended in [19] to the ARMAX system

$$y_n + \alpha_1 y_{n-1} + \cdots + \alpha_p y_{n-p} = \beta_1 u_{n-d} + \cdots + \beta_q u_{n-d-q+1} + \epsilon_n + c_1 \epsilon_{n-1} + \cdots + c_h \epsilon_{n-h}, \tag{21}$$

in which $d \ge 1$ represents the delay, without assuming $A(z)$ to be stable (i.e., to have zeros outside the unit circle). Central to this extension are analogues of (17), developed in [20], for various recursive estimators of θ in the ARMAX model.

Letting $\theta = (-\alpha_1, \cdots, -\alpha_p, \beta_1, \cdots, \beta_q, c_1, \cdots, c_h; y_0, \cdots, y_{1-p}, u_0, \cdots, u_{2-d-q}, \epsilon_0, \cdots, \epsilon_{1-h})^T$, the above ARMAX model can be written as a nonlinear stochastic regression model $y_n = f_n(\theta) + \epsilon_n$, in which f_n is $\mathcal{F}_{n-1}$-measurable and $\{\epsilon_n\}$ is a martingale difference sequence with respect to $\{\mathcal{F}_n\}$, cf. [21]. Analogues of (17) for least squares estimates in nonlinear stochastic regression models have recently been developed in [21] and [22], and we are currently studying their applications to adaptive control of nonlinear input-output systems of the form $y_n = g(y_{n-1}, \cdots, y_{n-p}, u_{n-d}, \cdots, u_{n-d-q+1}; \theta) + \epsilon_n$.

4. CONCLUSION

By incorporating adjustments for parameter uncertainties into certainty-equivalence rules, asymptotically optimal solutions have been developed for several classical stochastic adaptive control problems. In these problems, the optimal solutions are very simple when the parameters are known. It is therefore particularly tempting to preserve this simplicity by replacing the unknown parameters in these "fictitious" optimal solutions by their sequential estimates, as in the certainty-equivalence approach. Moreover, the alternative approach of putting a prior distribution on the unknown parameters leads to very complicated optimal stochastic control problems. On the other hand, a certainty-equivalence rule is only an "imitation" of the optimal rule that assumes knowledge of the system parameters, and how well it imitates the unrealizable optimal rule depends on how much information the current and past data provide about the parameter configurations that are relevant to the control objective. To illustrate this point, in the example of Section 2, we do not need to know the actual values of the parameters $\mu_1, \mu_2, \mu_3, \lambda$ for optimal control, but need only to know which of μ_1, μ_2, μ_3 is the largest. In the example of Section 3, we do not need to estimate the entire vector θ well, but only need reliable estimates of $\theta^T \psi_t$.

The essence of the uncertainty adjustments in Sections 2 and 3 and in the references cited there is to ensure that there is adequate information in the data to address questions on the parameters that are basic to the control objective. A subtle issue here is that we want adequate, but not too much, information. For instance, in the example of Section 2, we need to sample from Π_1 to learn about μ_1 but should sample a total number, up to stage n, of no more than $(1 + o(1))(\log n)/I(\mu_1, \max(\mu_2, \mu_3))$ observations if $\mu_1 < \max(\mu_2, \mu_3)$. For the adaptive control problem of the ARX system discussed in Section 3, to enhance the information content of the design so that one can estimate θ consistently, a key idea of [18] is to introduce occasional blocks of white-noise perturbations and to keep the total variance of such perturbations up to stage n within the order $o(\log n)$ so that the regret R_n can still have the logarithmic order given by (20).

In the stochastic adaptive control problems discussed herein, the parameters do not change with time. There has been much recent interest in adaptive control of ARMAX systems in which the parameters may undergo temporal changes. In these problems, one has to develop estimates of time-varying parameters and to assess the uncertainties in these estimates. In addition to parameter uncertainty, another basic issue in stochastic adaptive control is model uncertainty. The assumed model of system dynamics may turn out to be inadequate or inappropriate as the system evolves over time, and one should perform diagnostic tests on-line and make adjustments accordingly. These are some of the challenging new directions in stochastic adaptive control, to which the heuristic principle of certainty equivalence with uncertainty adjustments may again offer practical solutions that are nearly optimal.

REFERENCES

[1] T. L. Lai, "Information bounds, certainty equivalence and learning in asymptotically efficient adaptive control of time-invariant stochastic systems", in *Topics in Stochastic Systems, Modelling, Estimation and Adaptive Control* (L. Gerencser and P. E. Caines, Eds.), Springer-Verlag, 1991, pp. 335-368.

[2] T. L. Lai, "Adaptive treatment allocation and the multi-armed bandit problem", *Ann. Statist.*, vol. 15 (1987), pp. 1091-1114.

[3] T. L. Lai, "Asymptotic solutions of bandit problems", in *Stochastic Differential Systems, Stochastic Control Theory and Applications* (W. Fleming and P. L. Lions, Eds.), Springer-Verlag, 1988, pp. 275-292.

[4] J. C. Gittins, "Bandit processes and dynamic allocation indices", *J. Roy. Statist. Soc. Ser. B*, vol. 41 (1979), pp. 148-177.

[5] T. L. Lai and H. Robbins, "Asymptotically efficient adaptive allocation rules", *Adv. Appl. Math.*, vol. 6 (1985), pp. 4-22.

[6] T. L. Lai and H. Robbins, "Asymptotically optimal allocation of treatments in sequential experiments", in *Design of Experiments, Ranking and Selection* (T. J. Santner and A. C. Tamhane, Eds.), Marcel-Dekker, 1984, pp. 127-142.

[7] Y. S. Chow, H. Robbins and D. Siegmund, *Great Expectations: The Theory of Optimal Stopping.* Houghton Mifflin, 1972.

[8] T. L. Lai, "Nearly optimal sequential tests of composite hypotheses", *Ann. Statist.*, vol. 16 (1988), pp. 856-886.

[9] T. L. Lai, "Boundary crossing problems for sample means", *Ann. Probability*, vol. 16 91988), pp. 375-396.

[10] C. D. Fuh and T. L. Lai, "Asymptotically efficient adaptive control of Markov chains", Tech. Report, Department of Statistics, Stanford Univ., 1992.

[11] V. Borkar and P. Varaiya, "Adaptive control of Markov chains, I: Finite parameter set", *IEEE Trans. Automat. Contr.*, vol. AC-24 (1979), pp. 953-958.

[12] V. Borkar and P. Varaiya, "Identification and adaptive control of Markov chains", *SIAM J. Contr. Optimiz.*, vol. 20 (1982), pp. 470-489.

[13] R. Agrawal, D. Teneketzis and V. Anantharam, "Asymptotically efficient adaptive allocation schemes for controlled Markov chains: Finite parameter space", *IEEE Trans. Automat. Contr.*, vol. 34 (1989), pp. 1249-1259.

[14] K. J. Åström, "Theory and applications of adaptive control — A survey", *Automatica*, vol. 19 (1983), pp. 471-486.

[15] T. L. Lai and C. Z. Wei, "Least squares estimates in stochastic regression models with applications to identification and control of dynamic systems", *Ann. Statist.*, vol. 10 (1982), pp. 154-166.

[16] P. R. Kumar, "A survey of some results in stochastic adaptive control", *SIAM J. Contr. Optimiz.*, vol. 23 (1985), pp. 329-380.

[17] T. L. Lai, "Asymptotically efficient adaptive control in stochastic regression models", *Adv. Appl. Math.*, vol. 7 (1986), pp. 23-45.

[18] T. L. Lai and C. Z. Wei, "Asymptotically efficient self-tuning regulators", *SIAM J. Contr. Optimiz.*, vol. 25 (1987), pp. 466-481.

[19] T. L. Lai and Z. Ying, "Parallel recursive algorithms in asymptotically efficient adaptive control of linear stochastic systems", *SIAM J. Contr. Optimiz.*, vol. 29 (1991), pp. 1091-1127.

[20] T. L. Lai and Z. Ying, "Recursive identification and adaptive prediction in linear stochastic systems", *SIAM J. Contr. Optimiz.*, vol. 29 (1991), pp. 1061-1090.

[21] G. Zhu, "Least squares estimation and adaptive prediction in non-linear stochastic regression models with applications to time series and stochastic systems", Ph.D. Dissertation, Department of Statistics, Stanford Univ., 1992.

[22] T. L. Lai and G. Zhu, "Adaptive prediction in non-linear autoregressive models and control systems", *Statistica Sinica*, vol. 1 (1991), pp. 309-334.

Uniform Convergence of the Solutions to Riccati Equations Arising in Boundary/Point Control Problems[*]

I. Lasiecka and R. Triggiani
Department of Applied Mathematics
University of Virginia
Charlottesville, VA 22903

1. Introduction

This paper presents uniform convergence results for the solutions to a parametrized family of Algebraic Riccati Equations which arise in the context of control problems for systems described by partial differential equations with unbounded control actions. This setup includes the case of boundary or point control problems. The problem studied here is motivated by recent developments in the theory of adaptive control for distributed parameter systems [D-M-P], [D-G-P] where it is shown that the issue of uniform convergence of a family of Riccati operators is critical to this theory.

We shall formulate the problem in the context of abstract equations defined on Hilbert spaces. Let H, U and Z denote three Hilbert spaces. Let α be a real valued scaler or possibly vector parameter such that $|\alpha| \to 0$ and $|\alpha| < 1$ (without loss of generality). We are given: a family A_α of generators of C_0 – semigroups $e^{A_\alpha t}$ on H, such that

$$\rho(A_a) \subset [\omega_0, \infty] \text{ for some fixed } \omega_0 > 0 , \tag{1.1}$$

a family of control operators B_α (generally unbounded) such that

$$| \hat{A}_\alpha^{-1} B_\alpha |_{\mathcal{L}(U; H)} \leq M \text{ uniformly in } \alpha \text{ where } \hat{A}_\alpha \equiv -A_\alpha + \omega_0 I . \tag{1.2}$$

We consider the following family of control problems paremetrized by α:

$$\begin{cases} y_t(t) = A_\alpha y(t) + B_\alpha u(t) , \\ y(0) = y_0 \in H . \end{cases} \tag{1.3}$$

With (1.3) we associate the following Algebraic Riccati Equation

$$\text{(R.E.)} \quad \begin{cases} (A_\alpha^* P_\alpha x, y)_H + (P_\alpha A_\alpha x, y)_H + (R^* Rx, y)_H = (B_\alpha^* P_\alpha x, B_\alpha^* P_\alpha y)_U; \ x, y \in \mathcal{D}(A_\alpha) \\ \text{where } R \in \mathcal{L}(H; Z) . \end{cases}$$

Under suitable stabilizability/detectability conditions it is known (see [L-T.2]) that there exists a unique solution $P_\alpha \in \mathcal{L}(H)$ of equation (R.E.). The question addressed in this paper is: under which conditions imposed on A_α, B_α, R one obtains the uniform convergence as $|\alpha| \to 0$.

[*]This research was partially supported by the National Science Foundation under Grant DMS-8902811 and by the Air Force Office of Scientific Research under Grant DEF-89-0511.

$$|P_\alpha - P|_{\mathcal{L}(H)} \to 0 \; ; \tag{1.4}$$

$$|B_\alpha^* P_\alpha - B^* P|_{\mathcal{L}(H; U)} \to 0 \tag{1.5}$$

where we have adopted the notation $P \equiv P_0$; $B \equiv B_0$; $A \equiv A_0$ etc. A partial answer to this question has been given in [Ch-D-P] and [D-M-P]. However, the results in the papers cited above deal with the following restrictive situations

(i)   in the case of general C_0 semigroups, the control operators B_α are assumed to be *bounded:* $U \to H$ (see [Ch-D-P]);

(ii) in the case of analytic semigroups, only "mild" unboundedness of control operator is allowed i.e.: it is assumed that $R(\lambda_0, A^{-\gamma})B$ is bounded where $0 < \gamma < \frac{1}{2}$ (see [D-M-P]).

The main contribution of this paper is to dispense entirely with the above restrictions. We shall prove the *uniform* convergence of Riccati operators for general C_0 - semigroups with unbounded "admissible" control actions and in the case of analytic semigroups we will be able to treat the case of "fully unbounded" control operators i.e. when $\gamma < 1$. It should be noted that there is an *enormous* wealth of results devoted to a question of convergence of approximations (typically finite dimensional) of Riccati Equations (see [L-T] and references therein). However, most of these results are presented in the framework of finite dimensional approximations (with the specific properties placed over approximating subspaces) and hence not immediately readily applicable in the context of adaptive control. Ond the other hand, the techniques developed earlier for these problems (in particular [L-T.1]; [L.1]) are clearly adaptable to treat the issue of *uniform* convergence of a family of Riccati Equations. The purpose of this note is to provide a clear formulation/proof of these results in the context of adaptive control theory and to illustrate the findings with several examples arising from boundary/point control problems. The outline of the paper is as follows. Section 2 deals with the case of analytic semigroups. Subsection 2.1 provides the statement of the results, subsection 2.2 gives the proof and subsection 2.3 provides the examples. Section 3 treats the case of general C_0 semigroups.

2. Analytic semigroups

2.1. Statement of the results

Here we shall assume that for each α, A_α is a generator of analytic semigroup such that

$$|A_\alpha e^{A_\alpha t}|_{\mathcal{L}(H)} \leq C \frac{e^{w_0 t}}{t} \tag{A-1}$$

for some constants C, $w_0 > 0$ independent on α, and

$$|\hat{A}_\alpha^{-\gamma} B_\alpha|_{\mathcal{L}(U; H)} \leq C; \quad \gamma < 1 . \tag{A-2}$$

In order to guarantee the unique solvability of the Algebraic Riccati Equation we shall also assume: There exists $F \in \mathcal{L}(H ; U)$ such that the s.c. analytic semigroup $e^{(A + BF)t}$ (as guaranteed by (A-1)) is exponentially stable i.e.:

$$|e^{(A + BF)t}|_{L(H)} \leq M_F e^{-w_F t}; \quad w_F > 0 . \tag{A-3}$$

There exists $K \in L(Z; H)$ such that

$$|e^{(A + KR)t}|_{L(H)} \leq M_K e^{-w_K t}; \quad w_K > 0 . \tag{A-4}$$

It is well known (see [L-T]) that the conditions (A-1) - (A-4) imply that there exists unique solution P_α to (ARE) such that

$$(\hat{A}_\alpha^*)^\rho P_\alpha \in L(H) ; \quad \text{for any } \rho < 1 ; \tag{2.1}$$

and, is particular (see (A-2))

$$B_\alpha^* P_\alpha \in L(H; U) , \tag{2.2}$$

where $(B_\alpha u, v)_H = (u, B_\alpha^* v) \; \forall u \in U$: $v \in \mathcal{D}(B_\alpha^*) = \mathcal{D}(A_\alpha^*)$. In order to obtain convergence results for P_α we need to assume certain convergence properties for the generators A_α and the control operators B_α. We shall henceforth assume $\mathcal{D}(A_\alpha^*) = \mathcal{D}(A^*)$ and

$$|\hat{A}_\alpha^{*\gamma - 1} (A_\alpha^* - A^*)|_{L(\mathcal{D}(A^*);H)} \to 0 \quad \text{as } |\alpha| \to 0 , \tag{A-5}$$

$$|(B_\alpha^* - B^*)|_{L(\mathcal{D}(A^*); U)} \to 0 \quad \text{as } |\alpha| \to 0 . \tag{A-6}$$

Our main result is

Theorem 2.1

Assume (A-1) - (A-6). In addition assume that
(A-7i) either $R > 0$ or else $\hat{A}^{-1} K R: H \to H$ is compact;
(A-7ii) either $B^* \hat{A}^{*-1}: H \to H$ is compact or else $F: H \to U$ is compact.

Then as $|\alpha| \to 0$

$$|P_\alpha - P|_{L(H)} + |B_\alpha^* P_\alpha - B^* P|_{L(H; U)} \to 0 \tag{2.3}$$

$$|e^{(A_\alpha - B_\alpha B_\alpha^* P_\alpha)t}|_{L(H)} \leq C e^{-wt} \tag{2.4}$$

for some $C > 0$; $w > 0$ independent on $\alpha > 0$.

Remark 2.1

Assumptions (A-1)-(A-7) are very natural for the problem. Indeed hypothesis (A-1) is satisfied if A_α is uniformly coercive, (i.e. $(A_\alpha u, v)_H \leq C |u|_v |v|_v$ and $-(A_\alpha u, u)_H \geq \rho |u|_v^2 - C_0 |u|_H^2$.

Hypothesis (A-2) holds for all the cases of boundary or point control problems. Stabilizability detectability assumptions (A-3), (A-4) are typically satisfied with operators F and K which are of finite rank. This automatically implies that (A-7) holds.

Remark 2.2

Notice that Theorem 2.1 extends the results of [D-M-P] (related to uniform convergence of Riccaati operators) in several directions. Indeed, first of all and most importantly, γ in (A-2) is allowed now to range in $[0,1)$ rather than only in $[0, \frac{1}{2})$ (this prmits the treatment of several boundary/point control problems where the value of γ is strictly greater than $\frac{1}{2}$ (typically $\frac{3}{4} + \varepsilon$) see examples 2.3.1, 2.3.2). Second, the assumption of compactness of $\hat{A}^{-1}$ (in [D-M-P]), is now dispensed with. This generalization is important if one wants to treat cases of strongly damped wave/plate equations where the resolvent is *not* compact (see example 2.3.3). Third, the assumption on "uniform" (with respect to the parameter α) stability of the generators (see assumption (A-4) in [D-M-P]) is dispensed with. As far as stabilizability properties, all it is needed is stabilizability (not a-priori stability!) of a single process (A,B) (not of A_α, B_α). This, again, allow us to treat dynamics which are originally unstable and the only information required a priori is possibility of stabilization of original dynamics represented by (A, B). Notice also that in the special case when the control operators are given by$B_\alpha = A_\alpha \hat{A}^{-1} B$ (as in [D-M-P]) our approximating condition (A-6) is satisfied automatically. Indeed, $B_\alpha^* - B^* = (\hat{A}^{-1} B)^* [A_\alpha^* - A^*]$ and by (A-5)

$$|B_\alpha^* - B^*|_{\mathcal{L}(D(A^*); U)} \leq C |\hat{A}^{*-1} (A_\alpha^* - A^*)|_{\mathcal{L}(D(A^*); U)} \to 0.$$

2.2. Proof of Theorem 2.1

The proof of Theorem 2.1 follows very closely arguments used in the proof of Theorem 1.1 in [L-T.1] (whose statement implies all the assertions of Theorem 2.1). The main difference is that paper [L-T.1] deals with finite dimensional approximations A_h, B_h of the original operators A and B. However, if we restrict the framwork of [L-T.1] to $V_h \equiv H$ and $P_h = I$, then the approximating assumptions imposed on A_h (see (1.14), (1.15) p. 642 in L-T.1] follow from our assumptions (A-1)-(A-5). (We remark that the rate of convergence imposed on (1.15) of [L-T.1] is replaced now by the statement on the uniform (operators norm) convergence. This difference has no bearing on subsequent arguments). Similarly assumptions (1.3), (1.5), (1.6), (1.26)-(1.27) in [L-T.1] follow from our present hypotheses (A-2), (A-3), (A-4) and (A-7). The main difference is in the nature of assumptions imposed on the approximations B_α of B. The hypotheses (1.16) - (1.18) of [L-T.1] are more tailored toward finite dimensional approximations (like the inverse approximation property (1.16) typical for splines, etc.) rather than our present hypothesis (A-6) of Theorem 2.1 (which is in some sense more general). On the other hand, these hypotheses (1.16)-(1.18) are used only to obtain the fundamental "convergence" Lemma 3.1 in [L-T.1] p. 650. Thus, if we can prove the statements of this Lemma independently within our present framework, all the remaining arguments for the proof of Theorem 1.1 [L-T.1] (hence our Theorem 2.1) remain valid. Therefore, it suffices to prove the convergence results stated in Lemma 3.1 [L-T.1]. For the convenience of the reader, we shall provide its full statement.

Lemma 2.1

Assume (A-1) - (A-6) and let r, Q be arbitrary constants. Then there exists $w_0 > 0$ and the constant $C > 0$ (independent on α) such that as $|\alpha| \to 0$

$$\sup_{t \geq 0} t \, e^{-w_0 t} \, |B_\alpha^* \, e^{A_\alpha^* t} - B^* \, e^{A^* t} \, |_{\mathcal{L}(H; \, U)} \to 0 \, ; \tag{2.1}$$

$$\sup_{t \geq 0} e^{-w_0 t} \, |B_\alpha^* \, e^{A_\alpha^* T} - B^* e^{A^* t} \, |_{\mathcal{L}(\mathcal{D}(A^*); \, U)} \to 0 \, ; \tag{2.2}$$

$$\sup_{t \geq 0} t^{(1-Q)} \, e^{-w_0 t} \, |B_\alpha^* e^{A_\alpha^* T} - B^* \, e^{A^* t} \, |_{\mathcal{L} \, \mathcal{D}(\hat{A}^{*Q}); \, U)} \to 0 \tag{2.3}$$

$$|B_\alpha^* \, e^{A_\alpha^* t} |_{\mathcal{L}(H, \, U)} \leq C \, \frac{e^{w_0 t}}{t^\gamma} \, ; \tag{2.4}$$

$$\sup_{t \geq 0} t^{Q + (1-Q)\gamma} \, e^{w_0 t} | \, B_\alpha^* \, e^{A_\alpha^* T} - B^* \, e^{A^* t} |_{\mathcal{L}(H; \, U)} \to 0; \tag{2.5}$$

$$\sup_{t \geq 0} t^{(1-r)[Q + \gamma(1-Q)]} \, e^{w_0 t} |B_\alpha^* \, e^{A_\alpha^* t} - B^* \, e^{A^* t} |_{\mathcal{L}(\mathcal{D}(\hat{A}^{*r}); \, U)} \to 0 \, . \tag{2.6}$$

proof

Let Σ denote a closed triangular sector containing the axis $[-\infty, w_0]$ and delimited by the two rays $w_0 + \rho \, e^{\pm iQ}$ for some $\pi/2 < Q < 2\pi$. With the appropriate choice of Q, such sector contains the spectrum of A_α for all values of the parameter α (recall that constants C, w_0 in the hypothesis (A-1) are independent on α). Then the standard semigroup theory argument yields the following representation formula

$$B_\alpha^* \, e^{A_\alpha^* t} - B^* \, e^{A^* t} = \int_{\partial \Sigma} e^{\lambda t} \, [B_\alpha^* \, R(\lambda, A_\alpha^*) - B^* \, R(\lambda, A^*)] \, d\lambda \, . \tag{2.7}$$

By using the resolvent idendity we obtain

$$B_\alpha^* \, R(\lambda, A_\alpha^*) - B^* \, R(\lambda, A^*) = (B_\alpha^* - B^*) \, R(\lambda, A^*) + B_\alpha^* \, R(\lambda, A_\alpha^*) \, [A_\alpha^* - A^*] \, R(\lambda, A^*) \, . \tag{2.8}$$

Therefore

$$\int_{\partial \Sigma} |e^{\lambda t} \, (B_\alpha^* - B^*) \, R(\lambda, A^*) |_{\mathcal{L}(H; \, U)} \, d\lambda \leq \int_{\partial \Sigma} e^{|\lambda| t} \, |(B_\alpha^* - B^*) |_{\mathcal{L}(\mathcal{D}(A^*); \, U)} \, |A^* \, R(\lambda, A^*) |_{\mathcal{L}(H)} \, d\lambda$$

by (A-1) (in λ - version)

$$\leq c \, \frac{e^{w_0 t}}{t} \, |(B_\alpha - B^*) |_{\mathcal{L}(\mathcal{D}(A^*); \, U)} \, . \tag{2.9}$$

$$\int_{\partial \Sigma} |e^{\lambda t} \, B_\alpha^* \, R(\lambda, A_\alpha^*) \, [A_\alpha^* - A^*] \, R |\lambda, A^*) |_{\mathcal{L}(H; \, U)} \, d\lambda \leq \int_{\partial \Sigma} e^{|\lambda| t} \, |B_\alpha^* \, \hat{A}_\alpha^{*-\gamma} |_{\mathcal{L}(H; \, U)} \, |\hat{A}_\alpha^* \, R(\lambda, A_\alpha^*) |_{\mathcal{L}(H)}$$

$$|\hat{A}_\alpha^{*\gamma - 1} \, [A_\alpha^* - A^*] |_{\mathcal{L}(\mathcal{D}(A^*); \, H)} \cdot \, |A^* \, R(\lambda, A^*) |_{\mathcal{L}(H)} \, d\lambda$$

by (A-2) and (A-1)

$$\leq C \, |\hat{A}_\alpha^{*\gamma - 1} \, [A_\alpha^* - A^*] |_{\mathcal{L}(\mathcal{D}(A^*); \, H)} \int_{\partial \Sigma} e^{|\lambda| t} \, d\lambda \leq C \, \frac{e^{w_0 t}}{t} \, |\hat{A}_\alpha^* - A^*| |_{\mathcal{L}(\mathcal{D}(A^*); \, H)} \, . \tag{2.10}$$

Combining (2.7)-(2.10) and recalling the convergence hypotheses (A-5) and (A-6) yields the desired result in (2.1).

proof of 2.2. By using analyticity of A^* along the arguments as in [K.1]

$$\int_{\partial\Sigma} |e^{|\lambda|t}| |(B_\alpha^* - B^*) R(\lambda, A^*)|_{L(\mathcal{D}(A^*); U)} d\lambda$$

$$\leq |(B_\alpha^* - B^*)|_{L(\mathcal{D}(A^*); U)} \int_{\partial\Sigma} e^{|\lambda|t} |R(\lambda, A^*)|_{L(H)} d\lambda \leq C |B_\alpha^* - B^*|_{L(\mathcal{D}(A^*); U)}. \qquad (2.11)$$

$$\int_{\partial\Sigma} e^{|\lambda|t} |B_\alpha^* R(\lambda, A_\alpha^*) [A_\alpha^* - A^*] R(\lambda, A^*)|_{L(\mathcal{D}(A^*); U)}$$

$$\leq \int_{\partial\Sigma} e^{|\lambda|t} |B_\alpha^* \hat{A}_\alpha^{*-\gamma}|_{L(H; U)} |\hat{A}_\alpha^* R(\lambda, A_\alpha^*)|_{L(H)} |\hat{A}_\alpha^{*\gamma-1} [A_\alpha^* - A^*]|_{L(\mathcal{D}(A^*); H)} |R(\lambda, A^*)|_{L(H)} d\lambda$$

$$\leq C |\hat{A}_\alpha^{*\gamma-1} [A_\alpha^* - A^*]|_{L(\mathcal{D}(A^*); H)} \int_{\partial\Sigma} e^{|\lambda|t} |R(\lambda, A^*)|_{L(H)} d\lambda$$

$$\leq C |\hat{A}_\alpha^{*\gamma-1} [A_\alpha^* - A^*]|_{L(\mathcal{D}(A^*); H)}. \qquad (2.12)$$

Combining (2.11), (2.12) with (2.7) and (2.8) yields (2.2). (2.3) follows by interpolation from (2.1) and (2.2).

proof of 2.4 From (A-1), using interpolation and duality we obtain

$$|\hat{A}_\alpha^{*\gamma} e^{A_\alpha^* t}|_{L(H)} \leq C \frac{e^{w_0 t}}{t^\gamma}. \qquad (2.13)$$

From (H-2), by duality

$$|B_\alpha^* \hat{A}_\alpha^{\alpha-\gamma}|_{L(H; U)} \leq C. \qquad (2.14)$$

Hence

$$|B_\alpha^* e^{A_\alpha^* t}|_{L(H; V)} \leq |B_\alpha^* \hat{A}_\alpha^{*-\gamma}|_{L(H; V)} |\hat{A}_\alpha^{*\gamma} e^{A_\alpha^* t}|_{L(H)} \leq C \frac{e^{w_0 t}}{t^\gamma}$$

as desired.

proof of 2.5. From (2.4) and (A-1) (with (A-2) we obtain

$$\sup_{t \geq 0} t^\gamma e^{-w_0 t} |B_\alpha^* e^{A_\alpha^* t} - B^* e^{A^* t}|_{L(H; U)} \leq C. \qquad (2.15)$$

On the other hand from (2.1) (after "raising" it to power - Q) we obtain

$$\sup_{t \geq 0} t^{1-Q} e^{-w_0 t (1-Q)} |B_\alpha^* e^{A_\alpha^* t} - B^* e^{A^* t}|_{L(H; U)}^{1-Q} \to 0 \qquad (2.16)$$

Then "raising" (2.15) to power Q and multiplying the result by (2.16) yields the conclusion in (2.5). (2.6) follows by interpolation between (25) and (2.2). ∎

2.3 Examples

2.3.1 Heat equation with Dirichlet boundary control.

Let $\Omega \subset R^n$ be an open bounded domain with sufficiently smooth boundary Γ. In Ω, we consider

$$\begin{cases} y_t = (\alpha + 1) \Delta y + c^2 y & \text{in } (0, T] \times \Omega \equiv Q, \\ y(0, \cdot) = y_0 & \text{in } \Omega, \\ y|_{\Sigma} = u & \text{in } (0, T] \times \Gamma \equiv \Sigma. \end{cases} \tag{2.17}$$

Here the parameter $\alpha \to 0$ and without loss of generality we assume that $0 < \alpha \leq 1$. To put problem (2.17) into the abstract setting we introduce the following operators and spaces:

$$Z = H = L_2(\Omega); \quad U = L_2(\Gamma); \quad R = I;$$

$$A_\alpha h \equiv (\alpha + 1) \Delta h + c^2 h; \quad \mathcal{D}(A_\alpha) = H^2(\Omega) \cap H_0^1(\Omega); \quad B_\alpha u = -A_\alpha D u - c^2 \alpha D u$$

where D (Dirichlet map) is defined by

$$h = Dg \text{ iff } (\Delta + c^2) h = 0 \text{ in } \Omega \quad h|_\Gamma = g.$$

$$D: \text{ continuous } L_2(\Gamma) \to H^{1/2}(\Omega) \subset \mathcal{D}(\hat{A}^{1/4-\varepsilon}). \tag{2.18}$$

Verification of the assumptions of Theorem 2.1.

Hypothesis (A-1). It is enough to prove that A_α is uniformly coercive (see Remark 2.1). This, in turn, follows from the inequalities

$$(A_\alpha u, u)_{L_2(\Omega)} \geq |\nabla u|^2_{L_2(\Omega)} - c^2 |u|^2_{L_2(\Omega)},$$

and

$$(A_\alpha u, v)_{L_2(\Omega)} \leq 2|\nabla u|_{L_2(\Omega)} |\nabla u|_{L_2(\Omega)} + c^2 |u|_{L_2(\Omega)} |v|_{L_2(\Omega)}$$

so A_α is uniformly coercive with $V = H_0^1(\Omega)$.

Hypothesis (A-2) is satisfied with $\gamma = \frac{3}{4} + \varepsilon$ where $\varepsilon > 0$ can be taken arbitrarily small. Indeed, this follows from (2.18), since with $\gamma > \frac{3}{4}$

$$\hat{A}_\alpha^{-\gamma} B_\alpha = -\hat{A}_\alpha^{1-\gamma} D - \alpha c^2 \hat{A}_\alpha^{-\gamma} D \in \mathcal{L}(L_2(\Gamma); L_2(\Omega))$$

where we have used (2.18) together with the fact that $\mathcal{D}(\hat{A}_\alpha^Q) = \mathcal{D}(\hat{A}^Q); \, 0 \leq Q \leq 1$.

The stabilizability condition (A-3) holds in this case with a feedback operator which is finite rank (see [L-T.1] sect. 6.1). The detectability condition (A-4) holds automatically as $R = I$. We shall verify the validity of (A-5) and (A-6).

Hypothesis (A-5). Stronger conclusion holds:

$$|(A^* - A_\alpha^*)|_{\mathcal{L}(\mathcal{D}(A^*); H)} \to 0 \text{ as } \alpha \to 0. \tag{2.19}$$

Indeed, with $u \in \mathcal{D}(A^*) = \mathcal{D}(A) \subset H^2(\Omega)$ we obtain

$$|(A^* - A_\alpha^*) u|_{L_2(\Omega)} = \alpha \, |\Delta u|_{L_2(\Omega)} \leq \alpha \, |u|_{H^2(\Omega)} \tag{2.20}$$

which yields (2.19).

Hypothesis (A-6). Let $u \in \mathcal{D}(A^*) \subset H^2(\Omega)$

$$|B_\alpha^* u - B^* u|_{L_2(\Gamma)} \leq |D^*|_{\mathcal{L}(L_2(\Omega),\, L_2(\Gamma))} \, |A_\alpha^* u - A^* u|_{L_2(\Omega)} + c^2 \, \alpha \, |D^*|_{\mathcal{L}(L_2(\Omega);\, L_2(\Gamma))} \, |u|_{L_2(\Omega)}$$

$$\leq \alpha \, |D^*|_{\mathcal{L}(L_2(\Omega);\, L_2(\Gamma))} \, (1 + c^2) \, |u|_{H^2(\Omega)}$$

which implies (A-6).

Finally, conditions (A-7) are automatically satisfied as F is compact (finite rank) and R > 0. (Notice that we do not need to use the compactness of $\hat{A}^{-1}$.) Thus, all the required hypotheses (A-1)-(A-7) have been verified and the results of Theorem 2.1 apply.

Remark.

More general parabolic problem

$$y_t = \sum_{ij=1}^{n} \frac{\partial}{\partial x_j} \left(a_\alpha(x) \frac{\partial}{\partial x_i} \, y \right) + \sum_{i=1}^{n} b_\alpha(x) \frac{\partial}{\partial x_i} \, y + c_\alpha \, y$$

where a_α satisfy the usual ellipticity condition and

$$\sup_{x \in \Omega} \, [\, |a_\alpha(x) - a_0(x)| + |b_\alpha(x) - b_0(x)| + |c_\alpha(x) - c_0(x)| \,] \to 0 \quad \text{as } \alpha \to 0$$

can be treated in the same manner.

2.3.2. Structurally damped plate equation with boundary control.

Consider the following model of a plate equation in the deflection w (t, x), where $\rho > 0$ is any constant

$$\begin{cases} w_{tt} + (1 + \alpha) \, \Delta^2 w - \rho \Delta w_t = 0 & \text{in } Q \,; \\ w(0, \cdot) = w_0; \ w_t(0, \cdot) = w_1 & \text{in } \Omega \,; \\ w|_\Sigma = 0 \,; \\ \Delta w|_\Sigma = u. \end{cases} \tag{2.21}$$

To put problem (2.21) into the abstract setting we introduce the following spaces and the operators

$$Z = H \equiv [H^2(\Omega) \cap H_0^1(\Omega)] \times L_2(\Omega); \quad U = L_2(\Gamma); \quad R = I \,;$$

$$\mathcal{A}_\alpha h = (1 + \alpha) \, \Delta^2 h; \quad \mathcal{D}(\mathcal{A}_\alpha) = \{ h \in H^4(\Omega); \ h|_\Gamma = \Delta h|_\Gamma = 0 \} \,;$$

$$A_\alpha \equiv \begin{bmatrix} 0 & I \\ -\mathcal{A}_\alpha & -\rho\, \mathcal{A}^{\frac{1}{2}} \end{bmatrix} ;$$

$$B_\alpha u = \begin{bmatrix} 0 \\ (1+\alpha)\, \mathcal{A}^{\frac{1}{2}} Du \end{bmatrix}$$

where we recall $\mathcal{A}^{\frac{1}{2}}\, u = \Delta u;\ \ \mathcal{D}(\mathcal{A}^{\frac{1}{2}}) = H^2(\Omega) \cap H_0^1(\Omega).$

Verification of the assumptions.

Hypothesis A-1 This follows from the arguments of [Ch-T].

Hypothesis (A-2) is satisfied with $\gamma > \frac{3}{4}$. Since A is the direct sum of two normal operators on H,

$$\mathcal{D}((-A_\alpha^*)^Q) = \mathcal{D}((-A_\alpha)^Q) = \mathcal{D}(A_\alpha^{\frac{1}{2} + Q/2}) \times \mathcal{D}(\mathcal{A}^{Q/2}) = \mathcal{D}(\mathcal{A}^{\frac{1}{2} + Q/2}) \times \mathcal{D}(\mathcal{A}^{Q/2}) ;\ \ 0 \le Q \le 1. \tag{2.22}$$

Hence, by duality and closed graph theorem

$$\mathcal{A}_\alpha^{*-\gamma} \in \mathcal{L}(H;\ \mathcal{D}(\mathcal{A}^{\frac{1}{2} + \gamma/2} \times \mathcal{D}(\mathcal{A}^{\gamma/2})). \tag{2.23}$$

(with a norm uniform in α). On the other land, (A-2) is equivalent to

$$|B_\alpha^\alpha\, A_\alpha^{*-\gamma}|_{\mathcal{L}(H;\ U)} \le M. \tag{2.24}$$

where

$$B_\alpha^* v = (1+\alpha)\, D^*\, \mathcal{A}^{\frac{1}{2}}\, v_2;\ \ \ v = (v_1, v_2).$$

Thus, to establish (2.24), it is enough to verify that

$$|D^*\, \mathcal{A}^{\frac{1}{2}} v_2|_{L_2(\Gamma)} \le M\, |v_2|_{\mathcal{D}(\mathcal{A}^{\gamma/2})}. \tag{2.25}$$

From (2.18)

$$\mathcal{D}^*\, \mathcal{A}^{\frac{1}{2}(\frac{1}{4} - \varepsilon)} \in \mathcal{L}(L_2(\Omega); L_2(\Gamma)):\ \ \varepsilon > 0.$$

Hence for any $\varepsilon > 0$

$$|D^*\, \mathcal{A}^{\frac{1}{2}} v_2|_{L_2(\Gamma)} = |D^*\, \mathcal{A}^{\frac{1}{4} - \varepsilon}\, \mathcal{A}^{\frac{1}{2} - \frac{1}{4} + \varepsilon} v_2|_{L_2(\Gamma)} \le M\, |\mathcal{A}^{\frac{1}{4} + \varepsilon} v_2|_{L_2(\Gamma)}$$

which proves (2.24) with $\gamma > \frac{3}{4}$.

Hypothesis (A-5) We shall prove stronger conclusion

$$|A_\alpha^* - A^*|_{\mathcal{L}(\mathcal{D}(A^*), H)} \to 0\ \ \text{as } \alpha \to 0. \tag{2.26}$$

Indeed

$$A_\alpha^* - A^* = \begin{bmatrix} 0 & -I \\ \mathcal{A}_\alpha & -\rho\,\mathcal{A}^{\frac{1}{2}} \end{bmatrix} - \begin{bmatrix} 0, & -I \\ \mathcal{A}, & \rho\,\mathcal{A}^{\frac{1}{2}} \end{bmatrix} = \begin{bmatrix} 0 & 0 \\ \alpha\mathcal{A} & 0 \end{bmatrix}$$

with $u \in \mathcal{D}(A^*) = \mathcal{D}(\mathcal{A}) \times \mathcal{D}(\mathcal{A}^{\frac{1}{2}})$ we have

$$|(A_\alpha^* - A^*)\,u|_H = \alpha\,|\mathcal{A}u_1|_{L_2(\Omega)} = \le \alpha\,|u|_{\mathcal{D}(A^*)}$$

which proves (A-5).

Hypothesis (A-6) with $v = (v_1, v_2) \in \mathcal{D}(A^*)$, $(B_\alpha^* - B^*)v = \alpha\,D^*\,\mathcal{A}^{\frac{1}{2}}v_2$. Hence

$$|(B_\alpha^* - B^*)v|_{L_2(\Gamma)} \le \alpha\,|D^*\,\mathcal{A}^{\frac{1}{2}}v_2|_{L_2(\Gamma)} \le C\alpha\,|\mathcal{A}^{\frac{1}{2}}v_2|_{L_2(\Omega)} \le C\alpha\,|v|_{\mathcal{D}(A^*)}$$

where we have used (2.25).

Since e^{At} is exponentially stable, stabilizability (A-3) and detectability (A-7) condition hold automatically with $F = K = 0$. Assumption (A-7) holds as well, since $F = K = 0$. Thus all the assumptions of Theorem 2.1 have been verified.

2.3.3 Kelvin-Voigt plate equation with point control.

The Kelvin-Voigt model for a plate equation in the deflection $w(t, x)$ is

$$
\begin{aligned}
& w_{tt} + (1 + \alpha)\,\Delta^2 w + (1 + \alpha)\,\Delta^2 w_t = \delta(x - x^0)\,u(t) && \text{in } (0, T] \times \Omega = Q, \\
& w(0, \cdot) = w_0;\ w_t(0, \cdot) = w_1 && \text{in } \Omega, \\
& \Delta w|_\Sigma + (1 - \mu)\,B_1 w \equiv 0 && \text{in } (0, T] \times \Gamma = \Sigma, && (2.27) \\
& \frac{\partial \Delta w}{\partial \nu}\Big|_\Sigma + (1 - \mu)\,B_2 w \equiv 0 && \text{in } \Sigma,
\end{aligned}
$$

with $0 < \mu < \frac{1}{2}$ the Poisson modulus and $\rho > 0$ any constant, x^0 is an interior point of the open bounded $\Omega \subset R^n$, $n \le 2$. The boundary operators B_1 and B_2 are zero for $n = 1$, and for $n = 2$.

$$
\begin{aligned}
& B_1 w = 2\nu_1\,\nu_2\,w_{xy} - \nu_1^2\,w_{yy} - \nu_2^2\,w_{xx}; \\
& B_2 w = \frac{\partial}{\partial\tau}\,[(\nu_1^2 - \nu_2^2)\,w_{xy} + \nu_1\nu_2\,(w_{yy} - {}^\prime w_{xx})],
\end{aligned}
\qquad (2.28)
$$

where $\partial/\partial\tau$ is the tangential derivative, and $\nu = (\nu_1, \nu_2)$ outward normal.

Abstract setting. We introduce the nonnegative self-adjoint operator

$$\mathcal{A}_\alpha h = (1 + \alpha)\,\Delta^2 h, \qquad (2.29)$$

$$\mathcal{D}(\mathcal{A}_\alpha) = \{h \in H^4(\Omega):\ \Delta h + (1 - \mu)\,B_1 h|_\Gamma = 0;\ \frac{\partial \Delta h}{\partial \nu} + (1 - \mu)\,B_2 h|_\Gamma = 0\},$$

and select the following spaces and operators:

$$H = \mathcal{D}(\mathcal{A}^{\frac{1}{2}}) \times L_2(\Omega) = H^2(\Omega) \times L_2(\Omega);\quad U = R^1, \qquad (2.30)$$

$$A_\alpha = \begin{bmatrix} 0 & I \\ -\mathcal{A}_\alpha & \rho\mathcal{A}_\alpha \end{bmatrix}; \quad B_\alpha u = \begin{bmatrix} 0 \\ \delta(x-x')u \end{bmatrix}; \quad R = I \tag{2.31}$$

to obtain the abstract model (1.3).

Verification of Hypotheses

Hypothesis A-1. The operator A_α in (2.31) generates a s.c. contraction semigroup $e^{A_\alpha t}$ on H which moreover is uniformly (in α) analytic for $t > 0$. This is a special case of a much more general result in [Ch-T.2].

Hypothesis A-2. It is straightforward to verify that assumption (A-2) is satisfied with $\gamma = 1$. We require that

$$\begin{bmatrix} \mathcal{A}_\alpha^{-1} \delta(x-x^0) \\ 0 \end{bmatrix} \in H, \tag{2.32}$$

The Sobolev imbedding then yields that (2.32) holds true if $n \leq 3$. However, in order to verify assumption (A-2), which requires that γ should be < 1, the most elementary way is to check that assumption (A-2) holds in fact true with $\gamma = \frac{1}{2}$. In this case, we can in fact rely on the direct computation of $(-A)^{\frac{1}{2}}$ (for simplicity of notation, we take henceforth $\rho = 1$)

$$(-A_\alpha)^{-\frac{1}{2}} = \begin{bmatrix} (1) & \mathcal{A}_\alpha^{-\frac{1}{4}}(2I + \mathcal{A}_\alpha^{\frac{1}{2}})^{-\frac{1}{2}} \\ (2) & \mathcal{A}_\alpha^{-\frac{1}{4}}(2I + \mathcal{A}_\alpha^{\frac{1}{2}})^{-\frac{1}{2}} \end{bmatrix} \tag{2.33}$$

(where the entries $(1) = \mathcal{A}_\alpha^{-\frac{1}{4}}(2I + \mathcal{A}^{\frac{1}{2}})^{-\frac{1}{2}}$ and $(2) = \mathcal{A}_\alpha^{\frac{1}{4}}(2I + \mathcal{A}_\alpha^{\frac{1}{2}})^{-\frac{1}{2}}$ do not really count in the present analysis). We need to compute

$$(-\mathcal{A}_\alpha)^{-\frac{1}{2}} Bu = \begin{bmatrix} \mathcal{A}_\alpha^{-\frac{1}{4}}(2I + \mathcal{A}_\alpha^{\frac{1}{2}})^{-\frac{1}{2}} \delta(x-x^0) u \\ \mathcal{A}_\alpha^{-\frac{1}{4}}(2I + \mathcal{A}_\alpha^{\frac{1}{2}})^{\frac{1}{2}} \delta(x-x^0) u \end{bmatrix}. \tag{2.34}$$

From (2.34) we then readily see that $(-\mathcal{A}_\alpha)^{-\frac{1}{4}} Bu \in H = \mathcal{D}(\mathcal{A}^{\frac{1}{4}}) \times L_2(\Omega)$ provided (s): $\mathcal{A}_\alpha^{-\frac{1}{4}} \delta(x-x^0) \in L_2(\Omega)$. But $\mathcal{D}(\mathcal{A}_\alpha^{\frac{1}{2}}) = H^2(\Omega)$ (and, in fact, only $\mathcal{D}(\mathcal{A}_\alpha^{\frac{1}{2}}) \subset H^2(\Omega)$ suffices for the present analysis) so that condition (s) is satisfied provided $\delta(x-x^0) \in (H^2(\Omega))'$ (duality with respect to $L_2(\Omega)$): i.e., provided $H^2(\Omega) \subset C(\Omega)$, i.e., by Sobolev embedding provided $2 > \frac{n}{2}$, or $n < 4$, as desired. We have shown: Assumption (A-2) $(-A)^{-\gamma} B \in \mathcal{L}(U, Y)$ holds true for this problem with $n \leq 3$, and $\gamma = \frac{1}{2}$.

Stabilizability Condition (A-3). With $\mathcal{A}$ as in (2.31) the semigroup e^{At} is uniformly (exponentially) stable in $H/\mathcal{N}(A)$, where $\mathcal{N}(\mathcal{A})$ is the finite-dimensional nullspace of $\mathcal{A}$ [Ch-T] and thus (A-3) is automatically satisfied on this spce. For the eigenvalue $\lambda = 0$, we apply the same procedure as in the case of heat equation.

Detectability Condition (A-9). This is satisfied since in our case $R = 1$.

Hypothesis (A-5) W compute

$$A_\alpha^* - A^* = \begin{bmatrix} 0 & 0 \\ \alpha\mathcal{A}; & -\rho\alpha\mathcal{A} \end{bmatrix} ; \quad (\hat{A}^*)^{-1} = \begin{bmatrix} -\rho I ; \hat{\mathcal{A}}^{-1} \\ -I & 0 \end{bmatrix} .$$

Hence

$$|(A_\alpha^* - A^*)(\hat{A}^*)^{-1}|_{L(H)} \le \alpha \to 0 \text{ as } \alpha \to 0.$$

Assumption (A-6). Since $B_\alpha = B$, hypothesis (A-6) is automatically satisfied. Finally assumptions (A-7i) holds as $R > 0$ and (A-7ii) is satisfied since $B^*\hat{A}^{*-1}$ is compact. Indeed $\hat{A}^{-1}$ B is compact. This follows from (2.32); $\hat{\mathcal{A}}^{-1} \delta \in L(R; \mathcal{D}(\mathcal{A}^{\frac{1}{2}-}))$; and from the compactness of $\hat{\mathcal{A}}^{-1}$. ∎

3. General C_0 semigroups

3.1. Statement of the results

Let $e^{A_\alpha t}$ be a family of a C_0 generators. We shall assume:

$$|e^{A_\alpha t}|_{L(H)} \le C e^{w_0 t} \text{ for some positive constants } C > 0; \; w_0 > 0. \tag{B-1}$$

$$|e^{A_\alpha t} - e^{At}|_{L(H)} \to 0 \text{ for } t > 0 \text{ as } |\alpha| \to 0. \tag{B-2}$$

The operators $B_\alpha \in L(U; \mathcal{D}(A_\alpha^*)')$ (i.e. $\hat{A}_\alpha^{-1} B_\alpha \in L(U; H)$) are assumed to satisfy

$$\int_0^T |B_\alpha^* e^{A_\alpha^* t} x|_U^2 \, dt \le C_T |x|_H^2; \; x \in \mathcal{D}(A_\alpha^*) \supset \mathcal{D}(A^*). \tag{B-3}$$

$$\int_0^T |(B_\alpha^* e^{A_\alpha^* t} - B^* e^{A^* t}) x|_U^2 \, dt \le C(\alpha) |x|_H^2 \tag{B-4}$$

where $C(\alpha) \to 0$ as $|\alpha| \to 0$.

For every $y_0 \in H$ there exists $u \in L_2(0\infty; U)$ such that

$$J(u, y(u)) = \int_0^\infty |Ry|_Z^2 + |u|_U^2 \, dt < C |y_0|_H^2 \tag{B-5}$$

where $y = y(u)$ satisfies (1.3).

There exists an operator $K \in L(Z, H)$ such that

$$|e^{(A_\alpha^* + KR)t}|_{L(H)} \le C e^{-w_K t}; \; w_K > 0. \tag{B-6}$$

All the constants in (B-1)-(B-6) are required to be independent on the value of parameter $\alpha > 0$.

It was shown in [F-L-T] that the conditions (B-1), (B-3), (B-5), (B-6) imply the existence of a unique solution to the Riccati Equation (R.E.) with the following property

$$B_\alpha^* P_\alpha \in \mathcal{L}(\mathcal{D}(A_\alpha); U).$$ (3.1)

The main result of this section is

Theorem 3.1

(i) Assume (B-1)-(B-6). Then

$$|P_\alpha - P|_{\mathcal{L}(H)} \to 0 \quad \text{as} \quad |\alpha| \to 0.$$ (3.2)

(ii) If in addition we assume that

$$\int_0^\infty |R\, e^{\hat{A}_\alpha t}\, B_\alpha u\,|_Z dt \leq C\, |u|_U,$$ (3.3)

$$|R\, [e^{\hat{A}_\alpha t} B_\alpha - e^{\hat{A} t} B]\,|_{\mathcal{L}(U; L_1(0\infty; Z))} \to 0 \infty |\alpha| \to 0.$$ (3.4)

Then

$$|B_\alpha^* P_\alpha|_{\mathcal{L}(H;\, U)} \leq C,$$ (3.5)

$$|B_\alpha^* P_\alpha - B^* P|_{\mathcal{L}(H;\, U)} \to 0 \quad \text{as} \quad |\alpha| \to 0.$$ (3.6)

Remark 3.1. The main contribution of Theorem 3.1 is that it treats the case of unbounded control operators. In fact, Theorem 3.1 extends the earlier results of [Ch-D-P] which have been established for the bounded case.

Remark 3.2. If B is bounded then hypotheses (B-3), (B-4) follow directly from (B-1) and (B-2).

3.2. Proof of Theorem 3.1

Because of space limitations we shall prove the Theorem 3.1 in the case when w_0 in (B-1) is negative. The general case (of unstable generators) can be easily treated by combining the arguments (provided below) with those of [L.1]. To begin with we introduce the operator

$$L_\alpha: L_2(0\infty:\ U) \to L_2(0\infty;\ H) \quad \text{given by}$$ (3.7)

$$(L_\alpha u)(t) = \int_0^t e^{A_\alpha(t-s)}\, B\, u(s)\, d;$$ (3.8)

By the virtue of hypothesis (B-3) L_α is bounded from $L_2(0\infty:\ U) \to L_2(0\infty;\ H) \cap C\,(0\infty;\ H)$ with a bound uniform in $\alpha > 0$ (see [L-T]). The following formulas describing optimal control $u_\alpha^0\, y_\alpha^0$ are known [L-1]

$$R\, y_\alpha^0\, (t) = (I + R\, L_\alpha L_\alpha^*\, R^*)^{-1}\, R e^{A_\alpha(\cdot)}\, x,$$ (3.9)

$$P_\alpha\, x = \int_0^\infty e^{A_\alpha^* \tau}\, R^*\, R\, y_\alpha^0\, (\tau,\, x)\, d\tau$$ (3.10)

where by $y_\alpha^0(t, x)$ we denote the optimal trajectory y_α^0 originating at the time t with the initial condition $x \in H$.

Lemma 3.1 As $\alpha \to 0$

$$\sup_{|x|_H=1} |y_\alpha^0(\cdot, x) - y^0(\cdot, x)|_{L_2(0\infty; H)} \to 0,$$

(3.11)

$$\sup_{|x|_H=1} |u_\alpha^0(\cdot, x) - u^0(\cdot, x)|_{L_2(0\infty; U)} \to U.$$

(3.12)

To prove Lemma 3.1 we need the following auxiliary result

Proposition 3.1 As $\alpha \to 0$

(i) $\quad |L_\alpha - L|_{\mathcal{L}(L_2(0\infty; U); L_2(0\infty; H))} \to 0,$

(ii) $\quad |L_\alpha^* - L^*|_{\mathcal{L}(L_2(0\infty; H); L_2(0\infty; U))} \to 0$

Proof: It suffices to prove part (i).

$$|e^{A_\alpha^*(\cdot)} - e^{A^*(\cdot)}|_{\mathcal{L}(H; L_2(0\infty; H))} \to 0 \text{ as } \alpha \to 0.$$

(3.13)

Indeed

$$\int_0^\infty |(e^{A_\alpha^* t} - e^{A^* t})x|^2 dt \leq \int_0^T |e^{A_\alpha^* t} - e^{A^* t})x|^2 dt$$

$$+ \int_T^\infty |e^{A_\alpha^* t}|_{\mathcal{L}(H)} + |e^{A^* t}|_{\mathcal{L}(H)} dt \leq \int_0^T |(e^{A_\alpha t} - e^{A t})x|^2 dt + 2C e^{-w_0 T} |x|^2.$$

(3.14)

On the other hand for each finite T

$$\int_0^T |e^{A_\alpha^* t} - e^{A^* t})|^2_{\mathcal{L}(H)} dt \to 0$$

(3.15)

which follows from the assumptions (3.1), (3.2) and Lebesque Dominated Theorem. Selecting T large enough so $\forall \varepsilon > 0$, $2 C e^{-wT} < \dfrac{\varepsilon}{2}$ and α such that $\int_0^{T_0} \|e^{A_\alpha t} - e^{At}\| \leq \varepsilon/2$ yields the desired result in (3.13).

We shall show next that: there exists $\nu > 0$ such that

$$\sup_{|x|_H=1} \int_0^\infty |[B_\alpha^* e^{(A_\alpha^* + \nu)t} - B^* e^{(A^* + \nu)t}] x|^2_U dt \to 0.$$

(3.16)

Indeed, in view of (B-4) it is enough to prove that as $T \to \infty$

$$\sup_{|x|_H=1} \int_T^\infty |[B_\alpha^* e^{(A_\alpha^*+v)t} - B^* e^{(A^*+v)t}] x|_U^2 \, dt \to 0.$$

Lemma 3.1 of [L.2] shows that hypotheses (B-1), (B-3) imply that there exists $\gamma > 0$

$$\int_0^\infty |B_\alpha^* e^{(A_\alpha^*+v)t} x|_U^2 \, dt \le C \, |x|_U^2.$$

Hence by (B-3)

$$\int_T^\infty |B_\alpha^* e^{(A_\alpha^*+v)t} x|_U^2 = \int_0^\infty |B_\alpha^* e^{(A_\alpha^*+v)(s+T)} x|_U^2 \, dt \le \int_0^\infty |B_\alpha^* e^{(A_\alpha^*+v)s} e^{(A_\alpha^*+v)T} x|_U^2 \, dt$$

$$\le C \, |e^{(A_\alpha^*+v)T} x|_H^2 \le C \, e^{(w_0+v)T} \, |x|^2 \to 0 \text{ as } T \to \infty$$

which (together with (B-4)) proves (3.16).

Now we are ready to prove Proposition 3.1. Indeed

$$|(L_\alpha^* - L^*) f|_{L_2(0\infty;\, U)}^2 \le \int_0^\infty |\int_t^\infty (B_\alpha^* e^{A_\alpha^*(\tau-t)} - B^* e^{A^*(\tau-t)}) f(\tau) \, d\tau|^2 \, dt$$

$$\le \int_0^\infty \int_t^\infty |[(B_\alpha^* e^{(A_\alpha^*+v)(\tau-t)} - B^* e^{(A^*+v)(\tau-t)}] f(\tau)|^2 \, d\tau \int_0^\infty e^{-2v\tau} \, d\tau \, dt$$

$$\le C \, |f|_{L_2(0\infty;\, H)}^2 \, |B_\alpha^* e^{(A_\alpha^*+v)(\cdot)} - B^* e^{(A^*v)(\cdot)}|_{\mathcal{L}(H;\, L_2(0\infty;\, U))} \to 0 \text{ as } |\alpha| \to 0$$

where we have used (3.16) ∎

<u>Proposition 3.2.</u> The following operators are uniformly (in α) bounded.

$$|(I + R L_\alpha L_\alpha^* R^*)^{-1}|_{\mathcal{L}(L_2(0\infty;\, Z))} \le C \tag{3.17}$$

$$|(I + R L_\alpha L_\alpha^* R^*)^{-1} R L_\alpha|_{\mathcal{L}(L_2(0\infty;\, U) \to L_2(0\infty;\, Z))} \le C \tag{3.18}$$

$$|L_\alpha^* R^* (I + R L_\alpha L_\alpha^* R^*)^{-1}|_{\mathcal{L}(L_2(0\infty;\, Z) \to L_2(0\infty;\, U))} \le C. \tag{3.19}$$

<u>Proof.</u> (3.17) follow from the fact that $R L_\alpha L_\alpha^* R^*$ is positive selfadjoint. As for (3.18) we simply write $(I + R L_\alpha L_\alpha^* R^*)^{-1} R L_\alpha f = g$. Hence

$$(R L_\alpha f, g)_{L_2(0\infty;\, Z)} = |g|_{L_2(0\infty;\, Z)} + |L_\alpha^* R^* g|_{L_2(0\infty;\, U)}^2 \text{ and}$$

$$|g|_{L_2(0\infty;\, Z)}^2 + \frac{1}{2} |L_\alpha^* R^* g|_{L_2(0\infty;\, U)}^2 \le \frac{1}{2} |f|_{L_2(0\infty;\, U)}$$

which proves (3.18). (3.19) follows by duality from (3.18). ∎

Now we are in a position to prove Lemma 3.1. From (3.9)

$$R (y_\alpha^0 (\cdot , x) - y^0 (\cdot , x)) = (I + R L_\alpha L_\alpha^* R^*)^{-1} [R L_\alpha L_\alpha^* R^* - R L L^* R^*] (I + R L L^* R^*)^{-1}$$
$$R e^{A (\cdot)} x + (I + R L_\alpha L_\alpha^* R^*)^{-1} [R e^{A_\alpha (\cdot)} - R e^{A (\cdot)}] x \equiv I + II . \tag{3.20}$$

By the result (3.17) of Proposition 3.2 and (3.13) of Proposition 3.1 we obtain

$$|II|_{\mathcal{L} (H \to L_2(0\infty; Z))} \to 0 \text{ as } \alpha \to 0 . \tag{3.21}$$

As for the term I we decompose further

$$I = (I + R L_\alpha L_\alpha^* R^*)^{-1} R L_\alpha (L_\alpha^* - L^*) R^* (I + R L L^* R^*)^{-1} R e^{A (\cdot)} x + (I + R L_\alpha L_\alpha^* R^*)^{-1} .$$
$$R (L_\alpha - L) \cdot L^* R^* (I + R L L^* R^*)^{-1} R e^{A (\cdot)} x \equiv I_A + I_B . \tag{3.22}$$

By (3.17) and (3.18) of Proposition 3.2 and by Proposition 3.1 we infer

$$|I_A|_{\mathcal{L} (H \to L_2(0\infty; Z))} \to 0 \text{ as } \alpha \to 0 . \tag{3.23}$$

Similarly for I_B we evoke (3.17), (3.19) in Proposition 3.1. This yields

$$|I_B|_{\mathcal{L} (H \to L_2(0\infty; Z))} \to 0 \text{ as } \alpha \to 0 \tag{3.24}$$

Collecting the results in (3.22-3.24) we obtain

$$|I|_{\mathcal{L} (H \to L_2(0\infty; Z)} \to 0 \text{ as } \alpha \to 0$$

which together with (3.21) and (3.20) yields the conclusion (3.11) in Lemma 3.1. As for (3.12) it is enough to use optimality conditions which give

$$u_\alpha^0 (, x) - u^0 (, x) = L_\alpha^* R^* R y_\alpha^0 (, x) - L^* R^* R y^0 (\cdot , x) \tag{3.25}$$

together with the convergence results of Proposition 3.1 ∎

Proof of Theorem 3.1

By using the formula (3.10) we write

$$P_\alpha x - P x = \int_0^\infty (e^{A_\alpha^* \tau} - e^{A^* \tau}) R^* R y^0 (\tau, x) d\tau + \int_0^\infty e^{A_\alpha^* \tau} R^* R [y_\alpha^0 (\tau, x) - y^0 (\tau, x)] d\tau$$

$$= I + II . \tag{3.26}$$

From (B-1) and (3.11)

$$|II|_{\mathcal{L}(H)} \leq M_1 |R| \int_0^{\infty} e^{-w_1\tau} \sup_{|x|=1} |R [y_\alpha^0 (\tau, x) - y^0 (\tau, x)|_Z d\varsigma \leq \frac{M_1}{2w_1} |R| \sup_{|x|_H=1}$$

(3.27)

$$|R [y_\alpha^0 (\cdot, x) - y^0 (\cdot, x)]|_{L_2(0\infty; Z)} \to 0 \text{ as } \alpha \to 0.$$

$$|II|_{\mathcal{L}(H)} \leq |(e^{A_\alpha^*(\cdot)} - e^{A^*(\cdot)}) R^*|_{\mathcal{L}(Z \to L_2(0\infty; H))} \sup_{|x|_H=1} |R y^0 (\cdot, x)|_{L_2(0\infty; Z)} \to 0 \text{ as } \alpha \to 0 \quad (3.28)$$

where we have used (3.13) together with (3.9) and (3.17). Proof of part (i) of Theorem 3.1 is completed.

Proof of part (ii) of Theorem 3.1

Hypothesis (B-6) along with the argument as in [B-1] imply that there exists $w_1 > 0$ such that

$$|y_\alpha^0 (t, x)|_H \leq c e^{-w_1 t} |x|_H .$$

(3.29)

From (3.10), (B-5) and (3.3), (3.29)

$$(B_\alpha^* P_\alpha x, u)_U = (B^* \int_0^{\infty} e^{A_\alpha^*\tau} R^* R y_\alpha^0 (\tau, x) d\tau, u)_U = \int_0^{\infty} (R y_\alpha^0 (\tau, x), R e^{A_\alpha\tau} Bu)_Z d\tau$$

$$\leq C |R y_\alpha^0(\cdot, x)|_{C (0\infty; Z)} |R e^{A_\alpha(\cdot)} Bu|_{L_1 (0\infty; Z)} \leq C |x|_H |u|_U$$

where we have used (3.3) together with the exponential decay of $e^{A_\alpha t}$.

As for (3.6) we argue similarly

$$((B_\alpha^* P_\alpha - B^* P) x, u)_U = \int_0^{\infty} (R (y_\alpha^0 (\tau, x) - y^0 (\tau, x)), R e^{A_\alpha\tau} B_\alpha u)_Z d\tau$$

$$+ \int_0^{\infty} (R y^0 (\tau, x), R (e^{A_\alpha\tau} B_\alpha - e^{A\tau}B)u)_U \leq |R (y_\alpha^0(\cdot, x) - y^0(\cdot, x))|_{L_2(0\infty; Z)} \quad (3.30)$$

$$|R e^{A_\alpha(\cdot)} B_\alpha u|_{L_2(0\infty; Z)} + |R y^0 (\cdot, x)|_{C(0\infty; Z)} |R (e^{A_\alpha\tau}B_\alpha - e^{A\tau}B)u|_{L_1(0\infty, Z)}$$

By using the results of (3.11), (3.29) together with hypotheses (3.3) and (3.4) (and the exponential decay of $e^{A_\alpha t}$) we conclude from (3.30) that

$$|B_\alpha^* P_\alpha - B^* P|_{\mathcal{L}(H, U)} \to 0 \text{ as } \alpha \to 0 \quad \blacksquare$$

3.3. Example - wave equation with boundary control and perturbed boundary conditions.

Consider the wave equation

$$\begin{cases} y_{tt} = \Delta y - y_t & \text{in } Q \\ y\big|_\Gamma = \alpha\, g \int_\Omega f(x)\, y\,(x)\, dx + u & \text{in } \Sigma \\ y\,(t=0) = y_0; \;\; y_t(t=0) = y_1 & \text{in } \Omega\,. \end{cases} \qquad (3.31)$$

Here f (resp g) are given functions in $L_2(\Omega)$ (resp $(L_2(\Gamma))$) and the initial data y_0, y_1 are in $L_2(\Omega) \times H^{-1}(\Omega)$. To put problem (3.31) into abstract framwork we set

$$H \equiv L_2(\Omega) \times H^{-1}(\Omega); \;\; U \equiv L_2(\Gamma)$$

$$A \equiv \begin{bmatrix} 0 & I \\ \mathcal{A} & -I \end{bmatrix} \text{ where}$$

$$\mathcal{A}u \equiv \Delta u; \;\; \mathcal{D}\,(\mathcal{A}) = H^2(\Omega) \times H_0^1(\Omega)\,.$$

$$Bu = \begin{bmatrix} 0 \\ \mathcal{A}\,Du \end{bmatrix};$$

$$A_\alpha = A + \alpha\, B\, F$$

where $F \in \mathcal{L}\,(L_2(\Omega);\, L_2(\Gamma))$ is given by $Fy \equiv g \int_\Omega fy\, d\Omega$.

From the "trace regularity" results in [L-L-T] it follows that the condition (B-3) with $\alpha = 0$ is satisfied. Also, since e^{At} is exponentially stable hypotheses (B-1), (B-5), (B-6) *with $\alpha = 0$* hold as well. Our main task is to show that all these hypotheses hold with $0 < \alpha \le 1$. To accomplish this we may consider more general abstract equation which would encompass problem (3.30) as the special case. Indeed, we shall consider a family of abstract differential equations of the form

$$\begin{cases} y_t = Ay + \alpha\, BFy + Bu & \text{in } \mathcal{D}(A^*)' \\ y(0) = y_0 \in H\,. \end{cases} \qquad (3.32)$$

where $F \in \mathcal{L}\,(H;\, U)$ and we assume that

Hypotheses (B-1), (B.3), and (3.3), (3.4) are satisfied with $\alpha = 0$ (i.e. for unperturbed problem).   (C)

Theorem 3.2

Assume hypothesis (C). Then the conclusions of Theorem 3.1 apply to the problem (3.22).

Since in the case of the model (3.30) the conditions (B-1), (B-3), (B-5) and (B-6) are satisfied with $\alpha = 0$ we obtain

Corollary 3.1

Conclusions of Theorem 3.1 apply to the problem (3.30).

Proof of Theorem 3.2

By the results of [D-L-S] we know that for every value of the parametr α, A_α given by (3.31) generates a strongly continuous semigroup $e^{A_\alpha t}$ on H. Moreover, the adjoint operator A_α^* is given by

$$A_\alpha^* = A^* + \alpha \, F^* \, B^* \tag{3.33}$$

and we have $\mathcal{D}(A_\alpha^*) = \mathcal{D}(A^*)$. Notice however that in general $\mathcal{D}(A_{\alpha_1}) \neq \mathcal{D}(A_{\alpha_2})$; $\alpha_1 \neq \alpha_2$ and there is no relation between $\mathcal{D}(A_\alpha)$ and $\mathcal{D}(A)$. For simplicity of exposition we shall assume that w_0 in (B-1) (satisfied with $\alpha = 0$) is negative (this is not an essential assumption see sect. 3.1). To prove the Theorem 3.2 it suffices to verify the validity of (B-1)-(B-6) for $\alpha \neq 0$, $\alpha \rightarrow 0$. We shall start by proving

$$|e^{A_\alpha^* t}|_{\mathcal{L}(H)} \leq C \, e^{-w_1 t} \, ; \quad \text{for some } w_1 > 0 . \tag{3.34}$$

Notice that (3.34) automatically implies (B-1), (B-5) (with $\underline{u} = 0$) and (B-6) (with K=0). In order to prove (3.34) we shall show

$$\int_0^\infty |B^* \, e^{A_\alpha^* t} x|_U^2 \, dt \leq C \, |x|_H^2 \, ; \quad x \in \mathcal{D}(A_\alpha^*) = \mathcal{D}(A^*) . \tag{3.35}$$

Let $z_\alpha(t) \equiv e^{A_\alpha^* t} x$. Then by using the variation of parameter formula we obtain

$$z_\alpha(t) = e^{A^* t} x + \alpha \int_0^t e^{A^*(t-s)} F^* \, B^* \, z_\alpha(s) \, ds . \tag{3.36}$$

From the hypothesis (B-3) (with $\alpha = 0$) and from the result of Lemma 3.1 in [L-2] we know that there exists a constant $v > 0$ such that

$$\int_0^\infty |B^* \, e^{(A^* + v)t} x|_U^2 \, dt \leq C \, |x|_H^2 \, ; \quad x \in \mathcal{D}(A^*) \tag{3.37}$$

Applying operator B^* to both sides of (3.36) (which procedure is justified with $x \in \mathcal{D}(A^*)$ as $z_\alpha(t) \in \mathcal{D}(A^*) \subset \mathcal{D}(B^*)$) and using (3.37) gives:

$$\int_0^\infty |B^* z_\alpha(t)|_V^2 \, dt \leq \int_0^\infty |B^* e^{A^* t} x|_U^2 \, dt + \alpha^2 \int_0^\infty [\int_0^t B^* e^{A^*(t-s)} F^* B^* z_\alpha(s) \, ds]_U^2 \, dt$$

$$\leq C \, |x|_H^2 + \alpha \int_0^\infty dt \int_0^t |B^* \, e^{(A^* + v)(t-s)} F^* B^* z_\alpha(s)|_U^2 \, ds \cdot \int_0^t e^{-2v(t-s)} \, ds \leq C \, |x|_H^2$$

$$+ \frac{\alpha^2}{2v} \int_0^\infty ds \int_s^\infty |B^* \, e^{(A^* + v)(t-s)} F^* B^* z_\alpha(s)|_U^2 \, dt$$

by (3.37)

$$\leq C \, |x|_H^2 + C \frac{\alpha}{2v} \int_0^\infty |F^*|^2 \, |B^* z_\alpha(s)|_U^2 \, ds = C \, |x|_H^2 + \frac{C \alpha^2 \, |B^*|^2}{2v} \int_0^\infty |B^* Z_\alpha(s)|_U^2 \, ds$$

Selecting a sufficiently small α (i.e. $\dfrac{C\,\alpha^2/|F^*|^2}{2\nu} < 1$) yields (3.35).

Remark We notice that the inequality (3.35) on the finite time horizon (i.e. (B-3)) holds for every value of α. This can be proved by applying Contraction Mapping Principle to integral equation (3.36) (as in [D-L-S]).

To complete the proof of (3.34) we return to the integral equation (3.36). Applying inequality (3.35) we obtain

$$\int\limits_0^\infty |z_\alpha(t)|^2 \, dt \le C^2 \int\limits_0^\infty e^{-2w_0 t}\, dt \; |x|_H^2 + C^2\,\alpha^2 \int\limits_0^\infty |\int\limits_0^t e^{-w_0(t-s)}\; |B^* z_\alpha(s)|\, ds|^2\, dt \le C\, |x|_H^2 \, .$$

The final conclusion in (3.34) follows now from Datko's result (see [P.1]).

It remains to prove (B.-2) and (B-4). In fact, we shall prove that

$$|e^{A_\alpha t} - e^{At}|_{\mathcal{L}(H)} \le C_T \cdot \alpha \qquad \text{for } t \le T \tag{3.38}$$

$$\int\limits_0^T |B^*\,(e^{A_\alpha t} - e^{A^* t})\, x|_U^2 \, dt \le C_T\,\alpha^2\, |x|_H^2 \, . \tag{3.39}$$

Indeed with $z_\alpha(t) \equiv e^{A_\alpha t}\, x$ and $z(t) \equiv e^{A^* t}\, x$

$$z_\alpha(t) - z(t) = \alpha \int\limits_0^t e^{A^*(t-s)}\, F\, B^*\, z_\alpha(s)\, ds \, . \tag{3.40}$$

By (3.35) for $t \le T$

$$|z_\alpha(t) - z(t)|_H \le \alpha\, |B^* z_\alpha(\cdot)|_{L_2[0,t;\,U]} \le C\,\alpha\, |x|_H$$

which proves (3.38). As for (3.39) we return to (3.40) applying B^* to both sides of the equation. This yields

$$\int\limits_0^T |B^*\, [z_\alpha(t) - z(t)]|_U^2 \, dt \le \alpha^2 \int\limits_0^T t \int\limits_0^t |B^*\, e^{A^*(t-s)}\, F^*\, B^* z_\alpha(s)|_U^2\, ds\, dt$$

$$\le C_T \alpha^2 \int\limits_0^T \int\limits_s^T |B^*\, e^{A^*(t-s)}\, F^*\, B^*\, z_\alpha(s)\,|_U^2\, dt\, ds$$

by (B-3) and again (3.35)

$$\le C_T\,\alpha^2 \int\limits_0^T |F^*|^2\, |B^*\, z_\alpha(s)|_U^2\, ds \le C_T\,\alpha^2 |x|_H^2$$

as desired for (3.39) ∎

References

[B.1] A. V. Balakrishnan, Applied Functional Analysis, Springr Verlag.

[Ch-D-P] A. Chojnowska-Michalik, T. E. Duncan, B. Pasik-Duncan, Uniform operator continuity of the stationary Riccati equation in Hilbert space. *Applied Mathematics and Optimization*, to appear.

[Ch-T] S. Chen, R. Triggiani, Proof of exensions of two conjectures on structural damping for elastic systems, *Pacific J. Math.*, 136 (1989), 15-55.

[D-G-P] T. E. Duncan, B. Goldys, B. Pasik-Duncan, Adaptive control of linear stochastic evolution systems. Stochastics and Stochastics Reports 35 (1991a) 129-192.

[D-L-S] W. Doesch, I. Lasiecka, and W. Schappacher, Finite dimensional boundary feedback control problems for linear infinite dimensional systems, *Israel J. of Math.*, 51 (1985), 177-207.

[D-M-P] T. E. Duncan, B. Mastowski, B. Pasik-Duncan. Adaptive boundary control of linear stochastic distributed parameter systems.

[F-L-T] F. Flandoli, I. Lasiecka, and R. Triggiani, Algebraic Riccati equations with non-smoothing observation arising in hyperbolic and Euler-Brnouli equations, *Ann. Matem. Pura a Appl.*, Vol. CLiii (1988), 307-382.

[K.1] T. Kato, Perturbation theory of linear operators, Springer-Verlag, New York, 1966.

[L.1] I. Lasiecka, Approximations of the solutions of infinite dimensional Albebraic Riccati Equations with unbounded input operators, *Numerical Funct. Anal. and Optimiz.* Vol. 11, pp. 303-378 (1990).

[L.2] I. Lasiecka, Stabilization of hyperbolic and parabolic systems with nonlinearly perturbed boundary conditions. *Journal of Differential Equations* Vol. 75, No. 1, pp. 53-87 (1988).

[L-L-T] I. Lasiecka, J. L. Lions, and R. Triggiani, Non-homogeneous boundary value problms for second order hyperbolic oprators, *J. Mathem. Pure et Appl.* 65 (1986), 149-192.

[L-T] I. Lasiecka and R. Triggiani, Differential and Algebraic Riccati Equations with Applications to Boundary/Point Control Problems: Continuous Theory and Approximation Theory. Springer Verlag, Vol. 164 1991.

[L-T.1] I. Lasiecka and R. Triggiani, Numerical approximation for abstract systems modelled by analytic semigroups, and applications, *Mathematics of Computation* Vol. 57 No. 196, pp. 639-662.

[L-T.2] I. Lasiecka and R. Triggiani, Algebraic Riccati equations arising from systems with unbounded input - solution operator: applications to boundary control problems for wave and plate equations, *Journal of Nonlinear Analysis*.

[P.1] A. Pazy, Semigroups of operators and applications to partial differential equations, Springer-Verlag, 1983.

A (simple) perspective on adaptation and performance of adaptation mechanisms

Lennart Ljung
Department of Electrical Engineering
Linköping University
S-581 83 Linköping, Sweden

January 24, 1992

Abstract

Mechanisms for adapting models, filters, decisions, regulators and so on to changing properties of a system or a signal are of fundamental importance in many modern signal processing and control algorithms. In this chapter we give an overview of some basic set-ups and algorithms that are used for this. We pay special attention to the rationale behind the different algorithms, thus distinguishing between "optimal" algorithms and "ad hoc" algorithms. We also give an outline of the basic approaches to performance analysis of adaptive algorithms.

1   Introduction

Adaptation and adaptability are desired features in most systems' behaviour . In technical systems dealing with signal processing - in a broad sense - adaptive properties are manifested in such concepts as "adaptive control", "adaptive filtering", "adaptive prediction", and so on.

The main feature in any adaptation mechanisms is a *tracking facility*, which, explicitly or implicitly, tracks the time varying properties of the signal or system, to which we want to adapt.

Tracking a system's properties is always a question of critically evaluating the observation obtained from the process in question: Do they contain information about changes in the process or are they just dominated by random influations. Thus even in a non-mathematical setting, *adaptation and tracking is always characterised by a trade-off between tracking ability* (dare to believe signs of process changes in the measurements!) and *noise sensitivity* (don't get confused by random fluctuations!). We shall see this fundamental trade-off show up in various formalized ways in the course of this contribution.

One focus of our discussion will be how to translate certain assumptions about the system's behaviour and criteria for good tracking to optimal algorithms. We shall then also see that many common *ad hoc* algorithms can be interpreted as corresponding to certain assumptions about the system's behaviour.

Another focus of our discussion is to outline basic procedures for analytic performance evaluations of the various algorithms.

We shall mostly confine ourselves to the case where the underlying system or signal model can be formulated as a linear regression. See also the survey [14].

2 Optimal Algorithms for Tracking Drifting Parameters

We shall use the following linear regression signal model:

$$y(t) = \varphi^T(t)\theta + e(t) \tag{1}$$

where $\{y(t)\}$ and $\{\varphi(t)\}$ are observed signals. The vector θ contains the unknown parameters, which are to be estimated by the tracker.

The most common application of (1) in control and signal processing is when the regression vector $\varphi(t)$ consists of lagged outputs and inputs

$$\varphi^T(t) = (-y(t-1), \ldots, u(t-m)). \tag{2}$$

In this case (1) and (2) correspond to a linear difference relationship between the input and the output. In case there is no "input" signal $\{u(t)\}$ we have the well-known AR model for the signal $\{y(t)\}$.

We now assume that there is a true - and time varying - value $\theta_0(t)$ for the parameters and that these develop over time as a random walk. This means that the "true" description of the signals $\{y(t)\}$ and $\{\varphi(t)\}$ becomes

$$\theta_0(t) = \theta_0(t-1) + w(t) \tag{3}$$

$$y(t) = \theta_0^T(t)\varphi(t) + e(t). \tag{4}$$

We here assume $\{e(t)\}$ to be white Gaussian noise with variance $R_2(t)$, while $\{w(t)\}$ is white Gaussian noise with covariance matrix $R_1(t)$ independent of $\{e(t)\}$. It is then well known, see, e.g. Section 2.3 in [16] that the estimate $\hat{\theta}(t)$ that minimizes the conditional expectation, given past observations

$$\Pi(t) = E(\hat{\theta}(t) - \theta_0(t))(\hat{\theta}(t) - \theta_0(t))^T \tag{5}$$

(even in a matrix sense) is given by the Kalman filter:

$$\hat{\theta}(t) = \hat{\theta}(t-1) + K(t)\varepsilon(t) \tag{6}$$

$$\varepsilon(t) = y(t) - \varphi^T(t)\hat{\theta}(t-1) \tag{7}$$

where the gain vector $K(t)$ is given by

$$K(t) = \frac{P(t-1)\varphi(t)}{\hat{R}_2(t) + \varphi^T(t)P(t-1)\varphi(t)} \tag{8}$$

and the matrix $P(t)$ is updated according to

$$P(t) = P(t-1) - \frac{P(t-1)\varphi(t)\varphi^T(t)P(t-1)}{\hat{R}_2(t) + \varphi^T(t)P(t-1)\varphi(t)} + \hat{R}_1(t),$$

$$P(0) = P_0, \tag{9}$$

We have here used the notations $\hat{R}_1(t)$ and $\hat{R}_2(t)$ to indicate that the values used in the algorithm may very well differ from the true values $R_1(t)$ and $R_2(t)$. In the case $\hat{R}_1(t) \equiv R_1(t)$ and $\hat{R}_2(t) \equiv R_2(t)$, however, $\hat{\theta}(t)$ is the conditional expectation of $\theta_0(t)$, given the observations $\{u(k), y(k)\}$, $k \leq t$, and $P(t)$ is the conditional covariance matrix of the parameter estimation error.

Note also that if $R_1(t)$ is known then (6)-(9) is the optimal algorithm also for abrupt changes in θ_0. (Take $R_1(t) = 0$ except when a jump occurs, say for $t \in T_1$ take then $R_1(t) = R_1$.) However, this requires the time instants for the jumps to be known, not too realistic an assumption.

Remark. In fact, the problem of recursive parameter estimation can be seen as a special case of *non-linear filtering*. The parameters are then interpreted as states. There are consequently several important links to the wide literature on non-linear filtering. The reader may consult [16], Section 2.3 of [11], and [1] for some aspects of this. In the current context, though, the dynamics in (3) is linear, and, under Gaussian noise sources, the non-linear filtering problem specializes to a linear one.

In the algorithm (9) it follows that, after a transient, the size of $P(t)$ will be like the square root of $\hat{R}_1$. (This will be shown formally in (47) below.) For slowly changing systems, P will thus be small. To explicitly show this it is useful to scale P, so as to rewrite (6) as

$$\hat{\theta}(t) = \hat{\theta}(t-1) + \mu(t)P_t L(\varphi(t))(y(t) - \varphi^T(t)\hat{\theta}(t-1)) \tag{10}$$

We have here allowed a possible non-linear transformation L (such as normalization) of $\varphi(t)$. We shall regard (10) as the archetypical algorithm for adaptive parameter estimation. The link to (6)–(9) can be made explicit by associating

$$L(\varphi(t)) = \varphi(t)$$

$$P_t = \frac{1}{\mu(t)\hat{R}_2(t)}(P(t) - \hat{R}_1(t))$$

$$\mu^2(t) \approx ||\hat{R}_1(t)||$$

3 Some ad hoc Algorithms for Tracking Drifting Parameters

The basic formulation (3) and (4) with the optimal algorithm (6)-(9) is quite powerful. It can deal with both slowly drifting parameters and with sudden changes, by assigning proper values to the covariance matrix $\hat{R}_1(t)$ and the variance $\hat{R}_2(t)$. The main shortcoming is then that these values will rarely be known to the user. One approach to deal with this problem is to choose some ad hoc values for $\hat{R}_1(t)$. We will discuss two such ad hoc choices below.

3.1 The RLS Algorithm

A popular approach to deal with time-varying linear regressions is to minimize a weighted criterion

$$V_t(\theta) = \sum_{k=1}^{t} \beta(t,k)(y(k) - \theta^T \varphi(k))^2 \tag{11}$$

where

$$\beta(t,k) = \prod_{j=k+1}^{t} \lambda(j) \tag{12}$$

and where $|\lambda(j)| \leq 1$ is denoted the forgetting factor.

From, e.g. [16] we have that this is accomplished by the recursive least squares (RSL) algorithm, which is given by (6)-(8) with $K(t)$ chosen as

$$K(t) = \frac{P(t-1)\varphi(t)}{\lambda(t) + \varphi^T(t)P(t-1)\varphi(t)} \tag{13}$$

and

$$P(t) = \frac{1}{\lambda(t)}\left[P(t-1) - \frac{P(t-1)\varphi(t)\varphi^T(t)P(t-1)}{\lambda(t) + \varphi^T(t)P(t-1)\varphi(t)}\right]. \tag{14}$$

We note that this is a special case of (6)-(9), corresponding to the choices

$$\hat{R}_1(t) = \left(\frac{1}{\lambda(t)} - 1\right)$$

$$\times \left[P(t-1) - \frac{P(t-1)\varphi(t)\varphi^T(t)P(t-1)}{\lambda(t) + \varphi^T(t)P(t-1)\varphi(t)}\right] \approx \left(\frac{1}{\lambda(t)} - 1\right)P(t-1) \qquad (15)$$

$$\hat{R}_2(t) = \lambda(t).$$

(The approximation follows in the typical case where $||P|| << 1$)

For future use we also note that

$$P(t) = \left[\sum_{k=1}^{t} \beta(t,k)\varphi(k)\varphi^T(k)\right]^{-1}. \qquad (16)$$

The connection to the archetypical algorithm (10) is given by

$$\mu(t) = \left[\sum_{k=1}^{t} \beta(t,k)\right]^{-1} \qquad (17)$$

which gives

$$P_t = \frac{1}{\mu(t)\hat{R}_2(t)}(P(t) - \hat{R}_1(t)) =$$

$$\approx \frac{1}{\lambda(t)\mu(t)}(2 - \frac{1}{\lambda(t)})P(t) \approx \frac{1}{\mu(t)}P(t)$$

In the first approximation we set $P(t) \approx P(t-1)$ and in the second one we used $2\lambda - 1 \approx \lambda^2$ which holds for λ close to 1. This gives

$$P_t = [(\sum_{k=1}^{t} \beta(t,k)))^{-1} \cdot \sum_{k=1}^{t} \beta(t,k)\varphi(k)\varphi^T(k)]^{-1}$$

which shows that

$$P_t \approx [E\varphi(t)\varphi^T(t)]^{-1}$$

by a weighted sample sum approximation. Consequently, the normalization with μ makes P_t of a size that does not depend on λ. We shall later also use the expression

$$R(t) = \mu(t)P_t^{-1} \approx E\varphi(t)\varphi^T(t) \qquad (18)$$

3.2 The LMS Algorithm

Widrow's least mean squares algorithm (see, e.g. [27]), is a commonly used tool for adaptation. It is given by

$$L(t) = \mu\varphi(t) \tag{19}$$

The LMS algorithm can also be formulated in a normalized variant

$$L(t) = \frac{\mu\varphi(t)}{1 + \mu \mid \varphi(t) \mid^2}. \tag{20}$$

Again, we may verify that (6) and (20) is a special case of the basic algorithm (6)-(9) corresponding to

$$\hat{R}_1(t) = \mu^2 \frac{\varphi(t)\varphi^T(t)}{1 + \mu \mid \varphi(t) \mid^2} \tag{21}$$

$$\hat{R}_2(t) = 1 \tag{22}$$

$$P(0) = \mu \cdot I. \tag{23}$$

From which it follows $P(t) = \mu I$.

3.3 Estimating the Unknown Covariances

A more systematic approach to deal with the problem of unknown $R_1(t)$ and $R_2(t)$ values is of course to estimate them. We shall here discuss a few possibilities of this kind.

Let us consider the case where the parameters are slowly drifting and the values of $R_1(t) \equiv R_1$ and $R_2(t) \equiv R_2$ are nearly constant over extended periods of time. It is then feasible to devise efficient methods for estimating R_1 and R_2. Techniques for this go back to the literature on adaptive filtering. See, for example, [19], [22], [23] and [2]. [9] and [10] have developed this approach further and also tested the feasibility of such methods. The idea can be described as a least squares method applied to a linear regression model for the covariances. A variant is given in [24]. [10] also contains a survey of other approaches to estimate R_1 and R_2. See also [25] and [20] and [21].

Another common approach is to use the RLS algorithm (13) and (14) and adjust the size of the forgetting factor $\lambda(t)$. Several ways to do this can be conceived. [4] have devised one method that is based on monitoring the residual variance $\varepsilon^2(t)$, ($\varepsilon(t)$ defined in (7)). When this increases, $\lambda(t)$ is decreased. From (15) we see that methods to adjust $\lambda(t)$ can be seen as ways to estimate the "size" of $R_1(t)$, while direction information is neglected.

A third family of approaches that can be seen as adjustments or selections of $R_1(t)$ can be summarized under the name "directional forgetting". The prime idea is then to select $\hat{R}_1(t)$ in (9) not based on estimates of $R_1(t)$ but as a means to keep

$P(t)$ well conditioned. One interpretation is that we forget information only in the "direction" where the new one is obtained. Examples, of such strategies are given in [13], [8], [12] and [17].

4 Asymptotic Properties on the Decreasing Gain Case

The actual use of the adaptive algorithms is to track time-varying properties of a system or a signal. Still, a natural first question is to ask how well the algorithms are capable to handle a time invariant system. This corresponds to the special case $R_1(t) = \hat{R}_1(t) = 0$ in (9), (4) or $\lambda(j) \equiv 1$ in (12). A substantial part of [16] is devoted to such analysis, and we shall here only quote the bottom lines:

1. A recursive prediction error algorithm of which (6) is a special case will, as t tends to infinity, and as the gain tends to zero converge to a local minimum of the expected loss function

$$\bar{V}(\theta) = E\ell(\varepsilon(t,\theta),t) \tag{24}$$

i.e.

$$\hat{\theta}(t) \rightarrow \arg\min \bar{V}(\theta) \qquad \text{w.p 1 as } t \rightarrow \infty \tag{25}$$

2. If, in addition, the Gauss-Newton search direction is used, and asymptotically equal weighting is used ($\lambda(j) \equiv 1$) then the asymptotic accuracy

$$\bar{P} = \lim_{t\to\infty} tE(\hat{\theta}(t) - \theta_0)(\hat{\theta}(t) - \theta_0)^T$$

will be the same as for the corresponding off-line estimation method.

These asymptotic properties are thus the best one could ask for. It remains though to study how the algorihtms actually can cope with time varying systems. This is the question we turn to next.

5 Tracking Ability of the Algorithms

In the analysis of the tracking ability we will only study algorithms for linear regressions. We first develop an exact expression for the parameter error.

Let us consider the description (3)-(4) for the behaviour of the true system together with the generic parameter estimation algorithm (6) and (8)

$$\theta_0(t+1) = \theta_0(t) + \gamma w(t) \tag{26}$$

$$y(t) = \varphi^T(t)\theta_0(t) + e(t) \tag{27}$$

$$\hat{\theta}(t) = \hat{\theta}(t-1) + K(t)\varepsilon(t) \tag{28}$$

$$\varepsilon(t) = y(t) - \varphi^T(t)\hat{\theta}(t-1). \tag{29}$$

Introduce the parameter error

$$\tilde{\theta}(t) = \hat{\theta}(t) - \theta_0(t+1). \tag{30}$$

Remark. The variable γ is used to easily treat scaling of the parameter changes. The time indexing here may seem somewhat peculiar, but it will simplify the expressions to follow. From an expression for the covariance of $\tilde{\theta}(t)$ we can exactly derive, e.g. the covariance of $\hat{\theta}(t) - \theta_0(t)$. □

Then

$$\tilde{\theta}(t) = (I - K(t)\varphi^T(t))\tilde{\theta}(t-1) + K(t)e(t) - \gamma w(t). \tag{31}$$

The parameter error thus obeys a linear, time-varying difference equation. Notice that the $K(t)$ is always of the form

$$K(t) = P(t)\varphi(t) \tag{32}$$

for some matrix $P(t)$. Solving (31) gives

$$\tilde{\theta}(t) = \Phi(t,0)\tilde{\theta}(0) + \sum_{k=1}^{t} \Phi(t,k)[P(k)\varphi(k)e(k) - \gamma w(k)] \tag{33}$$

where

$$\Phi(t,k) = \prod_{j=k}^{t}(I - P(j)\varphi(j)\varphi^T(j)). \tag{34}$$

Expressions (31) and (33) form the basis for all analysis of the performance of the algorithm, and they hold for any sequences $\{\varphi(t)\}$, $\{e(t)\}$ and $\{w(t)\}$. The difficulty in the analysis lies in the complicated expression for $\Phi(t,k)$. Its properties depend entirely on the sequence $\{\varphi(t)\}$, but they are inherited in a fairly complicated way. We shall be interested in the properties of $\tilde{\theta}(t)$ as the gain $K(t)$ becomes small. We therefore write

$$K(t) = \mu P_t\varphi(t) \tag{35}$$

where μ is a positive scaling parameter (see (10)), and obtain

$$\tilde{\theta}(t) = (I - \mu P_t\varphi(t)\varphi^T(t))\tilde{\theta}(t-1) + \mu P_t\varphi(t)e(t) - \gamma w(t). \tag{36}$$

The quantity that we are interested in is the size of the error $\tilde{\theta}(t)$ as measured by the covariance matrix

$$\Pi(t) = E\tilde{\theta}(t)\tilde{\theta}^T(t). \tag{37}$$

Here expectation "E" is over $\{e(t)\}$, $\{w(t)\}$ as well as over any random components of $\{\varphi(t)\}$. The exact expression for $\Pi(t)$ follows a somewhat complex equation. Our goal is to show that $\Pi(t)$ is well approximated by $\hat{\Pi}(t)$, defined by

$$\hat{\Pi}(t) = (I - \mu\bar{P}_t Q(t))\hat{\Pi}(t-1)(I - \mu\bar{P}_t Q(t))^T$$

$$+ \mu^2 \bar{P}_t Q(t)\bar{P}_t \cdot R_2(t) + \gamma^2 R_1(t) \tag{38}$$

$$\hat{\Pi}(t_0) = \Pi(t_0). \tag{39}$$

Here

$$\bar{P}_t = EP_t \tag{40}$$

$$Q(t) = E\varphi(t)\varphi^T(t) \tag{41}$$

$$R_1(t) = Ew(t)w^T(t) \tag{42}$$

$$R_2(t) = Ee^2(t). \tag{43}$$

In essence, (38) is obtained from (36) by squaring it and applying expectation neglecting certain dependencies between random variables.

There are several possibilities to establish that Π and $\hat{\Pi}$ are close, and we refer to [14] and [15] for more details about this.

Let us briefly discuss the implications of the expression (38). There is a substantial amount of papers that discuss such implications, e.g. [26], [3], [18], and [5]. We shall only comment on the case of RLS with forgetting factor $\lambda = 1 - \mu$. This gives with

$$\bar{P}_t = \bar{P} = Q^{-1}$$

$$Q(t) = Q$$

$$\hat{\Pi}(t) = \hat{\Pi}(t-1) - 2\mu\hat{\Pi}(t-1) + \mu^2\hat{\Pi}(t-1) + \mu^2 Q^{-1} \cdot R_2 + \gamma^2 R_1 \tag{44}$$

As $t \to \infty$ we find that

$$\hat{\Pi}(t) \to \hat{\Pi}$$

where

$$\hat{\Pi} = \frac{1}{2}(\mu Q^{-1}R_2 + \frac{\gamma^2}{\mu}R_1) \tag{45}$$

(neglecting the term $\mu^2\hat{\Pi}$ which for small μ is of an order of magnitude less than the other terms).

This expression shows clearly the trade-off in the choice of step size (adaptation gain) μ (or forgetting factor $\lambda = 1 - \mu$). A small μ gives a small influence from the noise $\{e(t)\}$ in the term $\mu Q^{-1}R_2$ and a large tracking error from the term $\gamma^2/\mu R_1$ and vice versa for a large μ.

Other specific algorithms, such as LMS show similar trade-offs. We may note, in the general case, as t tends to infinity, that $\hat{\Pi}(t)$ will converge to the solution $\hat{\Pi}$ of

$$\bar{P}Q\hat{\Pi} + \hat{\Pi}Q\bar{P} = \mu\bar{P}Q\bar{P}R_2^0 + \frac{\gamma^2}{\mu}R_1^0 \tag{46}$$

(where we assume $\bar{P}$, Q, R_1 and R_2 to be time-invariant). If $P(t)$ obeys (9) and $\mathring{R}_1(t) = \mu^2\hat{R}_1$ is small and constant (or averages around such a value), similar arguments will show that $P(t) \approx \bar{\bar{P}}$ for small μ where

$$\bar{\bar{P}} = \bar{P} - \frac{\bar{P}Q\bar{P}}{\hat{R}_2 + \mu\bar{\bar{P}}Q} + \mu^2\hat{R}_1$$

If we scale $P(t)$ as in (32), (35),

$$P_t = \mu P(t)$$

we find that $P_t \approx \bar{P}$ which neglecting small terms ($tr\bar{\bar{P}}Q$ is negelected compared to R_2) is given by

$$\bar{P}Q\bar{P} = \hat{R}_1\hat{R}_2 \tag{47}$$

We refer to the references mentioned above for further discussion. In the nect section we shall develop expressions like (45) for the error in the estimated frequency functions of linear systems. These are more transparent in the general case.

6 Evaluation of the Error in the Frequency Domain

The expression for the mean square error that we derived in the previous section are somewhat implicit. In [6] and [7] explicit expressions for the mean square error of a corresponding transfer function estimate were derived. The results can be summarized as follows. Consider an FIR model, where $\varphi(t)$ contains only lagged inputs

$$y(t) = \varphi^T(t)\theta = \sum_{k=1}^{d} g_k u(t-k). \tag{48}$$

The corresponding transfer function then is

$$G(e^{i\omega}) = \sum_{k=1}^{d} g_k e^{ik\omega} = W_d^*(\omega)\theta \tag{49}$$

where

$$W_d(\omega) = [e^{i\omega} \cdots e^{di\omega}]^T \tag{50}$$

and where "*" denotes transpose and complex conjugate.

The mean square error of the transfer function estimate at frequency ω then is

$$\pi_d(\omega) = W_d^*(\omega)\hat{\Pi}W_d(\omega) \tag{51}$$

where $\hat{\Pi}$ is the mean square error matrix for the parameters, as derived in Section 7-10. The key properties to be used are as follows:

Let A and B be $d \times d$ Toeplitz-like matrices, that satisfy some regularity conditions, see [7]. We can then define the scalar functions $a(\omega)$ and $b(\omega)$ by

$$\frac{1}{d}W_d^*(\omega)AW_d(\omega) \rightarrow a(\omega) \text{ as } d \rightarrow \infty \tag{52}$$

and

$$\frac{1}{d}W_d^*(\omega)BW_d(\omega) \rightarrow b(\omega) \text{ as } d \rightarrow \infty. \tag{53}$$

Furthermore, it can be shown that

$$\frac{1}{d}W_d^*(\omega)ABW_d(\omega) \rightarrow a(\omega)b(\omega) \text{ as } d \rightarrow \infty \tag{54}$$

and

$$\frac{1}{d}W_d^*(\omega)A^{-1}W_d(\omega) \rightarrow \frac{1}{a(\omega)} \text{ as } d \rightarrow \infty. \tag{55}$$

When applying this operation to the covariance matrix

$$Q = E\varphi(t)\varphi^T(t)$$

with

$$\varphi^T(t) = (u(t-1), \ldots u(t-d))$$

(cf (41)) we get

$$\frac{1}{d}W_d^*(\omega)QW_d(\omega) \rightarrow \Phi_u(\omega) \text{ as } d \rightarrow \infty \tag{56}$$

where $\Phi_u(\omega)$ is the spectrum of the input $\{u(t)\}$.

We are now going to apply these results to the general expression (46) by evaluating

$$\bar{\pi}(\omega) = \lim_{d \rightarrow \infty} \frac{1}{d}\pi_d(\omega) = \lim_{d \rightarrow \infty} \frac{1}{d}W_d^*(\omega)\hat{\Pi}W_d(\omega) \tag{57}$$

as the order, d, of the FIR model (48) tends to infinity. For large order models we will thus have that the mean square error of the transfer function estimate at frequency ω is given by

$$\pi_d(\omega) \approx d \cdot \bar{\pi}(\omega). \tag{58}$$

Introduce

$$p(\omega) = \lim_{d \to \infty} \frac{1}{d} W_d^*(\omega) P W_d(\omega)$$

$$\hat{r}_1(\omega) = \lim_{d \to \infty} \frac{1}{d} W_d^*(\omega) \hat{R}_1 W_d(\omega) \qquad (59)$$

$$r_1^0(\omega) = \lim_{d \to \infty} \frac{1}{d} W_d^*(\omega) R_1^0 W_d(\omega).$$

(Recall that the normalization is such that actual parameter change covariance matrix is $\gamma^2 R_1^0$ and that the corresponding assumed covariance in (9) is $\mu^2 \hat{R}_1$).

From (47) we then find, by applying the limiting procedure to both members

$$p^2(\omega) \Phi_u(\omega) = \hat{r}_1(\omega) \qquad (60)$$

or

$$p(\omega) = \sqrt{\left(\frac{\hat{r}_1(\omega)}{\Phi_u(\omega)} \right)}. \qquad (61)$$

Similarly (46) gives

$$2p(\omega)\Phi_u(\omega)\bar{\pi}(\omega) = \mu R_2^0 \hat{r}_1(\omega) + \frac{\gamma^2}{\mu} r_1^0(\omega) \qquad (62)$$

or

$$\bar{\pi}(\omega) = \frac{1}{2} \sqrt{\left(\frac{\hat{r}_1(\omega)}{\Phi_u(\omega)} \right)} \left[\mu \cdot R_2^0 + \frac{\gamma^2}{\mu} \frac{r_1^0(\omega)}{\hat{r}_1(\omega)} \right]. \qquad (63)$$

Expressions (58) and (63) give an explicit and useful description of how the accuracy of the estimate varies with frequency and with the design variables $\hat{r}_1(\omega)$ and μ.

It is easy to explicitly minimize (63) with respect to these variables, and this gives, as it should (if $R_2^0 = 1$)

$$\hat{r}(\omega) = r_1^0(\omega), \quad \mu = \gamma. \qquad (64)$$

We also obtain for the LMS algorithm from (63) with $p(\omega) \equiv 1$

$$\bar{\pi}(\omega) = \frac{1}{2} \left[\mu \cdot R_2^0 + \frac{\gamma^2}{\mu} \frac{r_1^0(\omega)}{\Phi_u(\omega)} \right] \qquad (65)$$

and for the RLS algorithm

$$\bar{\pi}(\omega) = \frac{1}{2} \left(\mu \frac{R_2^0}{\Phi_u(\omega)} + \frac{\gamma^2}{\mu} \cdot r_1^0(\omega) \right). \qquad (66)$$

The results (63)-(66) thus describe how the basic recursive identification algorithm performs under small gain and under steady parameter drift. [7] contains a further discussion of these aspects.

7 Conclusions

We have outlined how to approach the problem of deriving or constructing adaptation algorithms for tracking time-varying systems. We have, among other things, stressed how the Kalman filter provides a natural starting point for the derivations. We have also stressed how common *ad hoc* approaches can be interpreted as special cases corresponding to specific assumptions about the behaviour of the true parameters.

The analysis of the tracking ability of adaptation algorithms is of foremost interest. We have shown the archetypical result where the true covariance matrix for the parameter error can be approximated by an expression that is simpler to study. This study brings out the basic trace-off between tracking ability and noise sensitivity. We have shown how this trade-off becomes especially explicit when evaluated in the frequency domain for linear systems and models.

References

[1] B. D. O. Anderson and J. B. Moore. *Optimal Filtering*. Prentice Hall, New Jersey, 1979.

[2] P. R. Belanger. Estimation of noise covariance matrices for a linear time-varying stochastic process. *Automatica*, 10:267–275, 1974.

[3] D.C. Farden. Tracking properties of adaptive signal processing algorithms. *IEEE Trans. Acoustics, Speech and Signal Processing*, ASSP-29(3):439–446, 1981.

[4] T. R. Fortesque, L. S. Kershenbaum, and B. F. Ydstie. Implementation of self-tuning regulators with variable forgetting factors. *Automatica*, 17:831–835, 1981.

[5] W. Gardner. Nonstationary learning characteristics of the LMS algorithms: A general study, analysis and critique. *IEEE Trans. of Circuits and Systems*, CAS-34(10):1199–1207, 1987.

[6] S. Gunnarsson. *Frequency domain aspects of modeling and control in adaptive systems*. PhD thesis, Department of Electrical Engineering, Linköping University, Linköping, Sweden, 1988.

[7] S. Gunnarsson and L. Ljung. Frequency somain tracking characteristics of adaptive algorithms. *IEEE Transactions on Acoustics Speech and Signal Processing*, ASSP-37:1072–1084, July 1989.

[8] T. Hägglund. *New estimation techniques for adaptive control.* PhD thesis, Department of Automatic Control, Lund University, Sweden., 1983.

[9] A. Isaksson. Identification of time-varying systems through adaptive kalman filtering. In *Preprints 10th IFAC World Congress*, pages 306–311, Munich, 1987.

[10] A. Isaksson. *On system identification in one and two dimensions with signal processing applications.* PhD thesis, Department of Electrical Engineering, Linköping University, 1988.

[11] A. Jazwinski. *Stochastic Process and Filtering Theory*, volume 64 of *Mathematics in Science and Engineering*. Academic Press, New York, 1970.

[12] R. Kulhavy. Restricted exponential forgetting in real-time identification. *Automatica*, (23):589–600, 1987.

[13] R. Kulhavy and M. Karny. Tracking of slowly varying parameters by directional forgetting. In *Proc. 9th IFAC World Congress*, pages 78–83, Budapest, 1984. Vol. X.

[14] L. Ljung and S. Gunnarsson. Adaptive tracking in system identification - a survey. *Automatica*, 26(1):7–22, 1990.

[15] L. Ljung and P. Priouret. A result of the mean square error obtained using general tracking algorithms. *Int. J. of Adaptive Control*, 5(4):231–250, 1991.

[16] L. Ljung and T. Söderström. *Theory and Practice of Recursive Identification.* MIT press, Cambridge, Mass., 1983.

[17] G. C. Goodwin M. E. Salgado and R. H. Middleton. Modified least squares algorithm incorporation resetting and forgetting. *Int. J. Control,*, 1988.

[18] O. Macchi and E. Eweda. Second-order convergence analysis of stochastic adaptive linear filtering. *IEEE Trans. Automatic Control*, AC-28(1):76–85, 1983.

[19] R. K. Mehra. On the identification of variances and adaptive kalman filtering. *IEEE Trans. Aut. Control*, AC-15(2):175–184, 1970.

[20] A. P. Sage and G. W. Husa. Adaptive filtering with unknown prior statistics. In *Proc. 1969 Joint Aut. Control Conf.*, pages 760–769, 1969.

[21] A. P. Sage and G. W. Husa. Algorithms for sequential adaptive estimation of prior statistics. In *Proc. 8th IEEE Symp. Adaptive Processes*, Pennsylvania State University, University Park., 1969.

[22] J. C. Shellenbarger. Estimation of covariance parameter for an adaptive kalman filter. In *Proc. National Electronics Conf.*, pages 698–702, 1966.

[23] J. C. Shellenbarger. A multivariance learning technique for improved dynamics system performance. In *Proc. National Electronics Conf.*, pages 146–151, 1967.

[24] J. G. Wang and Z. L. Deng. Simulation of a newly designed adaptive controller. In *IFAC Symp. on Simulation of Control Systems*, pages 93–196, Vienna, 1986.

[25] I. M. Weiss. A survey of discrete kalman-bucy filtering with unknown noise covariances. In *AIAA Guidance, Control and Flight Mechanics Conf.*, 1970.

[26] B. Widrow, J.M. McCool, M.G. Larimore, and C.R. Johnson Jr. Stationary and nonstationary learning characteristics of the lms adaptive filter. *Proceedings of the IEEE*, 64(8):1151–1162, 1976.

[27] B. Widrow and S. Stearns. *Adaptive Signal Processing*. Prentice-Hall, Englewood-Cliffs, 1985.

DISCOUNTED ESTIMATION AND DISCRETIZATION

IN ADAPTIVE CONTROL

Petr Mandl and Monika Laušmanová

Department of Probability and Mathematical Statistics

Charles University, Prague, Czechoslovakia

Linear autonomous systems are dealt with to the output of which undesirable oscillations are superposed. The oscillations are tracked using discounted least squares estimation in order to eliminate them by modulating the input signal. The performance is measured by means of the average of a quadratic function. An expansion of the criterion for vanishing discount rate is presented.

1. Introduction

Consider a system with transfer function

$$H(s) = \frac{N_0}{D_0 + D_1 s + \ldots + D_{q-1} s^{q-1} + s^q} = \frac{N_0}{D(s)} .$$

For the sake of simplicity the numerator is assumed to be a constant. Suppose that the output y of the system is subjected to slow oscillations which are to be reduced by adding a control signal u to the input x. In Laplace transform,

$$(1) \qquad y^* = H(s) (x^* + u^*) + \sin^* \phi t ,$$

where $\sin \phi t$ can be replaced by an arbitrary periodic function. ϕ is small. (1) means, interpreting s as the differentiation operator,

(2) $\qquad D(s)\, y \;=\; N_0\,(x + u) \;+\; \alpha(t)\, ,$

where

$$\alpha(t) = D(s)\, \sin \phi t = D_0 \sin \phi t + O(\phi)\, ,\quad \phi \to 0\, .$$

Let us have computer control of the system in mind , and let x and y be measured in discrete times with sampling interval δ and let u be constant in the interval $[k\delta,(k+1)\delta)$. We denote

(3) $\qquad x_k = x(k\delta),\quad y_k = y(k\delta),\quad u_k = u(k\delta),\quad \alpha_k = \alpha(k\delta).$

$\alpha(t)$ is to be estimated from the observations, and since the oscillations are slow it will be estimated as a constant by a procedure which - intuitively speaking - forgets the past. The least squares method with exponential discounting is the mostly used one among such procedures ([1],[3]).

To investigate the properties of estimation methods we have to introduce randomness and produce for (2) a discrete time model. There are several ways of defining a discrete time model for (2) ([2]). With regard to the Itô calculus for stochastic models we concentrate here on the backward Euler approximation. In terms of the z-transformation the transfer function of the discrete model is then

$$H\left(\frac{1 - z^{-1}}{\delta}\right) \;=\; \frac{n_0}{1 + d_1\, z^{-1} + \ldots + d_q\, z^{-q}}\, ,$$

and hence

(4) $\qquad y_t + d_1\, y_{t-1} + \cdots + d_q\, y_{t-q} \;=\; n_0\, (\, x_t + u_t\,) + \alpha_t.$

The discounted least squares estimate of α_t is the minimizer α^*_t of

(5) $(1 - \beta)\sum_{k=0}^{\infty} \beta^k\, (y_{t-k} + d_1\, y_{t-k-1} + \cdots - n_0\, (x_{t-k} + u_{t-k}) - \alpha)^2$

where $\beta = \exp(-\sigma\delta)$ is the discount factor. Consequently

(6) $\alpha^*_t = \beta\, \alpha^*_{t-1} + (1-\beta)\, (y_t + d_1\, y_{t-1} + \cdots - n_0\, (x_t + u_t)).$

The discrete time stochastic model is then obtained by introducing random variables into (4) which are interpreted as

Equation (2) defines a continuous time deterministic model. In distinction to the above said we shall transform it to a continuous time stochastic model before discretization. In the problem we consider it will be seen that the model yields what is expected from a continuous time description. Namely, explicit formulas giving better insight into the role of different parameters and good approximation to the discretized version.

2.Model Description

First we introduce white noise into (2) as random disturbance, and assume also that x is a random process. Let x be colored noise satisfying

$$dx = - k x \, dt + dW^q \, ,$$

where W^q is a Wiener process. Generally, x could be a component of a process satisfying a linear stochastic differential equation.

To obtain the state space model we introduce the state

(7) $\qquad X = (y \, , \, y', \ldots, \, y^{(q-1)}, \, x \,)' \, .$

Then

(8) $\qquad dX \;=\; \begin{pmatrix} 0 & 1 & 0 & \ldots & 0 & 0 \\ 0 & 0 & 1 & \ldots & 0 & 0 \\ \cdot & \cdot & \cdot & \ldots & \cdot & \cdot \\ \cdot & \cdot & \cdot & \ldots & & \\ 0 & 0 & 0 & \ldots & 1 & 0 \\ -D_0 & -D_1 & -D_2 & \cdots & -D_{q-1} & N_0 \\ 0 & 0 & 0 & \ldots & 0 & -k \end{pmatrix} X \, dt \;+\;$

$$+ \begin{pmatrix} 0 \\ 0 \\ \cdot \\ 0 \\ 1 \\ 0 \end{pmatrix} U \, dt \;+\; \begin{pmatrix} 0 \\ 0 \\ \cdot \\ 0 \\ 1 \\ 0 \end{pmatrix} \alpha \, dt \;+\; \begin{pmatrix} 0 \\ 0 \\ \cdot \\ 0 \\ dW^{q-1} \\ dW^q \end{pmatrix} .$$

Let us consider the general form of (8)

(9) $dX(t) = f\ X(t)\ dt + g\ U(t)\ dt + b\ \alpha(t)\ dt + dW(t)$.

X , U and α are of dimension n , m and p , respectively. f , g ,
b are matrices, f is stable. W is an n-dimensional Wiener
process with local covariance matrix h , i.e. $dW\ dW' = h\ dt$.

As stated in the introduction, the control U aims to
compensate b α(t) . To this purpose an estimate $\alpha^*(t)$ of α(t) is
constructed. The continuous time analogue of (5) is

(10) $\int_{\infty}^{t} \exp[\sigma(s-t)]\ [(\dot{X}(s)-fX(s)-gU(s)-b\alpha)'l(\dot{X}(s)-fX(s)-gU(s)-b\alpha)$

$$- \dot{X}(s)'l\ \dot{X}(s)]\ ds.$$

l is nonnegatively definite matrix such that

Q = b'l b

is nonsingular. σ > 0 is the discount factor. The undefined terms
$\dot{X}(s)'l\ \dot{X}(s)$ in the square bracket cancel, $\dot{X}(s)\ ds$ means dX(s).

The performance of the control U can be measured by means of
the average value R of a quadratic form X'r X , for example by
the square of the difference between the input and the output. r
is a nonnegatively definite matrix.

Let in (9) U and α be constant. Then

$$\frac{d}{dt}\ E\ X(t) = f\ E\ X(t) + g\ U + b\ \alpha .$$

Consequently, the average value $\overline{X}$ of X fulfils

$\overline{X} = - f^{-1}(\ g\ U + b\ \alpha\ )$.

Further

(11) $R = (\ g\ U + b\ \alpha\ )'f^{-1}'r\ f^{-1}(\ g\ U + b\ \alpha\ ) + trace(\ h\ r\ )$.

The minimum of (11) is attained when

(12) $g\ U = c\ \alpha$,

where in the case that $g'f^{-1}'r\ f^{-1}g$ is nonsingular

$$c = - g\ (\ g'f^{-1}'r\ f^{-1}g\ )^{-1}\ g'f^{-1}'r\ f^{-1}b .$$

To define the control $\alpha^*(t)$ is inserted into (12) instead of α,

$$g\ U(t)\ =\ -\ c\ \alpha^*(t)\ .$$

Thus from (9)

(13) $\qquad d\ X(t)\ =\ f\ X(t)\ dt\ +\ (\ b\ \alpha(t)\ -\ c\ \alpha^*(t)\ )\ dt\ +\ dW(t)$

obtains.

Minimizing (10) it follows

$$\alpha^*(t)\ =\ \sigma \int_{-\infty}^{t} \exp[\sigma(s-t)]\ Q^{-1}b'1\ (dX(s)\ -\ fX(s)\ ds\ +\ c\alpha^*(s)\ ds).$$

Differentiating and using $Q^{-1}b'1\ b = I$,

$$d\alpha^*(t)\ =\ \sigma\ Q^{-1}b'1\ (dX(t)\ -\ fX(t)dt\ +\ c\alpha^*(t)dt)\ -\ \sigma\ \alpha^*(t)dt,$$

(14) $\qquad d\alpha^*(t)\ =\ \sigma\ Q^{-1}b'1\ (dX(t)\ -\ fX(t)\ dt\ +\ (c-b)\ \alpha^*(t)\ dt),$

(15) $\qquad d\alpha^*(t)\ =\ \sigma\ Q^{-1}b'1\ (\ dW(t)\ +\ b\ (\ \alpha(t)\ -\ \alpha^*(t)\ )\ dt)\ =$

$$=\ \sigma\ Q^{-1}b'1\ dW(t)\ +\ \sigma\ (\ \alpha(t)\ -\ \alpha^*(t)\ )\ dt$$

Notice that the distribution of α^* does not depend on U .

Consider (14) for systems given by (8), and let in (10)

$$1\ =\ \begin{pmatrix} 0 & ,\ldots, & 0 & , & 0 \\ & \cdots\cdots\cdots\cdots \\ 0 & ,\ldots, & 0 & , & 0 \\ 0 & ,\ldots, & 1 & , & 0 \\ 0 & ,\ldots, & 0 & , & 1 \end{pmatrix}\ .$$

Then

$$b'1\ =\ (\ 0\ ,\ldots,\ 1\ ,\ 0\ )\ ,\qquad b'1\ b = 1\ ,\qquad c = b\ ,$$

and (14) reads

(16) $\quad d\alpha^*(t)\ =\ \sigma\ dX^{q-1}(t)\ +\ \sigma(D_0X^0(t)+\ldots+D_{q-1}X^{q-1}(t)-N_0X^q(t))dt.$

This is the continuous time equation for the estimate. If only the values of X^0 and X^q are sampled, to make use of (16) we have to set $dt = \delta$ and to replace the differentials of $X^0,\ldots,$ X^{q-1}, and α^* using the backward Euler approximation.

On the other hand, inserting $n_0u_t = -\alpha^*_{t-1}$ into (6) one obtains

$$\alpha^*_t\ =\ \alpha^*_{t-1}\ +\ (1-\beta)\ (y_t\ +\ d_1y_{t-1}\ +\ldots+\ n_0x_t)\ ,$$

which agrees with the approximation to (16) for

$$1 - \beta = 1 - \exp(-\sigma\delta) \sim \sigma\delta .$$

From (15) it follows

$$\alpha^*(t) = \sigma \int_{-\infty}^{t} \exp[\sigma(s-t)] \, \alpha(s) \, ds + \sigma \, Q^{-1}b'1 \int_{-\infty}^{t} \exp[\sigma(s-t)]dW(s).$$

This gives us the probability distribution of $\alpha^*(t)$.

Proposition 1. $\alpha^*(t)$ has normal distribution with the mean and the variance matrix

$$(17) \quad E\alpha^*(t) = \sigma \int_{-\infty}^{t} \exp[\sigma(s-t)]\alpha(s) \, ds, \ \operatorname{Var} \alpha^*(t) = \frac{\sigma}{2} Q^{-1}b'1h1bQ^{-1}.$$

As stated, $\alpha(t)$ has to represent slow oscillations. Slow means here with frequency comparable to the discount rate $\sigma \to 0+$. To this purpose we take a piecewise continuous periodic function $a(y)$, $-\infty < y < \infty$, and set

$$(18) \qquad \alpha(t) = a(\sigma t).$$

From (17) it follows then

$$(19) \qquad E\alpha^*(t/\sigma) = \int_{-\infty}^{t} \exp(y-t) \, a(y) \, dy \ = \bar{a}(t) .$$

3. Asymptotic Expansion

In Section 2 we introduced R , the average of X'r X to evaluate the efficiency of the control. Here we present an asymptotic formula for R as $\sigma \to 0+$. To this purpose we consider first the moments of X(t).

From (13) it follows

$$X(t) = \int_{-\infty}^{t} \exp[(t-s)f] \, (b\alpha(s)-c\alpha^*(s))ds + \int_{-\infty}^{t} \exp[(t-s)f] \, dW(s),$$

and with regard to (18), (19)

(20) $\quad EX(T/\sigma) = \int_{-\infty}^{T} \exp[(T-y)f/\sigma]\ (b\ a(y) - c\ \bar{a}(y))\ dy/\sigma\ =$

$$= f^{-1}(b\ a(T) - c\ \bar{a}(T)) + O(\sigma)\ ,$$

(21) $\quad Var\ X(T/\sigma)\ =\ S\ +\ O(\sigma)\ ,\quad \sigma\ ->\ 0+\ ,$

where S fulfils the equation

$$f\ S + S\ f' + h = 0\ .$$

S is the variance matrix of the process without oscillations.

From (20),(21) it follows

(22) $E\ X'(T/\sigma)\ r\ X(T/\sigma) = (ba(T)-c\bar{a}(T))'f^{-1}{}'r\ f^{-1}(ba(T)-c\bar{a}(T))\ +$

$$+\ trace(rS) + O(\sigma)\ .$$

First order expansion of (20),(21),(22) are presented in [5].

Let τ denote the period of a(y). To estimate R we have to take average of (22) over the period.

Proposition 2. It holds as $\sigma\ ->\ 0+$

$$R\ =\ \frac{1}{\tau}\int_{0}^{\tau}(ba(T)-c\bar{a}(T))'f^{-1}{}'r\ f^{-1}(ba(T)-c\bar{a}(T))\ dT\ +$$

$$+\ trace(rS) + O(\sigma)\ .$$

For

$$a(y)\ =\ \sin(2\pi\theta y)$$

one obtains

$$\bar{a}(y)\ =\ \frac{1}{1 + (2\pi\theta)^2}\ (\ \sin(2\pi\theta y) - 2\pi\theta\ \cos(2\pi\theta y)\ )\ .$$

Hence, for

$$æ\ =\ 1\ /\ \sqrt{(1+(2\pi\theta)^2)}\ ,\quad \mu = f^{-1}{}'r\ f^{-1}\ ,$$

(23) $\quad R = \dfrac{1}{2}\ (b-æc)'\mu\ (b-æc) + æ\ b'\mu\ c + trace(rS) + O(\sigma)\ .$

4.Example

Let (1) hold with

$$H(s) = f_2 / (s^2 + f_1 s + f_2) , \qquad \phi = 2\pi\theta\sigma .$$

Take the squared difference between the input and the output as criterion. Then, according to (8),

$$f = \begin{vmatrix} 0 & 1 & 0 \\ -f_2 & -f_1 & f_2 \\ 0 & 0 & -k \end{vmatrix} , \quad b = c = \begin{vmatrix} 0 \\ f_2 \\ 0 \end{vmatrix} , \quad h = \begin{vmatrix} 0 & 0 & 0 \\ 0 & h_2 & 0 \\ 0 & 0 & h_3 \end{vmatrix} ,$$

$$r = \begin{vmatrix} 1 & 0 & -1 \\ 0 & 0 & 0 \\ -1 & 0 & 1 \end{vmatrix} .$$

Moreover,

$$\alpha(t) = \sin \phi t + O(\sigma) ,$$

therefore (23) can be used.

Let there be discrete observation of the input and the output with sampling interval δ . Exact values of

$$(24) \qquad \lim_{t\to\infty} \frac{1}{t} \sum_{s=0}^{t-1} (X^1{}_s - X^3{}_s)^2 = R^* ,$$

were calculated by the method presented in [5], which reduces the problem to the investigation of a discrete linear system with state vector

$$(y_t, y_t', x_t, y_{t-1}, y_{t-2}, u_{t-1}) .$$

(24) was then compared to (23) giving

$$(25) \quad R^* \sim R_0 = \frac{1}{2} \frac{(2\pi\theta)^2}{1+(2\pi\theta)^2} + \frac{1}{2} (\frac{f_1{}^2 + f_2 + k\, f_1}{k\, f_1{}^2 + f_1 f_2 + k^2 f_1} h_3 + \frac{1}{f_1 f_2} h_2).$$

The first order expansion is

$$(26) \qquad R^* \sim R_0 + (\frac{4\,\phi^4 f_1}{(1+\phi^2)^2 f_2} + \frac{3\, h_2}{2\, f_2{}^2}) \sigma .$$

(25),(26) are explicit formulas obtained from a continuous time model, as promissed in the introduction.

For illustration some numerical results are presented in Table 1 where

$$f_1 = 20 \ , \ f_2 = 50 \ , \ k = 1 \ , \ h_2 = h_3 = 1 \ , \ \delta = 0.5 \ .$$

θ	σ	(24)	(25)	SIMULATIONS	
0.1	0.02	0.3077	0.3075	0.2934	0.2932
	0.05	0.3079		0.2856	0.2991
0.2	0.02	0.4725	0.4721	0.4402	0.4371
	0.05	0.4729		0.4712	0.4473

Table 1

The sample size N for the simulation is such that $\phi N\delta = 4\pi$.

If there are no oscillations then R = 0.1660 . For oscillations without damping R = 0.6660 . (26) differs from (25) for less then 10^{-4}.

References

[1] A.Aloneftis : Stochastic Adaptive Control. (Lecture Notes in Control and Information Sc.98.) Springer-Verlag, Berlin- -Heidelberg-New York 1987.

[2] K.J.Aström , B.Wittenmark : Computer Controlled Systems. Prentice-Hall, Englewood Cliffs 1984.

[3] K.J.Aström , B.Wittenmark : Adaptive Control. Adison-Wesley, Reading 1989.

[4] T.E.Duncan, P.Mandl, B.Pasik-Duncan : On exponentially discounted adaptive control. Kybernetika 26 (1990), 361-372.

[5] M.Laušmanová : Exponentially discounted estimates and oscillations in linear controlled systems. Kybernetika 28 (1992) to appear.

Model reference adaptive control of linear stochastic systems

Sean P. Meyn   Lyndon J. Brown

University of Illinois & the Coordinated Science Laboratory
1101 W. Springfield ave., Urbana, IL, 61801

Abstract

In this paper we describe new adaptive control designs and some new stability and optimality results for general delay, linear stochastic systems under adaptive control.

We develop a new indirect model reference adaptive control law which does not require the solution of a Diophantine equation. This control law coupled with the parameter estimation equations allows a representation of the closed loop system as

$$A_m y_{k+d} = B_m z_k + C_m \hat{v}_{k+d} + \psi_{k+d}$$

where A_m, B_m and C_m are the polynomials used to define the reference model, $\{z_k\}$ is a reference input sequence, $\{\hat{v}_{k+d}\}$ is an estimate of a disturbance process, and the term ψ_{k+d} can be bounded by a "small term" times the norm of a pseudo regression vector.

This expression greatly eases the stability and performance analysis of the closed loop system, and has potential for generalization to adaptive pole placement, and other control laws suitable for non-minimum phase systems.

1 Introduction

We develop in this paper a new class of indirect model reference adaptive control laws for linear stochastic systems. Our formulation eliminates the need to solve the ubiquitous Diophantine equation which results in significantly simpler implementation. Moreover, the control law allows a representation of the closed loop system in a form which greatly motivates and simplifies its analysis.

Typically, especially in deterministic theory, parameter projection is used to prevent the estimates from exiting a certain convex region of parameter space. This approach has several well-known drawbacks: for least squares algorithms, the required projection can be difficult to compute for complex regions and, for its implementation, this approach requires significant apriori information regarding the location of the "true" parameter. In this paper we avoid this approach and instead modify the control law directly whenever the parameter estimates approach an undesirable set of values. This modification of the control law is trivial to implement, and also significantly reduces the amount of apriori knowledge needed for its implementation.

In the ideal "white noise" case we show, even for general delay models, that the performance is optimal in the usual sample mean square sense. Our approach is general in nature and can be extended to provide proofs of stability for a wide class of parameter estimation algorithms, and even more general control laws.

The remainder of the paper is organized as follows: In Sections 2 and 3 we develop a new (non-adaptive) model reference control law which is based upon a reparametrization of the system model to a predictor form.

In Section 4 we extend this control law to the adaptive case, and we devise a modification of the control law which is used whenever the parameter estimate enters a "forbidden" region. Included in this section is an outline of the method used to prove our main stability and performance results.

Our main results are described in Section 5 where we consider time invariant linear models driven by disturbance processes modeled as martingale difference sequences.

In Section 6 we discuss some possible extensions of our results.

We now begin by developing a particular system description which will allow us to define the control laws analyzed in this paper.

2 System Description

In this paper we consider single input/single output general delay ARMAX models of the form

> **System Model**
>
> $$y_{k+d} = \sum_{i=1}^{l_0} a_0^i y_{k-i+d} + \sum_{i=0}^{m_0} b_0^i u_{k-i} + \sum_{i=1}^{n_0} c_0^i v_{k-i+d}^0 + v_{k+d}^0 \qquad (2.1)$$

where $b_0^0 \neq 0$, and the process $\{v_k^0\}$ is a martingale difference sequence satisfying for some $\beta > 2$ and all $k \in \mathbb{Z}_+$,

$$E[v_{k+1}^0 \mid \mathcal{V}_k] = 0 \quad \text{a.s.}$$
$$0 < E[|v_{k+1}^0|^\beta \mid \mathcal{V}_k] = \gamma_{v^0}^\beta < \infty \quad \text{a.s.} \qquad (2.2)$$

where $\mathcal{V}_k = \sigma\{v_0^0, \dots, v_0^k\}$ and γ_{v^0} is a deterministic constant. Then a procedure of Sin et. al. [11] may be used to transform the model (2.1) as follows: Write the model (2.1) in the polynomial form

$$A_0 y = z^d B_0 u + C_0 v^0 \qquad (2.3)$$

where z denotes the backwards shift operator.

Let F_0, F_1, G_0, G_1 be polynomials satisfying

$$F_0 A_0 + z^d G_0 = C_0$$
$$F_1 C_0 + z^d G_1 = 1$$

where $\deg F_0$, $\deg F_1 \leq d-1$. Multiplying both sides of (2.3) by the polynomial $F_0 F_1$, we obtain the system description

$$y = z^d (G_1 + F_1 G_0) y + z^d F_0 F_1 B_0 u + v - z^d G_1 v \qquad (2.4)$$

where $v = F_0 v^0$ is a d-step martingale difference sequence for which

$$E[v_{k+d} \mid \mathcal{V}_k] = 0 \quad \text{a.s.} \qquad (2.5)$$
$$0 < E[|v_{k+d}|^\beta \mid \mathcal{V}_k] \leq \gamma_v^\beta < \infty \quad \text{a.s.} \qquad (2.6)$$

where γ_v is a deterministic constant.

The system description (2.4) may be written in the regression form

$$y_{k+d} = \theta^T \varphi_k^0 + v_{k+d} \qquad (2.7)$$

where

$$\theta^T \triangleq (a^0, \dots, a^{l_1}, b^0, \dots, b^{m_1}, c^0, \dots, c^{n_1})$$

and

$$\varphi_k^{0T} = (y_k, \ldots, y_{k-\ell_1}, u_k, \ldots, u_{k-m_1}, v_k, \ldots, v_{k-n_1}).$$

The most important property of this final model is that, for any causal control law, the d-step optimal prediction of y_{k+d} given the observations $\{u_0, \ldots, u_k, y_0, \ldots y_k\}$ may be succinctly written as

$$E[y_{k+d} \mid \mathcal{Y}_k] = \theta^T \varphi_k^0 \tag{2.8}$$

where $\mathcal{Y}_k = \sigma\{y_0, \ldots, y_k, u_0, \ldots, u_k\}$.

The control laws developed in the next section are entirely based upon this formula.

3 Model Reference Control

Here we consider the following generalized model reference control objective: Find a sequence of controls $u_k \in \mathcal{Y}_k \triangleq \sigma\{y_0, \ldots, y_k, u_0, \ldots, u_{k-1}\}$, $k \in \mathbb{Z}_+$, which make the resulting closed loop system become

$$A_m y = z^d B_m z + C_m v \tag{3.1}$$

where A_m, B_m, C_m are given polynomials, $A_m(0) = 1$, and $\{z_k^*\}$ is a bounded $\mathcal{Y}_k$-adapted scalar reference input.

Clearly this is not possible for all polynomials (A_m, B_m, C_m); for example, we cannot have $A_m = 1$, and $C_m = 0$ since this would require exact predictions of future disturbances, which is ruled out by (2.5), (2.6).

The following result provides necessary and sufficient conditions under which the desired closed loop behavior is achievable.

For any polynomial $H(z) = \sum_{i=0}^{s} h_i z^i$, we let

$$H^0(z) \triangleq \sum_{i=0}^{s \wedge (d-1)} h_i z^i \text{ and } z^d H^1 \triangleq H - H^0.$$

Proposition 3.1 *Let $\{z_k^*\}$ be a $\mathcal{V}_k$-adapted reference sequence, and suppose that a $\mathcal{V}_k$-adapted input sequence $\{u_k\}$ exists which makes the output sequence satisfy (3.1). Then it is necessary that the relation*

$$C_m^0 = A_m^0 \tag{3.1}$$

be satisfied. Conversely, if (3.1) is satisfied, then the control law obtained by solving the equation

$$C_m(\theta^T \varphi_k) = [C_m - A_m]y_{k+d} + B_m z_k^* \tag{3.2}$$

for u_k results in the closed loop system (3.1), with $u_k \in \mathcal{V}_k$, $k \in \mathbb{Z}_+$.

We note that by our conditions on the model and on the control law we have $\mathcal{V}_k = \mathcal{Y}_k$, $k \geq 0$.

Proof Suppose that the closed loop system is of the form (3.1); we will demonstrate that (3.1) does hold.

From (3.1) we have

$$A_m^0 y_k - C_m^0 v_k = z^d B_m z_k^* + z^d C_m^1 v_k - z^d A_m^1 y_k \in \mathcal{V}_{k-d}$$

and since $y_k - v_k$ is also $\mathcal{V}_{k-d}$-measurable, this shows that

$$[A_m^0 - C_m^0]v_k \in \mathcal{V}_{k-d}, \qquad k \in \mathbb{Z}_+.$$

If $A_m^0 \neq C_m^0$, this implies that $v_n \in \mathcal{V}_{n-1}$ for all $n \in \mathbb{Z}_+$. It follows that the stochastic process $\{v_n\}$ is purely (non-linearly) deterministic (c.f. Caines [1]) which contradicts the whiteness property (2.5). Hence $A_m^0 = C_m^0$, which is the desired conclusion.

Conversely, under our assumption that $b_0^0 \neq 0$, the condition (3.1) immediately implies that equation (3.2) gives a control satisfying $u_k \in \mathcal{V}_k$, $k \in \mathbb{Z}_+$.

Substituting $\theta^T \varphi_k = y_{k+d} - v_{k+d}$ into (3.2) gives the closed loop system (3.1). $\qquad \square$

The proposition gives necessary and sufficient conditions under which a model is attainable, but it says nothing of what happens to the input sequence $\{u_k\}$. Indeed, the control law (3.2) is solved by inverting the polynomial B which, as in all model reference control laws, may result in explosion of the input sequence if B possesses zeros within the unit circle in $\mathbb{C}$. This fact prompts us to introduce a minimum phase condition.

Observe that we may compute the control law without solving a Diophantine equation as in [7,4, 10,12] as well as numerous other papers. This is a significant reduction in complexity of the control law, and we will see below that this simplification carries over to the adaptive control problem.

We conclude this section with an example to illustrate how a model may be chosen.

Let G and H be given polynomials with $G(0) = 1$, and suppose that we wish to minimize the *mean square model matching error*:

$$\lambda_k \triangleq \mathsf{E}\left[\left(y_{k+d} - \frac{H}{G}z_k^*\right)^2\right].$$

Then, since $y_{k+d} - \frac{H}{G}z_k^* - v_{k+d} \in \mathcal{V}_k$, we have by (2.5),

$$\lambda_k = \mathsf{E}\left[\left(y_{k+d} - \frac{H}{G}z_k^* - v_{k+d}\right)^2\right] + \mathsf{E}[v_{k+d}^2].$$

Assuming that G is Hurwitz, and ignoring the effect of initial conditions, this shows that the model matching error is minimized if the output follows the model

$$Gy_{k+d} = Hz_k^* + Gv_{k+d}.$$

This control problem clearly lies within the class of control strategies considered here. We also see that the matching condition (3.1) is trivially satisfied.

4 Model Reference Adaptive Control

We now generalize the control laws introduced in the previous section to the adaptive case, where the system parameters are not known apriori.

4.1 An indirect adaptive control law

It is desirable to describe the control law without using any detailed properties of an associated parameter estimator. We will just assume that a sequence $\{\hat{\theta}_k\}$ of parameter estimates are given, with $\hat{\theta}_k$ an estimate of θ, available at the time instant k. We also assume that estimates $\{\hat{v}_k\}$ of the disturbance process $\{v_k\}$ are given, where $\hat{v}_k$ is an estimate of v_k which is known at time k.

With these estimates at hand, an obvious adaptive version of (3.2) is to solve

$$C_m(\hat{\theta}_k^T\varphi_k) = [C_m - A_m]y_{k+d} + B_m z_k^* \tag{4.1}$$

for u_k, $k \in \mathbb{Z}_+$, where $\varphi_k^T = (y_k, \ldots, y_{k-l}, u_k, \ldots, u_{k-m}, \hat{v}_k, \ldots, \hat{v}_{k-n})$. While this control law is easily computed, and the analysis below shows that it is effective in terms of stability and performance under mild conditions, it has a few deficiencies which we will now explain and correct.

First of all, if $\deg C_m \geq d$, then this control uses the estimate $\varphi_{k-d-i}^T \hat{\theta}_{k-d-i}^T$ of $y_{k-i} - v_{k-i}$ for some values of $i \geq 0$, which is based upon information available at time k. Clearly y_{k-i} need not be estimated at time k, and in the estimators that we consider, $y_{k-i} - \hat{v}_{k-i}$ appears to be a more useful estimate of $y_{k-i} - v_{k-i}$ then is $\hat{\theta}_{k-d-i}^T\varphi_{k-d-i}$.

Another difficulty with (4.1) is that to apply this control algorithm it may be necessary to store past values of $\hat{\theta}_{k-i}^T\varphi_{k-i}$ (with $i \geq d$). Since past values of $\{\hat{v}_k\}$ are already stored in φ_k at time k, less storage is required if $y_{k-i} - \hat{v}_{k-i}$ is used to estimate $y_{k-i} - v_{k-i}$ for $i \geq 0$.

These considerations lead to the following control law: At time k, choose u_k so that

$$C_m^0(\hat{\theta}_k^T \varphi_k) + z^d C_m^1(y_{k+d} - \hat{v}_{k+d}) = [C_m - A_m]y_{k+d} + B_m z_k^*$$

which may also be written as

$$C_m^0(\hat{\theta}_k^T \varphi_k) = -A_m^1 y_k + C_m^1 \hat{v}_k + B_m z_k^*. \qquad (4.2)$$

This is essentially the control law considered in this paper, where we always assume that $A_m(0) = C_m(0) = 1$ and that $C_m^0 = A_m^0$. If $\hat{b}_k^0 \neq 0$, then (4.2) uniquely defines the control u_k as a function of variables which are known at time k.

The usual approach to this problem is to propose a linear control law, and then compute the appropriate parameters by solving a Diophantine equation. Here the solution of a Diophantine equation at each time instance is replaced by an application of the d-dimensional filter C_m^0 in (4.2).

4.2 Bounds on the input and output

To investigate the stability and performance of the resulting closed loop system we will exploit a simple formula which describes the error of the estimate $\hat{\theta}_k^T \varphi_k + \hat{v}_{k+d}$ of the output y_{k+d}.

For the stochastic gradient algorithm, extended least squares, the Kalman Filter based algorithm of [5,6] and many other parameter estimators we may obtain an expression of the form

$$y_{k+d} = \hat{\theta}_k^T \varphi_k + \hat{v}_{k+d} + \mu_{k+d}. \qquad (4.3)$$

Under very general assumptions, $|\mu_{k+d}| = |\varphi_k|\varepsilon_k$ where for the stochastic gradient algorithm

$$\sum_{i=0}^{\infty} \varepsilon_k^2 < \infty$$

and

$$\sum_{i=0}^{N} \varepsilon_k^2 = o\Big(\log \sum_{i=0}^{N} |\varphi_i|^2\Big), \qquad N \in \mathbb{Z}_+$$

for extended least squares.

Filtering (4.3) by C_m^0 and substituting the control law (4.2) gives

$$\begin{aligned} C_m^0 y_{k+d} &= -z^d A_m^1 y_{k+d} + z^d C_m^1 \hat{v}_{k+d} + B_m z_k^* \\ &\quad + C_m^0 \hat{v}_{k+d} + C_m^0 \mu_{k+d} \end{aligned}$$

From the matching condition and the definitions, this shows that

$$A_m y_{k+d} = B_m z_k^* + C_m \hat{v}_{k+d} + C_m^0 \mu_{k+d} \qquad (4.4)$$

Assuming that the zeros of A_m lie outside the closed unit disk in $\mathbb{C}$, this equation allows us to bound y_{k+d}^2 by a moving average of the bounded reference signal, the disturbance estimates, and the variables $\{\varepsilon_i^2 |\varphi_i|^2 : i \leq k\}$. This fact is the basis of the stability analysis performed in this paper.

To obtain stability, we also require that $\{u_k^2\}$ may be bounded by $\{y_k^2\}$ in some sense. The required bound is *almost* attained if we assume a minimum phase condition. However, if $\hat{b}_k^0$ is not prevented from approaching zero then the control law (4.1) can introduce large feedback gain which may result in lack of robustness or instability.

The most common approach to correcting this problem is to introduce a projection in the parameter estimation algorithm. Here we instead modify the control law directly to avoid high gain.

Fix a constant $\delta > 0$, let $(\delta_k, \mathcal{Y}_k)$ be an adapted sequence with $|\delta_k| \leq \delta$ for all k, and define the control action u_k through the formula

> Model Reference Adaptive Control Law
>
> Choose at time k the input u_k as the solution to the equation
>
> $$C_m^0(\hat{\theta}_k^T \varphi_k) + \delta_k u_k = -A_m^1 y_k + C_m^1 \hat{v}_k + B_m z_k^* \qquad (4.5)$$

We note that since $|\delta_k| \leq \delta$ for all $k \in \mathbb{Z}_+$, if the constant δ is small then (4.5) is a small perturbation of the control law (4.1).

Analogous modifications of the control law are used by Guo & Chen [3] and Middleton & Kokotovic [8] to prevent high gain.

Controlling the gain in this way is an obvious approach to keeping the control sequence bounded by the output and disturbance processes. It also ensures that the control law is always well defined. However, the introduction of $\{\delta_k\}$ destroys the formula (4.3), so that instead we have

$$A_m y_{k+d} = B_m z_k^* + C_m \hat{v}_{k+d} + C_m^0 \mu_{k+d} - \delta_k u_k. \qquad (4.6)$$

We will see that the introduction of this additional term has little effect on our results, besides somewhat complicating the analysis.

4.3 Performance

To measure performance, assume for concreteness that the goal is to make $y_{k+d} \cong \frac{B_m}{A_m} z_k^*$, so that we have $A_m = C_m$. In this case we may filter (4.6) to obtain

$$
\begin{aligned}
y_{k+d} &- \frac{B_m}{A_m} z_k^* - v_{k+d} \\
&= \hat{v}_{k+d} - v_{k+d} + \frac{A_m^0}{A_m}(\mu_{k+d}) - \frac{1}{A_M}(\delta_k u_k) + o(1)
\end{aligned} \qquad (4.7)
$$

where the term $o(1)$ is due to initial conditions, and decays to zero exponentially fast. Observe that the optimal mean square closed loop response is obtained when the left hand side is equal to zero.

For the estimators we consider, the accumulated error

$$\sum_{k=0}^{N} (v_k - \hat{v}_k)^2$$

grows very slowly, so that once stability is established, (4.7) may be used to bound the performance of the closed loop system.

In the next section we will use the formula (4.7) to obtain optimal performance for the adaptive control law based upon extended least squares.

5 Stability and Performance for Extended Least Squares

Here we prove that the control laws developed in the previous section are stabilizing and give optimal performance for the ELS (extended least squares) parameter estimation algorithm.

We consider the system model (2.1), transformed to the predictor form (2.7). Estimates $\{\hat{\theta}_k\}$ of the parameter θ defined in (2.7) are defined recursively by the ELS algorithm described here:

Extended Least Squares

$$\hat{\theta}_k = \hat{\theta}_{k-1} + \frac{P_{k-1}\varphi_{k-d}e_k}{1+\varphi_{k-d}^T P_{k-1}\varphi_{k-d}} \tag{5.1}$$

$$P_k = P_{k-1} - \frac{P_{k-1}\varphi_{k-d}\varphi_{k-d}^T P_{k-1}}{1+\varphi_{k-d}^T P_{k-1}\varphi_{k-d}}. \tag{5.2}$$

The apriori error e_k and the aposteriori error $\hat{v}_k$ are defined as

$$e_k \triangleq y_k - \hat{\theta}_{k-1}^T \varphi_{k-d} \tag{5.3}$$

$$\hat{v}_k \triangleq y_k - \hat{\theta}_k^T \varphi_{k-d}, \qquad k \in \mathbb{Z}_+, \tag{5.4}$$

and the pseudo-regression vector φ_k is defined as

$$\varphi_k^T \triangleq (y_k, \ldots, y_{k-\ell_1}, u_k, \ldots, u_{k-m_1}, \hat{v}_k, \ldots, \hat{v}_{k-n_1}), \tag{5.5}$$

We define the positive random variable r_k as

$$r_k = \text{trace } P_k^{-1}, \qquad k \in \mathbb{Z}_+, \tag{5.6}$$

and choose δ_k in the control law (4.5) as

$$\delta_k \triangleq \delta(\log^{\frac{1}{2}} r_k)^{-1}\text{sign}(\hat{b}_k^0)\mathbf{1}\big\{|\hat{b}_k^0| < \delta(\log^{\frac{1}{2}} r_k)^{-1}\big\}, \qquad k \in \mathbb{Z}_+.$$

where δ is a fixed positive number.

The following assumptions will be required in our main results:

A5.1 The zeros of the polynomial B_0 lie strictly outside of the closed unit disk in $\mathbb{C}$;

A5.2 The initial conditions are deterministic and $r_0 = \text{trace } P_0^{-1} \geq e$;

A5.3 The polynomial $C^{-1} - \frac{1}{2}$ is strictly positive real; that is, the zeros of the polynomial C lie outside the closed unit disk in $\mathbb{C}$, and

$$Re\big\{\frac{1}{C(e^{i\omega})} - \frac{1}{2}\big\} > 0, \quad \omega \in [0, 2\pi].$$

A5.4 The disturbance process $\{v_k\}$ is of the form $v_k = F_0 v_k^0$ where $F_0 = 1 + \sum_{i=1}^{d-1} f_i z^i$, and the process $\{v_k^0\}$ is bounded and satisfies (2.2);

A5.5 The zeros of the polynomial A_m lie strictly outside the closed unit disk in $\mathbb{C}$, $A_m^0 = C_m^0$, and $A_m(0) = C_m(0) = 1$.

The assumption that $\{v_k^0\}$ is a bounded sequence is unnecessary; however, it does considerably simplify the stability proof. To see how this condition may be relaxed, the reader is referred to [3].

The main results in this section show that the closed loop system is stable and that the performance is sample mean square optimal under the control law (4.5).

Theorem 5.1 *Suppose that conditions A5.1-A5.5 hold, and that the control (4.5) is applied to the system (2.1). Then we have with probability one,*

$$\limsup_{N \to \infty} \frac{1}{N} \sum_{k=1}^{N} (y_k^2 + u_k^2) < \infty \tag{5.7}$$

$$\lim_{N \to \infty} \frac{1}{N} \sum_{k=1}^{N} \Big(y_{k+d} - \frac{B_m}{A_m}z_k^* - \frac{C_m}{A_m}v_{k+d}\Big)^2 = 0. \tag{5.8}$$

These results extend the main results of [3] to general control laws, and to system models with arbitrary delay. Similar results are given for the unit delay case in [10].

The major new contribution of these results lies in the fact that the control algorithm considered here is far simpler than traditional indirect model reference adaptive control laws which have been analyzed in the literature. As a consequence of this property, the analysis of the resulting closed loop system is simplified considerably.

In previous work [6] we have considered the model reference adaptive control laws considered in this paper coupled with a non-vanishing gain parameter estimation algorithm which is based upon the Kalman filter. In these papers we show that the closed loop system is stable in the sense that the input-output process is bounded. These results hold for time-varying, general delay models whose time variations are "small in the mean".

The proof of Theorem 5.1 comprises the remainder of this section.

We will follow closely the proof outline given in Section 4. To begin, recall that we define $\hat{v}_k = y_k - \hat{\theta}_k^T \varphi_{k-d}$. It follows by iterating (5.1) that we may write

$$y_k = \hat{\theta}_{k-d}^T \varphi_{k-d} + \hat{v}_k + \sum_{i=0}^{d-1} \varphi_{k-d}^T P_{k-1-i} \varphi_{k-d-i} \hat{v}_{k-i}. \tag{5.9}$$

Filtering both sides of this equation by C_m^0, substituting the control law (4.5), and using Condition A5.5, we obtain

$$\begin{aligned}
A_m y_k &= B_m z_{k-d}^* + C_m \hat{v}_k \\
&\quad - \delta_{k-d} u_{k-d} \\
&\quad + C_m^0 \Big(\sum_{i=0}^{d-1} \varphi_{k-d}^T P_{k-1-i} \varphi_{k-d-i} \hat{v}_{k-i} \Big).
\end{aligned} \tag{5.10}$$

This formula is of the form (4.6), which enables us to directly apply the ideas introduced in Section 4.

Proposition 5.1 below is the foundation of the analysis performed in this section. Define for $k \in \mathbb{Z}_+$,

$$\tilde{\theta}_k \triangleq \theta - \hat{\theta}_k \tag{5.11}$$

$$\Upsilon_k \triangleq \varphi_{k-d}^T P_{k-1} \varphi_{k-d} \hat{v}_k^2. \tag{5.12}$$

The proof of Proposition 5.1 will require the following identity. The proof is standard (cf. [2,10]).

Lemma 5.1 *For all $k \in \mathbb{Z}_+$,*

$$C(v_k - \hat{v}_k) = \tilde{\theta}_k^T \varphi_{k-d}.$$

Proposition 5.1 *Under Conditions A5.3 and A5.4 the following inequalities hold for all $N \in \mathbb{Z}_+$ with probability one:*

$$\sum_{i=1}^{N}(\hat{v}_i - v_i)^2 = O(\log r_N) \tag{5.13}$$

$$\sum_{i=0}^{N}(\tilde{\theta}_i^T \varphi_{i-d})^2 = O(\log r_N) \tag{5.14}$$

$$|\tilde{\theta}_N|^2 = O\Big\{ \frac{\log(r_N)}{\lambda_{\min} P_N^{-1}} \Big\} \tag{5.15}$$

$$\sum_{i=1}^{N} \Upsilon_i = O(\log r_N) \tag{5.16}$$

Proof The bounds (5.13) and (5.15) are taken directly from Theorem 2.2 of [10]. From (5.13) and Lemma 5.1 we see that (5.14) also holds.

The result (5.16) follows from equations (2.46), (2.47) and (2.55) of [10]. □

The following simple result when combined with (5.10), and Proposition 5.1, will allow a Bellman-Gronwall Lemma based stability proof.

Define for each $k \in \mathbb{Z}_+$,

$$\Gamma_k = \text{trace}(P_k - P_{k+d}). \tag{5.17}$$

Lemma 5.2 *Under the conditions of Theorem 5.1, we have*

$$\Big(\sum_{i=0}^{d-1} \varphi_{j-d}^T P_{j-1-i} \varphi_{j-d-i} \hat{v}_{j-i}\Big)^2 \le d\Big(\sum_{i=0}^{d-1} \Upsilon_{j-i}\Big)(\Gamma_{j-d}|\varphi_{j-d}|^2 + 1)$$

Proof By the Cauchy-Schwarz inequality

$$\Big(\sum_{i=0}^{d-1} \varphi_{j-d}^T P_{j-1-i} \varphi_{j-d-i} \hat{v}_{j-i}\Big)^2$$

$$\le \Big(\sum_{i=0}^{d-1} \varphi_{j-d}^T P_{j-1-i} \varphi_{j-d}\Big)\Big(\sum_{i=0}^{d-1} \varphi_{j-d-i}^T P_{j-1-i} \varphi_{j-d-i} \hat{v}_{j-i}^2\Big) \tag{5.18}$$

By substituting the defining identity for P_j, given in (5.2), we obtain the following estimate:

$$\begin{aligned}
\varphi_{j-d}^T P_{j-1-i} \varphi_{j-d} &= \varphi_{j-d}^T P_j \varphi_{j-d} + \varphi_{j-d}(P_{j-1-i} - P_j)\varphi_{j-d} \\
&= \varphi_{j-d}^T\Big(P_{j-1} - \frac{P_{j-1}\varphi_{j-d}^T \varphi_{j-d} P_{j-1}}{1 + \varphi_{j-d}^T P_{j-1}\varphi_{j-d}}\Big)\varphi_{j-d} \\
&\quad + \varphi_{j-d}^T(P_{j-1-i} - P_j)\varphi_{j-d} \\
&\le 1 + \varphi_{j-d}^T(P_{j-1-i} - P_j)\varphi_{j-d}.
\end{aligned}$$

Also by (5.2), we have $P_{j-1-i} \le P_{j-d}$ for all $0 \le i \le d-1$. This together with the previous inequality shows that

$$\begin{aligned}
\varphi_{j-d}^T P_{j-1-i} \varphi_{j-d} &\le 1 + \varphi_{j-d}^T(P_{j-d} - P_j)\varphi_{j-d} \\
&\le 1 + \|P_{j-d} - P_j\||\varphi_{j-d}|^2.
\end{aligned}$$

This completes the proof as the operator norm $\| \cdot \|$ of a positive matrix is bounded by its trace. □

In the next result we demonstrate that $\{u_k^2\}$ may be bounded by $\{y_k^2\}$, so that for stability we may restrict our attention to the output process.

Lemma 5.3 *Under the conditions of Theorem 5.1, there exists $\rho_2 < 1$ such that*

$$u_k^2 = O\Big\{\sum_{j=0}^{k+d-1} \rho_2^{k-j} y_j^2 + \sum_{j=0}^{k} \rho_2^{k-j}\Big(\sum_{i=0}^{d-1} \Upsilon_{j-i+d}\Big)(\Gamma_j|\varphi_j|^2) + \log r_{k+d}\Big\} \tag{5.19}$$

$$u_k^2 = O\Big\{\log^2 r_k\Big(1 + \sum_{j=0}^{k+d-1} \rho_2^{k-j} y_j^2\Big)\Big\}. \tag{5.20}$$

Proof We begin with the identity,

$$b^0 u_k = \tilde{\theta}_{k+d}^T \varphi_k + y_{k+d} - \hat{v}_{k+d} + (b^0 u_k - \theta^T \varphi_k) \tag{5.21}$$

From this and (5.10) we see that

$$
\begin{aligned}
(b^0 + \delta_k) u_k &= \Big\{ \frac{B_m}{A_m} z_k^* + \frac{C_m}{A_m} \hat{v}_{k+d} - \Big(\frac{1}{A_m} - 1 \Big) \delta_k u_k \\
&\quad + \frac{C_m^0}{A_m} \Big(\sum_{i=0}^{d-1} \varphi_k^T P_{k+d-1-i} \varphi_{k-i} \hat{v}_{k+d-i} \Big) \Big\} \\
&\quad + \tilde{\theta}_{k+d}^T \varphi_k - \hat{v}_{k+d} + (b^0 u_k - \theta^T \varphi_k) + o(1). \tag{5.22}
\end{aligned}
$$

From Condition A5.1 and since $r_k \to \infty$ as $k \to \infty$ (cf. [9]) we have $|b^0 + \delta_k| \geq \frac{1}{2} |b^0| > 0$ for k sufficiently large.

From Proposition 5.1 we see that $\{\hat{v}_k\}$ and $\{\tilde{\theta}_{k+d}^T \varphi_k\}$ can be bounded as

$$\hat{v}_{k+d}^2 = O(\log r_{k+d}), \qquad (\tilde{\theta}_{k+d}^T \varphi_k)^2 = O(\log r_{k+d}).$$

These facts and the identity above show that

$$
\begin{aligned}
u_k^2 &= O\Big(\Big\{ \Big(\frac{1}{A_m} - 1 \Big) \delta_k u_k \Big\}^2 + \{ b_0 u_k - \theta^T \varphi_k \}^2 \\
&\quad + \Big\{ \frac{C_m^0}{A_m} \Big(\sum_{i=0}^{d-1} \varphi_k^T P_{k+d-1-i} \varphi_{k-i} \hat{v}_{k+d-i} \Big) \Big\}^2 \\
&\quad + \log r_{k+d} \Big).
\end{aligned}
$$

The first two terms on the right hand side of this bound do not include u_k, while the third term can be bounded using Lemma 5.2. From these observations, and stability of A_m, we see that for some $\rho_3 < 1$,

$$
\begin{aligned}
u_k^2 &= O\Big(\sum_{j=0}^{k-1} \rho_3^{k-j} u_j^2 + \sum_{j=0}^{k} \rho_3^{k-j} y_j^2 + \sum_{j=0}^{k} \rho_3^{k-j} \Big(\sum_{i=0}^{d-1} \Upsilon_{j-i+d} \Big) \big(\Gamma_j |\varphi_j|^2 + 1 \big) \\
&\quad + \log r_{k+d} \Big).
\end{aligned}
$$

The minimum phase condition A5.1, together with the assumption that the disturbance process is bounded, implies that for some $\rho_4 < 1$,

$$u_j^2 = O\Big(\sum_{i=0}^{j+d} \rho_4^{j-i} y_i^2 + 1 \Big).$$

This, Proposition 5.1, and the preceding bound completes the proof of (5.19).

The bound (5.20) follows directly from the control law (4.5), Proposition 5.1, and the minimum phase condition. □

Lemma 5.2, Proposition 5.1, and equation (5.10) show that y_k^2 is bounded by a moving average of terms which are of order $\log r_k$, and two terms, essentially Υ_k and Γ_k, times the square of the norm of the pseudo regression vector.

We saw in Proposition 5.1 that the sum of $\{\Upsilon_k\}$ can be bounded in terms of $\{\log r_k\}$. The following result provides an estimate to show that $\{\Gamma_k\}$ is in fact summable. The proof is obvious and is left to the reader.

Lemma 5.4 *For the estimator (5.1) (5.2) we have*

$$\sum_{k=0}^{\infty} \Gamma_k \leq \operatorname{trace}(P_0 + \cdots + P_{d-1}) < \infty.$$

□

Because this term appears frequently in our derivations, we will denote

$$f_k \triangleq \Big(\sum_{i=0}^{d-1} \Upsilon_{k-i+d}\Big)\Gamma_k. \tag{5.23}$$

We also define, for $\alpha < 1$, $k \in \mathbb{Z}_+$,

$$S_k(\alpha) \triangleq \sum_{i=0}^{k} \alpha^{k-i} y_i^2 \tag{5.24}$$

$$F_k \triangleq \sum_{i=0}^{k} \alpha^{k-i} f_i. \tag{5.25}$$

We now proceed to bound the output process. The following lemma allows a direct application of the Bellman-Gronwall Lemma.

Lemma 5.5 *Under the conditions of Theorem 5.1 we have for some $\alpha < 1$, $\beta < 1$, $\gamma < 1$,*

$$S_k(\beta) = O\Big(\sum_{j=1}^{k-1} \gamma^{k-j} H_j S_j(\beta) + \log^4 r_k\Big)$$

where

$$H_j \triangleq \delta_{j-d+1} + f_{j-d+1} + (F_{j-d+1}(\alpha))^2 \log^2 r_{j-d+1}.$$

Proof From (5.10) and stability of A_m, there exists $\rho_1 < 1$ such that

$$y_k^2 = O\Big(\sum_{j=1}^{k} \rho_1^{k-j}\big\{\delta_{j-d}^2 u_{j-d}^2 + (z_{j-d}^*)^2 + \hat{v}_j^2 \\ + \big(\sum_{i=0}^{d-1} \varphi_{j-d}^T P_{j-1-i}\varphi_{j-d-i}\hat{v}_{j-i}\big)^2\big\}\Big).$$

Since $\{z_k^*\}$ and $\{v_k\}$ are bounded sequences, since $\hat{v}_k^2 = O(|\hat{\theta}_k|^2 + 1) = O(\log r_k)$ from Proposition 5.1, and because the conclusions of Lemma 5.1 hold, it follows that

$$y_k^2 = O\Big(\sum_{j=1}^{k} \rho_1^{k-j}\delta_{j-d}^2 u_{j-d}^2 + \sum_{j=1}^{k} \rho_1^{k-j} f_{j-d}|\varphi_{j-d}|^2 + \sum_{j=1}^{k} \Upsilon_j + \log r_k\Big). \tag{5.26}$$

From Lemma 5.3 we have the bound

$$u_{j-d}^2 = O\Big(S_{j-1}(\rho_2) + \sum_{\ell=0}^{j} \rho_2^{j-\ell} f_{\ell-d}|\varphi_{\ell-d}|^2 + \log r_j\Big).$$

Substituting this into the preceding bound, and using $\sum_{j=1}^{k} \Upsilon_j = O(\log r_k)$, which follows from Proposition 5.1, gives

$$y_k^2 = O\Big(\sum_{j=1}^{k} \rho_1^{k-j}\delta_{j-d}^2\big\{S_{j-1}(\rho_2) + \sum_{\ell=0}^{j} \rho_2^{j-\ell} f_{\ell-d}|\varphi_{\ell-d}|^2\big\} \\ + \sum_{j=1}^{k} \rho_1^{k-j} f_{j-d}|\varphi_{j-d}|^2 + \log r_k\Big).$$

This implies that for any $\rho_0 > \max(\rho_1, \rho_2)$,

$$y_k^2 = O\Big(\sum_{j=0}^{k-1} \rho_0^{k-j}\delta_{j-d+1}^2 S_j(\rho_0) + \sum_{j=1}^{k} \rho_0^{k-j} f_{j-d}|\varphi_{j-d}|^2 + \log r_k\Big).$$

Applying Lemma 5.3 once more gives

$$|\varphi_{j-d}|^2 = O\Big(S_{j-1}(\rho_2) + \sum_{i=0}^{j} \rho_2^{j-i} f_{i-d}|\varphi_{i-d}|^2 + \log r_j\Big).$$

When substituted into the previous bound, this implies that

$$
\begin{aligned}
y_k^2 &= O\Big(\sum_{j=0}^{k-1} \rho_0^{k-j} \delta_{j-d+1}^2 S_j(\rho_0) + \sum_{j=1}^{k} \rho_0^{k-j} f_{j-d} S_{j-1}(\rho_0) \\
&\quad + \sum_{j=1}^{k} \rho_0^{k-j} f_{j-d}\Big\{\sum_{i=0}^{j} \rho_0^{j-i} f_{i-d}|\varphi_{i-d}|^2\Big\} + \log^2 r_k\Big)
\end{aligned}
\tag{5.27}
$$

where we have used here the bound $f_{i-d} = O(\log r_k)$, $i \le k$, which follows from Proposition 5.1.
Applying the second inequality in Lemma 5.3 gives

$$|\varphi_{i-d}|^2 = O(\log^2 r_{i-d} S_{i-1}(\rho_0) + \log^2 r_{i-d}).$$

Since $\{r_i\}$ is an increasing sequence and since $S_{i-1}(\rho) \le \rho^{-(j-i)} S_{j-1}(\rho)$ for $j \ge i$, the bound on φ_{i-d} can be transformed as

$$|\varphi_{i-d}|^2 = O(\log^2 r_{j-d} S_{j-1}(\rho)\rho^{-(j-i)} + \log^2 r_{j-d})$$

valid for any $\rho \ge \rho_0$. Substituting this into (5.27) gives, for any $\rho > \rho_0$,

$$
\begin{aligned}
y_k^2 &= O\Big(\sum_{j=0}^{k-1} \rho_0^{k-j} \delta_{j-d+1}^2 S_j(\rho) + \sum_{j=1}^{k} \rho_0^{k-j} f_{j-d} S_{j-1}(\rho) \cdot \\
&\quad + \sum_{j=1}^{k} \rho_0^{k-j} f_{j-d} \log^2 r_{j-d} S_{j-1}(\rho) F_{j-d}(\rho^0/\rho) + \log^4 r_k\Big).
\end{aligned}
$$

Using the bound $f_{j-d} \le F_{j-d}(\alpha)$, $\alpha < 1$, and taking the ρ-average of each side of this inequality gives the result with $\alpha = \rho^0/\rho$, $\beta = \rho$, and any $\gamma > \rho$. $\qquad\qquad\Box$

Lemma 5.6 *Under the conditions of Theorem 5.1 we have, for any $\epsilon > 0$,*

$$|\varphi_{n-d}|^2 = o(r_n^\epsilon), \qquad \text{a.s.} \quad n \in \mathbb{Z}_+.$$

Proof We saw in Lemma 5.5 that for an almost surely finite random variable $B \ge 1$,

$$\gamma^{-k} S_k(\beta) \le B \sum_{j=1}^{k-1} H_j \gamma^{-j} S_j(\beta) + \gamma^{-k} B \log^4 r_k.$$

Letting $X_k = \gamma^{-k} S_k(\beta)$, $h_k = BH_k$, and $J_k = B\gamma^{-k} \log^4 r_k$, we see from the Bellman Gronwall Lemma that for all $N \in \mathbb{Z}_+$,

$$S_{N+1}(\beta) \le \sum_{k=1}^{N}\Big(\prod_{j=k}^{N}(1 + BH_j)\Big)\gamma^{N+1-k} B \log^4 r_k + B \log^4 r_{N+1}.
\tag{5.28}$$

The main task is to bound the product in the right hand side of this inequality.
First we obtain a crude bound for $\{H_k\}$: since $\log r_j \ge 1$ for all j, and since we may assume without loss of generality that $B \ge 1$, we have

$$1 + BH_j \le (1 + B\delta_{j-d+1})(1 + BF_{j-d+1}(\alpha) \log r_{j-d+1})^2.$$

The second multiplicand may be bounded as follows: for any $0 < c < \infty$ we have

$$
\begin{aligned}
1 + BF_{j-d+1}(\alpha)r_{j-d+1} &= 1 + Br_{j-d+1}\sum_{\ell=0}^{j-d+1}\alpha^{j-d+1-\ell}\Big(\sum_{i=0}^{d-1}\Upsilon_{\ell-i+d}\Big)\Gamma_\ell \\
&\leq \Big(1 + c^{-1}B\log r_j\sum_{\ell=0}^{j-d+1}\alpha^{\frac{1}{2}(j-d+1-\ell)}\Gamma_\ell\Big) \\
&\quad \times\Big(1 + cd\sum_{\ell=0}^{j-d+1}\alpha^{\frac{1}{2}(j-d+1-\ell)}\Upsilon_{\ell+d}\Big).
\end{aligned}
$$

These bounds together with the estimate $1 + x \leq e^x$, $x \in \mathbb{R}$, gives

$$
\begin{aligned}
\prod_{j=k}^{N}(1 + BH_j) \leq \exp\Big\{ &B\sum_{j=k}^{N}\delta_{j-d+1} \\
&+ 2\sum_{j=k}^{N}c^{-1}B\log r_j\Big(\sum_{\ell=0}^{j-d+1}\alpha^{\frac{1}{2}(j-d+1-\ell)}\Gamma_\ell\Big) \\
&+ 2\sum_{j=k}^{N}cd\Big(\sum_{\ell=0}^{j-d+1}\alpha^{\frac{1}{2}(j-d+1-\ell)}\Upsilon_{\ell+d}\Big)\Big\}.
\end{aligned} \tag{5.29}
$$

From Proposition 5.1 we have

$$
\sum_{j=k}^{N}\sum_{\ell=0}^{j-d+1}\alpha^{\frac{1}{2}(j-d+1-\ell)}\Upsilon_{\ell+d} = O(\log r_{N+1}).
$$

Hence we may choose the (random) constant $c > 0$ so small that for any preassigned $\varepsilon_0 > 0$,

$$
2\sum_{j=k}^{N}cd\Big(\sum_{\ell=0}^{j-d+1}\alpha^{\frac{1}{2}(j-d+1-\ell)}\Upsilon_{\ell+d}\Big) \leq \varepsilon_0 \log r_{N+1} \qquad \text{a.s.} \qquad n \in \mathbb{Z}_+.
$$

The second term in the (5.29) can be bounded as follows: we may write

$$
\sum_{j=0}^{\infty}\sum_{\ell=0}^{j}\alpha^{\frac{1}{2}(j-\ell)}\Gamma_\ell = \frac{1}{1-\sqrt{\alpha}}\sum_{\ell=0}^{\infty}\Gamma_\ell < \infty.
$$

This shows that $\sum_{j=k}^{N}\sum_{\ell=0}^{j}\alpha^{\frac{1}{2}(j-\ell)}\Gamma_\ell = o(1)$ as $k \to \infty$.

Finally, since $\delta_k \to 0$ as $k \to \infty$ a.s., we see from (5.29) that for all $\varepsilon_0 > 0$, $\varepsilon_1 > 0$,

$$
\prod_{j=k}^{N}(1 + BH_j) = O(\exp(\varepsilon_1(N-k) + \varepsilon_0 \log r_{N+1})).
$$

This together with (5.28) easily implies that for all $\varepsilon > 0$,

$$
S_{N+1}(\beta) = O(r_{N+1}^\varepsilon).
$$

This last bound together with Lemma 5.3 completes the proof of the lemma. □

Finally, we may give the

Proof of Theorem 5.1

It follows from (5.10) and Lemma 5.2 that

$$y_k^2 = O\left[\sum_{j=1}^{k} \rho_1^{k-j}\left(\delta_{j-d}^2 u_{j-d}^2 + (z_{j-d}^*)^2 + \hat{v}_j^2 + \left(\sum_{i=0}^{d-1} \Upsilon_{j-i}\right)\left(\Gamma_{j-d}|\varphi_{j-d}|^2 + 1\right)\right)\right].$$

Hence, by summing from $k = 1$ to N and applying Lemma 5.6 and Proposition 5.1 we see that

$$\begin{aligned}
\sum_{k=1}^{N} y_k^2 &= o\left(\sum_{k=1}^{N} u_{k-d}^2\right) + O\left(N + \log r_N + \sum_{k=1}^{N}\left(\sum_{i=0}^{d-1} \Upsilon_{k-i}\right)\left(\Gamma_{k-d}|\varphi_{k-d}|^2 + 1\right)\right) \\
&= o\left(\sum_{k=1}^{N} u_{k-d}^2\right) + O\left(r_N^\epsilon \log r_N \left(\sum_{k=1}^{N} \Gamma_{k-d}\right) + \log r_N + N\right) \\
&= o\left(\sum_{k=1}^{N} u_{k-d}^2\right) + O(r_N^{2\epsilon} + N).
\end{aligned}$$

From the minimum phase condition A5.1 it follows that

$$\sum_{k=1}^{N} u_{k-d}^2 = O\left(\sum_{k=1}^{N} y_k^2 + N\right).$$

These two bounds show that

$$\sum_{k=1}^{N} u_k^2 = o\left(\sum_{k=1}^{N} u_k^2\right) + O(r_{N+d}^{2\epsilon} + N),$$

and hence that

$$r_{N+d} = O\left(\sum_{k=1}^{N}(u_k^2 + y_k^2)\right) = o(r_{N+d}) + O(N).$$

This implies that $r_N = O(N)$, which proves (5.7).

From (5.10) we see that

$$\begin{aligned}
\left(y_k - \frac{B_m}{A_m}z_{k-d}^* - \frac{C_m}{A_m}v_k\right) &= \frac{C_m}{A_m}(\hat{v}_k - v_k) \\
&\quad - \frac{1}{A_m}(\delta_{k-d}u_{k-d}) \\
&\quad + \frac{C_m^0}{A_m}\left(\sum_{i=0}^{d-1} \varphi_{k-d}^T P_{k-1-i}\varphi_{k-d-i}\hat{v}_{k-i}\right).
\end{aligned}$$

From this, stability of A_m, Lemma 5.6 (with $\epsilon = \frac{1}{2}$), and (5.7) we have

$$\begin{aligned}
\sum_{k=1}^{N}\left(y_k - \frac{B_m}{A_m}z_{k-d}^* - \frac{C_m}{A_m}v_k\right)^2 &= O\left(\sum_{k=1}^{N}(\hat{v}_k - v_k)^2\right) \\
&\quad + o(N) \\
&\quad + O\left(\sum_{k=1}^{N}\left(\sum_{i=0}^{d-1} \Upsilon_{k-i}\right)\Gamma_{k-d}|\varphi_{k-d}|^2\right) \\
&= O\left(\sum_{k=1}^{N}(\hat{v}_k - v_k)^2\right) \\
&\quad + o(N) \\
&\quad + O\left(N^{\frac{1}{2}}\sum_{k=1}^{N}\left(\sum_{i=0}^{d-1} \Upsilon_{k-i}\right)\Gamma_{k-d}\right).
\end{aligned}$$

Proposition 5.1 and (5.7) then imply that

$$\sum_{k=1}^{N}\Big(y_k - \frac{B_m}{A_m}z_{k-d}^* - \frac{C_m}{A_m}v_k\Big)^2 = O\Big(\log N + N^{\frac{1}{2}}(\log N)\sum_{k=1}^{N}\Gamma_{k-d}\Big) + o(N)$$
$$= o(N).$$

This expression may be equivalently written as

$$\lim_{N\to\infty}\frac{1}{N}\sum_{k=1}^{N}\Big(y_k - \frac{B_m}{A_m}z_{k-d}^* - \frac{C_m}{A_m}v_k\Big)^2 = 0,$$

which is (5.8).

6 Conclusions

In this paper we have presented several new techniques for constructing and analyzing adaptive controllers for linear systems.

These adaptive control laws have obvious generalizations to non-minimum phase systems: It appears that we may successfully construct a stabilizing indirect adaptive control law which does not rely on the solution of a Diophantine equation, even if the true system possesses unstable zeros. This is obvious in the non-adaptive case since in the model reference control design described in Section 3 it is entirely possible that the model parameters (A_m, B_m, C_m) may depend on the system parameters.

We also believe that a simple perturbation of the control law can be made, just as in the minimum phase case, so that stability and high performance can be guaranteed with minimal apriori knowledge.

Presently we are working towards such generalizations for both discrete and continuous time models, and we are also considering system models described by stochastic differential equations.

Acknowledgement This work supported in part by NSF grant # ECS 8910088 and University of Illinois Research Board grant # 1-2-69637. Lyndon Brown is supported in part by a Natural Science and Engineering Research Council of Canada Post Graduate Scholarship.

References

[1] P. E. CAINES, *Linear stochastic systems*, John Wiley & Sons, New York, NY, 1988.

[2] H. CHEN, *Recursive Estimation and Control for Stochastic Systems*, Wiley, New York, NY, 1985.

[3] L. GUO AND H. CHEN, *The Åström-Wittenmark self-tuning regulator revisited and ELS-based adaptive trackers*, IEEE Transactions on Automatic Control, 36 (1991), pp. 802–812.

[4] P. R. KUMAR, *Convergence of least-squares parameter estimate based adaptive control schemes*, IEEE Transactions on Automatic Control, 35 (1990), pp. 416–424.

[5] S. P. MEYN AND L. BROWN, *Adaptive control using a Kalman filter based estimation algorithm*, in Proceedings of the 29th Conference on Decision and Control, Honolulu, HI, December 1990, pp. 1432–1437.

[6] S. P. MEYN AND L. J. BROWN, *Model reference adaptive control of time varying & stochastic systems*. To appear in IEEE TAC, 1992.

[7] R. H. MIDDLETON AND G. C. GOODWIN, *Adaptive control of time-varying linear systems*, IEEE Transactions on Automatic Control, 33 (1988), pp. 150–155.

[8] R. H. MIDDLETON AND P. V. KOKOTOVIC, *Boundedness properties of simple indirect adaptive control systems*, in Proceedings of the 1991 American Control Conference, Boston, MA, June 1991, pp. 1216–1220.

[9] L. PRALY, S. LIN, AND P. R. KUMAR, *A robust adaptive minimum variance controller*, SIAM J. on Control and Optimization, 27 (1989), pp. 235–266.

[10] W. REN AND P. R. KUMAR, *Stochastic adaptive system theory: Recent advances and a reappraisal*. To appear in *Foundations of Adaptive Control*, edited by P. V. Kokotović, Springer-Verlag, 1991.

[11] K. S. SIN, G. C. GOODWIN, AND R. R. BITMEAD, *An adaptive d-step ahead predictor based on least squares*, IEEE, AC-25 (1980), pp. 1161–1164.

[12] K. S. TSAKALIS AND P. A. IOANNOU, *A new direct adaptive control scheme for time varying plants*, IEEE Transactions on Automatic Control, (1990).

Extended Least Squares Based Adaptive Control : Robustness Analysis*

Sanjeev M. Naik and P. R. Kumar
Coordinated Science Laboratory
University of Illinois, Urbana, IL 61801

Abstract

While the standard Least Squares parameter estimation algorithm is applicable for the control of linear stochastic systems in white noise, the Extended Least Squares algorithm is popularly used when the noise is colored. In this paper we examine whether this algorithm, designed to have good stochastic performance, is in fact robust to small unmodeled dynamics and bounded disturbances.

We establish robust boundedness of a one-step-ahead tracking adaptive control law based on a Weighted Extended Least-Squares type *non-interlaced* adaptation law with parameter projection. The nominal plant is assumed to be minimum phase.

1 Introduction

In this paper, we show that the popular Extended Least-Squares (ELS) based one-step ahead adaptive tracking algorithm is robust to both the presence of small unmodeled dynamics and violation of the stochasticity assumption on the noise. Specifically the noise is allowed to be any bounded sequence. The only modifications used are projection of the parameter estimates onto a convex, bounded set, and normalization by an "extended" regressor. The adaptive control algorithm is a weighted extended least-squares type *non-interlaced* adaptation law with projection. Assuming the <u>nominal</u> plant (i.e. without the small unmodeled dynamics) to be minimum-phase , we prove that all the signals in the closed-loop system are uniformly bounded.

In [2], Praly *et al.* have proved boundedness of an ELS algorithm which differs in two respects. First, it employs normalization of all the signals entering the update law by a specially constructed signal. The construction of this signal requires a priori knowledge about the nominal plant: namely the stability margin of the B polynomial. Also, it involves additional computation. Second, d *interlaced* algorithms are used, which further adds to the computational burden.

2 Problem Statement

Consider a class of systems of the form (z^{-1} is the unit-delay operator),

$$A(z^{-1})y(t) = z^{-d}B(z^{-1})u(t) + C(z^{-1})\omega(t), \tag{2.1}$$

where $A(z^{-1}) = 1 + a_1 z^{-1} + \ldots + a_{p'} z^{-p'}$, $B(z^{-1}) = b_0 + b_1 z^{-1} + \ldots + b_{q'} z^{-q'}$, and $C(z^{-1}) = 1 + c_1 z^{-1} + \ldots + c_{r'} z^{-r'}$, the coefficients being unknown. $B(z^{-1})$, $C(z^{-1})$ are assumed to have all their zeros within

*The research reported here has been supported in part by U.S.A.R.O. under Contract No. DAAL 03-91-G-0182, the JSEP under Contract No. N00014-90-J-1270, and by an International Paper Fellowship for the first author.

the open unit disk, and $\omega(t)$ is a white noise sequence. The true process under control, however, does not necessarily lie in the above class, but satisfies

$$A(z^{-1})y(t) = z^{-d}B(z^{-1})u(t) + C(z^{-1})w(t) + v'(t), \qquad (2.2)$$

where A, B are as earlier, $C(z^{-1}) = 1$, w is any bounded disturbance, and $v'(t)$ represents unmodeled dynamics, which is assumed to satisfy [1]

$$v'^2(t) \leq K_v m(t-1) + k_v, \qquad (2.3)$$

where $m(t)$ is defined by

$$m(t) = \sigma m(t-1) + K_y y^2(t) + K_u u^2(t) + K_3, m(0) > 0, \qquad (2.4)$$

with $0 < \alpha^{1+\delta} < \sigma < 1$, for some $0 \leq \delta < 1$ and $0 < \alpha < 1$ is such that all the zeros of $B(z^{-1})$ lie in the open disk $|z| < \alpha$.

It is desired to track a given bounded reference trajectory $y^*(t) \equiv r(t-d)$. Our aim is to study the behavior of an extended-least-squares (ELS) based one-step ahead adaptive tracking law designed under the assumption that the system belongs to the model class (2.1), when the process actually satisfies (2.2).

3 Parametrization and Adaptive Control Law

Define polynomials $F(z^{-1})$ (of degree $d-1$), and $G(z^{-1})$ through,

$$A(z^{-1})F(z^{-1}) + z^{-d}G(z^{-1}) = C(z^{-1}). \qquad (3.1)$$

Hence, by (2.2) and (3.1), we have $C[y(t) - y^*(t) - Fw(t)] = z^{-d}[BFu(t) + Gy(t) + (1-C)y^*(t+d) - y^*(t+d)] + Fv'(t)$. Define $\phi^T(t) := [u(t), \ldots, u(t-q+1), y(t), \ldots, y(t-p+1), y^*(t+d-1), \ldots, y^*(t+d-r)]^T$, $\theta := [\text{coefficients of } BF, \text{coefficients of } G, \text{coefficients of } (1-C)]^T$. We will assume that $\|\theta - \theta^0\| \leq M_0$ some θ^0, and that $b_0 \geq b_{min} > 0$, for some positive constants M and b_{min}. This yields

$$C[y(t) - y^*(t) - Fw(t)] = \phi^T(t-d)\theta - y^*(t) + Fv'(t). \qquad (3.2)$$

Motivated by this, and assuming $w(\cdot)$ to be a white noise sequence, one designs the certainty-equivalence adaptive control law

$$\phi^T(t)\hat{\theta}(t) = y^*(t+d), \qquad (3.3)$$

where $\hat{\theta}(t)$ is the estimate of θ at time t.

Assuming that the true process (2.2) is indeed in the class of systems (2.1), one usually uses the well-known ELS algorithm [1] to update the parameter estimate $\hat{\theta}(\cdot)$:

$$\hat{\theta}(t) = \hat{\theta}(t-1) + P(t)\phi(t-d)(y(t) - \phi^T(t-d)\hat{\theta}(t-1))$$
$$P(t)^{-1} = P(t-1)^{-1} + \phi(t-d)\phi^T(t-d).$$

A scheme to bound the condition number of the covariance matrix P is also generally employed.

Usually however, the true plant does not fit into the class of models (2.1). The robustness of ELS-based algorithms in this situation has received very little attention. Here, we establish the boundedness of all closed-loop signals using a *non-interlaced* weighted ELS algorithm where the signals entering the adaptation law need only be normalized by an "extended" regressor.

---

[1] All constants in this paper are positive, unless noted otherwise. K denotes a generic positive constant.

In [2], Praly et al. show that an *interlaced* ELS-type algorithm with parameter projection, where the signals entering the adaptation law are normalized by a specially constructed signal (which requires knowing the stability margin of the B-polynomial of the nominal plant), ensures bounded closed-loop signals.

We analyze the following parameter update law :

$$\widehat{\theta}''(t) = \widehat{\theta}(t-1) + a(t-1)P(t-1)\overline{\phi}(t-d)\overline{e}_a(t) \tag{3.4}$$

$$P'(t)^{-1} = \lambda(t-1)P(t-1)^{-1} + \overline{\phi}(t-d)\overline{\phi}^T(t-d), \tag{3.5}$$

$$\delta_1 I \geq P(-1) \geq \delta_0 I$$

$$P(t) = \begin{cases} P'(t) & \text{if } trace(P'(t)) \leq \delta_1 \\ P(t-1) & \text{otherwise}. \end{cases} \tag{3.6}$$

$$a(t-1) = [1 + \overline{\phi}^T(t-d)P(t-1)\overline{\phi}(t-d)]^{-1} \tag{3.7}$$

$$\widehat{\theta}'(t) = \widehat{\theta}''(t) + \max\{0, b_{\min} - \widehat{\theta}''_1(t)\}\frac{P_1(t)}{P_{11}(t)} \tag{3.8}$$

where $\quad \widehat{\theta}''_1(t) :=$ first component of the vector $\widehat{\theta}''(t)$ $\tag{3.9}$

$\qquad P_1(t) :=$ first column of the matrix $P(t)$ $\tag{3.10}$

and $\quad P_{11}(t) :=$ (1,1)th element of $P(t)$ $\tag{3.11}$

$$\widehat{\theta}(t) = \theta^0 + [\widehat{\theta}'(t) - \theta^0]\min\{1, \frac{M_0\delta_1}{\delta_0\|\widehat{\theta}'(t) - \theta^0\|}\} \tag{3.12}$$

$$e_a(t) = y(t) - \phi^T(t-d)\widehat{\theta}(t-1) \tag{3.13}$$

$$\overline{\phi}(t) = \frac{\phi(t)}{\sqrt{n(t)}}, \overline{e}_a(t) = \frac{e_a(t)}{\sqrt{n(t)}}, \text{ where} \tag{3.14}$$

$$n(t) = K_1\|\psi(t)\|^2 + K_2 \tag{3.15}$$

$$\psi(t) = [u(t),\ldots,u(t-q-d+2),y(t),\ldots,y(t-p-d+2),$$
$$y^*(t+d-1),\ldots,y^*(t+1-r)]^T, \tag{3.16}$$

together with the control law (3.3). We refer to $\psi(\cdot)$ as an "extended" regressor, and to $e_a(\cdot)$ as the "augmented error." Note that $\psi(t) \equiv \phi(t)$ in the unit delay case, i.e., when $d = 1$.

Furthermore, the sequence $\lambda(\cdot)$ is assumed to satisfy $0 < \frac{1}{2} + \epsilon < \lambda' \leq \lambda(t) \leq 1 - \frac{\delta_0}{K_1} < 1$ for some $0 < \epsilon < 1/2$. A common instance of this is $\lambda(t) \equiv \lambda$, $1/2 + \epsilon < \lambda \leq 1 - \delta_0/K_1$, which results in an exponentially weighted least-squares based algorithm.

Theorem 1.

The adaptive control law (3.3-3.16), when applied to the true process (2.2), ensures that all closed-loop signals are uniformly bounded.

The proof of this theorem will proceed through Sections 4-8. The essential idea is to construct a signal W which dominates all other signals, and which has a bounded growth-rate. This signal W is then shown to be bounded. The boundedness of all other signals then follows. The proof begins with some preliminary observations in Section 4. Next, in Section 5, a "switched system" is introduced to construct an intermediate dominating signal z (Such a "switched system" idea was originally used in [3] to study the robustness of a gradient-update based adaptive control algorithm, when applied to a unit-delay, minimum phase plant, to a restricted class of unmodeled dynamics.). In Section 6 we show that over certain finite intervals, z is comparable to the normalization n. Then, a "state error", e_c is defined and the model reference structure of the control law is exploited to show that e_c is generated by a stable system driven by the unmodeled dynamics, and a quantity depending on the parameter estimation error. Finally, the signal W is constructed from the signals z, and the state error e_c. A novel large-signal analysis over finite time-intervals is then shown to yield a contraction property for W. This combined with the bounded growth-rate of W then shows the boundedness of W, and hence of all closed-loop signals.

4 Preliminary Observations

Note that: (1) $\delta_0 I \le P(t) \le \delta_1 I$, $\forall t \ge 0$. (2) $\|\overline{\phi}(t)\|^2 \le 1/K_1$, and $\|P(t)\| \le \delta_1$ together imply $1 \ge a(t) \ge K_1/(\delta_1 + K_1)$, $\forall t \ge 0$. (3) The above scheme ensures $\|\widehat{\theta}(t)\| \le M$ where $M := \delta_1 M_0/\delta_0 + \|\theta^0\|$, and $\widehat{b}_0(t) \ge b_{\min} > 0$, $\forall t \ge 0$.

Next, define $e(t) := y(t) - y^*(t)$, the *output tracking error*. Applying the above-mentioned adaptive control law to the true process (2.2), using (3.2), and (3.3), and letting $\widetilde{\theta}(t) := \widehat{\theta}(t) - \theta$, we get

$$e(t) - Fw(t) = -\phi^T(t - d)\widetilde{\theta}(t - d) + Fv'(t),\tag{4.1}$$

Also, defining $v(t) = F(w(t) + v'(t))$, we have $y(t) = \phi^T(t - d)\theta + v(t)$, which gives, using (3.13),

$$e_a(t) = -\phi^T(t - d)\widetilde{\theta}(t - 1) + v(t).\tag{4.2}$$

It is also worth noting that since $e(t) = -\phi^T(t - d)\widetilde{\theta}(t - d) + v(t)$, we have

$$e_a(t) = s(t) + e(t),\text{ where } s(t) := \phi^T(t - d)(\widehat{\theta}(t - d) - \widehat{\theta}(t - 1)).\tag{4.3}$$

We note that the "augmented error" e_a is identical to the output tracking error e in the $d = 1$ case. Assume $|w(t)| \le K_w$. Using the definition of v, by (2.3), and (2.4) we get,

$$v^2(t) \le KK_v m(t - 1) + k_{vm}.\tag{4.4}$$

The following lemma will be useful. This result essentially states that the "difference" term $s(t)$ can be bounded in terms of the "augmented error" $e_a(t)$.

Lemma 1.

$$s^2(t) \le s_m^2(t) \ \le \ K \sum_{j=0}^{d-2} e_a^2(t - 1 - j),\tag{4.5}$$

$$\text{where } s_m(t) \ := \ \xi \sum_{j=0}^{d-2} |e_a(t - 1 - j)|/K_1,\tag{4.6}$$

$$\text{with } \xi \ := \ \sqrt{2\max\left((\delta_1/\delta_0)^2 - 1, 1\right)}.\tag{4.7}$$

Proof.

$$|s(t)| \ \le \ \sum_{j=0}^{d-2} \|\phi(t - d)\|\|\widehat{\theta}(t - 1 - j) - \widehat{\theta}(t - 2 - j)\|$$

$$\le \ \xi \sum_{j=0}^{d-2} \|\phi(t - d)\|\|\widehat{\theta}''(t - 1 - j) - \widehat{\theta}(t - 2 - j)\|$$

$$= \ \xi \sum_{j=0}^{d-2} \|\phi(t - d)\|\|a(t - 2 - j)P(t - 2 - j)$$

$$\cdot \overline{\phi}(t - d - 1 - j) \frac{e_a(t - 1 - j)}{\sqrt{n(t - 1 - j)}} \|$$

$$\leq \xi \sum_{j=0}^{d-2} |e_a(t - 1 - j)| / K_1 \,.$$

$\square$

5 Bounding signals

Define $z(\cdot)$ through the following "switched system":

$$
\begin{aligned}
z(t) &= I(t-1)(\sigma z(t-1) + K_y e_a^2(t) + K_u u^2(t-d) + K_3) \\
&\quad + (1 - I(t-1))(gz(t-1) + 2K_3)\,, \\
&\quad \text{where } 0 < \sigma < g < 1, z(0) > 0, \quad (5.1) \\
I(t-1) &= \quad 1 \quad \text{if } \sigma z(t-1) + K_y e_a^2(t) + K_u u^2(t-d) \\
&\qquad + K_3 \geq gz(t-1) + 2K_3, \\
&\qquad 0 \quad \text{otherwise}\,. \quad (5.2)
\end{aligned}
$$

Lemma 2.

(a) $m(t) \leq Kz(t) + K$.

(b) $z(t) \leq K_z z(t-1) + k_z$.

Proof.

(a) From the definition of $m(\cdot)$, we have

$$m(t) = \sigma^t m(0) + \sum_{j=0}^{t} \sigma^{t-j}[K_y y^2(j) + K_u u^2(j) + K_3]\,.$$

Note that $y^2(j) \leq 2e^2(j) + 2y^{*2}(j)$, and from the control law (3.3),

$$u(t) = \frac{[r(t) - \widehat{\theta}_u^T(t)\phi_u(t) - \widehat{\theta}_y^T(t)\phi_y(t) - \widehat{\theta}_{y1}(t)y(t) - \widehat{\theta}_{y^*}^T(t)\phi_{y^*}(t)]}{\widehat{b}_0(t)}\,, \quad (5.3)$$

where

$$
\begin{aligned}
\phi_u(t)^T &= (u(t-1), \dots, u(t-q+1))\,, \\
\phi_y(t)^T &= (y(t-1), \dots, y(t-p+1))\,,
\end{aligned}
$$

and

$$\phi_{y^*}(t)^T = (y^*(t+d-1), \dots, y^*(t+d-r))\,, \forall t \geq 0\,.$$

Also note that $\widehat{b}_0(t) \geq b_{min} > 0$, $\|\widehat{\theta}(t)\| \leq M$, and $y^{*2}(t) \leq k_{y^*} \ \forall t \geq 0$. The desired result then follows from Lemma 1, by using (5.3) repeatedly $(d-1)$ times, after noting that

$$z(t) \geq \sigma z(t-1) + K_y e_a^2(t) + K_u u^2(t-d) + K_3\,, \quad \text{which implies}$$

$$z(t) \geq \sigma^t z(0) + \sum_{j=0}^{t} \sigma^{t-j}[K_y e_a^2(t) + K_u u^2(j-d) + K_3]\,.$$

(b) The proof is straightforward and omitted. $\square$

Lemma 2(a) when combined with (4.4) gives

$$v^2(t) \leq KK_v z(t-1) + K \,, \forall t \geq 1\,. \quad (5.4)$$

Lemma 3.

$$u^2(t-d) \leq Kn(t-1) + K \qquad (5.5)$$

Proof. The control law (3.3) gives

$$u(t) = (r(t) - \hat{\theta}^T(t)\phi_I(t))/\hat{b}_0(t),$$

where $\phi_I(t)$ equals $\phi(t)$ except for the first component which is set to zero. The result then follows since $\|\phi_I(t)\|^2 \leq \|\psi(t+d-1)\|^2$. □

6 Comparing $n(t)$ and $z(t)$

First, note that we clearly have $n(t) \leq K_n z(t), \forall t \geq 0$.

Lemma 4.

(a) $I(t) = 1 \Rightarrow n(t) \geq K_{nz} z(t)$. (b) Consider a positive integer N'. Let t_1 be such that $I(t_1) = 1$. If $z(t) \geq L, \forall t \in [t_1 - N', t_1]$ where L is a large enough positive constant, then $\exists K_{vmax} > 0$ such that $\forall K_v \in [0, K_{vmax}]$, $n(t) > \delta(N')z(t), \forall t \in [t_1 - N', t_1]$, where $\delta(N') > 0$ is a positive constant which depends only on N'.

Proof.

(a) $I(t-1) = 1 \Rightarrow K_y e_a^2(t) + K_u u^2(t-d) \geq (g-\sigma)z(t-1) + K_3$. Hence, by Lemma 3, $K_y e_a^2(t) + Kn(t-1) \geq (g-\sigma)z(t-1) + (K_3 - K) =: \bar{g}z(t-1) + \overline{K} =: RHS$. If $n(t-1) \geq RHS/(2K)$, the result follows. So, suppose $n(t-1) \leq RHS/(2K)$. Then, $e_a^2(t) \geq RHS/(2K_y)$. Using (4.2) and (5.4), we get

$$
\begin{aligned}
2M^2\|\psi(t-1)\|^2 &\geq 2M^2\|\phi(t-d)\|^2 \geq e_a^2(t) - v^2(t) \\
&\geq [\bar{g}z(t-1) + \overline{K}]/(2K_y) - KK_v z(t-1) - K \\
&\geq (\bar{g}/(2K_y) - KK_v)z(t-1), \\
&\quad \text{assuming } \overline{K} \geq 2KK_y.
\end{aligned}
$$

(b) First we will bound the growth-rate of $n(t)/z(t)$, and then use this in a reversed time argument. Clearly, $z(t) \geq \sigma z(t-1), \forall t$. First note that

$$\|\psi(t)\|^2 = \sum_{j=0}^{q+d-2} u^2(t-j) + \sum_{j=0}^{p+d-2} y^2(t-j) + \sum_{j=1}^{r+d-1} y^{*2}(t+d-j).$$

Using the control law (3.3) gives,

$$
\begin{aligned}
u^2(t) &\leq Ky^2(t) + K\|\psi(t-1)\|^2 + K \\
&\leq K\|\psi(t-1)\|^2 + Kv^2(t) + K \\
\Rightarrow \|\psi(t)\|^2 &\leq K\|\psi(t-1)\|^2 + Kv^2(t) + K \\
\Rightarrow n(t) &\leq Kn(t-1) + Kv^2(t) + K.
\end{aligned}
$$

Using (5.4) gives $n(t) \leq Kn(t-1) + KK_v z(t-1) + K$. This gives

$$\frac{n(t)}{z(t)} \leq K\frac{n(t-1)}{z(t-1)} + KK_v + K/z(t).$$

$$\leq K\frac{n(t-1)}{z(t-1)} + (KK_v + K/L) =: K_a\frac{n(t-1)}{z(t-1)} + K_d$$

$$\Rightarrow \frac{n(t_1)}{z(t_1)} \leq K_a^{t_1-t'}[\frac{n(t')}{z(t')} + \sum_{j=t'+1}^{t_1} K_a^{t'-j}K_d], \forall t' \in [t_1 - N', t_1]$$

$$\leq K_a^{N'}[\frac{n(t')}{z(t')} + \frac{K_aK_d}{K_a-1}], \forall t' \in [t_1 - N', t_1]$$

$$\Rightarrow \frac{n(t')}{z(t')} \geq K_a^{-N'}\frac{n(t_1)}{z(t_1)} - \frac{K_aK_d}{K_a-1}$$

$$\geq K_a^{-N'}K_{nz} - \frac{K_aK_d}{K_a-1} =: \delta(N').$$

It should be noted that $\delta(N')$ can be made positive for any finite N' by restricting the size of the unmodeled dynamics to be small enough (i.e., making K_v small enough), and making L large enough. This concludes the proof. □

For later use, define $K_{nz}(N') := \min\{K_{nz}, \delta(N')\}$.

7 A Nonminimal System Representation

Recall from (4.1) that

$$y(t) = e(t) + r(t - d). \tag{7.1}$$

Furthermore, from the control law (3.3), we get, using (2.2), (3.1), (4.1), and the fact that $C(q^{-1}) = 1$,

$$\begin{aligned} u(t - d) &= -b'_1 u(t - d - 1) - b'_2 u(t - d - 2) - \ldots - b'_{q-1} u(t - d - q + 1) \\ &+ \frac{1}{b_0}(A(q^{-1})[e(t) - v(t) + r(t - d)]) \\ &- \frac{1}{b_0}(G(q^{-1})[w(t - d) + v'(t - d)]), \end{aligned} \tag{7.2}$$

where $b'_j := b_j/b_0$, $j = 1, \ldots, q - 1$.

From (7.1), and (7.2), defining $x(t) = [u(t - d), \ldots, u(t - d - q + 1), y(t), y(t - 1), \ldots, y(t - p)]$, we get

$$x(t) = A_c x(t - 1) + b_{c1}e(t) + b_{c2}(A(q^{-1})e(t)) + \ell(t), \tag{7.3}$$

where A_c is the following matrix, which is stable since $B(q^{-1})$ has all its roots strictly inside the unit circle.

$$A_c := \begin{bmatrix} J & 0 \\ 0 & H \end{bmatrix}$$

$$J = \begin{bmatrix} -b_1/b_0 \cdots & -b_{q-1}/b_0 \\ & & 0 \\ I & & \vdots \\ & & 0 \end{bmatrix}$$

$$H = \begin{bmatrix} 0 \cdots & 0 \\ & & 0 \\ I & & \vdots \\ & & 0 \end{bmatrix}$$

Furthermore, $b_{c1} := [0, \ldots, 0, 1, 0, \ldots, 0]^T$, $b_{c2} := [\frac{1}{b_0}, 0, \ldots, 0]^T$, and $\ell(t) := b_{c2}[(G(q^{-1})[w(t - d) + v'(t - d)]) + (A(q^{-1})[v(t) - r(t - d)])] + [0, \ldots, 0, 1, 0, \ldots, 0]^T r(t - d)$.

8 Ultimate Boundedness Analysis

Define $W(t) = k_e x^T(t) P x(t) + z(t)$ where $P = P^T > 0$ satisfies $A_c^T P A_c - P = -I$. Such a P exists since A_c is stable.

Now, $z(t) \leq g z(t-1) + K_y e_a^2(t) + K_u u^2(t-d) + 2K_3$, and $x(t) = A_c x(t-1) + b_{c1} e(t) + b_{c2}(A(q^{-1})e(t)) + \ell(t)$, which gives $W(t) \leq k_e [x^T(t-1)A_c^T + b_{c1}^T e(t) + b_{c2}^T(A(q^{-1})e(t)) + \ell^T(t)] P [A_c x(t-1) + b_{c1} e(t) + b_{c2}(A(q^{-1})e(t)) + \ell(t)] + g z(t-1) + K_y e_a^2(t) + K_u u^2(t-d) + 2K_3$.

This implies

$$
\begin{aligned}
W(t) \leq \ & k_e x^T(t-1)(P-I)x(t-1) \\
+ \ & 2k_e x^T(t-1)A_c^T P(b_{c1}e(t) + b_{c2}(A(q^{-1})e(t))) + 2k_e x^T(t-1)A_c^T P\ell(t) \\
+ \ & 2k_e b_{c1}^T P\ell(t)e(t) + 2k_e b_{c2}^T P\ell(t)(A(q^{-1})e(t)) \\
+ \ & k_e b_{c1}^T P b_{c1} e^2(t) + k_e \ell^T(t) P\ell(t) + k_e b_{c2}^T P b_{c2}(A(q^{-1})e(t))^2 \\
+ \ & g z(t-1) + K_y e_a^2(t) + K_u u^2(t-d) + 2K_3 \, .
\end{aligned}
$$

Now,

$$
\begin{aligned}
|u(t-d)| \leq \ & \tfrac{1}{b_{min}}(|\hat{\theta}_u^T(t-d)\phi_u(t-d)| + |\hat{\theta}_y^T(t-d)\phi_y(t-d)| \\
& + |\hat{\theta}_{y1}y(t-d)| + |\hat{\theta}_{y^*}^T(t-d)\phi_{y^*}(t-d)| + |r(t-d)|) \\
\leq \ & K\|x(t-1)\| + K|y(t-d)| + K \\
\leq \ & K\|x(t-1)\| + K \, .
\end{aligned}
$$

Also, since $|e(t)| \leq |e_a(t)| + |s(t)|$ and by Lemma 1, $|s(t)| \leq s_m(t)$, we have, after letting $\gamma_1 := \|A_c^T P b_{c1}\|$,

$$
\begin{aligned}
|2k_e x^T(t-1)A_c^T P b_{c1} e(t)| \leq \ & 2\gamma_1 k_e \|x(t-1)\|\|e(t)\| \\
\leq \ & 2\gamma_1 k_e \|x(t-1)\|(|e_a(t)| + s_m(t)) \\
\leq \ & \gamma_1 k_e[(\epsilon_1 + \epsilon_2)\|x(t-1)\|^2 \\
& + \tfrac{1}{\epsilon_1}e_a^2(t) + \tfrac{1}{\epsilon_2}s_m^2(t)] \, .
\end{aligned}
$$

Similarly, after letting $\gamma_2 := \|A_c^T P b_{c2}\|$, by (4.3) we have

$$
\begin{aligned}
|2k_e x^T(t-1)A_c^T P b_{c2}(A(q^{-1})e(t))| \leq \ & 2\gamma_2 k_e \|x(t-1)\|\|e(t)\| \\
\leq \ & 2\gamma_2 k_e \|x(t-1)\|(|A(q^{-1})e_a(t)| + |A(q^{-1})s(t)|) \\
\leq \ & \gamma_2 k_e[(\epsilon_3 + \epsilon_4)\|x(t-1)\|^2 \\
& + \tfrac{1}{\epsilon_3}(A(q^{-1})e_a(t))^2 + \tfrac{1}{\epsilon_4}(A(q^{-1})s(t))^2] \, .
\end{aligned}
$$

Next, after defining $\gamma_3 := \|A_c^T P\|$, we get,

$$
\begin{aligned}
|2k_e e_c^T(t-d)A_c^T P\ell(t)| \leq \ & 2\gamma_3 k_e \|e_c(t-d)\|\|\ell(t)\| \\
\leq \ & \gamma_3 k_e(\epsilon_5 \|e_c(t-d)\|^2 + \tfrac{1}{\epsilon_5}\|\ell(t)\|^2) \, ,
\end{aligned}
$$

and defining $\gamma_4 := \|P b_{c1}\|$, $\gamma_5 := \|P b_{c2}\|$, we get

$$
\begin{aligned}
|2k_e b_{c1}^T P\ell(t)e(t)| \leq \ & 2k_e \gamma_4 \|\ell(t)\|(|e_a(t)| + s_m(t)) \\
\leq \ & k_e \gamma_4((\epsilon_6 + \epsilon_7)\|\ell(t)\|^2 + \tfrac{1}{\epsilon_6}e_a^2(t) + \tfrac{1}{\epsilon_7}s_m^2(t)) \, ,
\end{aligned}
$$

and

$$
\begin{aligned}
|2k_e b_{c2}^T P\ell(t)(A(q^{-1})e(t))| \leq \ & 2k_e \gamma_5 \|\ell(t)\|(|A(q^{-1})e_a(t)| + |A(q^{-1})s(t)|) \\
\leq \ & k_e \gamma_5((\epsilon_8 + \epsilon_9)\|\ell(t)\|^2 + \tfrac{1}{\epsilon_8}(A(q^{-1})e_a(t))^2 + \tfrac{1}{\epsilon_9}(A(q^{-1})s(t))^2) \, .
\end{aligned}
$$

Finally, pick $0 < \gamma < 1$, such that $\gamma > \max\left\{1 - \frac{1}{\lambda_{\max}(P)}, g\right\}$, and ϵ_1, ϵ_2, ϵ_3, ϵ_4, ϵ_5 small enough, and k_e large enough so that $-1 + (1-\gamma)\lambda_{\max}(P) + (\epsilon_1 + \epsilon_2)\gamma_1 + (\epsilon_3 + \epsilon_4)\gamma_2 + \epsilon_5\gamma_3 + \frac{K}{k_e} \le 0$. This gives

$$W(t) \le \gamma W(t-1) + K_{ea}e_a^2(t) + K_{ea2}\sum_{j=0}^{q} e_a^2(t-j) + K_s s_m^2(t) + K_{s2}\sum_{j=0}^{q} s_m^2(t-j) + K_\ell\|\ell(t)\|^2 + 2K_3, \quad (8.1)$$

where $K_{ea} = [k_e(\frac{\gamma_1}{\epsilon_1} + \frac{\gamma_4}{\epsilon_6} + b_{c1}^T P b_{c1}) + K_y]$, $K_{ea2} = k_e(q+1)[\frac{\gamma_1}{\epsilon_3} + \frac{\gamma_5}{\epsilon_8} + b_{c2}^T P b_{c2}]$, $K_s = [k_e(\frac{\gamma_1}{\epsilon_2} + \frac{\gamma_4}{\epsilon_7})]$, $K_{s2} = k_e(q+1)[\frac{\gamma_2}{\epsilon_4} + \frac{\gamma_5}{\epsilon_9}]$, and $K_\ell = \left(\frac{\gamma_3}{\epsilon_5} + \gamma_4(\epsilon_6 + \epsilon_7) + \gamma_5(\epsilon_8 + \epsilon_9) + \lambda_{\max}(P)\right)k_e$.

From (8.1), since $z(t) \le W(t)$, $z(t) \ge \sigma z(t-1)$, and by Lemma 1, we have,

$$
\begin{aligned}
W(t) &\le (\gamma + K_{ea}\frac{e_a^2(t)}{z(t-1)} + K_{ea2}\sum_{j=0}^{q}\sigma^{-j+1}\frac{e_a^2(t-j)}{z(t-j)} \\
&\quad + KK_s\sum_{j=0}^{d-2}\sigma^{-j}\frac{e_a^2(t-1-j)}{z(t-1-j)} \\
&\quad + KK_{s2}\sum_{j=0}^{q}\sum_{i=0}^{d-2}\sigma^{-j-i}\frac{e_a^2(t-1-j-i)}{z(t-1-j-i)} \\
&\quad + K_\ell\frac{\|\ell(t)\|^2}{z(t-1)})W(t-1) + 2K_3 \\
&=: \ g(t)W(t-1) + 2K_3.
\end{aligned}
\quad (8.2)
$$

Lemma 5.

$$W(t) \le K_{wz}z(t) + k_{wz}.$$

Proof. This follows from directly from (5.4) using the fact that $y(t) = \phi^T(t-d)\theta + v(t)$, since $W(t) = k_e x^T(t)Px(t) + z(t)$, and $\|x(t)\|^2 \le y^2(t) + \sum_{j=0}^{q-1}u^2(t-d-j) + \sum_{j=1}^{p}y^2(t-j)$. $\square$

Corollary 6.

$$W(t) \le KW(t-1) + K.$$

Proof. This follows from Lemma 2(b) and Lemma 5. $\square$

Lemma 7.

Consider a time interval $[a,b]$ such that $W(t) \ge 2K_{wz}L$, $\forall t \in [a-d+1, b]$. Then for L large enough, (i) $z(t) \le W(t) \le 2K_{wz}z(t)$. (ii) If $I(t-1) = 0$, $\forall t \in [a,b]$, then $W(b) \le 2K_{wz}(g^{b-a} + \frac{2K_3}{(1-g)L})W(a)$. (iii) If for each $t \in [a,b]$ such that $I(t) = 0$, $\exists\, n' \in [0,N]$ such that $I(t+n') = 1$, then for some $0 < \beta < 1$, $W(b) \le K\exp[-\beta(b-a)]\left[1 + \frac{K}{L\gamma^{b-a}}\right]W(a)$.

Proof.

(i) This follows by choosing L large enough.

(ii) By (5.1), for $I(t) = 0$, $z(t) = gz(t-1) + 2K_3$, which implies that $z(b) \le g^{b-a}z(a) + 2K_3\sum_{j=a+1}^{b}g^{b-j} \le [g^{b-a} + \frac{2K_3}{(1-g)L}]W(a)$. Then using (i) completes the proof.

(iii) From (8.2), we have

$$
\begin{aligned}
\frac{W(b)}{W(a)} &\le \left(\prod_{j=a+1}^{b}g(j)\right)\left[1 + \frac{2K_3}{W(a)}\sum_{t=a+1}^{b}\left(\prod_{j=a+1}^{t}g(j)\right)^{-1}\right] \\
&\le \exp\left[\sum_{j=a+1}^{b}\ln g(j)\right]\left[1 + \frac{K_3}{K_{wz}L}\sum_{t=a+1}^{b}(\frac{1}{\gamma})^{t-a}\right]
\end{aligned}
\quad (8.3)
$$

Now,

$$\sum_{t=a+1}^{b} \ln g(t) \leq \sum_{t=a+1}^{b} g(t) - (b-a).$$

By (8.2), this gives

$$
\begin{aligned}
\sum_{t=a+1}^{b} \ln g(t) \leq \ & -(1-\gamma)(b-a) + K_{ea}\sum_{t=a+1}^{b}\frac{e_a^2(t)}{z(t-1)} \\
& + KK_{ea2}\sum_{j=0}^{q}\sigma^{-j+1}\sum_{t=a+1}^{b}\frac{e_a^2(t-j)}{z(t-j)} \\
& + KK_s\sum_{j=0}^{d-2}\sigma^{-j}\sum_{t=a+1}^{b}\frac{e_a^2(t-1-j)}{z(t-1-j)} \\
& + KK_{s2}\sum_{j=0}^{q}\sum_{i=0}^{d-2}\sigma^{-j-i}\sum_{t=a+1}^{b}\frac{e_a^2(t-1-j-i)}{z(t-1-j-i)} + K_\ell\sum_{t=a+1}^{b}\frac{\|\ell(t)\|^2}{z(t-1)}.
\end{aligned}
\tag{8.4}
$$

Since $z(t) \geq \sigma z(t-1)$, and since $z(t) \leq 2K_z z(t-1)$ for $z(t-1) \geq L$, we get,

$$
\begin{aligned}
\sum_{t=a+1}^{b} \ln g(t) \leq \ & -(1-\gamma)(b-a) + 2K_{ea}K_zK_n\sum_{t=a+1}^{b}\frac{e_a^2(t)}{n(t)} \\
& + KK_sK_n\sum_{j=0}^{d-2}\sigma^{-j}\sum_{t=a+1}^{b}\frac{e_a^2(t-1-j)}{n(t-1-j)} \\
& + KK_{ea2}K_n\sum_{j=0}^{q}\sigma^{-j+1}\sum_{t=a+1}^{b}\frac{e_a^2(t-j)}{n(t-j)} \\
& + KK_{s2}K_n\sum_{j=0}^{q}\sum_{i=0}^{d-2}\sigma^{-j-i}\sum_{t=a+1}^{b}\frac{e_a^2(t-1-j-i)}{n(t-1-j-i)} \\
& + \left[KK_v + \frac{K}{L}\right](1+\sigma^{-(d-1)})(b-a)
\end{aligned}
\tag{8.5}
$$

Now, by Lemma A.2 and Lemma 4,

$$\sum_{t=a+1}^{b}\frac{e_a^2(t)}{n(t)} \leq KM^2 + \frac{K(b-a)}{K_{nz}(N)}\left[KK_v + \frac{K}{L}\right].\tag{8.6}$$

Using (8.5) and (8.6), we get (upon choosing K_v small enough and L large enough),

$$\sum_{t=a+1}^{b}\ln g(t) \leq KM^2 - \beta(b-a), \quad \text{some } 0 < \beta < 1,$$

so that from (8.3),

$$\frac{W(b)}{W(a)} \leq \exp[-\beta(b-a)]\left[1 + \frac{K}{K_{wz}L\gamma^{b-a}(1-\gamma)}\right]\exp(KM^2).$$

$\square$

Contraction Lemma.

Pick $0 < \gamma^* < 1$. Then $\exists\, N, L$ large enough and K_{vmax} small enough, so that if $W(t) \geq 2K_{wz}L$, $\forall t \in [\ell-d, \ell+2N]$, then $\forall K_v \in [0, K_{vmax}]$,

$$W(\ell+2N) \leq \gamma^* W(\ell-d).$$

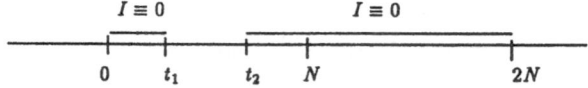

Figure 1: Illustration of Case 2

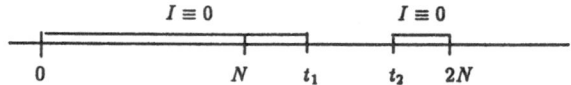

Figure 2: Illustration of Case 3

Proof. There are four cases:

Case 1: $(I(t) = 0, \forall t \in [\ell, \ell + 2N])$

From Lemma 7 (ii),

$$W(\ell + 2N) \leq 2K_{wz}\left(g^{2N} + \frac{2K_3}{(1-g)L}\right)W(\ell),$$

and by Corollary 6, $W(\ell) \leq KW(\ell-d)$. Hence, for N, L large enough and $K_{v\max}$ appropriately small, $W(\ell+2N) \leq \gamma^* W(\ell)$. Define $t_1 = \min\{t \in [0, 2N] : I(t+\ell) = 1\}$, $t_2 = \max\{t \in [0, 2N] : I(t+\ell) = 1\}$.

Case 2: $(0 \leq t_1 \leq t_2 \leq N)$

Using Lemma 7 (i) and (ii), we have

$$
\begin{aligned}
W(\ell + 2N) &\leq 2K_{wz}\left(g^{2N-t_2} + \frac{2K_3}{(1-g)L}\right)W(\ell + t_2) \\
&\leq 2K_{wz}\left(g^N + \frac{2K_3}{(1-g)L}\right)Ke^{-\beta t_2} \\
&\quad \cdot \left(1 + \frac{K}{L\gamma^{t_2}}\right)W(\ell) \\
&\leq 2K_{wz}K\left(g^N + \frac{2K_3}{(1-g)L}\right)\left(1 + \frac{K}{L\gamma^N}\right)W(\ell).
\end{aligned}
$$

Since $W(\ell) \leq KW(\ell - d)$ by Corollary 6, we have

$$W(\ell + 2N) \leq \gamma^* W(\ell - d),$$

where the last inequality holds for N, L large enough and $K_{v\max}$ appropriately small.

Case 3: $(N < t_1 < t_2 \leq 2N)$

Using Lemma 7 (i) and (ii), we have

$$
\begin{aligned}
W(\ell + 2N) &\leq 2K_{wz}\left(g^{2N-t_2} + \frac{2K_3}{(1-g)L}\right)W(\ell + t_2), \\
W(\ell + t_2) &\leq K\exp\left[-\beta(t_2 - N)\right]\left[1 + \frac{K}{L\gamma^{t_2-N}}\right]W(\ell + N),
\end{aligned}
$$

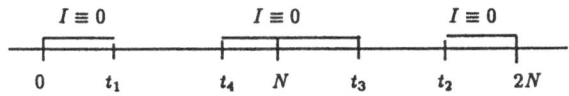

Figure 3: Illustration of Case 4

and

$$W(\ell + N) \le 2K_{wz}\left(g^N + \frac{2K_3}{(1-g)L}\right)W(\ell),$$

which implies

$$W(\ell + 2N) \le 4KK_{wz}^2\left(g^N + \frac{2K_3}{(1-g)L}\right)$$
$$\cdot\left(1 + \frac{2K_3}{(1-g)L}\right)\left(1 + \frac{K}{L\gamma^N}\right)W(\ell).$$

Since $W(\ell) \le KW(\ell - d)$ by Corollary 6, we have

$$W(\ell + 2N) \le \gamma^* W(\ell - d),$$

where the last inequality holds for N, L large enough and $K_{v\max}$ appropriately small.

Case 4: $(0 \le t_1 \le N < t_2 \le 2N)$

Define $t_3 := \min\{t \in [N, 2N] : I(t + \ell) = 1\}$, and $t_4 := \max\{t \in [0, N] : I(t + \ell) = 1\}$.

Case 4(a) $(t_3 - t_4 \le N)$

Using Lemma 7 (i) and (ii), we have

$$W(\ell + 2N) \le 2K_{wz}\left(g^{2N - t_2} + \frac{2K_3}{(1-g)L}\right)W(\ell + t_2),$$

and

$$W(\ell + t_2) \le K \exp\left[-\beta t_2\right]\left(1 + \frac{K}{L\gamma^{t_2}}\right)W(\ell),$$

which gives

$$W(\ell + 2N) \le 2KK_{wz}\left(1 + \frac{2K_3}{(1-g)L}\right)\exp(-\beta N)$$
$$\cdot\left(1 + \frac{K}{L\gamma^{2N}}\right)W(\ell).$$

Since $W(\ell) \le KW(\ell - d)$ by Corollary 6, we have

$$W(\ell + 2N) \le \gamma^* W(\ell - d),$$

where the last inequality holds for N, L large enough and $K_{v\max}$ appropriately small.

Case 4(b) $(t_3 - t_4 > N)$.

Using Lemma 7 (i) and (ii), we have

$$
\begin{aligned}
W(\ell + 2N) &\leq 2K_{wx}\left(g^{2N-t_2} + \frac{2K_3}{(1-g)L}\right)W(\ell + t_2), \\
W(\ell + t_2) &\leq K\exp(-\beta(t_2 - t_3))\left(1 + \frac{K}{L\gamma^{t_2-t_3}}\right)W(\ell + t_3), \\
W(\ell + t_3) &\leq 2K_{wx}\left(g^{t_3-t_4} + \frac{2K_3}{(1-g)L}\right)W(\ell + t_4),
\end{aligned}
$$

and

$$
W(\ell + t_4) \leq K\exp(-\beta t_4)\left(1 + \frac{K}{L\gamma^{t_4}}\right)W(\ell),
$$

which gives

$$
\begin{aligned}
W(\ell + 2N) &\leq 4K^2 K_{wx}^2\left(g^N + \frac{2K_3}{(1-g)L}\right) \\
&\quad \cdot \left(1 + \frac{K}{L\gamma^N}\right)^2\left(1 + \frac{2K_3}{(1-g)L}\right)W(\ell).
\end{aligned}
$$

Since $W(\ell) \leq KW(\ell - d)$ by Corollary 6, we have

$$
W(\ell + 2N) \leq \gamma^* W(\ell - d),
$$

where the last inequality holds for N, L large enough and $K_{v\max}$ appropriately small.

$\square$

Proof of Theorem 1. By Corollary 6, W has a bounded growth rate. The Contraction Lemma then proves that W is uniformly bounded. Since W bounds all other signals (through z and x), we conclude that all closed-loop signals are bounded. $\square$

9 Concluding Remarks

In this paper, we have proved uniform boundedness of an ELS-based adaptive tracking scheme subject to unmodeled dynamics, when the stochasticity assumption on the noise is violated. The nominal plant, without unmodeled dynamics, is assumed to be of minimum phase. We believe that to obtain good performance, a primary goal of adaptive control (as in tracking), it is necessary to design control laws on the basis of stochastic considerations. However, it is important to show that such algorithms are robust in a deterministic sense, as we have done in this paper.

Certain modifications to the adaptive law have been employed to achieve this; a weighted ELS update law is used where the signals entering the update were normalized by the norm of an extended regressor (and not by any specially constructed signal), and the parameter estimates are projected onto a compact set. We believe that it would be difficult to ensure boundedness without either of these two modifications.

The results presented in this paper extend the work reported in [4], where the robust boundedness of adaptively controlled continuous time plants in the presence of unmodeled dynamics and bounded disturbances is proved.

References

[1] G. C. Goodwin and K. S. Sin. Adaptive Filtering, Prediction and Control. Prentice-Hall, Inc., Englewood Cliffs, N.J., 1984.

[2] L. Praly, S. -F. Lin and P. R. Kumar. A robust adaptive minimum variance controller. *SIAM J. Control and Optimizn.*, Vol.27, No.2, pp. 235-266, 1989.

[3] B. E. Ydstie. Stability of discrete MRAC - revisited. *Systems & Control Letters*, vol. 13, pp. 429-438, 1989.

[4] S. M. Naik, P. R. Kumar, and B. E. Ydstie. "Robust continuous time adaptive control by parameter projection." *IEEE Transactions on Automatic Control*, Vol. 37, February 1992.

10 Appendix A

Lemma A.1 (Filtering).

Consider $w_{out}(t) = H(z^{-1})w_{in}(t)$, where $H(z^{-1})$ has all its poles within $|z| < \alpha$, where $\alpha^{1+\delta} < \sigma$, for some $0 \le \delta < 1$. If $w_{in}^2(t) \le Km(t) + K$, then $w_{out}^2(t) \le Km(t) + K$.

Proof. Let $h(t)$ denote the impulse response of $H(z^{-1})$. Then, letting $\alpha' := \alpha^{1-\delta}$, and $\alpha'' := \alpha^{1+\delta}$,

$$
\begin{aligned}
w_{out}^2(t) &= [\sum_{j=0}^{t} h(j)w_{in}(t-j)]^2 \le K(\sum_{j=0}^{t} \alpha^j |w_{in}(t-j)|)^2 \\
&= K(\sum_{j=0}^{t} \alpha'^{j/2}\alpha''^{j/2} |w_{in}(t-j)|)^2 \le K\sum_{k=0}^{t} \alpha''^k \sum_{j=0}^{t} \alpha'^j w_{in}^2(t-j) \\
&\le K\sum_{j=0}^{t} \alpha'^j m(t-j) + K
\end{aligned}
$$

Finally using $m(t) \ge \sigma^j m(t-j)$, we get,

$$
w_{out}^2(t) \le K\sum_{j=0}^{t} (\frac{\alpha'}{\sigma})^j m(t) + K,
$$

thereby concluding the proof. □

Lemma A.2.

Let $V(t) = \tilde{\theta}(t)^T P(t)^{-1}\tilde{\theta}(t)$. Then, $V(t) - V(t-1) \le -\beta_1 \bar{e}_a(t)^2 + \beta_2 \bar{v}(t)^2$ where

$$
\beta_1 = \frac{K_1}{(\delta_1 + K_1)} \min\left\{ \frac{(\delta_1 + 2K_1)}{(1+K_1)}\left(\lambda' - \frac{1}{2} - \epsilon\right), \frac{(\delta_1 + 2K_1)}{(1+K_1)} - \epsilon \right\},
$$

$$
\beta_2 = \max\left\{ 1 + \frac{(\delta_1 + K_1)(1-\lambda')^2}{K_1}\frac{1}{2\epsilon}, \frac{1}{\epsilon} \right\}.
$$

Proof. Define $\tilde{\theta}''(t) := \hat{\theta}''(t) - \theta$, and $\tilde{\theta}'(t) := \hat{\theta}'(t) - \theta$. Then, from (3.4) we have,

$$P(t)^{-1}\tilde{\theta}''(t) = P(t)^{-1}\tilde{\theta}(t-1) + a(t-1)P(t)^{-1}P(t-1)\overline{\phi}(t-d)\bar{e}_a(t).$$

There are two cases to consider.

<u>Case 1.</u> When $P(t) = P'(t)$, from (3.5) we have

$$
\begin{aligned}
P(t)^{-1}\tilde{\theta}''(t) &= \lambda(t-1)P(t-1)^{-1}\tilde{\theta}(t-1) + (\bar{v}(t) - \bar{e}_a(t))\overline{\phi}(t-d) \\
&+ \lambda(t-1)a(t-1)\bar{e}_a(t)\overline{\phi}(t-d) + (1 - a(t-1))\bar{e}_a(t)\overline{\phi}(t-d)
\end{aligned}
$$

So, $P(t)^{-1}\tilde{\theta}''(t) = \lambda(t-1)P(t-1)^{-1}\tilde{\theta}(t-1) + \{(\lambda(t-1)-1)a(t-1)\bar{e}_a(t) + \bar{v}(t)\}\overline{\phi}(t-d)$, which gives

$$\tilde{\theta}''^T(t)P(t)^{-1}\tilde{\theta}''(t) - V(t-1)$$
$$\leq \tilde{\theta}''(t)^T P(t)^{-1}\tilde{\theta}''(t) - \lambda(t-1)\tilde{\theta}(t-1)^T P(t-1)^{-1}\tilde{\theta}(t-1)$$
$$= \lambda(t-1)(\tilde{\theta}''(t) - \tilde{\theta}(t-1))^T P(t-1)^{-1}\tilde{\theta}(t-1)$$
$$+ \{(\lambda(t-1)-1)a(t-1)\bar{e}_a(t) + \bar{v}(t)\}\left[(\tilde{\theta}(t-1)^T\overline{\phi}(t-d))\right.$$
$$\left. + a(t-1)(\overline{\phi}(t-d)^T P(t-1)\overline{\phi}(t-d))\bar{e}_a(t)\right].$$

This implies

$$\tilde{\theta}''^T(t)P(t)^{-1}\tilde{\theta}''(t) - V(t-1)$$
$$\leq \lambda(t-1)a(t-1)(\bar{v}(t) - \bar{e}_a(t))\bar{e}_a(t)$$
$$+ \{[\lambda(t-1)-1]a(t-1)\bar{e}_a(t) + \bar{v}(t)\}[\bar{v}(t) - a(t-1)\bar{e}_a(t)]$$
$$\leq -[a(t-1)((1+a(t-1))\lambda(t-1) - a(t-1)) - \delta^2]\bar{e}_a(t)^2$$
$$+ \left(1 + \frac{a(t-1)^2(1-\lambda(t-1))^2}{\delta^2}\right)\bar{v}(t)^2,$$

the last inequality following after some algebraic manipulations, where $\delta^2 = \alpha_{min}(1 + \alpha_{min})\epsilon$, some $0 < \epsilon < 1/2$, where $\alpha_{min} := K_1/(\delta_1 + K_1)$. Then, since

$$a(t-1)((1+a(t-1))\lambda(t-1) - a(t-1)) - \delta^2$$
$$\geq \frac{K_1(\delta_1 + 2K_1)}{(\delta_1 + K_1)^2}(\lambda' - 1/2 - \epsilon), \text{ and}$$
$$\left(1 + \frac{a(t-1)^2[1-\lambda(t-1)]^2}{\delta^2}\right) \leq 1 + \frac{(1-\lambda')^2}{\epsilon}\frac{\delta_1 + K_1}{K_1},$$

we get $\tilde{\theta}''^T(t)P(t)^{-1}\tilde{\theta}''(t) - V(t-1) \leq -\beta_1 \bar{e}_a(t)^2 + \beta_2 \bar{v}(t)^2$.

<u>Case 2.</u> When $P(t) = P(t-1)$, we have,

$$P(t)^{-1}\tilde{\theta}''(t) = P(t-1)^{-1}\tilde{\theta}(t-1) + a(t-1)\overline{\phi}(t-d)\bar{e}_a(t).$$

This implies,

$$\tilde{\theta}''^T(t)P(t)^{-1}\tilde{\theta}''(t) - \tilde{\theta}^T(t-1)P(t-1)^{-1}\tilde{\theta}(t-1)$$
$$= a(t-1)[2\bar{v}(t)\bar{e}_a(t) - [1 + a(t-1)]\bar{e}_a^2(t)]$$
$$\leq -\frac{K_1}{\delta_1 + K_1}[\frac{\delta_1 + 2K_1}{\delta_1 + K_1} - \epsilon]\bar{e}_a^2(t) + \frac{1}{\epsilon}\bar{v}^2(t).$$

So, in either case, we have $\tilde{\theta}''^T(t)P(t)^{-1}\tilde{\theta}''(t) - V(t-1) \leq -\beta_1\bar{e}_a(t)^2 + \beta_2\bar{v}(t)^2$.

The remainder of the proof is similar to that of Lemma 3.2 of [2], and hence will be omitted. □

ON ROBUST SPECTRUM ASSIGNABILITY:
THERE ARE LIMITS TO POLE PLACEMENT THEOREM
FOR SYSTEMS WITH INFINITESIMAL PERTURBATIONS

Andrzej W. Olbrot
Department of Electrical and Computer Engineering
Wayne State University, Detroit, MI 48202

ABSTRACT
It is proved that, by using state feedback, it is not possible to place the poles of a linear time invariant system to the left of a certain critical negative real number if the system matrix is subject to uncertainties (perturbations) of a given (small) magnitude.

INTRODUCTION

The well-known pole placement theorem states that for any controllable linear finite dimensional time invariant system

(1) $d/dt\, x(t) = Ax(t) + Bu(t)$

there exists a linear constant state feedback

(2) $u(t) = Kx(t)$

such that the poles of the closed loop system (that is, the eigenvalues of the matrix $A + BK$) are located at arbitrarily preassigned positions [2]. In particular, the spectrum of the feedback system can be assigned in an arbitrary left half plane or, in other words, the solutions to (1) and (2) will decay to zero faster than a preassigned exponential function if the feedback matrix is chosen appropriately.

However, there are no results in the literature answering, at least partially, the following important question:

What regions in the complex plane are (robustly) spectrally assignable in the sense that the closed loop poles belong to an assigned region independent of all perturbations of system equations of a given class.

It is known that the power of feedback can overcome many types of uncertainties in the real system which means, mathematically, that the feedback system can maintain good performance, in particular: stability, under various kinds of perturbations of the mathematical model. In fact, for one dimensional systems (1), (2) even very large perturbations can be dominated by a sufficiently strong feedback in such a way that the exponential stability of any given decay rate is preserved (E.g., this is true when A and B vary in finite or infinite intervals such that the upper bound for A is known and B is nonzero). In general, it is very difficult to characterize spectrally assignable regions even for very simple perturbation models. However, it is possible to prove that, for systems of order two or more, the regions of primary interest in control, namely, left half planes, are not robustly assignable under infinitesimal perturbations. This is in sharp contrast with the case of first order systems and with the intuitive expectations stemming from the famous eigenvalue assignment theorem and the practical power of feedback. It would be reasonable to expect that for sufficiently small perturbations the eigenvalues of the feedback system can be shifted, by a sufficiently strong feedback, to any prespecified left half plane. However, this is not so, as described below, and this fact requires demystification of the famous pole placement theorem, already so popular in various undergraduate textbooks.

In a recent paper, [1], it was proved that there exists an unassignable left halfplane for systems with small perturbations in system matrices (A, B).

More precisely, assume (A_0, B_0) is controllable and let

$$(3) \qquad S = \{(A, B): \ d((A, B),(A_0, B_0)) < \epsilon_0\}$$

denote a family of system perturbations described by a fixed positive number ϵ_0 and a given distance function $d(.,.)$ in a suitable space of matrix pairs. The corresponding theorem is as follows:

THEOREM 1, [1]

Consider a family (3) corresponding to system (1) with one input and at least two state variables. Then for any $\epsilon_0 > 0$ (no matter how small) there exists a positive number r such that for any feedback matrix K placing the eigenvalues of $A_0 + B_0K$ to the left of $Re(s)$ $= -r$ there exists a pair (A,B) in the set S and an eigenvalue s_0 of $A + BK$ such that $Re(s_0) > -r$. Moreover, the pair (A,B) can be chosen of form (A_0, B), that is, with the perturbed matrix B_0 and not A_0.

General formulas for calculating the critical value of $r = r(\epsilon_0, A_0, B_0)$ are difficult to obtain. Some estimates, however, can be obtained rather easily. For instance, from [1] it follows that if

$$r^n > max ((det(-A_0), \epsilon_0^{-n/(n-1)})$$

then the left half plane $Re(s) < -r$ is not robustly assignable. The value of $det(A_0)$, equal to the product of all eigenvalues, affects the estimate of r only in case of stable systems (E.g., if all eigenvalues are real and equal $-s_0 < 0$ we get a natural inequality $r > s_0$).

For illustration, consider the following

EXAMPLE

Take the following two dimensional system

$$\frac{d}{dt} x_1 = x_2 - \epsilon u$$
$$\frac{d}{dt} x_2 = u$$

where ϵ is unknown but constrained to a known interval $[0, \epsilon_0]$. Consider a general state feedback

$$u = -(k_1 x_1 + k_2 x_2)$$

Our goal is to assign the eigenvalues of the feedback system in the left half plane $Re(s) < -r$, where $r > 0$ is given, independent of ϵ (robustly).

This goal is achieved if and only if

$$k_2 - \epsilon k_1 - 2r > 0$$

and

$$k_1 - r(k_2 - \epsilon k_1) + r^2 > 0$$

independent of ϵ. For $\epsilon = 0$, we obtain

$$k_1 = r^2 + rd_1 + d_2$$

and

$$k_2 = 2r + d_2$$

where d_1 and d_2 are two arbitrary positive numbers. For $\epsilon = \epsilon_0$, the inequalities above require

$$d_2(\epsilon_0 r - 1) > \epsilon_0(r^2 + d_1)$$

which produces contradiction if r is equal to $1/\epsilon_0$ or greater. The indicated value of r is the precise limit to our ability of exponential stabilization by state feedback in the presence of parameter perturbations.

MAIN RESULT

In this section we will describe a modification of Theorem 1 above by showing that similar restrictions to pole placement design are true when the matrix B of input gains is fixed and only the system matrix A is subject to variation. However, this holds only for systems of order $n > 2$.

THEOREM 2
Let

$$S = \{(A,B): A = A_0 + \delta A, B = B_0, \delta A \in N\}$$

be a family of controllable pairs of system matrices determined by a given neighborhood N of zero matrix. Assume that B_0 is a column and the size of A is $n > 2$. Then there exists a positive number r such that for any feedback matrix K placing the eigenvalues of $A_0 + B_0 K$ to the left of $Re(s) = -r$ there exists a pair (A,B) in the set S and an eigenvalue s_0 of $A + BK$ such that $Re(s_0) > -r$.

Proof:
It is well-known that any controllable pair (A_0, B_0) can be transformed by a change of coordinates and state feedback to the following Brunovsky canonical form

$$A_0 = \begin{bmatrix} 0 & 1 & 0 & \dots & 0 & 0 \\ 0 & 0 & 1 & \dots & 0 & 0 \\ & & \dots & \dots & & \\ & & \dots & \dots & & \\ 0 & 0 & 0 & \dots & 0 & 1 \\ 0 & 0 & 0 & \dots & 0 & 0 \end{bmatrix} \qquad B_0 = \begin{bmatrix} 0 \\ 0 \\ \cdot \\ \cdot \\ 0 \\ 1 \end{bmatrix}$$

so that we can assume, without loss of generality, the pair (A_0, B_0) in the form as above. With a feedback $u = -Kx$, $K = [k_1, k_2, ..., k_n]$, the unperturbed system eigenvalues are the roots of the polynomial

$$p(s) = \det(sI - A_0 + B_0K) = s^n + k_n s^{n-1} + ... + k_2 s + k_1$$

A necessary condition for $p(s)$ to have all roots in $Re(s) < -r$ is the positivity of all coefficients of

$$p(s-r) = s^n + d_n s^{n-1} + ... + d_2 s + d_1.$$

Hence, in particular,

(4) $\quad k_1 = r^n + d_n r^{n-1} + ... + d_2 r + d_1.$

Now, take δA as the matrix with all zero elements except for ϵ in the right upper corner where ϵ is a real number small enough to guarantee that the perturbed system belongs to the perturbation set S and such that $(-1)^n \epsilon < 0$. The corresponding perturbed characteristic polynomial of the feedback system can be written as

$$q(s) = p(s) + \epsilon k_1 s^{n-2}$$

and

$$q(s-r) = s^n + d_n s^{n-1} + (d_{n-1} + \epsilon k_1) s^{n-2} + ... + d_1 + \epsilon k_1 (-r)^{n-2}$$

Clearly, coefficients of $q(s-r)$ must be positive if the eigenvalues of the perturbed system matrix $A_0 + \delta A - B_0 K$ are to remain in the left half plane $Re(s) < -r$. In particular,

$$d_1 + \epsilon k_1 (-r)^{n-2} > 0.$$

Now, take r such that $\epsilon(-r)^{n-2} = -1$. Upon substitution of k_1 from (4), we obtain

$$r^n + d_n r^{n-1} + ... + d_2 r < 0,$$

a contradiction with positivity of r and d_i, $i = 2, ..., n$. This completes the proof.

FINAL REMARKS

An intriguing question is how common is the lack of robustness of exponential stabilization by state feedback. How small (fast) unmodelled dynamics affects the feedback design? How about small time delays? What if dynamic state feedback is allowed? While the unmodelled dynamics, changing the system order, is still under investigation, there are results available on small time delays [3, 4]. These results show exactly the same picture as above: The left half plane $Re(s) < -r$ is not assignable if

$$r > 1/h$$

where h is the maximum time delay expected as a perturbation of either the input signal ($u(t)$ replaced by $u(t-h)$) or in the state variables ($x(t)$ replaced by $x(t-h)$) in system (1),(2). The proof of the corresponding theorems can be obtained by utilizing the necessary stability conditions of Chebotarev and Meiman [5].

CONCLUSION

The results of this paper show a dramatic difference in the way we can vary by feedback the closed loop system poles under small perturbations: For systems with uncertain parameters there exists a left halfplane which is not assignable.

REFERENCES

[1] A. W. Olbrot and J. Cieslik, A qualitative bound in robustness of stabilization by state feedback, IEEE Trans. Autom. Control, Vol. 33, No. 12, Dec. 1988.

[2] M. Wonham, Linear Multivariable Control: A Geometric Approach, New York: Springer-Verlag 1979.

[3] A. W. Olbrot, The Pole Placement Theorem and Small Perturbations in Parameters: The Existence of an Unassignable Left Halfplane, 27th IEEE Conf. Decision and Control, Austin, TX, December 7-9, 1988.

[4] A. W. Olbrot, Nonrobustness of the Pole Placement Design under Small Perturbations in Time Delays, Proc. 1992 American Control Conference, Chicago, June 1992

[5] N. G. Chebotarev and N. N. Meiman, The Routh-Hurwitz problem for polynomials and entire functions, Trudy Mat. Inst. Steklov, Vol26, 1949.

ADAPTIVE CONTROL AND SELF-STABILIZATION

Miloje S. Radenkovic
Department of Electrical Engineering
University of Colorado at Denver
Denver, CO 80204

Anthony N. Michel
Department of Electrical Engineering
University of Notre Dame
Notre Dame, IN 46556

Abstract

This paper presents a new methodology for the global stability analysis, and consequently, for the design of robust deterministic and stochastic adaptive control, filtering and prediction. The methodology advocated herein represents a mathematical formalization of the self-stabilization mechanism which is a natural characteristic of every properly designed adaptive system. The effectiveness of the proposed approach is demonstrated by solving the robust deterministic and stochastic adaptive control problems. It is shown that very small algorithm gains μ, may produce very large signals in the adaptive loop, which are unacceptable for practical applications. The intensity of the admissible unmodelled dynamics does not depend on the algorithm design parameters and it is specified in terms of the corresponding H^∞ norm.

1 Introduction

In the seventies, significant progress was made in adaptive control theory, under the assumption that physical systems are described precisely by linear system models [1,2]. In these works it was assumed that the system parameters are unknown and that the relative degree and an upper bound on the order of the system are known. At the beginning of the eighties a disturbing fact was discovered, namely, that an adaptive controller designed for the case of a perfect system model in the presence of external disturbances or small modelling errors, can become unstable [3]. In order to guarantee stability, a variety of modifications of the algorithms originally designed for the ideal case have been proposed, such as σ and ε_1 modification, relative dead zone, signal normalization, projection of the parameter estimates and the like. An excellent unification of the existing robust deterministic adaptive control methods is developed in [4], where most of the significant works in the area of deterministic adaptive control are cited. In addition to the references cited in [4], we point to the work reported in [5,6,10], where the authors prove that the projection of the parameter estimates is, in fact, sufficient to guarantee global stability in the presence of small unmodelled dynamics. In the case of stochastic robust adaptive control there is no satisfactory theory explaining the behavior of the adaptive system in the presence of unmodelled dynamics. Attractive results have also been reported in [7], where the authors propose robust control based on the signal normalization philosophy, an approach inspired by the robust deterministic adaptive control results. In [8], robust stochastic adaptive control based on a stochastic approximation algorithm is proposed and global stability is established using Lyapunov function arguments.

The present paper attempts to provide a unified framework for a systematic and quantitative theory of deterministic and stochastic robust adaptive control, filtering and prediction. The contribution of the paper includes:

(i) A new methodology for the global stability analysis and consequently for the design of adaptive systems in the presence of unmodelled dynamics and external disturbances is developed. The present approach is quite natural and is based on the construction of corresponding Lyapunov functions for different periods of adaptation without using the Bellman-Gronwall Lemma. The proposed method makes it possible to treat robust deterministic and stochastic adaptive systems in a unique way, under formally identical assumptions.

(ii) It is shown that whenever, as a consequence of incorrect parameter estimates, an adaptive system becomes unstable, the adaptive algorithm will stabilize itself by generating correct parameter estimates. This self-stabilization mechanism and the possible occurrence of burstings are characterized analytically.

(iii) The proposed method provides a precise characterization of the class and size of the admissible unmodelled dynamics. Specifically, it is shown that the intensity of the tolerated unmodelled dynamics, expressed in terms of the corresponding H^∞ norm, does not depend on the design parameters of the adaptive system, and is identical to the case of non-adaptive robust control

2   Notation and Terminology

In this paper we shall consider simultaneously the robust deterministic and stochastic adaptive control problem for discrete time systems. Throughout the text we will assign the label DAC only to the relations which are applicable to Deterministic Adaptive Control. Similarly, to expressions applicable to the Stochastic Adaptive Control, we will assign the label SAC. For a discrete time function $x(t) : T \to \Re$, we define the following norm

$$n_x(t) = \left\{ \sum_{j=1}^{t} \lambda^{t-j} x(j)^2 \right\}^{1/2}, \quad 0 < \lambda \leq 1 \tag{2.1}$$

where T is the set of nonnegative integers. Throughout the text λ in (2.1) denotes a fixed number satisfying

$$0 < \lambda < 1, \quad \text{for DAC, and } \lambda = 1, \quad \text{for SAC} \tag{2.2}$$

H^∞ will denote the space of transfer functions $T(z)$ which are analytic and bounded outside and on the unit circle in the z plane. S^λ is the operator defined by $S^\lambda T(z) = T(\lambda^{1/2}z)$, for a fixed parameter λ, $0 < \lambda \leq 1$. $S^\lambda H^\infty$ is the space of transfer functions $T(z)$ such that $S^\lambda T(z) \in H^\infty$. In other words $T(z) \in S^\lambda H^\infty$, if $T(z)$ is analytic and bounded outside and on the circle $|z| = \lambda^{1/2}$ in the z plane. For $T(z) \in H^\infty$, the H^∞ norm is defined by

$$\| T(z) \|_{H^\infty} := \max_{|z|=1} |T(z)|. \tag{2.3}$$

Likewise, the norm of the $S^\lambda H^\infty$ space is defined by

$$\| T(z) \|_{H^\infty}^\lambda := \| S^\lambda T(z) \|_{H^\infty} = \max_{|z|=1} |T(\lambda^{1/2}z)|. \tag{2.4}$$

When performing majorizations, in order to account for initial conditions we will use nonnegative functions in the present paper

$$\xi_i(t) = c_i \lambda_\xi^t, 0 \le c_i < \infty \tag{2.5}$$

where

$$0 < \lambda_\xi < 1, \text{ for DAC, and } \lambda_\xi = 1, \text{ for SAC}. \tag{2.6}$$

3 Robust Ultimate Boundedness Theorem

In this section we will establish results which can be used in the analysis and consequently in the design of *robust adaptive control, filtering and prediction* algorithms.

Let us consider the following recursion for $t \ge 1$,

$$V(t+1) + \frac{S(t+1)}{\tilde{r}(t)} \le V(t) + \frac{S(t)}{\tilde{r}(t-1)} - 2\mu(1 - \frac{\mu}{2})\frac{z(t)^2}{\tilde{r}(t)}$$

$$+ \frac{2\mu(1-\mu)|z(t)| \cdot |\gamma(t)| + \mu^2\gamma(t)^2}{\tilde{r}(t)} + 2\mu\frac{d(t) - d(t-1)}{\tilde{r}(t)} + \frac{\beta(t)^2}{\tilde{r}(t)^2} \tag{3.1}$$

where $0 < \mu \le 1$ and all variables are real-valued with finite initial conditions.

Regarding relation (3.1) we introduce the following assumptions:

(A_1) $V(t) \in \Re^+$ and $V(t) \le d_0$ for all $t \ge 1$,

(A_2) for all $t \ge 0, \tilde{r}(t) > 0$ and $\tilde{r}(t) \ge \lambda\tilde{r}(t-1)$, where λ is defined by (2.2).

(A_3) $\tilde{r}(t) \le m(t)$ for $t \ge 1$, where

$$m(t) = \max\{C_\theta n_x(t-1)^2, g(t)\}, 0 < C_\theta < \infty \tag{3.2}$$

where $n_x(t)$ is defined by Eq. (2.1), when $x(t) = z(t)$, and $z(t) \in \Re$. In Eq. (3.2) $g(t) \in \Re^+$ and is given by

$$g(t) \le k_\theta + \xi_1(t), 0 < k_\theta < \infty, \text{ for } DAC, \tag{3.3}$$

$$\limsup_{t\to\infty} \frac{g(t)}{t} \le C_g, 0 \le C_g < \infty, \text{ for } SAC \tag{3.4}$$

(A_4) $n_\gamma(t) \le C_\gamma(\lambda)n_z(t) + n_\nu(t) + \xi_2(t), 0 < C_\gamma(\lambda) < \infty, t \ge 1$ where $n_\gamma(t)$ is defined by Eq. (2.1) when $x(t) = \gamma(t), \gamma(t) \in \Re$, while

$$\limsup_{t\to\infty} n_\nu(t)^2 \le \Sigma_{\nu D}^2 < \infty, \text{ for } DAC, \tag{3.5}$$

$$\limsup_{t\to\infty} \frac{1}{t}n_\nu(t)^2 \le \Sigma_{\nu S}^2 < \infty, \text{ for } SAC \tag{3.6}$$

where $n_\nu(t)$ is defined by Eq. (2.1), when $x(t) = \nu(t)$, for some sequence $\nu(t) \in \Re$.

(A_5) $\sum_{t=1}^\infty \beta(t)^2 / \tilde{r}(t)^2 \leq C_\beta < \infty$.

(A_6) $S(t) \in \Re^+, d(t) \in \Re^+$ and for $t \geq 1, S(t+1)/\tilde{r}(t) \leq C_S$, while $d(t)$ in (3.1) satisfies $d(t) \leq o[n_z(t)^2 + t]$.

In addition to the above assumptions the following holds:

$$S(t) = S_0(t)I_A, d(t) = d_0(t)I_A, \beta(t) = \beta_0(t)I_A \tag{3.7}$$

where the indicator function I_A is given by

$$I_A = \begin{cases} 0, & \text{for } DAC \\ 1, & \text{for } SAC \end{cases} \tag{3.8}$$

and $S_0(t) \in \Re^+, d_0(t) \in \Re^+$ and $\beta_0(t) \in \Re$.

For the purpose of future analysis, let us define the following function

$$\begin{aligned} W(t+1) &= \mu \sum_{j=1}^t \lambda^{t-j} \left\{ [1 - \frac{\mu}{2} + (1-\mu)C_\gamma(\lambda) + \frac{\mu}{2}C_\gamma(\lambda)^2]z(j)^2 - 2(1-\mu)|z(j)| \parallel \gamma(j)| \right. \\ &\quad \left. - \mu\gamma(j)^2 - 2[d(j) - d(j-1)] \right\}. \end{aligned} \tag{3.9}$$

Note that relation (3.1) describes the evolution of the Lyapunov type function $V(t)$ for which we are interested in determining the l_2^λ stability of the variable $z(t)$. The term $\gamma(t)$ tends to destabilize the convergence of the sequence $V(t)$ and the l_2^λ stability of the sequence $z(t)$. In adaptive control, filtering, and prediction problems, $\gamma(t)$ usually describes the effects of the unmodelled dynamics and external disturbances. The intensity of the unmodelled dynamics is quantified by the constant $C_\gamma(\lambda)$ defined by assumption (A_4). It turns out that the behavior of the sequence $W(t)$ is crucial for the convergence of the recursive scheme (3.1) and consequently for the l_2^λ stability of the sequence $z(t)$. Throughout this paper, the sequence $W(t)$ will be referred to as a *bursting function*. The reason for this terminology, as will be seen, can be found in the role played by the function $W(t)$ in determining the stability (or instability) of the recursive scheme (3.1). As will be seen in the subsequent analysis, in the time intervals where $W(t) \leq 0$, the stability of the sequence $z(t)$ follows trivially; however, in these time-intervals the Lyapunov function $V(t) + \frac{S(t)}{\tilde{r}(t-1)}$ may be increasing (diverging), thus giving rise to bursting of the sequence $z(t)$. As a consequence of this phenomenon, the function $W(t)$ becomes positive, forcing the function $V(t+1) + [S(t+1) + W(t+1)]/\tilde{r}(t)$ to converge, thus stabilizing $z(t)$.

The intensity of the admissible unmodelled dynamics will be specified by the following assumption

(A_7) $\rho_1 = 1 - \frac{\mu}{2} - (1-\mu)C_\gamma(\lambda) - \frac{\mu}{2}C_\gamma(\lambda)^2 > 0$.

The convergence properties of recursion (3.1) will be formulated in the following theorem:

Theorem 3.1: (Robust Ultimate Boundedness Theorem)

Let the assumptions $(A_1) - (A_7)$ hold. Then for all $t \geq 1$ we have

$$DAC : \limsup_{t\to\infty} n_z(t)^2 \leq \max\left\{ \Sigma_{\gamma D}^2, \left(\frac{\lambda}{C_\theta} + \frac{d_0}{\mu\rho_1} \right) k_\theta \right\} e^{\frac{d_0 C_\theta}{\lambda\mu\rho_1}} \tag{3.10}$$

where

$$\Sigma^2_{\gamma D} = \max \left\{ \frac{16[1 - \mu + \mu C_\gamma(\lambda)]^2}{\rho_1^2} ; \frac{2\mu}{\rho_1} \right\} \Sigma^2_{\nu D} \tag{3.11}$$

and

$$SAC : \lim_{t \to \infty} \sup_t \frac{1}{t} n_z(t)^2 \leq \max \left\{ \Sigma^2_{\gamma S}, \left(\frac{1}{C_\theta} + \frac{d_1'}{\mu \rho_1} \right) C_g \right\} e^{\frac{d_1' C_\theta}{\mu \rho_1}} \tag{3.12}$$

where

$$\Sigma^2_{\gamma S} = \max \left\{ \frac{16[1 - \mu + \mu C_\gamma(1)]^2}{\rho_1^2} ; \frac{2\mu}{\rho_1 - \rho_0} \right\} \Sigma^2_{\nu S} \tag{3.13}$$

where $0 < \rho_0 << \rho_1$, and d_1' in (3.12) is given by

$$d_1' = d_0 + C_S + C_\beta. \tag{3.14}$$

The constants $C_\theta, k_\theta, \Sigma_{\nu D}, \Sigma_{\nu S}, C_S, C_\beta, C_g$ and ρ_1 are defined by assumptions $(A_1) - (A_7)$.

Proof: The statements of the theorem will be derived by considering the behavior of the bursting function $W(t)$ given by Eq. (3.9) and its effects on the convergence of the recursive scheme (3.1). Let us define the sequences τ_k and $\sigma_k, k \geq 1$ as follows

$$1 \stackrel{\Delta}{=} \tau_1 < \sigma_1 < \tau_2 < \sigma_2 < \cdots < \tau_k < \sigma_k < \tau_{k+1} < \cdots, \tag{3.15}$$

so that

$$W(t+1) \leq 0 \text{ for } t \in Q_k \text{ and } W(t+1) > 0 \text{ for } t \in T_k, \tag{3.16}$$

where the time intervals T_k and Q_k are defined by

$$Q_k = [\tau_k, \sigma_k), \ T_k = [\sigma_k, \tau_{k+1}), \ k \geq 1. \tag{3.17}$$

If $W(2) > 0$, we set $\tau_1 = 0, \sigma_1 = 1$ and Q_k is defined for $k \geq 2$. If $W(t+1) > 0$ for all $t \geq 1$ we define $\sigma_1 = 1$ and $\tau_2 = +\infty$. In the case when $W(t+1) \leq 0$ for all $t \geq 1$, we set $\tau_1 = 1$ and $\sigma_1 = +\infty$.

Next we analyze the case when in (3.15) for all finite k, we have $\tau_k < \infty$ and $\sigma_k < \infty$. Let us first consider the time intervals Q_k. Since $W(t+1) \leq 0$ for $t \in Q_k$, from Eq. (3.9) and assumption (A_4) we obtain

$$\begin{aligned} \rho_1 n_z(t)^2 &\leq 2[1 - \mu + \mu C_\gamma(\lambda)]n_z(t)[n_\nu(t) + \xi_2(t)] \\ &+ \mu[n_\nu(t) + \xi_2(t)]^2 + 2d(t), \quad t \in Q_k \end{aligned} \tag{3.18}$$

where ρ_1 is defined by assumption (A_7). The above relation implies that for $t \in Q_k$

$$\begin{aligned} \rho_1 n_z(t)^2 &\leq 2 \max \left\{ 2[1 - \mu + \mu C_\gamma(\lambda)]n_z(t)[n_\nu(t) + \xi_2(t)]; \right. \\ &; \left. \mu[n_\nu(t) + \xi_2(t)]^2 + 2d(t) \right\}. \end{aligned} \tag{3.19}$$

Using the fact that in the case of $DAC, d(t) = 0$ (Eq. (3.7)), we obtain from (3.19) and (3.5)

$$DAC : \lim_{k \to \infty} \sup_k \sup_{t \in Q_k} n_z(t)^2 \leq \Sigma^2_{\gamma D} \tag{3.20}$$

where $\Sigma_{\gamma D}$ is given by Eq. (3.11). At the same time from (3.19), by using relation (3.6) and Assumption (A_6), we derive,

$$SAC : \limsup_{k \to \infty} \sup_k \sup_{t \in Q_k} \frac{1}{t} n_z(t)^2 \leq \Sigma_{\gamma S}^2 \tag{3.21}$$

where $\Sigma_{\gamma S}$ is defined by Eq. (3.13).

Let us analyze the time intervals $T_k, k \geq 1$, where the bursting function $W(t+1) > 0$. Using the definition of $W(t+1)$ (Eq. (3.9)) and assumption (A_2), we obtain from (3.1)

$$V(t+1) + \frac{S(t+1) + W(t+1)}{\tilde{r}(t)} \leq V(t) + \frac{S(t) + W(t)}{\tilde{r}(t-1)} - \mu\rho_1 \frac{z(t)^2}{\tilde{r}(t)} + \frac{\beta(t)^2}{\tilde{r}(t)^2} \tag{3.22}$$

where ρ_1 is given by assumption (A_7). After summation from $t = \sigma_k + 1$ to $N < \tau_{k+1}$, we obtain from (3.22)

$$V(N+1) + \frac{S(N+1) + W(N+1)}{\tilde{r}(N)} \leq V(\sigma_k+1) + \frac{S(\sigma_k+1) + W(\sigma_k+1)}{\tilde{r}(\sigma_k)}$$
$$- \mu\rho_1 \sum_{t=\sigma_k+1}^{N} \frac{z(t)^2}{\tilde{r}(t)} + \sum_{t=\sigma_k+1}^{N} \frac{\beta(t)^2}{\tilde{r}(t)^2}. \tag{3.23}$$

Note that from (3.1) it follows that

$$V(\sigma_k+1) + \frac{S(\sigma_k+1)}{\tilde{r}(\sigma_k)} \leq V(\sigma_k) + \frac{S(\sigma_k)}{\tilde{r}(\sigma_k-1)} - 2\mu(1-\frac{\mu}{2})\frac{z(\sigma_k)^2}{\tilde{r}(\sigma_k)}$$
$$+ \frac{2\mu(1-\mu)|z(\sigma_k)| \cdot |\gamma(\sigma_k)| + \mu^2\gamma(\sigma_k)^2}{\tilde{r}(\sigma_k)}$$
$$+ 2\mu\frac{d(\sigma_k) - d(\sigma_k-1)}{\tilde{r}(\sigma_k)} + \frac{\beta(\sigma_k)^2}{\tilde{r}(\sigma_k)^2} \tag{3.24}$$

and from Eq. (3.9) it follows that

$$W(\sigma_k+1) = \mu[1 - \frac{\mu}{2} + (1-\mu)C_\gamma(\lambda) + \frac{\mu}{2}C_\gamma(\lambda)^2]z(\sigma_k)^2 \tag{3.25}$$
$$- 2\mu(1-\mu)|z(\sigma_k)||\gamma(\sigma_k)| - \mu^2\gamma(\sigma_k)^2 - 2\mu[d(\sigma_k) - d(\sigma_k-1)] + \lambda W(\sigma_k).$$

Using the fact that by definition (3.16), $W(\sigma_k) \leq 0$, we obtain after substituting (3.24) and (3.25) into (3.23),

$$V(N+1) + \frac{S(N+1) + W(N+1)}{\tilde{r}(N)} \leq V(\sigma_k) + \frac{S(\sigma_k)}{\tilde{r}(\sigma_k-1)}$$
$$- \mu\rho_1 \sum_{t=\sigma_k}^{N} \frac{z(t)^2}{\tilde{r}(t)} + \sum_{t=\sigma_k}^{N} \frac{\beta(t)^2}{\tilde{r}(t)^2} \tag{3.26}$$

for $N \in T_k, k \geq 1$. From the above relation and assumptions $(A_1), (A_5)$ and (A_6) we obtain for $N \in T_k$

$$\mu\rho_1\Sigma_{t=\sigma_k}^{N}z(t)^2/\tilde{r}(t) \leq d_1 \tag{3.27}$$

where the constant d_1 is given by

$$
\begin{aligned}
DAC &: \quad d_1 = d_0, \quad (S(t) = 0, \beta(t) = 0) \\
SAC &: \quad d_1 = d_0 + C_S + C_\beta.
\end{aligned}
\tag{3.28}
$$

Next, we will show that relation (3.27) together with assumption (A_3) are crucial for establishing the stability in the time intervals T_k, $k \geq 1$. Let us define the integers $p_{ik} \in T_k$ and $l_{ik} \in T_k$ be defined as follows

$$
p_{0k} < l_{1k} < p_{1k} < \cdots < l_{ik} < p_{ik} < l_{(i+1)k} < \cdots
\tag{3.29}
$$

so that for $m(t)$ given by Eq. (3.2), the following holds

$$
m(t) = C_\theta n_z(t-1)^2 \text{ for } t \in L_{ik}, \; i \geq 1, k \geq 1
\tag{3.30}
$$

and

$$
m(t) = g(t) \text{ for } t \in D_{ik}, \; i \geq 1, k \geq 1
\tag{3.31}
$$

where the time intervals L_{ik} and D_{ik} are defined by

$$
L_{ik} = [p_{(i-1)k}, l_{ik}) \text{ and } D_{ik} = [l_{ik}, p_{ik}).
\tag{3.32}
$$

It is obvious that $l_{ik} < \tau_{k+1}$ and $p_{ik} < \tau_{k+1}$ for $i \geq 0$. If $C_\theta n_z(\sigma_k-1)^2 \geq g(\sigma_k)$, then $p_{0k} = \sigma_k$. If $C_\theta n_z(\sigma_k - 1)^2 < g(\sigma_k)$, we set $p_{0k} = 0$, $l_{1k} = \sigma_k$ and the intervals L_{ik} are defined for $i \geq 2$. In the case when $C_\theta n_z(t-1)^2 < g(t)$ for all $t \in T_k$, we define $p_{0k} = 0$, $l_{1k} = \sigma_k$ and $p_{1k} = \tau_{k+1}$. If $C_\theta n_z(t-1)^2 \geq g(t)$ for all $t \in T_k$ we set $p_{0k} = \sigma_k$ and $l_{1k} = \tau_{k+1}$.

Note that from (3.2) and (3.31) we can obtain

$$
n_z(t-1)^2 \leq g(t)/C_\theta \text{ for } t \in D_{ik}, \; i \geq 1, k \geq 1.
\tag{3.33}
$$

On the other hand, relations (3.27), (3.2) and (3.31) imply that for $t \in D_{ik}$,

$$
z(t)^2 \leq \frac{d_1}{\mu \rho_1} \tilde{r}(t) \leq \frac{d_1}{\mu \rho_1} g(t).
\tag{3.34}
$$

Since $n_z(t)^2 = z(t)^2 + \lambda n_z(t-1)^2$, relations (3.33) and (3.34) yield

$$
n_z(t)^2 \leq \left(\frac{\lambda}{C_\theta} + \frac{d_1}{\mu \rho_1} \right) g(t), \; t \in D_{ik}.
\tag{3.35}
$$

From the above relation we obtain in view of (3.3) and (3.4)

$$
DAC : \limsup_{k \to \infty} \sup_k \sup_i \sup_{t \in D_{ik}} n_z(t)^2 \leq \left(\frac{\lambda}{C_\theta} + \frac{d_1}{\mu \rho_1} \right) k_\theta
\tag{3.36}
$$

and

$$
SAC : \limsup_{k \to \infty} \sup_k \sup_i \sup_{t \in D_{ik}} \frac{1}{t} n_z(t)^2 \leq \left(\frac{1}{C_\theta} + \frac{d_1}{\mu \rho_1} \right) C_g
\tag{3.37}
$$

where the constant d_1 is specified by Eq. (3.28). Next, we consider the time intervals $L_{ik} \subset T_k$ for $i \geq 1, k \geq 1$. From (3.27), by assumption (A_3) and Eq. (3.30) we can obtain

$$R_{ik} = \sum_{t=p_{(i-1)k}}^{N} \frac{z(t)^2}{n_z(t-1)^2} \leq \frac{d_1}{\mu\rho_1} C_\theta \tag{3.38}$$

for $N \in L_{ik}$. From the above relation it follows that

$$
\begin{aligned}
R_{ik} &= \sum_{t=p_{(i-1)k}}^{N} \frac{n_z(t)^2 - \lambda n_z(t-1)^2}{n_z(t-1)^2} = \sum_{t=p_{(i-1)k}}^{N} \frac{\lambda}{\lambda n_z(t-1)^2} \int_{\lambda n_z(t-1)^2}^{n_z(t)^2} dx \\
&\geq \lambda \sum_{t=p_{(i-1)k}}^{N} \int_{\lambda n_z(t-1)^2}^{n_z(t)^2} \frac{dx}{x} = \lambda \sum_{t=p_{(i-1)k}}^{N} \left\{ \log n_z(t)^2 - \log \lambda n_z(t-1)^2 \right\} \\
&= \lambda \log \frac{n_z(N)^2}{n_z(p_{(i-1)k} - 1)^2} + \lambda[N - p_{(i-1)k}] \log \frac{1}{\lambda}.
\end{aligned}
\tag{3.39}
$$

Relations (3.38) and (3.39) imply that

$$n_z(N)^2 \leq e^{\frac{d_1}{\lambda\mu\rho_1}C_\theta} n_z(p_{(i-1)k} - 1)^2 \tag{3.40}$$

for $N \in L_{ik}$. Since $p_{(i-1)k} - 1 \in D_{(i-1)k}$ we can obtain from (3.40), (3.36) and (3.37) for $i \geq 2$,

$$DAC : \limsup_{k \to \infty} \sup_{k} \sup_{i \geq 2} \sup_{t \in L_{ik}} n_z(t)^2 \leq k_\theta \left(\frac{\lambda}{C_\theta} + \frac{d_1}{\mu\rho_1} \right) e^{\frac{d_1}{\lambda\mu\rho_1}C_\theta} \tag{3.41}$$

and

$$SAC : \limsup_{k \to \infty} \sup_{k} \sup_{i \geq 2} \sup_{t \in L_{ik}} \frac{1}{t} n_z(t)^2 \leq \left(\frac{1}{C_\theta} + \frac{d_1}{\mu\rho_1} \right) C_g e^{\frac{d_1}{\mu\rho_1}C_\theta}. \tag{3.42}$$

Let us evaluate $n_z(t)$ in the time intervals $L_{1k}, k \geq 1$. If $p_{0k} = \sigma_k$, we derive from (3.40), (3.20) and (3.21)

$$DAC : \limsup_{k \to \infty} \sup_{k} \sup_{t \in L_{1k}} n_z(t)^2 \leq e^{\frac{d_1}{\lambda\mu\rho_1}C_\theta} \Sigma_{\gamma D}^2 \tag{3.43}$$

and

$$SAC : \limsup_{k \to \infty} \sup_{k} \sup_{t \in L_{1k}} \frac{1}{t} n_z(t)^2 \leq e^{\frac{d_1}{\mu\rho_1}C_\theta} \Sigma_{\gamma S}^2 \tag{3.44}$$

where d_1 is given by Eq. (3.28). In the case $p_{0k} = 0$ and $l_{1k} = \sigma_k$, time intervals L_{ik} are defined for $i \geq 2$. In (3.39), the case $n_z(N)^2 \leq n_z(p_{(i-1)k} - 1)^2$ is trivial, and it is covered by relation (3.40).

It is not difficult to see that (3.20), (3.21), (3.36), (3.37), (3.41), (3.42), (3.43), and (3.44), establish the statements of the theorem. Thus the theorem is proved. $\square$

4 Robust Adaptive Control

The power of the resultrs established in the previous section will now be demonstrated by solving the robust adaptive control problem. Let us consider the following discrete-time SISO system with unmodelled dynamics

$$
\begin{aligned}
A(q^{-1})y(t+1) &= B(q^{-1})[1 + \Delta_1(q^{-1})]u(t) \\
&+ A(q^{-1})\Delta_2(q^{-1})u(t) + [1 + \Delta_3(q^{-1})]w(t+1)
\end{aligned}
\tag{4.1}
$$

where $\{y(t)\}, \{u(t)\}$ and $\{w(t)\}$ are output, input and disturbance sequences, respectively, while q^{-1} represents the unit delay operator. The polynomials $A(q^{-1})$ and $B(q^{-1})$ describe the nominal system model and are given by

$$
A(q^{-1}) = 1 + a_1 q^{-1} + \cdots + a_{n_A} q^{-n_A}, B(q^{-1}) = b_0 + b_1 q^{-1} + \cdots + b_{n_B} q^{-n_B}, (b_0 \neq 0). \tag{4.2}
$$

In Eq. (4.1) $\Delta_i(q^{-1}), i = 1, 2$ denote multiplicative and additive system perturbations. The dynamics of complex and unstructured external disturbances are represented by the operator $\Delta_3(q^{-1})$. The transfer functions $\Delta_i(z^{-1}), i = 1, 2, 3$, are causal and stable.

In (4.1) $\{w(t)\}$ is the external disturbance satisfying:

(S_1) $DAC : |w(t)| \leq k_w < \infty$

$SAC : w(t)$ is a stochastic process defined on the underlying probability space $\{\Omega, \mathcal{F}, P\}$, and

$$
\begin{aligned}
E\{w(t+1)|\mathcal{F}_t\} &= 0, E\{w(t+1)^2|\mathcal{F}_t\} = \sigma_w^2 < \infty, \\
E\{|w(t+1)|^{2+\eta}|\mathcal{F}_t\} &\leq k_1 < \infty, \eta > 0 \ (a.s.)
\end{aligned}
$$

where $\mathcal{F}_t$ is the σ-algebra generated by $\{w(1), \ldots, w(t)\}$.

We will stabilize system (4.1) by designing the adaptive controller so that for a given reference signal the following functional criterion is minimized:

$$
DAC : J = (y(t) - y^*(t))^2, SAC : J = \lim_{N \to \infty} \frac{1}{N} \sum_{t=1}^{N} (y(t) - y^*(t))^2 \tag{4.3}
$$

where we assume that

(S_2) $\{y^*(t)\}$ is a bounded deterministic sequence, i.e., $|y^*(t)| \leq m_1 < \infty$

Note that the system model (4.1) can be written in the form

$$
\eta(t) = \theta_0^T \phi(t) - y^*(t+1) + \gamma(t) \tag{4.4}
$$

where

$$
\begin{aligned}
\theta_0^T &= [-a_1, \ldots, -a_{n_A}; b_0, b_1, \ldots, b_{n_B}] \tag{4.5} \\
\phi(t)^T &= [y(t), \ldots, y(t - n_A + 1); u(t), \ldots, u(t - n_B)] \tag{4.6} \\
\eta(t) &= y(t+1) - y^*(t+1) - I_A w(t+1) \tag{4.7}
\end{aligned}
$$

and

$$
\gamma(t) = \Delta_0(q^{-1})u(t) + [\Delta_3(q^{-1}) + (1 - I_A)]w(t+1) \tag{4.8}
$$

where

$$\Delta_0(q^{-1}) = B(q^{-1})\Delta_1(q^{-1}) + A(q^{-1})\Delta_2(q^{-1}). \tag{4.9}$$

We will use the following adaptive control law

$$\hat{\theta}(t)^T \phi(t) = y^*(t+1) \tag{4.10}$$

where $\hat{\theta}(t)$ is an estimate of θ_0. From Eqs. (4.4) and (4.10) it follows that the closed-loop adaptive system is given by

$$\eta(t) = -z(t) + \gamma(t), z(t) = \tilde{\theta}(t)^T \phi(t), \tilde{\theta}(t) = \hat{\theta}(t) - \theta_0. \tag{4.11}$$

From (4.10) it is not difficult to obtain

$$B(q^{-1})u(t) - q[A(q^{-1}) - 1]y(t) = y^*(t+1) - z(t). \tag{4.12}$$

Combining Eqs. (4.7), (4.11) and (4.12) yields

$$B(q^{-1})u(t) = A(q^{-1})[-z(t) + y^*(t+1)] + [A(q^{-1}) - 1][I_A w(t+1) + \gamma(t)]. \tag{4.13}$$

Substituting $u(t)$ from Eq. (4.13) into Eq. (4.8), we obtain

$$\begin{aligned}\{B(q^{-1}) &- \Delta_0(q^{-1})[A(q^{-1}) - 1]\}\gamma(t) = -\Delta_0(q^{-1})A(q^{-1})[z(t) - y^*(t+1)] \\ &+ \{\Delta_0(q^{-1})[A(q^{-1}) - 1]I_A + [\Delta_3(q^{-1}) + 1 - I_A]B(q^{-1})\}w(t+1).\end{aligned} \tag{4.14}$$

Concerning system model (4.1) we assume

(S_3) A lower bound $\lambda_0, 0 < \lambda_0 < 1$, for which the zeros of $B(z^{-1})$ and the poles of the transfer functions

$$D_1(z^{-1}) = \Delta_0(z^{-1})A(z^{-1})/\{B(z^{-1}) - \Delta_0(z^{-1})[A(z^{-1}) - 1]\}$$

and

$$D_2(z^{-1}) = \{\Delta_0(z^{-1})[A(z^{-1}) - 1]I_A + [\Delta_3(z^{-1}) + 1 - I_A]B(z^{-1})\}/\{B(z^{-1}) - \Delta_0(z^{-1})[A(z^{-1}) - 1]\},$$

z_j, satisfy $|z_j|^2 \le \lambda_0$, is known.

Based on the available prior information related to the nominal system model, we introduce the following assumption:

(S_4) The compact convex set Θ^0 which contains θ_0, the sign of b_0 and a lower bound $b_{0,\min}$ of $|b_0|$, are known. Without loss of generality we assume that $b_0 > 0$ and $b_{0,\min} > 0$.

For the estimation of θ_0 we propose the following algorithms

$$\hat{\theta}(t+1) = \mathcal{P}\left\{\hat{\theta}(t) + \frac{\mu}{\tilde{r}(t)}\phi(t)[y(t+1) - y^*(t+1)]\right\}, 0 < \mu \le 1 \tag{4.15}$$

where $\mathcal{P}\{\cdot\}$ projects orthogonally onto Θ^0, so that $\mathcal{P}\{\hat{\theta}\} \in \Theta^0$ for all $\hat{\theta} \in \Re^{n_A + n_B + 1}$, and there exists a finite constant d_0 so that $\| \hat{\theta}(t) - \theta_0 \|^2 \le d_0 < \infty$, and $\hat{b}_0(t) \ge b_{0,\min} > 0$ for all $t \ge 1$. The algorithm gain sequence is given by

$$DAC : \tilde{r}(t) = n_0 + n_\phi(t)^2, \ 0 < n_0 < \infty \tag{4.16}$$

where

$$n_\phi(t)^2 = \lambda n_\phi(t-1)^2 + \| \phi(t) \|^2, \lambda_1 + \lambda_0 < \lambda < 1, 0 < \lambda_1 << \lambda_0 \qquad (4.17)$$

where λ_0 is defined by assumption (S_3),

$$SAC : \tilde{r}(t) = 2r(t) \qquad (4.18)$$

or

$$SAC : \tilde{r}(t) = \max\{2 \max_{1 \le \tau \le t} \| \phi(\tau) \|^2, r(t)^{1-\varepsilon}\}, 0 < \varepsilon < \frac{1}{2} \qquad (4.19)$$

where

$$r(t) = r(t-1) + \| \phi(t) \|^2 + 1, \; r(0) > 1. \qquad (4.20)$$

Let us define the following H^∞ norms

$$C_{AB}(\lambda) = \| \frac{A(z)}{B(z)} \|_{H^\infty}^\lambda, C_A(\lambda) = \| \frac{A(z)-1}{B(z)} \|_{H^\infty}^\lambda, C_\gamma(\lambda) = \| \frac{\Delta_0(z)A(z)}{B(z) - \Delta_0(z)[A(z)-1]} \|_{H^\infty}^\lambda$$

$$C_w(\lambda) = \| \frac{\Delta_0(z)[A(z)-1]I_A + [\Delta_3(z)+1-I_A]B(z)}{B(z) - \Delta_0(z)[A(z)-1]} \|_{H^\infty}^\lambda \qquad (4.21)$$

Throughout the following text, the values of important constants used in bounds will be precisely specified. Certain constants whose exact values are unimportant and which do not depend on $C_\gamma(\lambda)$ and $C_w(\lambda)$ will be denoted by C_i, $i = 1, 2, 3, \cdots$. Constants whose exact values are unimportant but which depend on $C_\gamma(\lambda)$ and $C_w(\lambda)$ and are such that their values decrease as the $C_\gamma(\lambda)$ and $C_w(\lambda)$ decrease, will be denoted by $\bar{C}_i$, $i = 1, 2, 3, \cdots$.

Throughout, all constants are positive, unless otherwise noted.

The following lemma will be useful for future reference.

Lemma 4.1: Let the assumptions (S_3) and (S_4) hold. Then

1) $n_\gamma(t) \le C_\gamma(\lambda)n_z(t) + n_\nu(t) + \xi_3(t), n_\nu(t) = C_\gamma(\lambda)n_{y^*}(t+1) + C_w(\lambda)n_w(t+1)$ $\quad (4.22)$
where the constants $C_\gamma(\lambda)$ and $C_w(\lambda)$ are defined by Eq. (4.21). The norms $n_\gamma(t), n_z(t), n_{y^*}(t)$ and $n_w(t)$ are given by Eq. (2.1) when $x(t) = \gamma(t), x(t) = z(t), x(t) = y^*(t)$ and $x(t) = w(t)$, respectively.

2) $n_\phi(t)^2 \le C_{\phi 1}(\lambda)n_z(t-1)^2 + C_{\phi 2}(\lambda)n_{y^*}(t+1)^2 + C_{\phi 3}(\lambda)n_w(t+1)^2 + \xi_4(t)$ $\quad (4.23)$

where $n_\phi(t)$ is defined by Eq. (4.17), and the constants $C_{\phi_i}(\lambda), i = 1, 2, 3$ are given by

$$C_{\phi 1}(\lambda) = 4C_1(\lambda)\{[C_{AB}(\lambda) + C_A(\lambda)C_\gamma(\lambda)]^2 + [1 + C_\gamma(\lambda)]^2\} \qquad (4.24)$$

$$C_{\phi 2}(\lambda) = \lambda^{-1}C_{\phi 1}(\lambda) + 2C_2(\lambda)/b_{0,\min}^2, C_{\phi 3}(\lambda) = 4C_1(\lambda)[I_A + C_w(\lambda)]^2[1 + C_A(\lambda)^2] \qquad (4.25)$$

where

$$C_1(\lambda) = \left(1 + \frac{2f_\theta C_2(\lambda)}{b_{0,\min}^2}\right)C_2(\lambda), C_2(\lambda) = \sum_{i=0}^{\bar{n}} \lambda^{-i}, \; \bar{n} = \max(n_A, n_B) \qquad (4.26)$$

while f_θ is the upper bound of $\| \hat{\theta}(t) \|^2 \le f_\theta$. The H^∞ norms $C_{AB}(\lambda), C_A(\lambda), C_\gamma(\lambda)$ and $C_w(\lambda)$ are defined by Eq. (4.21).

Proof: The proof of the lemma is given in [11].

The admissible unmodelled dynamics quantified by the constant $C_\gamma(\lambda)$ which is defined by Eq. (4.21). We assume that

$$(S_5) \quad \rho_1 = 1 - \frac{\mu}{2} - (1-\mu)C_\gamma(\lambda) - \frac{\mu}{2}C_\gamma(\lambda)^2 = [1 - C_\gamma(\lambda)]\{1 - \frac{\mu}{2}[1 - C_\gamma(\lambda)]\} > 0.$$

4.1 Deterministic Adaptive Control

From Eqs. (4.15), (4.7) and (4.11) it is not difficult to obtain

$$V(t+1) \leq V(t) - 2\mu(1 - \frac{\mu}{2})\frac{z(t)^2}{\bar{r}(t)} + \frac{2\mu(1-\mu)|z(t)| \cdot |\gamma(t)| + \mu^2\gamma(t)^2}{\bar{r}(t)} \tag{4.27}$$

where $z(t)$ and $\tilde{\theta}(t)$ are given by Eq. (4.11) and

$$V(t) = \| \tilde{\theta}(t) \|^2, \tag{4.28}$$

From (3.7) we conclude that the relations (3.1) and (4.27) are identical and therefore global stability results can be established simply by using Theorem 3.1.

Theorem 4.1: Let the assumptions $(S_1) - (S_5)$ hold. Then the DAC algorithm (4.10), (4.15)-(4.17), provides

1)
$$\limsup_{t \to \infty} \sum_{j=1}^{t} \lambda^{t-j}(\tilde{\theta}(j)^T\phi(j))^2 \leq \Sigma_d^2 \tag{4.29}$$

where

$$\Sigma_d^2 = \max\left\{\Sigma_{\gamma D}^2, \left(\frac{\lambda}{C_\theta} + \frac{d_0}{\mu\rho_1}\right)k_\theta\right\}e^{\frac{d_0 C_\theta}{\lambda\mu\rho_1}} \tag{4.30}$$

where $\Sigma_{\gamma D}$ is given by Eq. (3.11), where

$$\Sigma_{\nu D} = \frac{1}{1-\lambda}[m_1 C_\gamma(\lambda) + k_w C_w(\lambda)]^2 \tag{4.31}$$

while $C_\gamma(\lambda)$ is defined by Eq. (4.21) and ρ_1 is given by assumption (S_5). In Eq. (4.30), k_θ and C_θ are defined by

$$k_\theta = \frac{2}{1-\lambda}[m_1^2 C_{\phi 2}(\lambda) + k_w^2 C_{\phi 3}(\lambda)] + 2n_0, \quad C_\theta = 2C_{\phi 1}(\lambda) \tag{4.32}$$

where $C_w(\lambda)$ is given by Eq. (4.21), while the constants $C_{\phi i}(\lambda), i = 1, 2, 3$ are defined by Eqs. (4.24)-(4.25), respectively. In (4.30) d_0 is the upper bound for $\| \hat{\theta}(t) - \theta_0 \|^2$, and n_0 in Eq. (4.32) is defined by (4.16).

2) $\limsup_{t \to \infty} \sum_{j=1}^{t} \lambda^{t-j}[y(j+1) - y^*(j+1)]^2 \leq \{[1 + C_\gamma(\lambda)]\Sigma_d + \Sigma_{\nu D}\}^2.$ \hfill (4.33)

3) $\limsup_{t \to \infty} n_\phi(t)^2 \leq \lambda^{-1}C_{\phi 1}(\lambda)\Sigma_d^2 + \frac{k_\theta}{2}.$ \hfill (4.34)

Proof: In order to apply Theorem 3.1 from Section 3, we need to verify that assumptions $(S_1) - (S_5)$ imply assumptions $(A_1) - (A_7)$ introduced in Section 3. Assumption (A_1) is ensured by the projection in the estimation algorithm (4.15). Assumption (A_2) follows directly from the definition of $\tilde{r}(t)$ given by Eq. (4.16). Assumption (A_3) is a consequence of statement 2) of Lemma 4.1. Specifically, from (4.23) we can derive

$$n_0 + n_\phi(t)^2 \leq \max\{C_\theta n_z(t-1)^2, k_\theta + \xi_{10}(t)\} \tag{4.35}$$

where the constants C_θ and k_θ are defined by Eq. (4.32). Assumption (A_4) follows from statement (1) of Lemma 4.1. From Eq. (4.22) it is obvious that $\Sigma_{\nu D}$ in (3.5) is given by Eq. (4.31). Assumption (A_7) is equivalent to assumption (S_5). Assumptions (A_5) and (A_6) do not apply to DAC. Finally, applying the Robust Ultimate Boundedness Theorem (Theorem 3.1), statement (4.29) follows directly. From Eq. (4.11) and (4.22) we obtain

$$n_\eta(t) \leq [1 + C_\gamma(\lambda)]n_z(t) + n_\nu(t) + \xi_3(t) \tag{4.36}$$

where $n_\eta(t)$ is given by Eq. (2.1), when $x(t) = \eta(t)$. Definition (4.7) and Eq. (4.11), together with relations (4.29) and (4.36), imply the second statement of the theorem. Statement (3) of the theorem follows from (4.29) and (4.23). Thus the theorem is proved. □

4.2 Stochastic Adaptive Control

From Eqs. (4.7), (4.11), (4.15) and (4.18) (or(4.19)) we can obtain

$$V(t+1) \leq V(t) \quad - \quad 2\mu(1 - \frac{\mu}{2})\frac{z(t)^2}{\tilde{r}(t)} + \frac{2\mu(1-\mu)|z(t)| \cdot |\gamma(t)| + \mu^2\gamma(t)^2}{\tilde{r}(t)}$$
$$+ \quad 2\mu\frac{\tilde{\theta}(t)^T\phi(t)w(t+1)}{\tilde{r}(t)} + 2\mu^2\frac{\| \phi(t) \|^2}{\tilde{r}(t)^2}w(t+1)^2 \tag{4.37}$$

where $z(t)$ is defined by Eq. (4.11) and $V(t)$ is given by (4.28). By the Local Martingale Convergence Theorem we derive

$$S(t+1) = 2\mu\{d(t) - \sum_{j=1}^{t} \tilde{\theta}(j)^T\phi(j)w(j+1)\} > 0, t \geq 1 \quad (a.s.) \tag{4.38}$$

where

$$d(t) = C_d r(t)^{1-\epsilon}, 0 < C_d < \infty, \quad 0 < \epsilon < 1/2 \tag{4.39}$$

From (4.37) and Eq. (4.38) we obtain

$$V(t+1) \quad + \quad \frac{S(t+1)}{\tilde{r}(t)} \leq V(t) + \frac{S(t)}{\tilde{r}(t)} - 2\mu(1 - \frac{\mu}{2})\frac{z(t)^2}{\tilde{r}(t)}$$
$$+ \quad \frac{2\mu(1-\mu)|z(t)| \cdot |\gamma(t)| + \mu^2\gamma(t)^2}{\tilde{r}(t)} + 2\mu\frac{d(t) - d(t-1)}{\tilde{r}(t)} + \frac{\beta(t)^2}{\tilde{r}(t)^2} \tag{4.40}$$

where

$$\beta(t)^2 = 2\mu^2 \| \phi(t) \|^2 w(t+1)^2. \tag{4.41}$$

Obviously relations (3.1) and (4.40) are identical, and global stability of the SAC algorithms will be established by using Theorem 3.1. Let us verify that assumptions $(S_1) - (S_5)$ imply assumptions $(A_1) - (A_7)$ which are required for Theorem 3.1. Assumption (A_1) is satisfied as a consequence of the projection in (4.15). Assumption (A_2) follows from the definition of the sequence $\tilde{r}(t)$ given by Eq. (4.18), or Eq. (4.19). If we set $\lambda = 1$ in the first statement of Lemma 4.1, it is not difficult to see that assumption (A_4) is satisfied (a.s.), where from Eq. (4.22) we derive that $\Sigma_{\nu S}$ in (3.6) is given by

$$\Sigma_{\nu S}^2 = [m_1 C_\gamma(1) + \sigma_w C_w(1)]^2 \tag{4.42}$$

Assumption (A_5) is satisfied as a consequence of the Martingale Convergence Theorem and assumption (S_2), i.e., $\Sigma_{t=1}^\infty \beta(t)^2/\tilde{r}(t)^2 \leq C_\beta < \infty$ where $\beta(t)$ is defined by Eq. (4.41). Note that (4.38) implies $S(t+1)/\tilde{r}(t) \leq 4\mu C_d$ (a.s.). From Eq. (4.39), and statement (2) of Lemma 4.1 (for $\lambda = 1$) we conclude that $d(t) \leq o(n_z(t)^2 + t)$, by which assumption (A_6) is verified. Assumption (A_7) is identical to assumption (S_5) for $\lambda = 1$. Only assumption (A_3) is left to be proved.

A. Stochastic Adaptive Control when $\tilde{r}(t)$ is defined by Eq. (4.18)

From statement (2) of Lemma 4.1 and Eq. (4.18), it follows that assumption (A_3) is satisfied, where the constant C_θ and C_g are given by

$$C_\theta = 4C_{\phi 1}(1) \quad \text{and} \quad C_g = 4[m_1^2 C_{\phi 2}(1) + \sigma_w^2 C_{\phi 3}(1) + 1] \tag{4.43}$$

where $C_{\phi i}(1)$, $i = 1, 2, 3$ are defined by Eqs. (4.24)-(4.25), respectively.

After applying Theorem 4.1 the following global stability result is established:
Theorem 4.2: Let assumptions $(S_1) - (S_5)$ hold for $\lambda_0 = 1$ and $\lambda = 1$. Then the stochastic adaptive control algorithm (4.10), (4.15) and (4.18) provides:

1) $\lim_{N \to \infty} \sup_N \dfrac{1}{N} \sum_{t=1}^N (\tilde{\theta}(t)^T \phi(t))^2 \leq \Sigma_1^2$ (a.s.) \hfill (4.44)

where

$$\Sigma_1^2 = \max \left\{ \Sigma_{\gamma S}^2, \left(\frac{1}{C_\theta} + \frac{d_1}{\mu \rho_1} \right) C_g \right\} e^{\frac{d_1}{\mu \rho_1} C_\theta} \tag{4.45}$$

where $\Sigma_{\gamma S}$ is given by Eq. (3.13), while $\Sigma_{\nu S}$ in (3.13) is defined by Eq. (4.42) and ρ_1 is given by assumption (S_5) for $\lambda = 1$. In (4.45) the constants C_θ and C_g are defined by Eq. (4.43), while d_1 is given by (3.2S).

2) $\lim_{N \to \infty} \sup_N \dfrac{1}{N} \sum_{t=1}^N [y(t) - y^*(t)]^2 \leq \{\sigma_w + [1 + C_\gamma(1)]\Sigma_1 + \Sigma_{\nu S}\}^2$ (a.s.) \hfill (4.46)

3) $\lim_{N \to \infty} \sup_N \dfrac{1}{N} \sum_{t=1}^N \| \phi(t) \|^2 \leq C_{\phi 1}(1)\Sigma_1^2 + C_{\phi 2}(1)m_1^2 + C_{\phi 3}(1)\sigma_w^2$ (a.s.) \hfill (4.47)

Proof: The first statement of the theorem follows directly from Theorem 3.1. The second statement of the theorem follows from (4.36) (when $\lambda = 1$), and relation (4.44). Statement (3) of the theorem is obtained from (4.23) and (4.44). Thus the theorem is proved. □

B. Stochastic Adaptive Control when $\tilde{r}(t)$ is defined by Eq. (4.19)

Let us verify assumption (A_3). From statement (2) of Lemma 4.1 we derive

$$\max_{1 \leq \tau \leq t} \| \phi(\tau) \|^2 \leq \frac{C_{\phi 1}(\lambda)}{1 - \lambda} \max_{1 \leq \tau \leq t} z(\tau - 1)^2 + C_6 \max_{1 \leq \tau \leq t} \omega(\tau + 1)^2 \tag{4.48}$$

Relations (4.19) and (4.48) give

$$\tilde{r}(t) \leq \max \left\{ C_\theta \sum_{j=1}^{t-1} z(j)^2, o(t) \right\} \quad (a.s.) \tag{4.49}$$

where we have used the fact that $\max_{1 \leq \tau \leq t} w(\tau)^2 \leq o(t)$ $(a.s.)$, which follows from assumption (S_1) by applying the Markov inequality together with the conditional Borel Cantely Lemma [9]. In Eq. (4.49) the constant C_θ is given by

$$C_\theta = 4 \max \left\{ \frac{C_{\phi 1}(\lambda)}{1 - \lambda}; C_{\phi 1}(1)^{1-\epsilon} \right\}. \tag{4.50}$$

Note that the relation (4.49) establishes assumption (A_3) in the case where $C_g = 0$. After applying Theorem 4.1, the following result is obtained:

Theorem 4.3: Let assumptions $(S_1) - (S_5)$ hold where λ satisfies (4.17). Then the SAC algorithm (4.10), (4.15) and (4.19) yields

1) $\lim\limits_{N \to \infty} \sup\limits_N \dfrac{1}{N} \sum\limits_{t=1}^{N} (\tilde{\theta}(t)^T \phi(t))^2 \leq \Sigma_2^2 \quad (a.s)$ \hfill (4.51)

where

$$\Sigma_2^2 = \max \left\{ \frac{16[1 - \mu + \mu C_\gamma(1)]^2}{\rho_1^2}; \frac{2\mu}{\rho_1 - \rho_0} \right\} \Sigma_{\nu S}^2 e^{\frac{d_1 C_\theta}{\mu \rho_1}} \tag{4.52}$$

where $\Sigma_{\nu S}$ is defined by Eq. (4.42), C_θ is given by (4.50) and ρ_1 is defined by assumption (S_5).

2) $\lim\limits_{N \to \infty} \sup\limits_N \dfrac{1}{N} \sum\limits_{t=1}^{N} [y(t) - y^*(t)]^2 \leq \{\sigma_w + [1 + C_\gamma(1)]\Sigma_2 + \Sigma_{\nu S}\}^2 \quad (a.s.)$ \hfill (4.53)

3) $\lim\limits_{N \to \infty} \sup\limits_N \dfrac{1}{N} \sum\limits_{t=1}^{N} \| \phi(t) \|^2 \leq C_{\phi 1}(1)\Sigma_2^2 + C_{\phi 2}(1)m_1^2 + C_{\phi 3}(1)\sigma_w^2 \quad (a.s.)$ \hfill (4.54)

Proof: Since assumptions $(S_1) - (S_5)$ imply assumptions $(A_1) - (S_7)$, Theorem 3.1 can be applied, to yield the first statement of the theorem. In the proof of (4.51) we used the fact that in (3.4), $C_g = 0$, which follows from relation (4.49). The second statement of the theorem is obtained from (4.36) (when $\lambda = 1$) and (4.51). Statement (3) of the theorem follows from (4.51) and from statement (2) of Lemma 4.1 for $\lambda = 1$. Thus the theorem is proved. $\qquad \square$

Remark 4.1: From Theorems 4.1, 4.2, and 4.3, it is obvious that small algorithm gain μ may result in unacceptably large input and output signals. Specifically, from (4.30), (4.45) and (4.52) it follows that

$$\lim_{\mu \to 0} \Sigma_d = +\infty, \lim_{\mu \to 0} \Sigma_1 = +\infty \text{ and } \lim_{\mu \to 0} \Sigma_2 = +\infty \tag{4.55}$$

which means that for small μ. tracking error may diverge. From Eqs. (4.30), (4.45) and (4.52) it is obvious that there exists an optimal value μ_0 for which the established upper bounds are minimal. Similarly, from Assumption (S_5) it follows that the restriction on the tolerated unmodelled dynamics is given by the relation $C_\gamma(\lambda) < 1$. Clearly, the magnitude of the unmodelled dynamics is specified precisely and does not depend on the algorithm gain μ, as is usually the case in the literature on robust adaptive control.

5 Self-stabilization Mechanism in Adaptive Control

The results presented in Section 3 show that in the presence of unmodelled dynamics and external disturbances, the adaptive control algorithm possesses not only self-tuning but also a self-stabilization property. The latter means the following: whenever, as a consequence of the incorrect parameter estimates the adaptive system becomes unstable, the adaptive algorithm will stabilize itself by generating correct parameter estimates. During its operation, the adaptive controller passes through two phases characterized by the time intervals Q_k and T_k, defined by Eq. (3.17). In the time intervals Q_k, the bursting function $W(t + 1) \leq 0$, which implies the stability of the input and output signals for $t \in Q_k$. From (3.1) it is clear that in these time-intervals no characterization of the function $V(t + 1) + S(t + 1)/\tilde{r}(t)$ can be made. This function may diverge, thereby generating drifts of the parameter estimates. As a consequence of this, controller parameters may escape from the set of stabilizing controllers and the adaptive system will become unstable. Accordingly, the time intervals Q_k correspond to the *drift* phase of the adaptive algorithm. As time progresses, the norm $n_z(t)$ of the residual $z(t)$ becomes larger than $n_\gamma(t)$ and the bursting function $W(t + 1)$ given by Eq. (3.9) becomes positive. From relation (3.16) it is obvious that these periods of operation of the adaptive system correspond to the time intervals $T_k, k \geq 1$. Therefore, drift of the paraameter estimates in the time intervals Q_k gives rise to the bursting phenomenon. The behavior of the Lyapunov function $V_1(t+1) = V(t+1)+[S(t+1)+W(t+1)]/\tilde{r}(t)$ is described by the relation (3.22) and it is clear that $V_1(t + 1)$ decreases for $t \in T_k$. From (3.26) it also follows that in the time intervals T_k, fast adaptation takes place. As a consequence of this, the parameter estimates reenter the set of stabilizing controllers. It is obvious that the time-intervals $T_k, k \geq 1$, correspond to the *self-stabilization* phase of the adaptive system.

6 Conclusion

In this paper, a novel approach to the global stability analysis and design of robust adaptive systems is presented. The proposed methodology is based on the concept of the bursting function and Lyapunov theory, and is simple, elegant and quite natural. Using formally identical assumptions, the result established in Section 3 treats the global stability analysis of robust deterministic and stochastic systems in a unified way. The power of the Robust Ultimate Boundedness Theorem is demonstrated in the case of the robust deterministic and stochastic adaptive control problems. For the sake of clarity, the case of unit system delay is treated, and in the case of the stochastic problem statement, the ARX form of the nominal system model is assumed. If the system delay is assumed to be known, then the extension of the presented results in the case of general system delay is straightforward. In [9], it is shown that stochastic nominal system models having the ARMAX form with non-positive real noise dynamics can be treated by the same methodology. In a similar way, the application of the results established in Section 3 for robust adaptive filtering and prediction problems, can be accomplished without special difficulties.

References

[1] I. D. Landau, *Adaptive Control - The Model Reference Approach*, Marcel Dekker, New York, 1979.

[2] G. C. Goodwin and K. S. Sin, *Adaptive Filtering, Prediction and Control*, Englewood Cliffs, N. J.: Prentice-Hall, 1984.

[3] C. E. Rohrs, L. Valavani, M. Athans, and G. Stein, "Analytical Verification of Undesirable Properties of Direct Model Reference Adaptive Control Algorithm", *Proc. 20th IEEE Conf. Decision Contr.*, pp. 1272-1284, San Diego, CA, Dec. 1981.

[4] P. A. Ioannou and J. Sun, "Theory and Design of Robust Direct and Indirect Adaptive-Control Schemes", *Int. J. Control*, Vol. 47, pp. 775-813, 1988.

[5] B. E. Ydstie, "Stability of Discrete MRAC-revisited", *Systems & Control Letters*, Vol. 13, pp. 429-438, 1989.

[6] S. M. Naik, P. R. Kumar and B. E. Ydstie, "Robust Continuous Time Adaptive Control by Parameter Projection", submitted to *IEEE Trans. on Autom. Control*.

[7] L. Praly, S. F. Lin and P. R. Kumar, "A Robust Adaptive Minimum Variance Controller", *SIAM Journal on Control and Optimization*, Vol. 27, No. 2, pp. 235-266, 1989.

[8] M. S. Radenkovic and A. N. Michel, "Verification of the Self-Stabilization Mechanism in Robust Stochastic Adaptive Control Using Lyapunov Function Arguments", accepted for publication in *SIAM Journal on Control and Optimization*.

[9] W. F. Stout, *Almost Sure Convergence*, Academic Press, New York, 1974.

[10] B. E. Ydstie, "Stability of the Direct Self-tuning Regulator", in *Foundations of Adaptive Control*, Ed. by P. Kokotovic, Springer Verlag, 1991.

[11] M. S. Radenkovic and A. N. Michel, "Robust Adaptive Systems and Self-stabilization", Technical Report 91-74, Notre Dame University.

The Convergence of Self-tuning Feedback Control for Linear Stochastic Systems*

Wei Ren

Department of Electrical Engineering and Computer Science
University of California
Berkeley, CA 94720

P. R. Kumar
Coordinated Science Laboratory
University of Illinois
Urbana, IL 61801

Abstract

We consider adaptive control of linear stochastic systems, i.e., the control of unknown linear systems subject to stochastic disturbances whose spectra are also unknown. We examine the basic convergence issues, including the convergence of adaptive controllers and parameter estimates as well as the convergence of input and output. Despite over a decade of effort, previous works in this area are very much fragmented. Relatively complete convergence results are available only for adaptive minimum variance control of unit delay systems.

In this paper we propose the generalized certainty equivalence approach to stochastic adaptive control, where the estimates of disturbance innovations as well as parameter estimates are utilized. Based on this, the self-optimality of adaptive minimum variance controllers using an indirect approach and the stochastic gradient algorithm is established for general delay systems. Then we show that the self-optimality implies the self-tuning of adaptive controllers in general, by exhibiting the convergence of the parameter estimates to the null space of a certain covariance matrix and by characterizing the null space. The role of the system disturbance in providing an "internal excitation" is delineated. Finally we determine the exact order of external excitation required in order for the parameter estimates to converge to the true parameter. As a special case, it is found that for systems with white noise and delay greater than one, the parameter estimate is strongly consistent even without any external excitation,

*The research reported here has been supported by the U.S.A.R.O. under Contract No. DAAL-03-88-K-0046 and the Joint Services Electronics Program under Grant No. N00014-90-J-1270

and that for systems with white noise and unity delay, a nonzero constant set-point provides sufficient excitation for parameter consistency.

Of pedagogical interest is a deterministic reduction viewpoint we adopt in which all relevant properties of stochastically modeled disturbances are characterized deterministically by some long term average properties. Readers more familiar with deterministic theory may well find this viewpoint to be more enlightening with respect to understanding the goals and results of stochastic adaptive system theory.

1   Introduction

The goals of control are stabilization, command tracking and disturbance attenuation. When plants and disturbance models are known, various control laws can be designed to achieve these goals. Adaptive control is concerned with the case when the plants and disturbance models are unknown and when control is based on on-line modeling.

Previous works on adaptive control mostly fall into either the "deterministic school" or the "stochastic school." The former considers disturbance-free[1] plant and is mostly concerned with stabilization and command tracking, and robust stability with respect to bounded disturbance and unmodeled dynamics. And the latter assumes a stochastic model for disturbance, and hence can potentially achieve disturbance attenuation in addition to stabilization and command tracking, by making use of the estimated spectrum of the disturbance. Modeling the disturbance also allows the possibility of delineating the role of disturbance in providing natural "internal excitation" to adaptation in the feedback loop.

The so-called certainty equivalence approach to adaptive control is to couple a recursive parameter estimation scheme with an appropriate control design which maps the estimated plant and disturbance parameters to controller parameters. Clearly, the most desirable property of a certainty equivalent adaptive control is whether it will behave as if plant and disturbance parameters were known. More specifically, the following asymptotic properties are of basic interest.

i) Stability: is the closed-loop system stable, say in a mean square sense?

---

[1]A plant subject to periodic disturbance can be modeled as a disturbance-free system with uncontrollable modes on the stability boundary

ii) Convergence of input and output (self-optimality): Do input and output converge in the mean square sense to what they would be if the system were known.

iii) Convergence of adaptive controllers (self-tuning property): Does the adaptive controller self-tune almost surely.

iv) Convergence of parameter estimate: Does the parameter estimate converge to the true parameter?

Due to difficulty in analysis, existing convergence results for stochastic adaptive control are very much fragmented and incomplete despite over a decade of effort. Relatively complete convergence results are available only for adaptive minimum variance control (MV) of unit delay systems. The self-optimality of adaptive MV control of unit delay systems is shown by Goodwin, Ramadge and Caines [1] for stochastic gradient (SG) algorithm, by Sin and Goodwin [2], Zhang [3], and Chen [4] for modified least squares (MLS) algorithm, and by Guo and Chen [5] for extended least squares (ELS) algorithm. The self-tuning property of adaptive MV control of unit delay systems using SG algorithm is established by Becker, Kumar and Wei [6] and Kumar and Praly [7]. The self-optimality of adaptive MV control of general delay systems using SG algorithm is established in [8] by using a direct approach, i.e., by reparameterizing the system in the form of the MV controller, and by employing an interlaced multiple recursion for parameter estimation. Partial results using an indirect approach are reported by Fuchs [9]. A fairly complete set of convergence results for adaptive model reference control based on least squares algorithm has been obtained recently for general delay systems, however with the restricive assumptions that the noise is i.i.d gaussian and that the true parameter of the system lies outside certain set of zero Lebeque measure.

Very recently we have been able to obtain a set of relatively complete convergence results for both indirect and direct approaches using non-interlaced SG and ELS algorithms. While the direct approach is considered in Ren and Kumar [10], this paper focuses on the indirect approach for general delay-colored noise systems using the SG algorithm. Specifically, we analyze the almost sure convergence of adaptive controllers and parameter estimates as well as the mean square convergence of input and output. We first establish the mean square convergence of input and output, which implies self-optimality. Then we show the almost sure convergence of the parameter estimates to the null space of a certain covariance matrix. The self-tuning of adaptive controllers is shown by characterizing the null space as the intersection of two linear varieties, defined by excitations due to the external command signal and the system disturbance, which may we refer to as *external and internal excitations*, respectively. It is found that adaptive minimum variance regulators self-tune even without any external excitation. Finally we determine the exact order of the external excitation required in order for the parameter estimates to converge to the true parameter. As a special case, it is found that for systems with white noise and delay greater than one, the parameter estimate is consistent even without any external excitation, and that for systems with white noise but unity delay, a nonzero constant set-point provides sufficient excitation for parameter consistency.

Of pedagogical interest is a deterministic reduction viewpoint we adopt in which all relevant properties of stochastically modeled disturbance are characterized deterministically as some long term average properties. Some readers may find that this is a more convenient point of view to adopt.

The rest of the paper is organized as follows. The deterministic reduction viewpoint of stochastic disturbances is introduced in Section 2. In Section 3, we introduce the generalized certainty equivalence approach to stochastic adaptive control. The self-optimality of adaptive minimum variance control for general delay systems is established in Section 4. Section 5 considers the issues of self-tuning and parameter consistency.

2 A Deterministic Reduction of Stochastic Disturbances

As remarked earlier, one of the key concerns of stochastic adaptive control is to deal with the stochastically modeled disturbances. However, it is possible to fully dispense with all probabilistic assumptions, and adopt a completely deterministic model of such disturbances.

To start, let us consider the key properties satisfied by a "white noise" sequence $w(t)$. A probabilistic model for "white noise" is a stochastic process which satisfies the following three assumptions.

i) $w(t)$ is a martingale difference sequence with respect to an increasing sequence of σ-algebras $\{\mathcal{F}_t\}$, i.e.,

$$E[w(t) \mid \mathcal{F}_{t-1}] = 0 \text{ a.s., } \forall t .$$

ii) $E[w^2(t)|\mathcal{F}_{t-1}] = \sigma^2 > 0$ a.s., $\forall t$.

iii) $\sup_t E[|w(t)|^\alpha|\mathcal{F}_{t-1}] < +\infty$ a.s. for some $\alpha > 2$.

The key consequences of these assumptions are given in the following lemma, see $[11, 12, 6]$

Lemma 1. *Let f_t be a $\mathcal{F}_t$-measurable sequence. If assumptions i) and iii) above are satisfied, then*

a) $\sum_{t=1}^{N} f_{t-1} w(t) = o\left(\sum_{t=1}^{N} f_{t-1}^2\right) + O(1)$ a.s.[2]

b) $\sum_{t=1}^{N} f_{t-1} w(t)$ *converges a.s. on the event* $\left\{\sum_{t=1}^{\infty} f_{t-1}^2 < \infty\right\}$,

c) $\sum_{t=1}^{N} |f_{t-1}| w_t^2 = O\left(\sum_{t=1}^{N} |f_{t-1}|\right)$ *a.s. on the event* $\{\sup_t |f_t| < \infty\}$.

If, in addition, $w(t)$ satisfies assumption ii) above, then

d) $\frac{1}{N} \sum_{t=1}^{N} (w^2(t) - \sigma^2) = O(N^{-\delta})$ *a.s., where* $0 < \delta < \frac{\alpha-2}{\alpha}$,

e) $\lim \frac{1}{N} \sum_{t=1}^{N} f_{t-1}^2 = 0$ *a.s. on the event*

$$\{\lim \frac{1}{N} \sum_{t=1}^{N} f_{t-1}^2 w_t^2 = 0 \text{ and } \sup_t |f_{t-1}| < \infty\}.$$

In the above, f_t is a signal which depends only on $w(k)$, $k \leq t$, and on any other signal supposed to be "uncorrelated" with $\{w(t)\}$. Properties a) and b) essentially capture the fact that $w(t)$ is mean zero, and moreover is "uncorrelated" with f_{t-1}. Since the value of $f_{t-1} w(t)$

---

[2]Here and in the sequel, we write $\alpha_t = o(\beta_t)$ if $\alpha_t/\beta_t \to 0$, $\alpha_t = O(\beta_t)$ if $\sup_t \alpha_t/\beta_t < \infty$, and $\alpha_t \sim \beta_t$ if $\alpha_t = O(\beta_t)$ and $\beta_t = O(\alpha_t)$.

is therefore as "likely" to be positive as negative, in forming the sum $\sum_{t=1}^{N} f_{t-1} w(t)$ there are likely to be many cancellations, thus rendering it small in comparison with the energy, $\sum_{t=1}^{N} f_{t-1}^2$, in f. Property c) essentially states that since $w^2(t)$ has a bounded average value, by assumption iii), the sum $\sum_{t=1}^{N} |f_{t-1}| w^2(t)$ grows at the same rate as $\sum_{t=1}^{N} |f_{t-1}|$. Finally properties d) and e) show that w is a natural non-negligible unpredictable excitation.

The above sample path properties capture the essence of a white noise. Thus, instead of starting with a set of probabilistic assumptions, one could dispense with all stochastic assumptions, and simply suppose that the noise has the properties (a-e) above, and all the results in this paper would continue to hold.

In this paper we consider the single input, single output linear system described by

$$A(q^{-1})y(t) = q^{-d}B(q^{-1})u(t) + v(t), \tag{1}$$

where q^{-1} is the backward shift operator, i.e., $q^{-1}y(t) = y(t-1)$,

$$
\begin{aligned}
A(q^{-1}) &= 1 + a_1 q^{-1} + \cdots + a_p q^{-p} \\
B(q^{-1}) &= b_0 + b_1 q^{-1} + \cdots + b_h q^{-h}, \tag{2}
\end{aligned}
$$

$\{y(t)\}$ and $\{u(t)\}$ are the output and input of the system, and $\{v(t)\}$, the disturbance. To capture the frequency spectral property of the disturbance $v(t)$, we will model it as a moving average of a "white noise" sequence satisfying the sample path properties of Lemma 1. Thus

$$v(t) = C(q^{-1})w(t), \tag{3}$$

where

$$C(q^{-1}) = 1 + c_1 q^{-1} + \cdots + c_r q^{-r}.$$

The shift structure contained in the above moving average representation (3), and Lemma 1 a)-c), will provide us with certain auto-correlation property which we can exploit for disturbance attenuation using feedback control. On the other hand, Lemma 1 d) and e) will provide us with certain "persistency of excitation" properties which we will find useful.

3 The Generalized Certainty Equivalence Approach

A general linear control law is usually of the form

$$R(q^{-1})u(t) = S(q^{-1})y(t) + T(q^{-1})y^*(t), \tag{4}$$

where $y^*(t)$ is the command signal, $R, S,$ and T are polynomials in q^{-1} and are generated from the model parameter (A, B, C) by some design mapping $\mathcal{F}$,

$$(R, S, T) = \mathcal{F}(A, B, C), \tag{5}$$

The traditional certainty equivalence adaptive control is to replace at each time instant (R, S, T) in (4) by $(\widehat{R}_t, \widehat{S}_t, \widehat{T}_t)$ which is produced by

$$(\widehat{R}_t, \widehat{S}_t, \widehat{T}_t) = \mathcal{F}(\widehat{A}_t, \widehat{B}_t, \widehat{C}_t), \tag{6}$$

where $(\widehat{A}_t, \widehat{B}_t, \widehat{C}_t)$ is the estimated model parameter at time t.

In view of the fact that the parameter estimator will automatically produce as a by-product the estimate $\widehat{w}(t)$ of the disturbance innovations $w(t)$, we propose to consider instead the control law of a more general form

$$R(q^{-1})u(t) = S(q^{-1})y(t) + T(q^{-1})y^*(t) + M(q^{-1})w(t), \tag{7}$$

where $R, S,$ and T are polynomials in q^{-1}, $M(q^{-1})$ is a stable rational transfer function in q^{-1}, and (R, S, T, M) are prescribed by some design mapping $\mathcal{F}_G$,

$$(R, S, T, M) = \mathcal{F}_G(A, B, C). \tag{8}$$

The *generalized certainty equivalence approach* to adaptive control which we propose is to replace in (7), at each time instant, $w(t)$ by $\widehat{w}(t)$, and (R, S, T, M) by $(\widehat{R}_t, \widehat{S}_t, \widehat{T}_t, \widehat{M}_t)$ produced by

$$(\widehat{R}_t, \widehat{S}_t, \widehat{T}_t, \widehat{M}_t) = \mathcal{F}_G(\widehat{A}_t, \widehat{B}_t, \widehat{C}_t). \tag{9}$$

It can be shown that a MV control law in the form of (7) for the system (1,3) is given by

$$B(q^{-1})R'(q^{-1})u(t) = S(q^{-1})y(t) + y^*(t) + M(q^{-1})w(t), \tag{10}$$

where R', S, and M are the polynomial solutions of the following equations,

$$A(q^{-1})R'(q^{-1}) - q^{-d}S(q^{-1}) = 1 \tag{11}$$

$$F(q^{-1}) - q^{-d}M(q^{-1}) = C(q^{-1})R'(q^{-1}) \quad \text{with } F(q^{-1}) = \sum_{i=0}^{d-1} f_i q^{-i}, f_0 = 1. \tag{12}$$

It is instructive to compare the MV control law given by (10-12) with the following standard MV control law in the form of (4) (see [13]),

$$B(q^{-1})F(q^{-1})u(t) = G(q^{-1})y(t) + C(q^{-1})y^*(t), \tag{13}$$

where F and G are the polynomial solution of the following equation,

$$A(q^{-1})F(q^{-1}) - q^{-d}G(q^{-1}) = C(q^{-1}) \quad \text{with } F(q^{-1}) = \sum_{i=0}^{d-1} f_i q^{-i}, f_0 = 1. \tag{14}$$

It is now easy to show that if $w(t)$ in (10) is generated by the following "linear observer,"

$$C(q^{-1})w(t) = A(q^{-1})y(t) - q^{-d}B(q^{-1})u(t), \tag{15}$$

then the control law given by (10-12) is in fact identical to that given by (13-14). Of course, when (10) is used as the underlying control law of the generalized certainty equivalence adaptive control described earlier, $w(t)$ will be generated by the parameter estimator, which may be regarded as a nonlinear observer, thus eliminating the need for the "redundant" linear observer (15).

4 Self-Optimality

In this section, we establish the self-optimality of the adaptive minimum variance control based on the generalized certainty equivalence approach and (10). The following stochastic gradient (SG) algorithm is employed for parameter estimation,

$$\theta(t) = \theta(t-1) + \frac{\phi(t-1)}{r(t-1)}(y(t) - \phi^T(t-1)\theta(t-1)) \tag{16}$$

$$r(t-1) = r(t-2) + \|\phi(t-1)\|^2, \quad r(-1) = 1, \tag{17}$$

where $\theta(t)$ is the estimate of $\theta_0 := [a_1, \ldots a_p, b_0, \ldots, b_h, c_1, \ldots, c_r]^T$, and

$$\phi(t-1) := [-y(t-1), \ldots, -y(t-p), u(t-d), \ldots, u(t-d-h), \hat{w}(t-1), \ldots, \hat{w}(t-r)]^T,$$

and

$$\hat{w}(t) := y(t) - \phi^T(t-1)\theta(t)$$

The following theorem summarizes the useful asymptotic properties of the SG algorithm when applied to the system (1,3). For convenience, let us define

$$\phi_0(t-1) := [-y(t-1), \ldots, -y(t-p), u(t-d), \ldots, u(t-d-h), w(t-1), \ldots, w(t-r)]^T,$$

$$
\begin{aligned}
R(N) &:= I + \sum_{t=1}^{N} \phi(t)\phi^T(t) \\
R_0(N) &:= I + \sum_{t=1}^{N} \phi_0(t)\phi_0^T(t) \\
r_0(N) &:= \text{trace}\left(R_0(N)\right) \\
\tilde{\theta}(t) &:= \theta(t) - \theta_0.
\end{aligned}
$$

Theorem 1. *Consider the SG algorithm defined by (16) and (17) and applied to the AR-MAX model (1,3). Assume that $C(q^{-1})$ is Hurwitz and satisfies*

$$\min_{|z|=1} \text{Re}[C(z)] > 0. \tag{18}$$

Then,

$$\|\tilde{\theta}(t)\| \text{ converges a.s.} \tag{19}$$

$$\sum_{t=1}^{\infty} \|\theta(t) - \theta(t-k)\|^2 < \infty, \text{ a.s., } \forall k < \infty \tag{20}$$

$$\sum_{t=1}^{+\infty} \frac{(\hat{v}(t) - v(t))^2}{r(t-2)} < \infty, \text{ a.s.} \tag{21}$$

$$r(t) \sim r_0(t) \text{ a.s.} \tag{22}$$

$$\sum_{t=1}^{\infty} \frac{(y(t) - \phi^T(t-1)\theta(t-k) - v(t))^2}{r(t-1)} < \infty, \text{ a.s., } \forall k < \infty. \tag{23}$$

Moreover, if

$$r_0(N) \to \infty, \tag{24}$$

then

$$\frac{1}{r(N-1)}\|R(N-1)\tilde{\theta}(N)\| \to 0 \quad \text{a.s.} \tag{25}$$

Proof. See [14, 15].

The following theorem establishes the self-optimality of adaptive MV control.

Theorem 2. *Consider the adaptive MV control based on the parameter estimator (16-17) and the MV control law (10-12), and using the generalized certainty equivalence approach (9).*

Suppose that

A1) $B(q^{-1})$ is Hurwitz, and $b_0 \neq 0$

A2) $\min_{|z|=1} Re\left[C(z)\right] > 0$

A3) $y^*(t)$ is bounded.

A4) The distribution of $(w(0),\ldots,w(t))$ is absolutely continuous with respect to the Lebesgue measure for $\forall t \geq 0$.

Then

$$\lim_{N\to\infty} \frac{1}{N}\sum_{t=1}^{N}[y(t)-y^0(t)]^2 \tag{26}$$

where $y^0(t) := y^*(t) + F(q^{-1})w(t)$ is the optimum output. Further, let $u^0(t)$ be the optimum input corresponding to $y^0(t)$, i.e., it satisfies the following equation

$$A(q^{-1})y^\circ(t) = q^{-d}B(q^{-1})u^\circ(t) + C(q^{-1})w(t). \tag{27}$$

Then

$$\lim_{N\to\infty} \frac{1}{N}\sum_{t=1}^{N}[u(t)-u^0(t)]^2 = 0 \quad \text{a.s.} \tag{28}$$

Proof. For details of the proof, we refer the reader to [14, 15].

5 The Convergence of Adaptive Controllers and Parameter Estimates

In the previous section, we have established the self-optimality of adaptive MV control. In this section we consider the issue of self-tuning and parameter consistency and show that self-optimality implies self-tuning in general.

Define $\phi^o(t)$ and $R^o(t)$ as $\phi_0(t)$ and $R_0(t)$, using $y^o(t)$ and $u^o(t)$ in place of $y(t)$ and $u(t)$, respectively. Then from (26) and (28), we have

$$\frac{1}{N}\sum_{t=0}^{N} \|\phi^o(t) - \phi(t)\|^2 \to 0 \text{ a.s.,}$$

and

$$\frac{1}{N}\|R^o(N) - R(N)\| \to 0 \text{ a.s.}$$

From (26,28 and Lemma 1, $r_0(N) \to \infty$. It then follows from (25) that

$$\frac{1}{N}\|R^o(N-1)\tilde{\theta}(N)\| \to 0 \text{ a.s.}$$

For simplicity, assume that the ensemble correlations of the command signal $y^*(t)$ exist. Then $\frac{R^o(N-1)}{N}$ converges. Let

$$\Phi := \lim_{N\to\infty} \frac{R^o(N-1)}{N} .$$

It can then be shown that $\tilde{\theta}(t)$ converges almost surely to the null space of Φ. The following theorem summarizes the above development and further characterizes the null space of Φ.

Theorem 3. *Consider the indirect adaptive MV control using the SG algorithm. Let the assumptions of Theorem 2 hold, and suppose that the ensemble correlations of $\{y^*(t)\}$ exist. Then*

$$\tilde{\theta}(t) \to \mathcal{N}(\Phi) \text{ a.s. }$$

Further, $\forall \tilde{\theta} \in \mathcal{N}(\Phi)$, using the obvious notation, let $\hat{A}$, $\hat{B}$, $\hat{C}$, $\hat{F}$ and $\hat{G}$ be the appropriate polynomials corresponding to the parameter vector $\tilde{\theta} + \theta_o$. Then

$$F(q^{-1}) = \hat{F}(q^{-1}), \tag{29}$$

$$B(q^{-1})\widehat{G}(q^{-1}) = \widehat{B}(q^{-1})G(q^{-1}), \tag{30}$$

and

$$\frac{1}{N}\sum_{t=0}^{N}\left((\widehat{B}C - B\widehat{C})y^*(t)\right)^2 \to 0. \tag{31}$$

Moreover, if $B(q^{-1})$ and $G(q^{-1})$ do not have a common factor, and $\{y^*(t)\}$ is persistently exciting of order $\ell_p \geq \min(\deg C(q^{-1}) - \deg F(q^{-1}) + 1, \deg A(q^{-1}) + 1)$, then

$$\theta(t) \to \theta_0 \text{ a.s.}$$

Proof. Since $\tilde{\theta} \in \mathcal{N}(\Phi)$, we have

$$\frac{1}{N}\sum_{t=1}^{N}((A - \widehat{A})y^\circ(t) + q^{-d}(\widehat{B} - B)u^\circ(t) + (\widehat{C} - C)w(t))^2 \to 0 \text{ a.s.}$$

Therefore,

$$\frac{1}{N}\sum_{t=1}^{N}((A - \widehat{A})y^*(t) + q^{-d}(\widehat{B} - B)u^\circ(t) + ((A - \widehat{A})F + \widehat{C} - C)w(t))^2 \to 0. \tag{32}$$

Since $u^\circ(t)$ is $\mathcal{F}_t$-measurable, it follows as in [6, 7] that the first d coefficients of $(A - \widehat{A})F + \widehat{C} - C$ have to be zero, which yields (29).

Multiplying (32) by $B(q^{-1})$ and substituting (27) into (32), we obtain

$$\frac{1}{N}\sum_{t=0}^{N}\left[(\widehat{B}A - B\widehat{A})y^*(t) + (B(\widehat{C} - \widehat{A}F) - \widehat{B}(C - AF))w(t)\right]^2 \to 0 \text{ a.s.}$$

Hence,

$$\frac{1}{N}\sum_{t=0}^{N}\left((\widehat{B}A - B\widehat{A})y^*(t)\right)^2 \to 0 \tag{33}$$

and

$$\frac{1}{N}\sum_{t=0}^{N}\left((B(\widehat{C} - \widehat{A}F) - \widehat{B}(C - AF))w(t)\right) \to 0 \text{ a.s.} \tag{34}$$

So

$$B(\widehat{C} - \widehat{A}F) = \widehat{B}(C - AF), \tag{35}$$

which is equivalent to (30). If B and G do not have a common factor, then

$$\widehat{B} = \lambda B, \quad \widehat{G} = \lambda G$$

for some scalar λ.

If $\ell_p \geq \deg A + 1$, from the above and (33), we obtain $\lambda A - \hat{A} = 0$. It then follows from (35) that $\lambda = 1$, $\hat{A} = A$, and $\hat{C} = C$.

From (35), we obtain

$$(\hat{B}A - B\hat{A})F = \hat{B}C - B\hat{C}.$$

Therefore,

$$\lambda A - \hat{A} = \frac{\lambda C - \hat{C}}{F}$$

It follows that $\deg(\lambda A - \hat{A}) \leq \deg C - \deg F$. Hence if $\ell_p \geq \deg C - \deg F + 1$, we can proceed similarly to conclude parameter consistency. This completes the proof.

Remarks

i) The above method for studying parameter convergence and self-tuning property can be easily extended to other situations, e.g., adaptive model reference control based on the ELS and MLS algorithms, once the input and output convergence in the mean square sense is established.

ii) The richness of the signals in the feedback loop is due to both external and internal excitations. While Equation (33) defines a linear variety to which the limit of the parameter estimates is constrained due to the external excitation, Equation (34) defines that due to the internal excitation. When the internal excitation is not accounted for, the minimum order of the external excitation required for parameter consistency would be $\deg A + \deg B + 1$, which is much larger than that required in Theorem 3.

iii) Consider for instance the white noise case, i.e., when $C(q^{-1}) = 1$, and suppose $\deg F = d - 1$, which holds generically, then for $d \geq 2$, parameter consistency occurs even without any external excitation; for $d = 1$, parameter consistency occurs for nonzero constant reference tracking, i.e., set-point regulation.

iv) In the case of direct adaptive MV *regulation* of systems with general delay, based on the SG algorithm using a priori estimates, the parameters of $C(q^{-1})$ need not be estimated, and it can then be shown that if there is no overparameterization, the limit set $\Phi + \theta_o$ of $\theta(t)$ is a line passing through the origin and the true parameter. The fact that $\|\tilde{\theta}(t)\|$ converges can then be used to establish the convergence of $\theta(t)$ to a particular point on the line. This result then generalizes to the case of general delay the result of Becker, Kumar and Wei [6], which establishes the convergence of the parameter estimates to a random multiple of the true parameter for the unit delay case.

6 Concluding Remarks

With recent development reported in this paper and in [10, 5, 14], we are now near the end of the crusade for convergence and performance analysis of self-tuning minimum variance prediction and model reference control, except for a special case involving ELS, general delay and indirect scheme. However, the performance of adaptive control involving time varying stochastic systems and short-term memory algorithms still defies rigorous analysis and remains a significant open problem.

References

[1] G. C. Goodwin, P. J. Ramadage, and P. E. Caines, "Discrete time stochastic adaptive control," *SIAM J. Control Optimization*, vol. 19, pp. 829–853, 1981. *SIAM J. Control and Optimization*, vol. 20(6):893.

[2] K. Sin and G. Goodwin, "Stochastic adaptive control using a modified least squares algorithm," *Automatica*, vol. 18, pp. 315–321, 1982.

[3] Z. You-hong, "Stochastic adaptive control and prediction based on a modified least squares–the general delay-colored noise case," *IEEE Transactions*, vol. 27, pp. 1257–1260, 1982.

[4] H. F. Chen, "Recursive system identification and adaptive control by use of the modified least squares algorithm," *SIAM J. Control Optimization*, vol. 22, pp. 758–776, 1984.

[5] L. Guo and H. F. Chen, "The Åström-Wittenmark's self-tuning regulator revisited and ELS-based adaptive trackers," *IEEE Transactions*, vol. AC-36, no. 7, pp. 802–812, 1991.

[6] A. Becker, P. R. Kumar, and C. Z. Wei, "Adaptive control with the stochastic approximation algorithm: Geometry and convergence," *IEEE Transactions on Automatic Control*, vol. AC-30, pp. 330–338, April 1985.

[7] P. R. Kumar and L. Praly, "Self-tuning trackers," *SIAM Journal on Control and Optimization*, vol. 25, pp. 1053–1071, July 1987.

[8] G. C. Goodwin, K. S. Sin, and K. K. Saluja, "Stochastic adaptive control and prediction-the general delay-colored noise case," *IEEE Transactions*, vol. AC-25, pp. 946–949, 1980.

[9] J. J. Fuchs, "Indirect stochastic adaptive control: The general delay–colored noise case," *IEEE Transactions on Automatic Control*, vol. AC-27, no. 2, pp. 470–472, 1982.

[10] W. Ren and P. R. Kumar, "Interlacing is not necessary for adaptive control," in *Proceedings of the 1991 American Control Conference, Boston, MA, June 26-28, 1990*.

[11] Y. S. Chow, "Local convergence of martingales and the law of large numbers," *Annals of Mathematical Statistics*, vol. 36, pp. 552–558, 1965.

[12] T. L. Lai and C. Z. Wei, "Least squares estimate in stochastic regression with applications to identification and control of dynamic systems," *Annals of Mathematical Statistics*, vol. 10, pp. 154–166, 1982.

[13] G. C. Goodwin and K. S. Sin, *Adaptive Filtering, Prediction and Control*. Englewood Cliffs, NJ: Prentice-Hall, 1984.

[14] W. Ren and P. R. Kumar, "Stochastic adaptive system theory: Recent advances and a reappraisal." To appear in *Foundations of Adaptive Control*, P. V. Kokotović. Ed. Springer Verlag, 1991.

[15] W. Ren, *Stochastic Adaptive System Theory for Identification, Filtering, Prediction and Control*. Urbana, IL: Ph.D Dissertation, University of Illinois, August 1991.

Some aspects of Robustness in Stochastic and Adaptive Control

Wolfgang J. Runggaldier

Dipartimento di Matematica Pura ed Applicata

Università di Padova

35131 - Padova, Italy

ABSTRACT : We consider a family of stochastic control models with partial state observation that are driven by noise disturbances supposed to model reality more closely than white Gaussian noise (WGN). Under some assumptions it is shown that, for small values of the indexing parameter, these models are close in a suitable sense to an ideal limit model with linear dynamics and WGN. It is furthermore shown that nearly optimal controls for the limit model remain nearly optimal also in the prelimit models of the given family and that, is a "limiter" is used in the observations, the corresponding value of the objective function is little sensitive to the tails of the observation noise distribution.

INTRODUCTION

To solve real world problems one first needs to build a model. The problems are then solved with respect to the model and the natural question arises of how good is the solution derived for the model when applied to reality ? This question is of course outside the realm of mathematics, but one can cope with the issue by designing solution methods that perform satisfactorily also when applied to situations that are "not too far" from the assumed model. Following standard terminology, we shall refer to such methods as *"robust methods"*. Sometimes the robustness of a solution method depends also on the model that is being used and a clever choice of a model may enhance the robustness of the corresponding solution method. Models with such a property will be referred to as *"robust models"*.

For stochastic problems such as filtering and control, a crucial element when building a model are the driving noise disturbances. For mathematical convenience, but not exclusively for this reason, these noise disturbances are generally assumed to be white Gaussian (WGN). In reality, the disturbances are never exactly WGN, but may be close (in the sense of convergence in distribution) to WGN. In addition, real noise distributions may have fatter tails than a Gaussian. Such fatter tails often produce what by analogy to Statistics we may call "outliers" and they may cause poor performance of standard (even robust) solution methods. To further robustify the methods, in Statistics one uses a preliminary trimming of the data with the purpose of cutting off, or at least neutralizing, the outliers (see e.g. [4]). A related device is known from engineering practice, namely the use of a "limiter" (hard or soft) in the observations (see e.g.[3], [6]).

In the present paper we investigate some issues related to robust methods and robust models in the context of stochastic control with partial state observation. We consider in fact a family of stochastic control models with partial state observation ((1.1)-(1.3) below), where the driving noises are close in distribution to WGN and where, to enhance robustness, the possibility is also given of using a "limiter" in the observations. In applications we may think of the "real model" as corresponding to one element in the given family. Under some assumptions we show that the models in the family are "close" in a suitable sense to an ideal limit model with linear dynamics and driving noises that are WGN. It is then shown that nearly optimal controls for the limit model, provided they possess a Lipschitz property, are robust, namely they remain nearly optimal also for the (prelimit) models in the given family. In addition to showing robustness for nearly optimal controls, in the present paper we also show that, with a suitable choice of a limiter in the observations, one can enhance their robustness in the sense that the corresponding value of the objective function is very little sensitive to the tails of the observation noise distribution.

The results of this paper are based on previous results, obtained in a filtering context in [8] and [13] (for related results in a filtering context see also [7], [10], [17]). The approach itself falls within the area of "diffusion approximations" and related techniques wich have a wide range of applicability; see e.g. the contributions to this Conference by Krichagina, Sethi and Taksar, namely [5], [18], [20]; see also [21] for a related robustness study in a specific adaptive stochastic control model.

In section 1 we describe our family of prelimit stochastic control models with partial observation of the state; we also formulate our assumptions and state the basic result of the paper, whose proof is deferred to section 3. In section 2 we recall some results from [13] for the filtering problem associated to our prelimit stochastic control models with partial observation of the state. In section 3 we derive our main results by extending to the control case the filtering results mentioned in section 2. We finally remark that the results of this paper can easily be extended to adaptive stochastic control and control problems in discrete time (for such an extension in a filtering context see [14]).

1. DESCRIPTION OF THE MODELS, ASSUMPTIONS AND BASIC RESULT

On a given probability space $(\Omega, \mathcal{F}, P)$ consider the following family, parametrized by ε, of stochastic control models with partial observation of the state

$$\dot{x}_t^\varepsilon = A\, x_t^\varepsilon + u_t + \frac{1}{\varepsilon}\, \psi_{t/\varepsilon^2} \tag{1.1}$$

$$\dot{y}_t^\varepsilon = \frac{1}{\varepsilon}\, H(\,\varepsilon\, C\, x_t^\varepsilon + \phi_{t/\varepsilon^2}\,) \quad ; \quad y_0^\varepsilon = 0 \tag{1.2}$$

$$J^\varepsilon(u) = E\left\{\int_0^T r_t(x_t^\varepsilon, u_t)\, dt\right\} \tag{1.3}$$

where x_t^ε, y_t^ε are the state and observation processes respectively that for simplicity we assume to be scalar valued; $\psi = (\psi_t)$ and $\phi = (\phi_t)$ are strictluy stationary and ergodic stochastic processes that for simplicity we assume to be independent ; u_t is the control action at time t taking values in a compact set $U \subset \mathcal{R}$; $H(.)$ is a (nonlinear) function, the so-called "limiter". The criterion is to minimize, for a fixed finite horizon T, the objective function $J^\varepsilon(u)$ over the admissible controls that are the measurable functions

$$u_t = u_t(y^\varepsilon) \tag{1.4}$$

which, for $t \leq T$, depend on y_s^ε with $s \leq t$.

In line with [13] we make the following assumptions where $\mathcal{F} = (\mathcal{F}_t)_{t \in \mathcal{R}}$ is the filtration generated by x_0^ε as well as by the processes ψ and ϕ, and which satisfies the general conditions.

<u>A.1.</u> For some $p > 2$ and with $\| \alpha \|_p = (E |\alpha|^p)^{1/p}$ we have

$$\| \psi_0 \|_p < \infty , \quad E (\psi_0) = 0 , \quad \int_0^\infty \| E (\psi_t | \mathcal{F}_0) \|_p \, dt \; < \; \infty$$

<u>A.2.</u> The (limiter) function $H = H(x)$ is twice continuosly differentiable and its derivatives are uniformly bounded, i.e. for some $L > 0$

$$| H'(x) | \leq L \;\; , \;\; | H''(x) | \leq L$$

<u>A.3.</u> For some $p > o$ we have

$$\| H(\phi_0) \|_p < \infty , \quad E (H(\phi_0)) = 0 , \quad \int_0^\infty \| E (H(\phi_t) | \mathcal{F}_0) \|_p \, dt \; < \; \infty$$

<u>A.4.</u> The admissible controls satisfy a Lipschitz condition in the sense that there exists a nonnegative function $L(s)$ with $L(s) \leq L$ such that

$$| u_t(y) - u_t(y') | \leq \int_0^t L(s) | y_s - y_s' | \, ds \tag{1.5}$$

<u>A.5.</u> The cost functions $r_t(x,u)$ are bouded and Lipschitz in the sense that there exists $M > 0$ such that, with L as before, we have for all $t \leq T$

$$| r_t(x,u) - r_t(x',u') | \leq \min [M, L |x\text{-}x'|] + \min [M, L |u\text{-}u'|]$$

Remark 1.1.

i) A classical limiter function H is the so-called "soft limiter" which, for a given $k > 0$, takes the form

$$H(x) = \begin{cases} -1 & \text{for } x < -k \\ x/k & \text{for } -k \le x \le k \\ +1 & \text{for } x > k \end{cases}$$

This function H does not directly satisfy A.2; by smoothing out the corners at $x = \pm k$, one can however obtain a limiter function satisfying A.2, whose practical effect is the same as that of the limiter in (1.6).

ii) Considering admissible controls that, besides (1.4), satisfy also (1.5), may appear to be restrictive. If however the purpose is not so much to obtain an optimal control, but rather a nearly optimal control (this is in fact the purpose below), then an a-priori restriction to Lipschitz controls becomes acceptable (see e.g. [9; Thm.4]).

iii) Assumption A.5 allows to include locally Lipschitz cost functions, provided they are truncated if necessary (e.g. truncated quadratic costs).

Under the given assumptions we may use results from [13] that allow model (1.1), (1.2) to be rewritten as

$$x_t^\varepsilon = x_0^\varepsilon + \int_0^t A\, x_s^\varepsilon\, ds + \int_0^t u_s(y^\varepsilon)\, ds + M_t^\varepsilon \tag{1.7}$$

$$y_t^\varepsilon = \int_0^t E\{ H'(\phi_0)\}\, C\, x_s^\varepsilon\, ds + Q_t^\varepsilon + q_t^\varepsilon \tag{1.8}$$

where

$$M_t^\varepsilon = \frac{1}{\varepsilon}\int_0^t \psi_{s/\varepsilon 2}\, ds\ , \qquad Q_t^\varepsilon = \frac{1}{\varepsilon}\int_0^t H\,(\phi_{s/\varepsilon 2})\, ds \tag{1.9}$$

$$q_t^\varepsilon = \int_0^t [\, H'(\phi_{s/\varepsilon 2}) - E\{ H'(\phi_0)\}\,]\, C\, x_s^\varepsilon\, ds + \int_0^t \gamma_s^\varepsilon\, ds \tag{1.10}$$

with

$$\gamma_t^\varepsilon = \varepsilon^{-1} H (\varepsilon C x_t^\varepsilon + \phi_{t/\varepsilon^2}) - \varepsilon^{-1} H(\phi_{t/\varepsilon^2}) - H'(\phi_{t/\varepsilon^2}) C x_t^\varepsilon \tag{1.11}$$

Furthermore, it follows from [13] that for $\varepsilon \to 0$

$$M_t^\varepsilon \Rightarrow B W_t \quad , \quad Q_t^\varepsilon \Rightarrow D_H V_t \tag{1.12}$$

where $\Rightarrow$ denotes convergence in distribution, W_t, V_t are independent standard Wiener processes, and

$$B = 2 \int_0^\infty E (\psi_t \psi_0) \, dt \quad , \quad D_H = 2 \int_0^\infty E (H(\phi_t) H(\phi_0)) \, dt \tag{1.13}$$

Finally, by analogy to (6.9)-(6.13) in [13] one can show that

$$P - \lim_{\varepsilon \to 0} \sup_{t \leq T} |q_t^\varepsilon| = 0 \tag{1.14}$$

which, together with (1.12), implies

$$Q_t^\varepsilon + q_t^\varepsilon \Rightarrow D_H V_t \tag{1.15}$$

In conclusion, letting

$$C_H = E \{ H'(\phi_0) \} C \quad , \quad N_t^\varepsilon = Q_t^\varepsilon + q_t^\varepsilon \tag{1.16}$$

we have that the pair $(x_t^\varepsilon, y_t^\varepsilon)$ admits the representation

$$x_t^\varepsilon = x_0^\varepsilon + \int_0^t A x_s^\varepsilon \, ds + \int_0^t u_s(y^\varepsilon) \, ds + M_t^\varepsilon \tag{1.17}$$

$$y_t^\varepsilon = \int_0^t C_H x_s^\varepsilon \, ds + N_t^\varepsilon \tag{1.18}$$

where, for $\varepsilon \to 0$,

$$M_t^\varepsilon \Rightarrow B W_t \quad , \quad N_t^\varepsilon \Rightarrow D_H V_t \tag{1.19}$$

and where we furthermore assume that

$$x_0^\varepsilon \Rightarrow x_0 \quad , \quad E x_0^\varepsilon \to E x_0 \tag{1.20}$$

with x_0 being a Gaussian random variable, independent of W_t and V_t.

Consider then the following "limit" control model with linear dynamics and noise disturbances given by standard Wiener processes

$$x_t = x_0 + \int_0^t A \, x_s \, ds + \int_0^t u_s(y) \, ds + B \, W_t \qquad (1.21)$$

$$y_t = \int_0^t C_H \, x_s \, ds + D_H \, V_t \qquad (1.22)$$

$$J(u) = E \left\{ \int_0^T r_t(x_t, u_t) \, dt \right\} \qquad (1.23)$$

where, as before, $u_t(y)$ satisfies (1.4), (1.5).

The main result of this paper, to be proved in section 3, is

Theorem 1.1. We have

$$\lim_{\varepsilon \to 0} | J^\varepsilon(u) - J(u) | = 0$$

uniformly with respect to the admissible controls u satisfying (1.4), (1.5) ■

This theorem immediately implies the following robustness result

Corollary 1.1. If u^δ is a δ - optimal control for the limit model (1.21)-(1.23) within the class of admissible controls (1.4),(1.5), then, for sufficiently small ε, u^δ is 2δ - optimal also for the prelimit models (1.1) - (1.3). ■

Methods to obtain δ - optimal controls for models of the type (1.21) - (1.23) can be found e.g. in [11], [2], [15] and in [16] for the adative control case in discrete time.

So far we have obtained robustness of the nearly optimal controls for the "ideal" model (1.21) - (1.23) in the sense that they remain nearly optimal also for more "realistic" models of the type (1.1)-(1.3) and this is in line with the common notion of robustness in stochastic control. In section 3 we will however also show how models with a limiter in the observations may enhance robustness, in particular with respect to the tails of the observation noise distribution. To this effect, in the next section 2 we recall some results from [13] concerning the filtering problem associated with model (1.1) - (1.2).

2. ROBUSTNESS RESULTS IN A FILTERING CONTEXT

Consider the family $(x_t^\varepsilon, y_t^\varepsilon)$ of partially observed processes, where x_t^ε satisfies (1.1) for $u_t \equiv 0$ and is the unobserved component, while y_t^ε satisfies (1.2) and is the observable component. Assume also that A.1 - A.3 hold. Given this family $(x_t^\varepsilon, y_t^\varepsilon)$, consider the problem of constructing a linear filter estimate $\hat{x}_t^\varepsilon$ of x_t^ε, given the observations $\{ y_s^\varepsilon, s \le t \}$, which has the smallest possible mean square error. To this effect recall from [13] (see also proposition 3.1 below) that under the assumptions A.1 - A.3

the pair $(x_t^\varepsilon, y_t^\varepsilon)$ converges, for $\varepsilon \to 0$, in distribution to the limit pair (x_t, y_t) which satisfies (1.21),(1.22) for $u_s(y) \equiv 0$ and which is of the familiar linear-Gaussian type. The most natural way to obtain a good linear filter estimate for x_t^ε, based on $\{\, y_s^\varepsilon,\ s \le t\,\}$, is then to consider the Kalman-Bucy filter for the limit pair (x_t, y_t), which in fact minimizes the mean square error, and to apply it to the generic prelimit pair $(x_t^\varepsilon, y_t^\varepsilon)$. In this way we obtain the following linear filter

$$\dot{\hat{x}}_t^\varepsilon = A\, \hat{x}_t^\varepsilon + K_t^H [\, \dot{y}_t^\varepsilon - C_H\, \hat{x}_t^\varepsilon\,] \qquad , \qquad \hat{x}_0^\varepsilon = E\{\, x_0^\varepsilon\,\} \tag{2.1}$$

$$K_t^H = P_t^H\, C_H\, D_H^{-2} \tag{2.2}$$

$$\dot{P}_t^H = 2A\, P_t^H + B^2 - (P_t^H)^2\, C_H^2\, D_H^{-2} \qquad , \qquad P_0^H = E\{\, (X_0 - E\, X_0)^2\,\} \tag{2.3}$$

and from the convergence in distribution $(x_t^\varepsilon, y_t^\varepsilon) \Rightarrow (x_t, y_t)$ as well as from the fact that $\hat{x}_t^\varepsilon$ is a linear functional of y_t^ε and that $E\, x_0^\varepsilon \to E\, x_0$, it follows that we have the joint convergence in distribution

$$(x_t^\varepsilon, \hat{x}_t^\varepsilon, y_t^\varepsilon) \Rightarrow (x_t, \hat{x}_t, y_t) \tag{2.4}$$

where $\hat{x}_t = E\{\, x_t \mid y_s,\ s \le t\,\}$ is the Kalman-Bucy filter estimate for the limit pair (x_t, y_t).

A first result (see [8]) can now be shown concerning the robustness of the Kalman-Bucy filter corresponding to the limit model. More precisely, letting $N(x; m, V)$ denote the Gaussian distribution with mean m and variance V, we have the following theorem, whose proof is a rather immediate consequence of (2.4) and the fact that $\int f(x)\, dN(x; \hat{x}_t, P_t^H)$ minimizes the mean square error among the estimates of $f(x_t)$ that are based on $\{\, y_s,\ s \le t\,\}$.

Theorem 2.1. ([8]). Let $f(x)$ be a continuous and bounded function and $F_t(y)$ a continuous and bounded functional of $y_s,\ s \le t$. Then

$$\lim_{\varepsilon \to 0} E\{\, [\, f(x_t^\varepsilon) - F_t(y^\varepsilon)\,]^2\,\} \ge \lim_{\varepsilon \to 0} E\{\, [\, f(x_t^\varepsilon) - \int f(x)\, dN(x; \hat{x}_t^\varepsilon, P_t^H)\,]^2\,\} \qquad \blacksquare$$

The theorem essentially states that the Kalman-Bucy filter for the limit pair (x_t, y_t), when applied to the generic prelimit pair $(x_t^\varepsilon, y_t^\varepsilon)$, nearly minimizes the mean square error among all the estimators of $f(x_t^\varepsilon)$ that are continuous and bounded functionals of the past observations $y_s^\varepsilon,\ s \le t$ ($f(x)$ has to be continuous and bounded as well). This result corresponds to the robustness property of the nearly optimal controls for the ideal limit model (1.21)-(1.23) mentioned in the previous section.

Notice now that, under a uniform integrability condition (in practice this can be obtained by means of a truncation; see [13] also for a weaker requirement), from (2.4) we have

$$\lim_{\varepsilon \to 0} E\{\, [\, x_t^\varepsilon - \hat{x}_t^\varepsilon\,]^2\,\} = E\{\, [\, x_t - \hat{x}_t\,]^2\,\} = P_t^H \tag{2.5}$$

On the other hand, the filter variance P_t^H for the limit pair (x_t, y_t) satisfies the Riccati equation (2.3), from which it is immediately seen that the magnitude of P_t^H depends on the "signal-to-noise ratio" C_H^2 / D_H^2, which in turn depends, see (1.13),(1.16), on the structure of the function H as well as on the properties of the process ϕ_t. More precisely, given two limiter functions $H_1(.), H_2(.)$ we have that

$$\text{If} \quad C_{H_1}^2 / D_{H_1}^2 > C_{H_2}^2 / D_{H_2}^2, \quad \text{then} \quad P_t^{H_1} < P_t^{H_2} \tag{2.6}$$

Notice that, by (2.5), the implication (2.6) holds (with $\leq$ instead of $<$ in the right hand side) for small values of ε also for the mean square error $E\{ | x_t^\varepsilon - \hat{x}_t^\varepsilon |^2 \}$ of the filter $\hat{x}_t^\varepsilon$ in (2.1).

Given the essentially linear structure of the models (1.1),(1.2) and the optimality property of the Kalman-Bucy filter in linear-Gaussian models, it is intuitively clear (for a formal investigation see [13; Ch.3]) that, if ϕ_t is a Gaussian process, then the smallest value of the mean square error of $\hat{x}_t^\varepsilon$ in (2.1) is obtained by choosing the limiter function H to be the identity function, i.e. no limiter is best. If however the observation noise distribution has fat tails, then the mean square error of $\hat{x}_t^\varepsilon$ naturally increases; property (2.6) combined with (2.5) then gives a possibility to reduce,for small values of ε, the mean square error of $\hat{x}_t^\varepsilon$ by a suitable choice of the limiter function. In other words, prelimit models (1.1),(1.2), that allow for a limiter in the observations, are robust models in the sense that they enhance the robustness of the filter (2.1)-(2.3) by allowing to keep the mean square error of $\hat{x}_t^\varepsilon$ small also when the observation noise distribution has fatter tails than a Gaussian. For a theoretical investigation of this issue see, besides [13], also [12]. Simulation results for corresponding time discretized models are reported in [14], where the process ϕ_t generating the observation noise disturbances is assumed to be given by

$$\phi_t = \alpha \gamma_t + (1-\alpha) \gamma_t^{2q-1} \tag{2.7}$$

with $\alpha \in [0,1]$, q a positive integer, and γ_t the stationary Gaussian process

$$\gamma_t = \gamma_0 e^{-t/2} + \int_0^t e^{-(t-u)/2} d\beta_u \tag{2.8}$$

where β_t is a standard Wiener process independent of ψ_t. By its definition it is easily seen that the tails of the distribution of ϕ_t become heavier with increasing values of q. The simulations in [14] show that, without limiter in the observations, the Kalman-Bucy filter for the limit model performs well also in the prelimit models when the value of q does not exceed 3; for values of q greater than or equal to 3, the mean square error of the filter in the prelimit models (namely of $\hat{x}_t^\varepsilon$ in (2.1)) increases very rapidly with q, but can be kept almost constant with respect to q, if a soft limiter of the type of (1.6) is being used.

3. EXTENSION OF THE ROBUSTNESS RESULTS TO A CONTROL SETTING

The purpose of this section is twofold : First we prove the main result of this paper, namely theorem 1.1; then, by analogy to the second part of the previous section, we discuss the enhancement of robustness obtained from the use of a limiter in the observations.

Let $(x_t^\varepsilon, y_t^\varepsilon)$ be given by (1.1),(1.2) and (x_t, y_t) by (1.21),(1.22). With M_t^ε as in (1.9) and N_t^ε as in (1.16), BW_t and $D_H V_t$ as in (1.12) and x_0 as in (1.20) we have

Proposition 3.1. Under the assumptions A.1 - A.4, and independently of the choice of the control function u(.) satisfying (1.4),(1.5), there exists a constant K_1 such that a.s.

$$\sup_{t \leq T} \left[\, |x_t^\varepsilon - x_t| + |y_t^\varepsilon - y_t| \, \right] \leq K_1 \{ \, |x_0^\varepsilon - x_0| + \sup_{t \leq T} [\, |M_t^\varepsilon - BW_t| + |N_t^\varepsilon - D_H V_t| \,] \, \}$$

Proof : Putting $z_t^\varepsilon := [x_t^\varepsilon, y_t^\varepsilon]$, $z_t := [x_t, y_t]$, $\Gamma_t^\varepsilon := [M_t^\varepsilon, N_t^\varepsilon]$, $\Gamma_t := [BW_t, D_H V_t]$, and using in $\mathcal{R}^2$ the norm $\| z \| = |x| + |y|$, we may rewrite (1.17),(1.18) and (1.21),(1.22) as

$$z_t^\varepsilon = z_0^\varepsilon + \int_0^t F_s(z^\varepsilon) \, ds + \Gamma_t^\varepsilon \tag{3.1}$$

$$z_t = z_0 + \int_0^t F_s(z) \, ds + \Gamma_t \tag{3.2}$$

respectively, where the functional $F_s(.)$ satisfies

$$\| F_t(z^\varepsilon) - F_t(z) \| \leq \int_0^t \Lambda(s) \, \| z_s^\varepsilon - z_s \| \, ds \tag{3.3}$$

for a given nonnegative $\Lambda(s) \leq \Lambda$, independent of the choice of u(.). We then have a.s.

$$\| z_t^\varepsilon - z_t \| = \| z_0^\varepsilon - z_0 \| + \int_0^t \| F_s(z^\varepsilon) - F_s(z) \| \, ds + \| \Gamma_t^\varepsilon - \Gamma_t \| \leq$$

$$\| z_0^\varepsilon - z_0 \| + \int_0^t \int_0^s \Lambda(u) \, \| z_u^\varepsilon - z_u \| \, du \, ds + \| \Gamma_t^\varepsilon - \Gamma_t \| \leq$$

$$\| z_0^\varepsilon - z_0 \| + T \int_0^t \Lambda(s) \, \| z_s^\varepsilon - z_s \| \, ds + \| \Gamma_t^\varepsilon - \Gamma_t \| \tag{3.3}$$

from which, by Gronwall-Bellman's inequality and for appropriate constants K_1, K_2, K_3

$$\| z_t^\varepsilon - z_t \| = K_2 \| z_0^\varepsilon - z_0 \| + \| \Gamma_t^\varepsilon - \Gamma_t \| + \int_0^t T\Lambda(s) \, \| \Gamma_s^\varepsilon - \Gamma_s \| \exp\left[\int_s^t T\Lambda(u)du\right] ds \leq$$

$$K_2 \| z_0^\varepsilon - z_0 \| + \| \Gamma_t^\varepsilon - \Gamma_t \| + K_3 \int_0^t \| \Gamma_s^\varepsilon - \Gamma_s \| ds \leq$$

$$K_1 \left[\| z_0^\varepsilon - z_0 \| + \sup_{t \leq T} \| \Gamma_t^\varepsilon - \Gamma_t \| \right] \qquad \text{a.s.} \qquad (3.4)$$

and this implies the statement of the proposition. ∎

Proposition 3.2. Under the assumptions A.1 - A.5 we have for a suitable $K > 0$ and for all control functions $u(.)$ satisfying (1.4),(1.5)

$$| J^\varepsilon(u) - J(u) | \leq$$

$$E \min\left\{ 2MT \; ; \; K \left[| x_0^\varepsilon - x_0 | + \sup_{t \leq T} \left(| M_t^\varepsilon - BW_t | + | N_t^\varepsilon - D_H V_t | \right) \right] \right\}$$

$$(3.5)$$

Proof : We have, using also proposition 3.1,

$$| J^\varepsilon(u) - J(u) | \leq E \int_0^T | r_t(x_t^\varepsilon, u_t(y^\varepsilon)) - r_t(x_t, u_t(y)) | \, dt \leq$$

$$E \int_0^T \min [M; L | x_t^\varepsilon - x_t |] \, dt + E \int_0^T \min [M; L | u_t(y^\varepsilon) - u_t(y) |] \, dt \leq$$

$$E \min [M; L \int_0^T | x_t^\varepsilon - x_t | dt] + E \int_0^T \min [M; L^2 \int_0^t | y_s^\varepsilon - y_s | ds] \, dt \leq$$

$$E \min\left\{ 2MT; (TL^2+1) \int_0^T (| x_t^\varepsilon - x_t | + | y_t^\varepsilon - y_t |) \, dt \right\} \leq$$

$$E \min\left\{ 2MT; TK_1(TL^2+1) \left[| x_0^\varepsilon - x_0 | + \sup_{t \leq T} \left(| M_t^\varepsilon - BW_t | + | N_t^\varepsilon - D_H V_t | \right) \right] \right\}$$

Proof of theorem 1.1 : Let (see (3.5))

$$\eta^\varepsilon := \min \left\{ 2MT; \; K\left[\; | \; x_o^\varepsilon - x_o | + \sup_{t \leq T} \left(\; | \; M_t^\varepsilon - BW_t | + | \; N_t^\varepsilon - D_H V_t | \; \right) \right] \right\} \tag{3.6}$$

From the weak convergence $x_o^\varepsilon \Rightarrow x_o$, $M_t^\varepsilon \Rightarrow BW_t$, $N_t^\varepsilon \Rightarrow D_H V_t$, using Skorokhod imbedding [19], we have that there exists a suitable probability space $(\Omega, \mathcal{F}, \tilde{P})$ on which

$$\eta^\varepsilon \Rightarrow 0 \quad , \quad \tilde{P} \text{ - a.s.} \tag{3.7}$$

On the other hand, for $\varepsilon < \varepsilon_o$, we have

$$0 < \eta^\varepsilon \leq 2MT \quad , \quad \tilde{P} \text{ - a.s.} \tag{3.8}$$

It follows that

$$\lim_{\varepsilon \to 0} E(\eta^\varepsilon) = \lim_{\varepsilon \to 0} \tilde{E}(\eta^\varepsilon) = 0 \tag{3.9}$$

which, together with (3.5) gives the statement of theorem 1.1. ∎

Coming to the second part of this section, notice first that in many cases the optimal minimal value of $J(u)$ in the limit model (1.21)-(1.23) is an increasing function of the variance of the Kalman-Bucy filter for (1.21),(1.22) (with $u_t \equiv 0$). This holds true e.g. if $r_t(x,u)$ are quadratic functions of their arguments (see e.g. [1; Ch.8.6]). On the other hand, we have from section 2 that, without limiter in the observations, the mean square error of the filter $\hat{x}_t^\varepsilon$ in (2.1) grows with increasingly fatter tails of the distribution of ϕ_t; furthermore, by (2.5), this mean square error of $\hat{x}_t^\varepsilon$ is close to the variance P_t^H of the filter $\hat{x}_t$ associated with the limit model (1.21)-(1.23). From this and the fact that, by theorem 1.1, $J^\varepsilon(u)$ is close to $J(u)$ for small values of ε, it follows that, although a δ - optimal control for the limit model (1.21)-(1.23) is 2δ - optimal for small ε also in the prelimit models (1.1)-(1.3), the corresponding value of $J^\varepsilon(u)$ may become rather large when the observation noise distribution has fat tails. Using however the robust models (1.1)-(1.3) with the limiter in the observations, we know from section 2 that we can keep the mean square error of $\hat{x}_t^\varepsilon$ in (2.1) small also when the observation noise distribution has fat tails; this in turn enhances the robustness of the δ - optimal controls for the limit model (1.21)-(1.23) in the sense that, with a suitable choice of the limiter, the corresponding value of $J^\varepsilon(u)$ can be kept small also with fat tails in the observation noise distribution.

REFERENCES

[1] Åström, K.J., *Introduction to Stochastic Control*, Academic Press, 1970.

[2] Bensoussan, A. and Runggaldier, W., An approximation method for stochastic control problems with partial observation of the state - a method for constructing ε- optimal controls, *Acta Applicandae Mathematicae* **10** (1987), 145-170.

[3] Davenport, W.B., Signal to noise ratios in band pass limiters, *J. Appl. Phys.* **24** (1953), 720-727.

[4] Huber, P., *Robust Statistics*, J.Wiley, New York, 1981.

[5] Krichagina, E., Production control in a failure-prone manufacturing system : Diffusion approximation and asymptotic optimality, This Conference.

[6] Kushner, H.J., *Approximation and Weak Convergence Methods for Random Processes, With Applications to Stochastic Systems Theory*, MIT Press, Cambridge, Mass., 1984.

[7] Kushner, H.J. and Huang, H., Approximation and limit results for nonlinear filters with wide bandwidth observation noise, *Stochastics* **16** (1986), 65-96.

[8] Kushner, H.J. and Runggaldier, W.J., Filtering and control for wide bandwidth noise driven systems, *IEEE-Transactions* AC-32 (1987), 123-133.

[9] Kushner, H.J. and Runggaldier, W.J., Nearly optimal state feedback controls for stochastic systems with wideband noise disturbances, *SIAM J. on Control and Optimization* **25** (1987), 298-315.

[10] Kushner, H.J. and Ramachandran, K.M., Nearly optimal singular controls for wideband noise driven systems, *SIAM J. on Control and Optimization* **26** (1988), 569-591.

[11] Kushner, H.J., Numerical methods for stochastic control problems in continuous time, *SIAM J. on Control and Optimization* **28** (1990), 999-1048.

[12] Liptser, R.Sh. and Lototski, S.V., Diffusion approximation and robust Kalman filter, submitted to *Journal of Mathematical Systems, Estimation and Control*.

[13] Liptser, R.Sh. and Runggaldier, W.J., On diffusion approximation for filtering, *Stoch. Processes and Their Applications* **38** (1991), 205-238.

[14] Runggaldier, W.J. and Bonollo,M., Robust techniques for combined filtering and parameter estimation, in : *Control and Dynamic Systems* (C.T.Leondes,ed.), Academic Press, to appear.

[15] Runggaldier, W.J. and Stettner, L., On the construction of nearly optimal strategies for a general problem of control of partially observed diffusions, *Stochastics and Stochastics Reports* **37** (1991), 15-47.

[16] Runggaldier, W.J. and Zane, O., Approximations for discrete-time adaptive control : construction of ε- optimal controls, *Math. Control Signals Systems* 4 (1991), 269-291.

[17] Schick, I.C., Robust recursive estimation of the state of a discrete-time stochastic linear dynamic system in the presence of heavy-tailed observation noise, Ph.D.Thesis, Massachusetts Institute of Technology, 1990.

[18] Sethi, S.P., Hierarchical investment and production decisions in stochastic manufacturing systems : Asymptotic optimality and error bounds, This Conference.

[19] Skorokhod, A.V., Limit theorems for stochastic processes, *Theory of Probab. and its Applications* 1 (1956), 262-290.

[20] Taksar, M.I., On the optimal control of a queueing system, This Conference.

[21] Zhang,Q., Controlled diffusions with rapidly oscillating unknown parameter processes, This Conference.

CUMULANT MINIMIZATION AND ROBUST CONTROL*

Michael K. Sain
Department of Electrical Engineering
University of Notre Dame
Notre Dame, Indiana
USA 46556

Chang-Hee Won
Department of Electrical Engineering
University of Notre Dame
Notre Dame, Indiana
USA 46556

B. F. Spencer, Jr.
Department of Civil Engineering and Geological Sciences
University of Notre Dame
Notre Dame, Indiana
USA 46556

ABSTRACT

The use of higher order spectral information has recently attracted a great deal of attention in the estimation literature. Emphasis there has been placed upon the cumulants of the random variable in question. A related development in the stochastic control literature has to do with risk sensitive optimal control. Glover has pointed out the connections between the risk sensitive control problem and H_∞ methods. In this paper we show the relation between a natural approximation to the risk-sensitive problem, by means of series expansion, and the problem of minimizing linear combinations of cumulants of the performance function. Special attention is given to the linear combination of cumulants one and two, the mean and variance respectively. Using the theory for this minimal cost variance formulation, we present numerical results for a single degree-of-freedom building excited by an earthquake process.

I. INTRODUCTION

Over the last quarter century, several investigators have suggested that the field of control theory was expanding exponentially in space and time. They pointed out that it was not just a question of the field having various subfields. Indeed, the subfields themselves were further subdivided by fascinating alternative viewpoints and methodologies. Professors found a mini-research problem, for example, in the selection of a book for a first graduate course in control systems. Did you wish to study linear multivariable control? Well, there were the state-space methods and the frequency-domain methods, the linear quadratic methods and the geometric methods, and so forth. The books that were written tended to emphasize one viewpoint or method; space to present them all would have been too costly—if a thread of unity could have been pulled through. During some years, then, an overall synthesis of the area seemed like a pipe dream.

But things appear to be changing. A rally has taken place around the subject of robust control, which after all lies at the heart of the subject, and is easily detected in the original master references. One way to think about this area is by means of optimization problems placed in terms of various normed spaces and induced operators. Here we shall take the alternative of highlighting three major players on the field: H_∞ methods, deterministic dynamical game theory, and stochastic control. Intuitively, of course, there are overlaps in these concepts. Thus, for instance, one may have a stochastic game. What we have in mind, however, is the same problem having the capability to be posed and solved alternatively by each of the various players. This is a remarkable advance, and we will describe one aspect of it in the sequel here. It hints at the possibility of a grand synthesis, in which the central problems could be firmly identified and posed, with the alternative

*This work has been supported in part by the National Science Foundation under Grant BCS 90-06781, and in part by the Frank M. Freimann Chair in Electrical Engineering at the University of Notre Dame. The ideas herein were first announced at the Stochastic Theory and Adaptive Control Workshop, University of Kansas, September 26–28, 1991.

mathematical approaches highlighting different views of the solution. Investigators are welcoming this coalescence, which it is hoped will lead to a new plateau and re-grouping for a more final attack on the essential feedback questions.

The present paper addresses the problem area of risk sensitive optimal control, which has been discussed in book form recently by Whittle [1]. It has been shown by Glover and Doyle [2] that there is a precise connection between H_∞ control and risk sensitivity. We will elaborate upon this point shortly. The connection, on the other hand, between H_∞ control and deterministic dynamic games is more pandemic; and we refer the reader to the new book by Basar and Bernhard [3]. It seems that the origins of the risk sensitive optimal control idea originate with the work of Jacobson [4] in 1973.

Specifically, we wish to bring out the connection between performance cumulant minimization and risk sensitive optimal control. The idea of controlling performance variance was suggested to the first author by R. A. Rohrer after a seminar at the Coordinated Science Laboratory of the University of Illinois, Urbana, in 1963. Subsequently, under the support and direction of J. B. Cruz, Jr., it was developed [5] into a dissertation in 1965, appearing shortly thereafter [6] in journal form. Work on this theory continued until 1978, from which it is clear that at least a five-year overlap occurred with the beginnings of risk sensitive control. However, the two investigations proceeded independently, perhaps because the key linking idea was that of a cumulant, hardly a common term in control theory at that time.

If one thinks of expanding the risk sensitive performance function in a series, then each of the terms in the series corresponds to a cumulant; and control of cumulants may be regarded as control of individual terms, or linear combinations thereof, in that series. In particular, minimal variance control addresses the first two terms in the series. Thus, although the two problems can stand alone, under the proper circumstances they approximate each other.

We begin the detailed discussion with the notion of risk sensitive optimal control.

II. RISK SENSITIVE OPTIMAL CONTROL

Recently, the notion of risk sensitive optimal control has been receiving increased attention in the literature. For a summary and additional reading, the reader is referred to [7]. We shall wish to discuss the history of evolution of the concept, but initially we want to sketch the basic mathematical framework.

Assume that a performance function

$$J = \int_0^{t_F} j(x(t), u(t))dt + j_F(x(t_F)) \tag{1}$$

is given, and that it is a random variable because of the dynamical relationship that exists between $x(t)$ and $u(t)$ over the interval $[0, t_F]$. For example, such a relationship might arise by means of a stochastic differential equation. The risk sensitive performance index is

$$J_{RS} = -\frac{2}{\theta} \log\{E_k \exp\left(-\frac{\theta}{2}J\right)\}, \tag{2}$$

where θ is a small parameter and the subscript k on E denotes expectation based upon a control law k generating the control action $u(t)$ from the state $x(t)$ or from a measurement history arising from that state.

Recall now that the characteristic function of a random variable J is given by

$$\phi(s) = E\exp(-sJ). \tag{3}$$

Actually, (3) is sometimes called the *first* characteristic function of J. In any event, every student is familiar with the series expansion for $\phi(s)$, which takes the form

$$1 + \sum_{i=1}^{\infty} \frac{(-1)^i}{i!}\alpha_i s^i, \tag{4}$$

where α_i denotes the ith moment of J and where we have assumed for convenience that all moments exist. Suitable adjustments can be made otherwise. Corresponding to (3) and (4), there is a *second characteristic function* $\psi(s)$ defined by

$$\psi(s) = \log \phi(s) = \sum_{i=1}^{\infty} \frac{(-1)^i}{i!} \beta_i s^i, \tag{5}$$

in which the $\{\beta_i\}$ are known as the *cumulants*, or sometimes the *semi-invariants*, of J. Once again, we have assumed that all the cumulants exist. This is done only for convenience. If they do not, we employ a remainder. Historically, the idea of cumulants seems to be a standard part of statistical theory. In the 1960s, only β_1, the mean, and β_2, the variance, were in common use in stochastic control theory. More recently, with the rise of higher-order methods in signal processing, the term is becoming increasingly visible in control literature.

If we now compare (2) and (3) with (4), it is seen that

$$J_{RS} = \left(-\frac{2}{\theta}\right) \left\{ \sum_{i=1}^{\infty} \frac{(-1)^i}{i!} \beta_i(J,\ k) \left(\frac{\theta}{2}\right)^i \right\}, \tag{6}$$

in which we have employed the notation $\beta_i(J,\ k)$ to refer to the cumulants of J with respect to the control law k. Up to second order, (6) becomes

$$J_{RS} = \beta_1(J,\ k) - \frac{\theta}{4}\beta_2(J,\ k) + O(\theta^2) \tag{7}$$

$$= E_k J - \frac{\theta}{4}\text{VAR}_k J + O(\theta^2). \tag{8}$$

Thus we see that, for small θ, minimization of J_{RS} relates closely to the minimization of a linear combination of cost mean and cost variance.

Now consider (8). If $\theta > 0$, then the uncertainty associated with J being a random variable decreases the performance cost through $\text{VAR}_k J$. This is called the optimistic or "risk seeking" case. We shall not ordinarily be interested in this case. If $\theta < 0$, on the other hand, $\text{VAR}_k J$ increases the performance cost. This is the so-called pessimistic or "risk aversive" case. It corresponds to the typical problem of interest in robust control. For completeness, we mention that "risk neutrality" occurs when θ equals zero. This is just the classical minimum average performance problem of stochastic control theory. Accordingly, we may view the use of cost variance as a move away from "statistical risk neutrality".

In the preceding section, we mentioned that the risk sensitive optimal control history originated in 1973 with the work of Jacobson [4]. That investigation was carried out for the "linear exponential quadratic Gaussian" control problem. For such a study, Jacobson set up a linear dynamical relationship between $x(t)$ and $u(t)$, in the manner

$$\dot{x}(t) = Ax(t) + Bu(t) + w(t), \tag{9}$$

in which $w(t)$ is Gaussian white noise, and formulated J_{RS} by choosing

$$J = \int_0^{t_F} (x'(t)Qx(t) + u'(t)Ru(t))dt + x'(t_F)Px(t_F) \tag{10}$$

where t_F is a fixed final time. Matrices P, Q, and R are symmetric positive semi-definite; and R has an inverse. The superscript ($'$) denotes transposition. The remarkable thing about the optimal controller that solves this version, which assumes perfect observation of $x(t)$, is that the control law k is linear in x. While it is clear that this should be the case when $w(t) \equiv 0$, it is not at all transparent that this should be the case when we look at (6), with all its higher-order statistics. Additional insight as to how this happens will be given in a later section, when we discuss the nature of the performance cumulants.

Additional work by Speyer, Deyst, and Jacobson [8] in 1974 considered the case of measuring $x(t)$ in additive noise. An interesting solution is obtained, in terms of the entire set of state measurements up to the present. To reduce this situation to a smaller sufficient statistic, the authors assumed

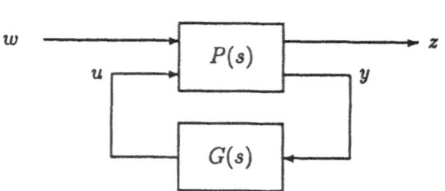

Figure 1: Diagram Setup for H_∞-Connection

zero plant noise. Then, in 1976, Speyer [9] considered the noisy measurement case again, but with zero instantaneous state weighting. Kumar and van Schuppen [10] in 1981 gave a solution in continuous time with zero plant noise. Then, in the same year, Whittle [11] used a certainty-equivalance derivation in discrete time for a general solution. Four years later, in 1985, Bensoussan and van Schuppen [12] presented an alternate approach to a general solution in continuous time. The development of a risk sensitive maximum principle for the case of partially observed states, and with a view toward easier-to-calculate costates, is the subject of [7], as mentioned earlier. It is based upon observations in [13] and [1], in 1986 and 1990, respectively.

Especially attractive in these events is the possible extension to nonlinear problems by means of large-deviation theory [7]. See also Fleming and McEneaney [14], in which viscosity solution methods to treat risk sensitive optimal control problems are discussed relative to the nonlinear system case.

III. THE H_∞-CONNECTION

Glover and Doyle [2] gave in 1988 a connection between a class of H_∞ problems and risk sensitivity. This paper considered the discrete-time case. Subsequently, Glover [15] developed a similar line of reasoning for the continuous-time case. Here we wish just to present a brief sketch of the idea presented in [15].

Consider a linear, time-invariant feedback system as shown in Figure 1. A basic H_∞ control problem is to find the controller $G(s)$ so that the closed-loop system is internally stable and such that the transfer function $T_{zw}(s)$ from w to z satisfies

$$\|T_{zw}(s)\|_\infty < \gamma \tag{11}$$

where γ is a suboptimal bound. The set of all solutions can be parameterized; and one unique choice is that solution which minimizes the entropy integral

$$-\frac{\gamma^2}{2\pi} \int_{-\infty}^{\infty} \log \left| \det \left(I - \frac{1}{\gamma^2} T_{zw}^* T_{zw} \right) \right| d\omega, \tag{12}$$

in which superscript (*) denotes conjugate transposition. Then this entropy integral can be shown to differ from

$$J_T(\gamma) = \frac{\gamma^2}{T} \log E \left\{ \exp \left(\frac{1}{2\gamma^2} V_T \right) \right\} \tag{13}$$

by terms of order $(1/T)$, where

$$V_T = \int_{-T}^{T} z'(t) z(t) dt, \tag{14}$$

provided that $z(t)$ is a stationary Gaussian process whose spectral density has H_∞-norm less than γ^2. In the steady state case, then, minimizing the risk sensitive form (13) is identical to minimizing the entropy form (12).

IV. CUMULANT STRUCTURE

In 1971, Liberty [16] studied the characteristic functions of integral quadratic forms, such as (10), over linear systems, such as (9). With respect to the second characteristic function, whose expansion yields the cumulants, he showed that every cumulant is the sum of two terms, one of which is independent of the mean of the initial state, and one of which is quadratic in the mean of the initial state. It had, of course, been known for years that the first cumulant, or cost mean, was quadratic in mean initial state. Moreover, it was implicit in the results of [6], which led to linear controls, that the second cumulant, or cost variance, was quadratic in mean initial state. Liberty seems, however, to have been the first to report this property for the complete set of statistical quantities. When applied on a cost-to-go basis, of course, with current measurements of the state, this insight is in complete agreement with the results of Jacobson [4], who found a linear controller solution. In subsequent years, Liberty and Hartwig studied the problem of generating the cumulants in the time domain, from differential equations. This work was eventually reported [17] in 1976.

Specifically, Liberty and Hartwig studied the system of equations

$$\dot{x}(t) = A(t)x(t) + C(t)w(t) \tag{15}$$

with a Gaussian initial condition and $w(t)$ as in (9). Cumulants were examined for the variate

$$\int_{t_0}^{t_F} x'(t)Q(t)x(t)dt + x'(t_F)Px(t_F). \tag{16}$$

The kth cumulant takes the explicit form

$$(k-1)!2^{k-1}\sum_{i=1}^{\infty}\lambda_i^k + k!2^{k-1}\sum_{i=1}^{\infty}m_i^2\lambda_i^{k-1}, \tag{17}$$

where m_i is the mean of x_i in an orthonormal expansion

$$x(t) \sim \sum_{i=1}^{\infty} x_i\phi_i(t) \tag{18}$$

and the λ_i are associated eigenvalues. In [16], it was shown that each m_i^2 is quadratic in $Ex(t_0)$; see also [17]. There are similarities between [16, 17] and [15], with respect to representations of the cost variable.

V. MINIMAL VARIANCE CONTROL

The basic idea of minimal variance control [5, 6] is to minimize

$$J_{MV} = \text{VAR}_k J \tag{19}$$

while satisfying a constraint

$$E_k J = M. \tag{20}$$

By means of a multiplier μ, corresponding to the constraint (20), one can form the function

$$J_{MV} = \mu(E_k J - M) + \text{VAR}_k J, \tag{21}$$

which for minimization amounts to

$$\hat{J}_{MV} = \mu E_k J + \text{VAR}_k J. \tag{22}$$

The natural relation of $\hat{J}_{MV}$ to J_{RS} occurs through the identity

$$\mu\theta + 4 = 0. \tag{23}$$

Minimal Variance Control		Risk Sensitive Control	
[5]	1965		
[6]	1966		
[20]	1968		
[16], [18]	1971		
		1973	[4]
		1974	[8]
[17]	1976	1976	[9]
[19]	1978		
		1981	[10], [11]
		1985	[12]
		1986	[13]
		1990	[1]
[this paper]	1991	1991	[14], [17]

Figure 2: A Time-Line Comparison of MV and RS.

When θ is small, which means that we are nearly at the minimum average cost problem, it is expected that minimal variance controllers and risk sensitive controllers should give good approximations to one another. Of course, when θ is small, then μ is large; and μ corresponds intuitively to multiplier "effort" to meet the constraint (20). Notice that the use of $\hat{J}_{MV}$ to approximate J_{RS} corresponds to finding an exact solution to an approximate problem. On the other hand, $\hat{J}_{MV}$ can also be solved even when μ is small. In this case, it is a stand-alone problem and appears to offer a risk-aversive-like control scheme, even though θ is not small.

In addition to the references already cited, we should also point out [18], in which a Riccati solution to $\hat{J}_{MV}$ minimization is developed for the case

$$u(t) = k(t, \; x(0)). \tag{24}$$

A sketch of this approach is given in the next section. Also, in 1978, Liberty and Hartwig [19] reported on solving $\hat{J}_{MV}$ in the feedback case with noisy state measurements. Finally, [20] examines the minimal cost variance concept for problems of estimation. A linear display of the references is sketched in Figure 2.

VI. THE RICCATI SOLUTION FOR $\hat{J}_{MV}$ [18]

An introductory MV problem is that of controlling the linear system

$$\dot{x}(t) = A(t)x(t) + B(t)u(t) + w(t) \tag{25}$$

according to the performance measure

$$J = \int_0^{t_F} [x'(t)Qx(t) + u'(t)Ru(t)] \, dt + x'(t_F)Px(t_F) \tag{26}$$

where t_F is a fixed final time. The noise $w(t)$ is assumed to be zero-mean Gaussian with essentially white characteristics, that is

$$E\{w(t)w'(\sigma)\} = S\delta(t - \sigma). \tag{27}$$

The state $x(t) \in R^n$ and the control action $u(t) \in R^m$. Matrices P, Q, R, and S are all symmetric and positive semidefinite with R also having an inverse. The superscript ($'$) denotes transposition, and $x(0)$ is assumed to be known exactly—a convenience more than an essential factor.

The minimal variance control problem is to find a class of functions $\{u(t)\}_M$ on $[0, t_F]$ so that

$$E_k J = M \tag{28}$$

for prespecified M (which must of course be at least as great as the well-known minimum average cost) and within that class to choose a $u_M^0(t)$ such that

$$\text{VAR}_k J \tag{29}$$

is minimum. It can be shown [16] that this problem has a unique solution in an appropriate Hilbert space setting. The computational solution to the minimal variance-control problem, first presented in [18], is an outgrowth of original work in [6] and of the work of Baggeroer [21].

The solution is defined by the $4n$ differential equations

$$\dot{z}(t) = A(t)z(t) - \frac{1}{2\mu}B(t)R^{-1}B'(t)\rho(t) \tag{30}$$

$$\dot{\rho}(t) = -A'(t)\rho(t) - 2\mu Q z(t) - 8Q v(t) \tag{31}$$

$$\dot{v}(t) = A(t)v(t) + S y(t) \tag{32}$$

$$\dot{y}(t) = -A'(t)y(t) - Q z(t) \tag{33}$$

together with the $4n$ boundary conditions

$$z(0) = x(0) \tag{34}$$

$$\rho(t_F) = 2\mu P z(t_F) + 8P v(t_F) \tag{35}$$

$$v(0) = 0 \tag{36}$$

$$y(t_F) = P z(t_F) \tag{37}$$

and the control action relationship

$$u(t) = \frac{-1}{2\mu}R^{-1}B'(t)\rho(t). \tag{38}$$

Strictly speaking, these $4n$ functions should be designated by an optimum superscript (^{o}), but it is more convenient to omit such notation because only these optimum functions are hereafter considered. The variable $z(t)$ is the mathematical expectation of $x(t)$; the variable $\rho(t)$ corresponds closely to the usual costate variable of optimal control theory, since it is the variable which enforces the differential equation constraint between $z(t)$ and $u(t)$. The variables $v(t)$ and $y(t)$ are introduced via the method of Baggeroer to reduce the integro-differential equation of [6]. It should be noted that M has been suppressed in the solution equations; this is a consequence of the greater convenience afforded by using instead the multiplier μ corresponding to the constraint. No difficulty therefore arises because there is a smooth one-one inverse relationship between μ and M. The positive variable μ becomes large as M becomes small and vice-versa. By means of the following identities,

$$M = z'(t_F)P z(t_F) + \int_0^{t_F} [z'(t)Q z(t) + \frac{1}{4\mu^2}\rho'(t)B(t)R^{-1}B'(t)\rho(t)]\,dt + K_m \tag{39}$$

$$\text{VAR}_k J = 4[z'(t_F)P v(t_F) + \int_0^{t_F} [z'(t)Q v(t)]\,dt] + K_v \tag{40}$$

tables relating μ, M, and $\text{VAR}_k J$ can be constructed. Moreover, the smooth inverse relationship between μ and M can be used to set up recursive algorithms for computing the μ corresponding to a particular M; this μ then leads to the entire problem solution. K_m and K_v are the constant mean and variance of a random variable not influenced by $u(t)$.

The performance-measure variance, as expressed above, takes on an inner product form in an appropriate space of functions. As such it is not immediately clear that it is nonnegative, as a variance should be. We now bring it into a new form which is clearly nonnegative and in the process gives added insight into the nature of the variable $y(t)$.

The time derivative of the function $y'(t)v(t)$ becomes

$$\frac{d}{dt}[y'(t)v(t)] = -z'(t)Qv(t) + y'(t)Sy(t).$$ (41)

As a consequence, integration yields

$$y'(t_F)v(t_F) - y'(0)v(0) = -\int_0^{t_F} z'(t)Qv(t)\,dt + \int_0^{t_F} y'(t)Sy(t)\,dt$$ (42)

or just

$$z'(t_F)Pv(t_F) + \int_0^{t_F} z'(t)Qv(t)\,dt = \int_0^{t_F} y'(t)Sy(t)\,dt$$ (43)

so that

$$\mathrm{VAR}_k J = 4\int_0^{t_F} y'(t)Sy(t)\,dt + K_v.$$ (44)

Since K_v is not affected by $u(t)$, this establishes that $y(t)$ is the principal contributor to what might be termed the controlled part of the performance-measure variance.

Suppose, for example, that the following matrices describe the plant, noise covariance, and performance-measure weightings:

$$A = \begin{bmatrix} -1 & 1 \\ 0 & -1 \end{bmatrix} \qquad B = \begin{bmatrix} 0 \\ 1 \end{bmatrix}$$

$$P = \begin{bmatrix} 0 & 0 \\ 0 & 0 \end{bmatrix} \qquad Q = I \qquad R = I$$

$$S = \begin{bmatrix} 0.5 & 0 \\ 0 & 0.5 \end{bmatrix} \qquad t_F = 3.0 \qquad \text{and } x(0) = \begin{bmatrix} 4 \\ 3 \end{bmatrix}.$$

Then the corresponding values for the controlled part of the mean and variance are given below.

μ	$M-K_m$	$\mathrm{VAR}_k J-K_v$
0.1	30.50	3.46
0.35	22.04	5.02
1.00	18.83	6.88

The equations above contain the information necessary to construct the solution to minimal variance control problems. The question of the existence of a unique solution to these equations is related to an indefinite Riccati equation, which arises as a fundamental feature of the solution.

Considered as one set of $4n$ differential equations, the solution has an overall state matrix $\bar{A}$ described by

$$\bar{A}(t) = \begin{bmatrix} \bar{A}_{11}(t) & \bar{A}_{12}(t) \\ \bar{A}_{21}(t) & \bar{A}_{22}(t) \end{bmatrix} = \begin{bmatrix} A(t) & 0 & -\frac{1}{2\mu}B(t)R^{-1}B'(t) & 0 \\ 0 & A(t) & 0 & S \\ -2\mu Q & -8Q & -A'(t) & 0 \\ -Q & 0 & 0 & -A'(t) \end{bmatrix}.$$ (45)

Associated with $\bar{A}(t)$ is the transition matrix $\Phi(t, 0)$, with which a corresponding partition is given, namely,

$$\Phi(t, 0) = \begin{bmatrix} \Phi_{11}(t, 0) & \Phi_{12}(t, 0) \\ \Phi_{21}(t, 0) & \phi_{22}(t, 0) \end{bmatrix}.$$ (46)

Because of the mixed boundary conditions a question of the existence of a unique solution on $[0, t_F]$ comes forward. It is a straightforward calculation to show that necessarily

$$\left\{ \begin{bmatrix} 2\mu P & 8P \\ P & 0 \end{bmatrix} \Phi_{12}(t_F, 0) - \Phi_{22}(t_F, 0) \right\} \begin{bmatrix} \rho(0) \\ y(0) \end{bmatrix}$$
$$= \left\{ \Phi_{21}(t_F, 0) - \begin{bmatrix} 2\mu P & 8P \\ P & 0 \end{bmatrix} \Phi_{11}(t_F, 0) \right\} \begin{bmatrix} z(0) \\ v(0) \end{bmatrix}. \tag{47}$$

With the obvious definitions, this can be rewritten as

$$F(t_F, 0) \begin{bmatrix} \rho(0) \\ y(0) \end{bmatrix} = E(t_F, 0) \begin{bmatrix} z(0) \\ v(0) \end{bmatrix}. \tag{48}$$

Now $\rho(0)$ and $y(0)$ may not be uniquely defined in terms of $z(0)$ and $v(0)$. In order to examine this question of uniqueness consider the matrix differential system

$$\dot{X}(t) = \bar{A}_{11}(t)X(t) + \bar{A}_{12}(t)Y(t) \tag{49}$$

$$\dot{Y}(t) = \bar{A}_{21}X(t) + \bar{A}_{22}(t)Y(t) \tag{50}$$

with the boundary conditions

$$X(0) = I \tag{51}$$

$$Y(t_F) = \begin{bmatrix} 2\mu P & 8P \\ P & 0 \end{bmatrix} X(t_F). \tag{52}$$

Suppose that there exists a matrix $M(t)$ on $[0, t_F]$ which satisfies the Riccati matrix differential equation

$$\dot{M}(t) + M(t)\bar{A}_{11}(t) - \bar{A}_{22}(t)M(t) - \bar{A}_{21} + M(t)\bar{A}_{12}M(Lt) = 0 \tag{53}$$

and the boundary condition

$$M(t_F) = \begin{bmatrix} 2\mu P & 8P \\ P & 0 \end{bmatrix}. \tag{54}$$

Under appropriate assumptions on system parameter matrices, this matrix $M(t)$ is unique. It is straightforward to establish that $M(t)X(t)$ is a solution for $Y(t)$ on $[0, t_F]$. Moreover, once $X(t)$ is specified on $[0, t_F]$, $Y(t)$ is uniquely specified on $[0, t_F]$. Consequently, every solution pair $X(t)$, $Y(t)$ is related on the interval $[0, t_F]$ by the equation

$$Y(t) = M(t)X(t). \tag{55}$$

As a result, then, $Y(0) = M(0)$ uniquely determines $Y(0)$. Therefore, if we have a solution $M(t)$ on $[0, t_F]$, then the solution to our $4n$ differential equations is unique.

This Riccati equation and its boundary condition have two features which distinguished them from other equations of a similar nature arising early in LQG control. The first of these features was that $M(t)$ is indefinite at its boundary condition; the constant term in its differential equation displays the same property. The second feature is that $M(t)$ is not symmetric. A result of Maguire's [22], however, eliminates this difficulty by making it evident that $M_{12}(t) = 8M'_{21}(t)$ on $[0, t_F]$. This result can be used to eliminate asymmetry in the problem by defining $\hat{v}(t) = 8v(t)$.

VII. AN EARTHQUAKE EXAMPLE

We consider in this section a numerical example based upon an earthquake excitation to a single degree-of-freedom building model. The building is represented by [23]

$$\dot{x} = \begin{bmatrix} 0 & 1 \\ -\omega_0^2 & -2\zeta\omega_0 \end{bmatrix} x + \begin{bmatrix} 0 \\ \dfrac{-4k_c \cos\alpha}{m} \end{bmatrix} u + \begin{bmatrix} 0 \\ -1 \end{bmatrix} \ddot{x}_g, \tag{56}$$

where x_1 is floor position and x_2 is floor velocity, and with the constants

$$k_c = 2124\frac{\text{lb}}{\text{in}}, \quad m = 16.69\frac{\text{lb-sec}^2}{\text{in}}, \quad \zeta = 0.0124, \quad \alpha = \frac{36\pi}{180}\text{ rad}, \quad \omega_0 = 6.94\pi\frac{\text{rad}}{\text{sec}}.$$

The earthquake is modelled by [24]

$$\dot{x}_e = \begin{bmatrix} 0 & 1 \\ -\omega_g^2 & -2\zeta_g\omega_g \end{bmatrix} x_e + \begin{bmatrix} 0 \\ 1 \end{bmatrix} w, \tag{57}$$

$$\ddot{x}_g = \begin{bmatrix} -\omega_g^2 & -2\zeta_g\omega_g \end{bmatrix} x_e, \tag{58}$$

with the constants

$$\zeta_g = 0.65, \quad \omega_g = 18.85\frac{\text{rad}}{\text{sec}}. \tag{59}$$

The spectral constant for the Gaussian white noise w is

$$S_0 = 0.7204\frac{\text{in}^2}{\text{sec}^3}. \tag{60}$$

The cost function is chosen to be

$$J = \int_0^{t_F} \left\{ \begin{bmatrix} x \\ xe \end{bmatrix}' \begin{bmatrix} 7934 & 0 & 0 & 0 \\ 0 & 0 & 0 & 0 \\ 0 & 0 & 0 & 0 \\ 0 & 0 & 0 & 0 \end{bmatrix} \begin{bmatrix} x \\ xe \end{bmatrix} + k_c u^2 \right\} dt. \tag{61}$$

Nominal t_F is 1.0 second. See Figure 3 for a sketch of the plant.

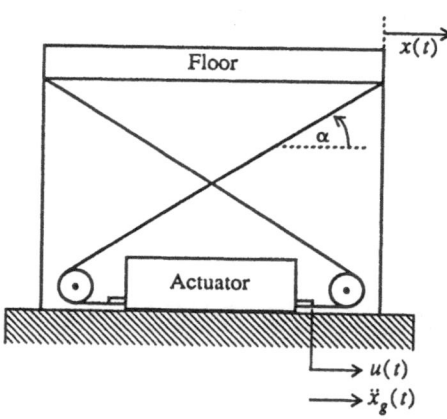

Figure 3: Schematic Diagram for Single Degree of Freedom Structure

Numerical simulations have been carried out to study a number of the basic minimal variance relationships in this example. Figure 4 shows the effect of terminal time t_F upon cost mean, denoted M, and cost variance, denoted VAR. The symbol x is used for M, while o denotes VAR. The curve is a cross section for $\mu = 1$. The initial condition was obtained by simulating the earthquake excitation for an initial time segment, and selecting the resulting state vector. Figure 5 depicts the interconnection between cost mean and cost variance. The upper graph is a family of curves identified with multiples of $x(0)$. Each curve, for a fixed $x(0)$, has μ as a parameter. The direction of increasing μ is indicated on the graph. The lower graph may be used to determine the size of μ for given pairs M and VAR. Figure 6 presents two graphs showing the effects of building model damping ζ and natural frequency ω_o on M and VAR. Notice that the cost mean and variance have large slopes in the neighborhood of the building nominal parameters.

VIII. CONCLUSIONS

In this paper, we have reviewed the question of stochastic control by cumulant minimization. It has been indicated how such an approach relates to risk sensitive optimal control, and therefore to robust control by H_∞ methods. When a focus is placed upon the first two cumulants, cost mean and variance, we outlined a minimal variance design procedure based upon Riccati methods and techniques similar to those used by Baggeroer. For solutions near to minimum average cost, minimal variance methods and risk sensitive methods approximate each other. However, minimal variance methods do not require this assumption. An application has been presented to the case of a building model excited by an earthquake. The results indicate relatively high performance mean and variance sensitivity at nominal building damping and stiffness.

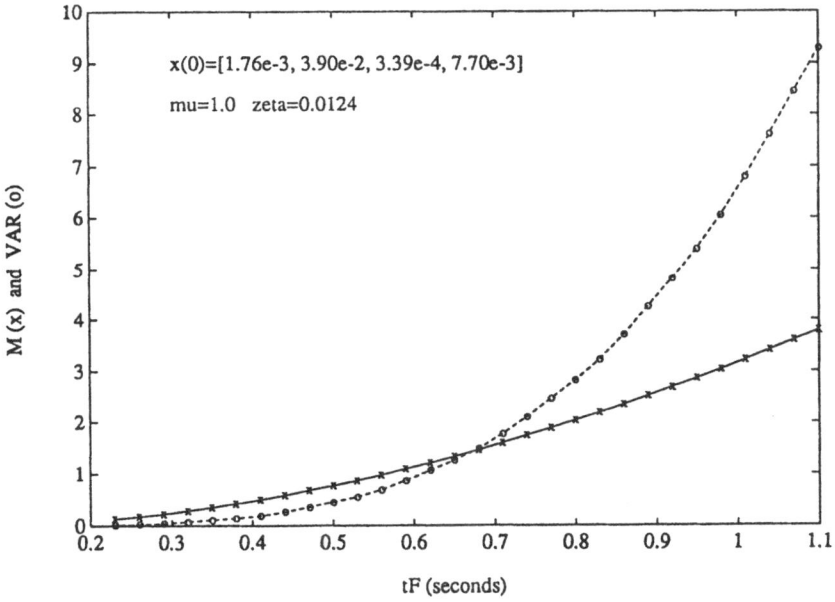

Figure 4: Cost Mean and Variance vs. Terminal Time

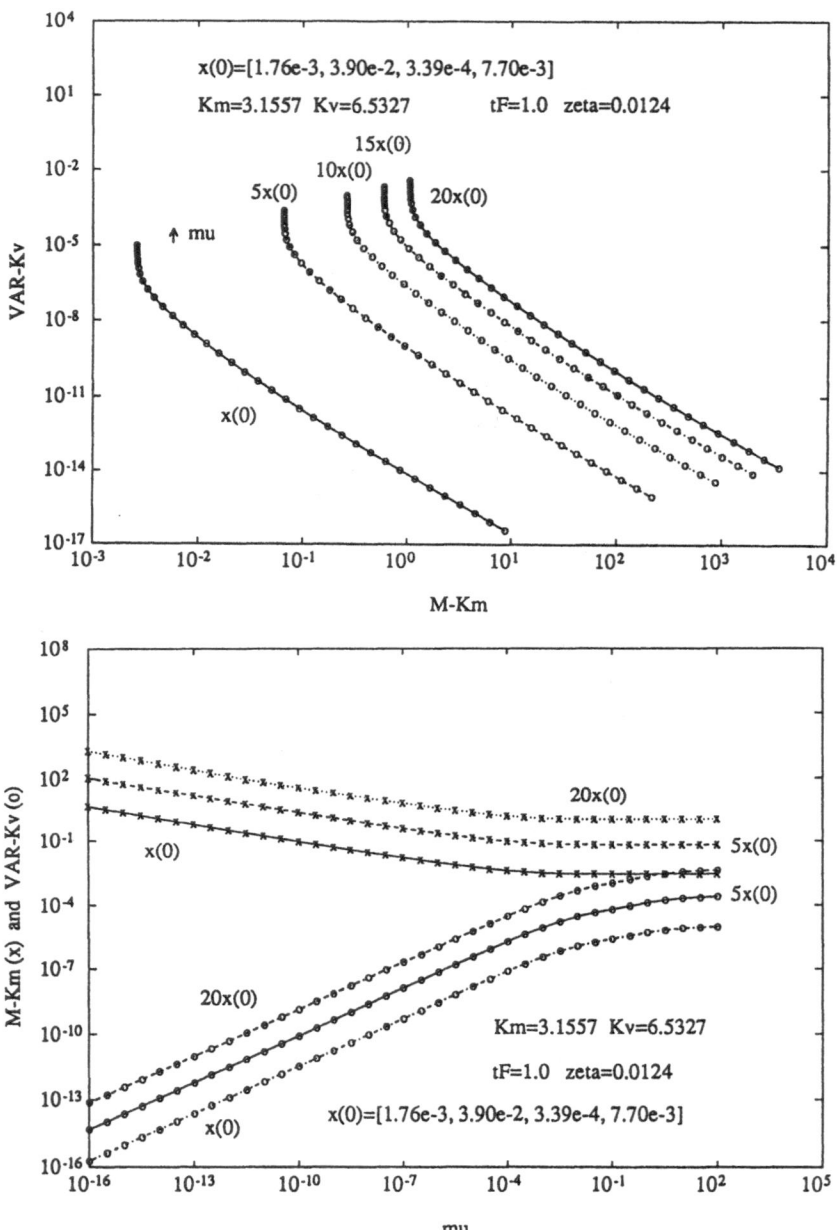

Figure 5: Cost Mean and Variance as a Function of $x(0)$ and μ

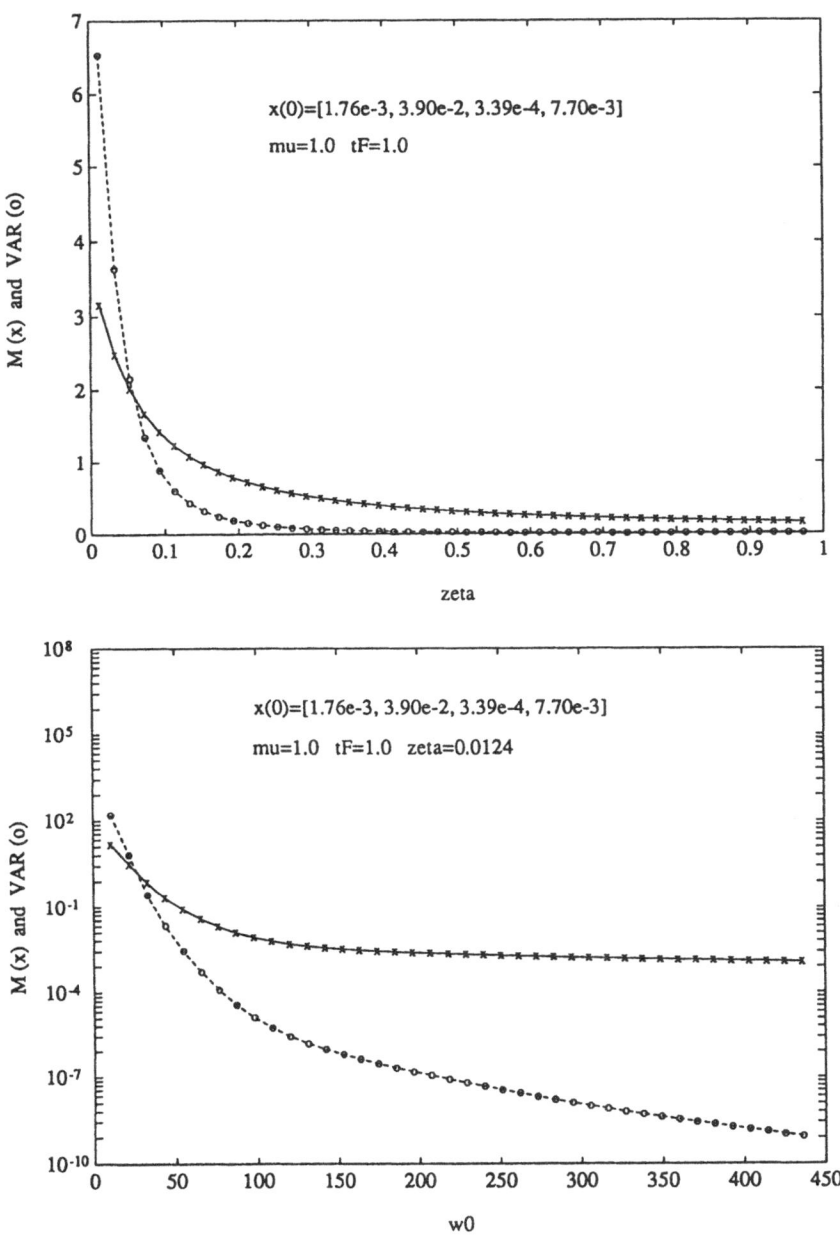

Figure 6: Cost Mean and Variance vs. Damping and Natural Frequency

REFERENCES

1. Peter Whittle, *Risk Sensitive Optimal Control.* New York: John Wiley & Sons, 1990.

2. K. Glover and J. C. Doyle, "State-Space Formulae for All Stabilizing Controllers that Satisfy an H_∞-Norm Bound and Relations to Risk Sensitivity," *Systems and Control Letters,* Volume 11, pp. 167–172, 1988.

3. Tamer Basar and Pierre Bernhard, H_∞-*Optimal Control and Related Minimax Design Problems.* Boston: Birkhäuser, 1991.

4. D. H. Jacobson, "Optimal Stochastic Linear Systems with Exponential Performance Criteria and Their Relationship to Deterministic Differential Games," *IEEE Transactions on Automatic Control,* AC-18, pp. 124–131, 1973.

5. Michael K. Sain, "On Minimal-Variance Control of Linear Systems with Quadratic Loss," *Ph.D Thesis,* Department of Electrical Engineering and Coordinated Science Laboratory, University of Illinois, Urbana, Illinois, January 1965.

6. M. K. Sain, "Control of Linear Systems According to the Minimal Variance Criterion—A New Approach to the Disturbance Problem," *IEEE Transactions on Automatic Control,* AC-11, No. 1, pp. 118–122, January 1966.

7. P. Whittle, "A Risk-Sensitive Maximum Principle: The Case of Imperfect State Observation," *IEEE Transactions on Automatic Control,* AC-36, No. 7, pp. 793–801, July 1991.

8. Jason L. Speyer, John Deyst, and David H. Jacobson, "Optimization of Stochastic Linear Systems with Additive Measurement and Process Noise Using Exponential Performance Criteria," *IEEE Transactions on Automatic Control,* AC-19, No. 4, pp. 358–366, August 1974.

9. J. L. Speyer, "An Adaptive Terminal Guidance Scheme Based on an Exponential Cost Criterion with Application to Homing Missile Guidance," *IEEE Transactions on Automatic Control,* AC-21, pp. 371–375, 1976.

10. P. R. Kumar and J. H. van Schuppen, "On the Optimal Control of Stochastic Systems with an Exponential-of-Integral Performance Index," *Journal of Mathematical Analysis and Applications,* Volume 80, pp. 312–332, 1981.

11. P. Whittle, "Risk-Sensitive Linear/Quadratic/Gaussian Control," *Advances in Applied Probability,* Volume 13, pp. 764–777, 1981.

12. A. Bensoussan and J. H. van Schuppen, "Optimal Control of Partially Observable Stochastic Systems with an Exponential-of-Integral Performance Index," *SIAM Journal on Control and Optimization,* Volume 23, pp. 599–613, 1985.

13. P. Whittle and J. Kuhn, "A Hamiltonian Formulation of Risk-Sensitive Linear/ Quadratic/Guassian Control," *International Journal of Control,* Volume 43, pp. 1–12, 1986.

14. W. Fleming and W. McEneaney, "Risk Sensitive Optimal Control," *Workshop on Stochastic Theory and Adaptive Control,* University of Kansas, September 26–28, 1991.

15. K. Glover, "Minimum Entropy and Risk-Sensitive Control: The Continuous Time Case," *Proceedings 28th IEEE Conference on Decision and Control*, pp. 388–391, December 1989.

16. S. R. Liberty, "Characteristic Functions of LQG Control," Ph.D. Dissertation, Department of Electrical Engineering, University of Notre Dame, August 1971.

17. S. R. Liberty and R. C. Hartwig, "On the Essential Quadratic Nature of LQG Control-Performance Measure Cumulants," *Information and Control*, Volume 32, Number 3, pp. 276–305, 1976.

18. M. K. Sain and S. R. Liberty, "Performance Measure Densities for a Class of LQG Control Systems," *IEEE Transactions on Automatic Control*, AC-16, Number 5, pp. 431–439, October 1971.

19. S. R. Liberty and R. C. Hartwig, "Design-Performance-Measure Statistics for Stochastic Linear Control Systems," *IEEE Transactions on Automatic Control*, AC-23, Number 6, pp. 1085–1090, December 1978.

20. M. K. Sain and C. R. Souza, "A Theory for Linear Estimators Minimizing the Variance of the Error Squared," *IEEE Transactions on Information Theory*, IT-14, Number 5, pp. 768–770, September 1968.

21. A.B. Baggeroer, "A State-Variable Approach to the Solution of Fredholm Integral Equations," *IEEE Transactions on Information Theory*, IT-15, pp. 557–570, September 1969.

22. M.V. Maguire, "On a Nonsymmetric Riccati Equation of Stochastic Control," M.S. Thesis, Department of Electrical Engineering, University of Notre Dame, Notre Dame, Indiana, 1971.

23. L. L. Chung, A. M. Reinhorn, and T. T. Soong, "Experiments on Active Control of Seismic Structures," *Journal of Engineering Mechanics*, Volume 114, Number 2, pp. 241–256, February 1988.

24. J. N. Yang and M. J. Lin, "Building Critical-Mode Control: Nonstationary Earthquake," *Journal of Engineering Mechanics*, Volume 109, Number 6, pp. 1375–1389, December 1983.

Hierarchical investment and production decisions in stochastic manufacturing systems*

S.P. Sethi; M. Taksar; and Q. Zhang†

October 9, 1991

Abstract

This paper presents an asymptotic analysis of hierarchical investment and production decisions in a manufacturing system with machines subject to breakdown and repair. The demand facing the system is assumed to be a given constant. The production capacity can be increased by purchasing a new machine at a fixed cost at some time in the future. The control variables are a pair of a Markov time to purchase the new machine and a production plan. The rate of change in machine states is assumed to be much larger than the rate of discounting of costs. This gives rise to a limiting problem in which the stochastic machine availability is replaced by the equilibrium mean availability. The value function for the original problem converges to the value function of the limiting problem. Moreover, three different methods are developed for constructing controls for the original problem from the optimal controls of the limiting problem in a way which guarantees their asymptotic optimality. The convergence rate of the value function for the original problem to that of the limiting problem is also found. This helps in providing error estimates for the constructed asymptotically optimal controls.

1 Introduction

Consider a firm that must satisfy a given constant demand for its product over time so as to minimize its discounted cost of investment, production, and inventory/shortage. Suppose that the firm has an existing machine, which is failure-prone with given rates of breakdown and repair. When in working order, it has a unit production capacity and when broken down, it has zero capacity. Assume that the demand for the firm's product is higher than the average production capacity of the existing machine. However, the firm has some initial inventory of its product to absorb the excess demand for a few initial periods. It is obvious that the firm must increase its production capacity at some future time. For this purpose, the firm has an option to purchase a new machine, identical to the existing machine, at a fixed given cost in order to double its average production capacity. This, of course, relies on the firm having sufficient repair capacity to handle two machines even when they are both broken down during some time interval.

---

*This work was supported in parts by the NSERC Grant A4619, URIF, and the Manufacturing Research Corporation of Ontario.

†Faculty of Management, University of Toronto, Toronto, Ontario, Canada M5S 1V4

‡Department of Applied Mathematics and Statistics, SUNY at Stony Brook, Stony Brook, NY 11794

It is possible to formulate the problem of the firm as a stochastic optimal control problem in which the decision variables are the time to purchase the new machine and the rate of production over time. We shall refer to this problem as the original problem (facing the firm). Such a problem is formulated as a simple example in §2 as a special case of the more general model treated in this paper. Even this simple example is quite hard to solve as it involves machine purchase time, which is a stopping time to be determined.

Moreover in practice, the investment or the capacity expansion decisions and the production decisions are carried out at different levels of hierarchy that exist within the firm. In the case of two levels, upper (corporate) and lower (operational), the machine purchase decision is usually carried out at the upper level, while the production decisions lie in the lower level's domain.

The important and the obvious question that arises is whether there is a two-level decision making procedure that is simpler than solving the original problem and is, at the same time, a good approximation to the optimal solution of the original problem. The theory developed in this paper answers the question in the affirmative under reasonable assumptions. Indeed, it is possible to develop several different two-level procedures that accomplish the task. One such two-level procedure can be described as follows. The upper level solves a deterministic problem, termed the limiting problem, obtained by replacing random capacities by their averages. The solution of this limiting problem yields the purchase date for the new machine as well as an average production plan. The upper level releases an order to have the new machine installed at that date and informs the lower level of this decision. With regards to the production plan, it is clear that the average production plan is not feasible for the original stochastic problem. However, it may be easy to construct a feasible production plan at the lower level that takes into account the information regarding the date at which the new machine will be available. Moreover, it is possible to show that this two-level procedure results in a solution that is nearly optimal for the original stochastic problem.

The general model considered in this paper replaces the single existing machine in the example by a finite state Markov process representing the existing system capacity over an infinite horizon. Purchase of the new machine at some future time is replaced by an expansion of the existing capacity at that time. This results in an enhanced capacity process represented by another finite state Markov process having a larger average capacity than the existing one. The cost of production and inventory/shortage is assumed to be either jointly convex or separable with linear production cost and convex inventory/shortage cost. It is assumed that the rate of breakdown and repair events is much larger than the rate discounting of costs; see Lehoczky, Sethi, Soner, and Taksar [1] for a detailed discussion of this point. It is under this assumption that we are able to prove that two-level decision procedures developed in the paper provide asymptotically optimal solution to the original stochastic control problem as the rates of breakdown and repair events become very large or, in other words, approach infinity.

In this paper, we extend the problem studied in [1] and [3] by incorporating the optimal stopping time of the capacity expansion event in the stochastic optimal control problem, in which only the optimal production plan was needed to be determined. Our extension results in a dynamic problem in which both the optimal stopping time and the optimal production plan are determined simultaneously by two homogeneous HJB equations: the equation dealing with the dynamics of the system before the capacity expansion and the equation concerning the dynamics after the capacity expansion. Then, by using these two equations, we define what is called a switching set, on which the solutions of the two equations coincide; a more detailed description on this point will be given in §4. This switching set will be used to determine the optimal stopping time at which to expand the capacity. Owing to the complexities in the dynamics of the system, the exact optimal solution for such a system is very difficult to obtain. In order to reduce the complexity in the manner of the averaging method mentioned above, we make use of the idea of hierarchical control. Namely, we derive a limiting control problem, which is simpler to solve than the original problem. This limiting problem is obtained by replacing the stochastic machine availability processes before and after the capacity expansion event by their respective average total capacities and by appropriately modifying the objective function. From its solution, we construct an approximately optimal control of the original, more complex, problem.

The plan of the paper is as follows. In §2, we formulate our model of the manufacturing system under consideration and the related optimization problem. We discuss some elementary properties of the associated value function in §3. In §3, we define the limiting control problem and show that the value function of our problem converges to the value function of the limiting problem as the oscillating rate of the machine capacity goes to infinity. Then in §4, we study the convergence rates of the value functions and construction of the asymptotically optimal controls by means of an asymptotic analysis on the capacity process. We discuss a method of constructing controls for the original problem starting from the solution of the limiting problem. We show that these controls are asymptotically optimal. §7 concludes the paper. The proofs of results are omitted and referred to [4].

2   Problem formulation

We consider a stochastic manufacturing system with the inventory/backlog or surplus $x_t \in R^n$ and production rate $u_t \in R^n$ that satisfy

$$\dot{x}_t = u_t - z, \ x_0 = x, \tag{2.1}$$

where $z \in R^n$ denotes the constant rates of demand and x is the initial surplus level. We assume $u_t \geq 0$ and for some positive vector $p^0 = (p_1^0, \cdots, p_n^0)$ such that $p^0 \cdot u_t \leq \alpha^\epsilon(t)$ where $\alpha^\epsilon(t)$ is a stochastic production capacity process with ϵ as a small parameter in the characterization of the capacity process to be precisely specified later. Moreover, the specification of $\alpha^\epsilon(t)$ involves the purchase of some given additional capacity at some time τ, $0 \le \tau \le \infty$ at a cost of K, where $\tau = \infty$ means not to purchase it at all. Therefore, our control variable is a pair $(\tau, u.)$ of a Markov time $\tau \ge 0$ and a production process $u.$ over time.

We consider the cost function J^ϵ defined by

$$J^\epsilon(x, \alpha, \tau, u.) = E[\int_0^\infty e^{-\rho t} G(x_t, u_t) dt + K e^{-\rho \tau}], \tag{2.2}$$

where $\alpha^\epsilon(0) = \alpha$ is the initial capacity and $\rho > 0$ is the discount rate. The problem is to find an admissible control $(\tau, u.)$ that minimizes $J^\epsilon(x, \alpha, \tau, u.)$.

We take $U = \{(u_1, \cdots, u_n) \ge 0 : p_1^0 u_1 + \cdots + p_n^0 u_n \le 1\}$, where $p^0 = (p_1^0, \cdots, p_n^0) \ge 0$. Then U is a compact convex subset of R^n. U will be used later in defining admissible controls.

Define $\alpha_1^\epsilon(t)$ and $\alpha_2^\epsilon(t)$ as two Markov processes with state spaces $\mathcal{M}_1 = \{0, 1, \cdots, m_1\}$ and $\mathcal{M}_2 = \{0, 1, \cdots, m_1 + m_2\}$, respectively. Here, $\alpha_1^\epsilon(t) \ge 0$ denotes the existing production capacity process and $\alpha_2^\epsilon(t) \ge 0$ denotes the capacity process of the system if it were to be supplemented by the additional new capacity at time $t = 0$.

Let $\mathcal{F}_1(t)$ and $\mathcal{F}_2(t)$ denote the filtrations generated by $\alpha_1^\epsilon(t)$ and $\alpha_2^\epsilon(t)$, respectively, i.e., $\mathcal{F}_1(t) = \sigma\{\alpha_1^\epsilon(s) : s \le t\}$ and $\mathcal{F}_2(t) = \sigma\{\alpha_2^\epsilon(t) : s \le t\}$.

We define a new process $\alpha^\epsilon(t)$ as follows: For each $\mathcal{F}_1(t)$-Markov time $\tau \ge 0$,

$$\alpha^\epsilon(t) = \begin{cases} \alpha_1^\epsilon(t) & \text{if } t < \tau \\ \alpha_2^\epsilon(t - \tau) & \text{if } t \ge \tau \end{cases}, \tag{2.3}$$

such that $\alpha^\epsilon(\tau) = \alpha_2^\epsilon(0) := \alpha_1^\epsilon(\tau) + m_2$. Here m_2 denotes the maximum additional capacity resulting from the addition of the new capacity.

We make the following assumptions on the running cost function G and the random processes $\alpha_1^\epsilon(t)$ and $\alpha_2^\epsilon(t)$.

A1) There exist convex functions $h(x)$ and $c(u)$ such that $G(x, u) = h(x) + c(u)$. For all x, x', u, u', there exist constants C_g and k_g such that $0 \le h(x) \le C_g(1 + |x|^{k_g})$,

$$|h(x) - h(x')| \le C_g(1 + |x|^{k_g} + |x'|^{k_g})|x - x'|, \text{ and } |c(u) - c(u')| \le C_g|u - u'|.$$

A2) $\alpha_1^\epsilon(t) \in \mathcal{M}_1$ and $\alpha_2^\epsilon(t) \in \mathcal{M}_2$ are Markov processes with generators $\epsilon^{-1}Q_1$ and $\epsilon^{-1}Q_2$, respectively, where $Q_1 = (q_{ij}^{(1)})$ and $Q_2 = (q_{ij}^{(2)})$ are matrices such that $q_{ij}^{(k)} \ge 0$ if $i \ne j$ and $q_{ii}^{(k)} = -\sum_{i \ne j} q_{ij}^{(k)}$ for $k = 1, 2$. Moreover, Q_1 and Q_2 are both irreducible.

Let $\mathcal{F}(t)$ denote the filtration generated by $\alpha^\epsilon(t)$, i.e., $\mathcal{F}(t) = \sigma\{\alpha^\epsilon(t) : s \le t\}$ and let $\alpha \cdot U = \{\alpha u : u \in U\}$ for any $\alpha \in R^1$.

Definition. We say that a control $(\tau, u.)$ is admissible if 1) τ is an $\mathcal{F}_1(t)$-Markov time; 2) u_t is $\mathcal{F}(t)$ adapted and $u(t) \in \alpha^\epsilon(t) \cdot U$ for $t \ge 0$.

We use $\mathcal{A}_I$ to denote the set of all admissible controls $(\tau, u.)$. Then the problem is:

$$\mathcal{P}(\mathrm{I}) : \begin{cases} \min_{(\tau, u.) \in \mathcal{A}_I} & J^\epsilon(x, \alpha, \tau, u.) \\ \text{s.t.} & \dot{x}_t = u_t - z, \ x_0 = x. \end{cases}$$

We write $v^\epsilon(x, \alpha)$, the value function, to be the minimum cost on $\mathcal{A}_I$, i.e.,

$$v^\epsilon(x, \alpha) = \inf_{(\tau, u.) \in \mathcal{A}_I} J^\epsilon(x, \alpha, \tau, u.). \tag{2.4}$$

Let us give a simple one dimensional example to illustrate the model we have just formulated. **A simple example.** Let us assume that the existing capacity consists of one machine of unit maximum capacity, i.e., $m_1 = 1$. Assume further that the purchase of an identical machine at a cost of K is under consideration at some time $\tau \geq 0$. The problem is to find the optimal time of purchase as well as the optimal production simultaneously. More specifically, we consider the following problem:

$$\begin{aligned} \text{minimize} \quad & J^\epsilon(x, \alpha, \tau, u.) = E[\int_0^\infty e^{-\rho t}|x_t|dt + Ke^{-\rho \tau}] \\ \text{s.t.} \quad & \dot{x}_t = u_t - z, \ u_t \in \mathcal{A}_I, \ x_0 = x. \end{aligned} \tag{2.5}$$

We take $0 < z \leq 1$, $U = [0, 1]$, $\mathcal{M}_1 = \{0, 1\}$, and $\mathcal{M}_2 = \{0, 1, 2\}$. We assume further that

$$Q_1 = \begin{bmatrix} -1 & 1 \\ 1 & -1 \end{bmatrix} \text{ and } Q_2 = \begin{bmatrix} -1 & 1 & 0 \\ 1 & -2 & 1 \\ 0 & 1 & -1 \end{bmatrix}.$$

The specification of Q_1 and Q_2 in our example represents the following situation. The existing machine has the breakdown rate of ϵ^{-1} and the system has sufficient capacity to repair it at the rate of ϵ^{-1}. Thus, the machine is in working order half the time on average. Furthermore, upon the addition of an identical machine, there is sufficient repair capacity in the system to repair each machine at rate of ϵ^{-1} even when both machine are down.

This example will be used throughout the rest of the paper to illustrate the theory as it develops. Furthermore, in view of this simple example, we will use the terms existing capacity and existing machine and the terms new (or additional) capacity and new machine interchangeably.

It follows immediately from (2.4) that

$$v^\epsilon(x, \alpha) = \inf_\tau \inf_{u.} J^\epsilon(x, \alpha, \tau, u.).$$

This means that we can optimize over all production rates first for any fixed τ, and then search for the Markov time τ that is optimal.

We now define an auxiliary value function $v_a^\epsilon(x, \alpha)$ to be the optimal cost with $\tau = 0$ a.s., i.e.,

$$v_a^\epsilon(x, \alpha + m_2) = \inf_{(0, u.) \in \mathcal{A}_I} J^\epsilon(x, \alpha, 0, u.) \text{ for all } \alpha \in \mathcal{M}_1.$$

For convenience, we shall call the pair $(v^\epsilon, v_a^\epsilon)$ as the value functions of the problem.

On account of the definition of $\alpha^\epsilon(t)$ in (2.3), it follows that

$$v_a^\epsilon(x, \alpha + m_2) \geq v^\epsilon(x, \alpha), \text{ for } \alpha \in \mathcal{M}_1. \tag{2.6}$$

3 Limiting control problem

In this section, we first state the elementary properties of the value functions $(v^\epsilon, v^\epsilon_a)$.

Theorem 3.1. $(v^\epsilon(x, \alpha), v^\epsilon_a(x, \alpha))$ *are locally Lipschitz convex functions and are the viscosity solutions to the following dynamic programming equations.*

$$\min\left\{ \min_{u \in \alpha \cdot U}[(u - z) \cdot \nabla v^\epsilon(x, \alpha) + G(x, u)] + \epsilon^{-1} Q_1 v^\epsilon(x, \cdot)(\alpha) - \rho v^\epsilon(x, \alpha), \right.$$
$$\left. v^\epsilon_a(x, \alpha + m_2) - v^\epsilon(x, \alpha) \right\} = 0, \tag{3.1}$$

for any $\alpha \in \mathcal{M}_1$ *and*

$$\min_{u \in \alpha \cdot U}[(u - z) \cdot \nabla v^\epsilon_a(x, \alpha) + G(x, u)] + \epsilon^{-1} Q_2 v^\epsilon_a(x, \cdot)(\alpha) - \rho(v^\epsilon_a(x, \alpha) - K) = 0, \tag{3.2}$$

for any $\alpha \in \mathcal{M}_2$.

We next consider the asymptotic behavior of the system (2.1) and (2.4). We show that the system (2.1) with random capacity due to unreliable machines can be simplified and reduced to a deterministic capacity system. In a large measure, this is accomplished by showing that there exists a value function $v(x)$ and an auxiliary value function $v_a(x)$ of some system, to be determined, such that $(v^\epsilon(x, \alpha), v^\epsilon_a(x, \alpha)) \to (v(x), v_a(x))$ for all (x, α) as $\epsilon \to 0$.

Let $\nu^{(1)} = (\nu^{(1)}_0, \nu^{(1)}_1, \cdots, \nu^{(1)}_{m_1})$ and let $\nu^{(2)} = (\nu^{(2)}_0, \nu^{(2)}_1, \cdots, \nu^{(2)}_{m_1+m_2})$ denote the equilibrium distributions of Q_1 and Q_2, respectively. Then $\nu^{(1)}$ and $\nu^{(2)}$ are the only positive solutions to

$$\nu^{(1)} Q_1 = 0 \text{ and } \sum_{i=0}^{m_1} \nu^{(1)}_i = 1,$$
$$\nu^{(2)} Q_2 = 0 \text{ and } \sum_{i=0}^{m_1+m_2} \nu^{(2)}_i = 1. \tag{3.3}$$

We now define a limiting problem. We first define control sets for the limiting problem. Let $U_1 = \{(u_0, \cdots, u_{m_1}) : \text{ such that } u_i \in i \cdot U\}$ and $U_2 = \{(u_0, \cdots, u_{m_1+m_2}) : \text{ such that } u_i \in i \cdot U\}$. Then $U_1 \subset R^{n \times (m_1+1)}$ and $U_2 \subset R^{n \times (m_1+m_2+1)}$.

Definition. We use $\mathcal{A}_{II}$ to denote the set of the following controls (*admissible controls* for a limiting problem): 1) τ is a deterministic time; 2) There exist deterministic u_t such that for $t < \tau$, $u_t = (u_0(t), \cdots, u_{m_1}(t)) \in U_1$ and for $t \geq \tau$, $u_t = (u_0(t), \cdots, u_{m_1+m_2}(t)) \in U_2$.

Let

$$J(x, \tau, u.) = \int_0^\tau e^{-\rho t} \sum_{i=0}^{m_1} \nu^{(1)}_i G(x_t, u_i(t)) dt + \int_\tau^\infty e^{-\rho t} \sum_{i=0}^{m_1+m_2} \nu^{(2)}_i G(x_t, u_i(t)) dt + e^{-\rho \tau} K$$

and let

$$\bar{u}_t = \begin{cases} \sum_{i=0}^{m_1} \nu^{(1)}_i u_i(t) & \text{if } t < \tau \\ \sum_{i=0}^{m_1+m_2} \nu^{(2)}_i u_i(t) & \text{if } t \geq \tau \end{cases}. \tag{3.4}$$

We define the following optimal control problem (limiting problem):

$$\mathcal{P}(\text{II}) : \begin{cases} \min_{(\tau, u.) \in \mathcal{A}_{II}} & J(x, \tau, u.) \\ \text{s.t.} & \dot{x}_t = \bar{u}_t - z, \ x_0 = x. \end{cases} \tag{3.5}$$

Remark. Note that the dimension of the control space in $\mathcal{P}(II)$ is $n \times (1 + m_1 + m_2)$. However, if $c(u)$ is linear in u, then the dimension of the control space in $\mathcal{P}(II)$ can be reduced to n. In fact, if there exists a vector $p \geq 0$ such that $c(u) = p \cdot u$, then the cost function of $\mathcal{P}(II)$ can be written as follows:

$$J(x,\tau,u.) = \int_0^\infty e^{-\rho t}(h(x_t) + p \cdot \bar{u}_t)dt + e^{-\rho \tau}K,$$

where $\bar{u}_t$ is defined in (3.4). Then, we can treat $\bar{u}_t \in R^n$ as the control variable of the limiting problem $\mathcal{P}(II)$.

Theorem 3.2. i) $(v(x), v_a(x))$ are the only viscosity solutions to the following equations:

$$\min\Big\{\min_{u \in U_1}[(\sum_{i=0}^{m_1} \nu_i^{(1)}u_i - z) \cdot \nabla v(x) + \sum_{i=0}^{m_1} \nu_i^{(1)}G(x,u_i)] - \rho v(x), v_a(x) - v(x)\Big\} = 0, \tag{3.6}$$

$$\min_{u \in U_2}[(\sum_{i=0}^{m_1+m_2} \nu_i^{(2)}u_i - z) \cdot \nabla v_a(x) + \sum_{i=0}^{m_1+m_2} \nu_i^{(2)}G(x,u_i)] - \rho(v_a(x) - K) = 0. \tag{3.7}$$

ii) $(v(x), v_a(x))$ are the value functions for $\mathcal{P}(II)$, i.e.,

$$v(x) = \inf_{(\tau,u.)\in\mathcal{A}_{11}} J(x,\tau,u.) \quad and \quad v_a(x) = \inf_{(0,u.)\in\mathcal{A}_{11}} J(x,0,u.).$$

iii) $|v^\epsilon(x,\alpha) - v(x)| + |v_a^\epsilon(x,\alpha) - v_a(x)| \leq C(1 + |x|^{k_9})\sqrt{\epsilon}$.

Remark. If $c(u) = p \cdot u$, then for any $r \in R^n$,

$$\sum_{i=0}^{m_1} \nu_i^{(1)} \min_{u \in i\cdot U}[(u - z) \cdot r + G(x,u)] = \min_{u \in \bar{\alpha}_1 \cdot U}[(u - z) \cdot r + G(x,u)]$$

$$\sum_{i=0}^{m_1+m_2} \nu_i^{(2)} \min_{u \in i\cdot U}[(u - z) \cdot r + G(x,u)] = \min_{u \in \bar{\alpha}_2 \cdot U}[(u - z) \cdot r + G(x,u)],$$

where $\bar{\alpha}_1$ and $\bar{\alpha}_2$ are the equilibrium means of $\alpha_1^\epsilon(t)$ and $\alpha_2^\epsilon(t)$, respectively, i.e., $\bar{\alpha}_1 = \sum_{i=0}^{m_1} i\nu_i^{(1)}$ and $\bar{\alpha}_2 = \sum_{i=0}^{m_1+m_2} i\nu_i^{(2)}$. Therefore, the dynamic programming equations will be reduced to

$$\min\Big\{\min_{u \in \bar{\alpha}_1 \cdot U}[(u - z) \cdot \nabla v(x) + G(x,u)] - \rho v(x), v_a(x) - v(x)\Big\} = 0,$$

$$\min_{u \in \bar{\alpha}_2 \cdot U}[(u - z) \cdot \nabla v_a(x) + G(x,u)] - \rho(v_a(x) - K) = 0.$$

The admissible controls in $\mathcal{A}_{11}$ will be $(\sigma, u.)$ such that $u_t \in \bar{\alpha}_1 \cdot U$ if $t < \sigma$ and $u_t \in \bar{\alpha}_2 \cdot U$ if $t \geq \sigma$.

4 Asymptotically optimal controls and error bounds

In this section we discuss a method to construct asymptotically optimal controls based on the controls for the limiting problems $\mathcal{P}(II)$.

Construction of nearly optimal controls. Let $(\sigma, u.^0) \in \mathcal{A}_{11}$ denote any admissible control for the limiting problem $\mathcal{P}(II)$ where $u_t^0 = \begin{cases} (u_0(t), \cdots, u_{m_1}(t)) \in U_1 & \text{if } t < \sigma \\ (u_0(t), \cdots, u_{m_1+m_2}(t)) \in U_2 & \text{if } t \geq \sigma \end{cases}$. Note that here σ denotes a deterministic (calendar) time. We take

$$u_t^\epsilon = \begin{cases} \sum_{i=0}^{m_1} \chi_{\{\alpha^\epsilon(t)=i\}}u_i(t) & \text{if } t < \sigma \\ \sum_{i=0}^{m_1+m_2} \chi_{\{\alpha^\epsilon(t)=i\}}u_i(t) & \text{if } t \geq \sigma \end{cases}. \tag{4.1}$$

Then, the constructed control $(\sigma, u.^\epsilon) \in \mathcal{A}_\text{I}$ is apparently admissible for $\mathcal{P}(\text{I})$.

Remark. By calendar in the parentheses above, we mean a time that can be marked on a calendar. An example would be to buy the new machine on Jan. 15, 2000. Alternatively, a deterministic time can be also expressed by a time at which a certain deterministic trajectory enters a specified deterministic set known as the switching set; see §5.

Remark. If $c(u) = p \cdot u$, then the control u_t^ϵ is expected to have a simpler form. In fact, if we take $(\sigma, \bar{u}.)$ to be an admissible control for $\mathcal{P}(\text{II})$. Then $\bar{u}_t \in \bar{\alpha}_1 \cdot U$ for $t < \sigma$ and $\bar{u}_t \in \bar{\alpha}_2 \cdot U$ for $t \geq \sigma$. Let

$$u_i(t) = \begin{cases} i(\bar{\alpha}_1)^{-1}\bar{u}_t & \text{if } t < \sigma \\ i(\bar{\alpha}_2)^{-1}\bar{u}_t & \text{if } t \geq \sigma \end{cases}.$$

Then $u_i(t) \in i \cdot U$. Hence, the control constructed in (4.1) is equal to

$$u_t^\epsilon = \begin{cases} \sum_{i=0}^{m_1} \chi_{\{\sigma^\epsilon(t)=i\}} u_i(t) & \text{if } t < \sigma \\ \sum_{i=0}^{m_1+m_2} \chi_{\{\sigma^\epsilon(t)=i\}} u_i(t) & \text{if } t \geq \sigma \end{cases} = \begin{cases} \alpha^\epsilon(t)(\bar{\alpha}_1)^{-1}\bar{u}(t) & \text{if } t < \sigma \\ \alpha^\epsilon(t)(\bar{\alpha}_2)^{-1}\bar{u}(t) & \text{if } t \geq \sigma \end{cases}.$$

We will see in Theorem 4.1 that $(\sigma, u.^\epsilon)$ will be asymptotically optimal for $\mathcal{P}(\text{I})$ provided $(\sigma, \bar{u}.)$ is near optimal for $\mathcal{P}(\text{II})$.

Theorem 4.1. *Let $(\sigma, u.^0) \in \mathcal{A}_\text{II}$ be an ϵ-optimal control for the limiting problem $\mathcal{P}(\text{II})$ and let $(\sigma, u^\epsilon.) \in \mathcal{A}_\text{I}$ be the control constructed in Method I (cf. (4.1)). Then, $(\sigma, u^\epsilon.)$ is asymptotically optimal with error bound $\sqrt{\epsilon}$, i.e.,*

$$|J^\epsilon(x, \alpha, \sigma, u^\epsilon.) - v^\epsilon(x, \alpha)| \leq C(1 + |x|^{k_9})\sqrt{\epsilon}.$$

Remark. The significance of ii) in Theorem 4.1 is that the corporate level (upper level) management only has to solve an upper level problem (simpler problem) and obtain a solution $(\sigma, u.^0)$, while the lower level (operational level) management simply uses $(\sigma, u.^\epsilon)$, a scaled version of $(\sigma, u.^0)$ and obtains a near optimal solution for the original problem $\mathcal{P}(\text{I})$.

Let us now define the Markov times for $\mathcal{P}(\text{II})$ as $\bar{\tau} = \inf\{t : x_t \in S\}$. The control can now be defined as follows: Choose u_t to be optimal for $t < \bar{\tau}$ and choose u_t to be optimal for $t \geq \bar{\tau}$. More precisely, for any deterministic time $\tau \geq 0$ (τ will be taken as $\bar{\tau}$ shortly), let

$$u_\tau^*(t, x) = \begin{cases} \operatorname{argmin}\{[\sum \nu_i^{(1)} u_i \cdot \nabla v(x) + \sum \nu_i^{(1)} G(x, u_i)] : u \in U_1\}, & \text{if } t < \tau \\ \operatorname{argmin}\{[\sum \nu_i^{(2)} u_i \cdot \nabla v_a(x) + \sum \nu_i^{(2)} G(x, u_i)] : u \in U_2\}, & \text{if } t \geq \tau. \end{cases}$$

Theorem 5.2. *Assume that $c(u)$ is second differentiable with $\frac{\partial^2}{\partial u^2} c(u) \geq c_0 I_{n \times n} > 0$. Furthermore, there exist constants C and $k_1 > 0$ such that*

$$|h(x + y) - h(x) - \nabla h(x) \cdot y| \leq C(1 + |x|^{k_1})|y|^2.$$

Let $u^*(t) = u_\tau^*(t, x_t)$. Then $(\bar{\tau}, u^*(\cdot))$ is optimal for the limiting problem $\mathcal{P}(\text{II})$.

A simple example (cont.). In example (2.5), the value function $v(x)$ is the only viscosity solution to the following dynamic programming equations:

$$\min\left\{\min_{u\in\bar{\alpha}_1}{}_{\cdot U}[(u-z)\cdot\nabla v(x) + |x|] - \rho v(x), v_a(x) - v(x)\right\} = 0,$$

$$\min_{u\in\bar{\alpha}_2}{}_{\cdot U}[(u-z)\cdot\nabla v_a(x) + |x|] - \rho(v_a(x) - K) = 0.$$

To easy the exposition, we assume $\bar{\alpha}_1 - z < 0$.

Let x^* be a number defined as follows:

$$x^* = \begin{cases} z\rho^{-1}\log[|\bar{\alpha}_1 - z|/(\rho^2 K)]\ (\geq 0) & \text{if } 0 < K\rho^2 \leq |\bar{\alpha}_1 - z| \\ (\bar{\alpha}_1 - z)\rho^{-1}\log[(\bar{\alpha}_2 - \bar{\alpha}_1 - \rho^2 K)/(\bar{\alpha}_2 - z)]\ (\leq 0) & \text{if } |\bar{\alpha}_1 - z| < \rho^2 K < \bar{\alpha}_2 - \bar{\alpha}_1 \end{cases}.$$

Then the solutions of the equations can be written in terms of x^*.

$$v(x) = \begin{cases} (\bar{\alpha}_1 - z)\rho^{-2}e^{-\rho x/z} + x/\rho + z\rho^{-2}[e^{-\rho x/z} - 1] & \text{if } x \geq \max\{0, x^*\} \\ -x/\rho + |\bar{\alpha}_1 - z|\rho^{-2} & \text{if } x^* < x < \max\{0, x^*\} \\ (\bar{\alpha}_2 - z)\rho^{-2}[e^{\rho x/(\bar{\alpha}_2 - z)} - \rho x/(\bar{\alpha}_2 - z) - 1] + K & \text{if } x \leq x^* \end{cases}$$

$$v_a(x) = \begin{cases} z\rho^{-2}[e^{-\rho x/z} + \rho x/z - 1] + K & \text{if } x \geq 0 \\ (\bar{\alpha}_2 - z)\rho^{-2}[e^{\rho x/(\bar{\alpha}_2 - z)} - \rho x/(\bar{\alpha}_2 - z) - 1] + K & \text{if } x < 0 \end{cases}.$$

Recall that $S = \{x : v_a(x) = v(x)\}$. Thus, in this example,

$$S = \begin{cases} (-\infty, \infty) & \text{if } K = 0 \\ (-\infty, x^*] & \text{if } 0 < K < (\bar{\alpha}_2 - \bar{\alpha}_1)\rho^{-2} \\ \emptyset & \text{if } K \geq (\bar{\alpha}_2 - \bar{\alpha}_1)\rho^{-2} \end{cases}.$$

Let

$$\bar{u}_t = \begin{cases} \bar{\alpha}_1(1 - \text{sgn}\,\bar{x}_t)/2 & \text{if } t < \sigma \\ \bar{\alpha}_2(1 - \text{sgn}\,\bar{x}_t)/2 & \text{if } t \geq \sigma, \end{cases}$$

where

$$\dot{\bar{x}}_t = \bar{u}_t - z, \quad x_0 = x \text{ and } \sigma = \inf\{t : \bar{x}_t \in S\}. \tag{4.2}$$

It is not difficult to see that (4.2) has unique solutions. Therefore, by Theorem 5.2, $(\sigma, \bar{u}.)$ is optimal for $\mathcal{P}(\text{II})$.

Let u_t^ϵ denote the constructed control. Then, $u_t^\epsilon = \alpha^\epsilon(t)(1 - \text{sgn}\,\bar{x}_t)/2$. Theorem 4.1 yields that $(\sigma, u^\epsilon.)$ is asymptotically optimal for $\mathcal{P}(\text{I})$ with error bound $\sqrt{\epsilon}$.

In particular, if $K = 0$, then $\sigma = 0$, i.e., to buy the new machine immediately as the machine is available *gratis*.

If $K \geq (\bar{\alpha}_2 - \bar{\alpha}_1)\rho^{-2}$, then $\sigma = \infty$, i.e., not to buy it at all, because the machine is too expensive. An immediate interpreation of this point can be given as follows: Suppose $\bar{x}_0$ is very negative. Then buying a machine helps to reduce the cost of inventory at

$$\int_0^\infty e^{-\rho t}(\bar{\alpha}_2 - \bar{\alpha}_1)t\,dt = (\bar{\alpha}_2 - \bar{\alpha}_1)\rho^{-2}.$$

However, if this is smaller than K, the cost of a new machine, i.e., $(\bar{\alpha}_2 - \bar{\alpha}_1)\rho^{-2} \leq K$, it is certainly better off not to buy any new machine at all.

References

[1] Lehoczky, J., Sethi, S.P., Soner, H.M., and Taksar, M., An asymptotic analysis of hierarchical control of manufacturing systems under uncertainty, *Mathematics of Operations Research*, Vol. 16, No. 3, pp. 596-608, (1991).

[2] Sethi, S.P. and Zhang, Q., Asymptotic optimality in hierarchical control of manufacturing systems under uncertainty: State of the art, forthcoming in *Proceedings of the International Conference on Operations Research 1990*, Vienna, Austria, August 28-31, (1990).

[3] Sethi, S.P. and Zhang, Q., Hierarchical production planning in dynamic stochastic manufacturing systems: Asymptotic optimality and error bounds, submitted to *SIAM Journal on Control and Optimization*, (1991).

[4] Sethi, S.P., Taksar, M., and Zhang, Q., Hierarchical investment and production decisions in stochastic manufacturing systems: Asymptotic optimality and error bounds, submitted to *Mathematics of Operations Research*, (1991).

[5] Sethi, S.P., Zhang, Q., and Zhou, X.Y., Hierarchical controls in stochastic manufacturing systems with convex costs, submitted to *Journal of Optimization Theory and Applications*, (1991).

[6] Soner, H.M., Optimal control with state-space constrraint II, *SIAM Journal on Control and Optimization*, Vol. 24, No. 6, (1986).

Guaranteed Performance Regions for Multi-User Markov Models

NAHUM SHIMKIN and ADAM SHWARTZ [1]

Department of Electrical Engineering
Technion – Israel Institute of Technology
Haifa 32000, ISRAEL

Abstract: A user facing a multi-user resource-sharing system considers a vector of performance measures (e.g. response times to various tasks). Acceptable performance is defined through a set in the space of performance vectors. Can the user obtain a (time-average) performance vector which approaches this desired set? We consider the worst-case scenario, where other users may, for selfish reasons, try to exclude his vector from the desired set. For a controlled Markov model of the system, we give a sufficient condition for approachability, and construct appropriate policies. Under certain recurrence conditions, a complete characterization of approachability is then provided for convex sets. The mathematical formulation leads to a theory of approachability for stochastic games. A simple queueing example is analyzed to illustrate the applicability of this approach.

1 Introduction

Consider entering a multi-user resource-sharing system, for example a computer system. The objective is to guarantee acceptable service level for yourself, for example a fast response time of the terminal, adequate computation speed and reasonable delay at the printer queue. Naturally, a somewhat larger delay at the printer would be acceptable if we could gain in the response time. This tradeoff is modeled by defining a set in the performance space—in this example $\mathbb{R}^3$—which we wish to approach.

We model the dynamics of the system as a controlled Markov chain, where each user exerts some control. We make no assumptions on the behavior of the other users. The question is: for a given set in the performance space, can we guarantee that the time-averaged performance vector will (in the long run) fall into this set, even if the other users are doing their best to obstruct us (worst-case)? Or, can a group of malicious users exclude our performance from approaching this set?

Since we are considering a worst-case scenario, we may as well assume that we are facing a single "opponent", whose goals may conflict with ours. This framework can also be used to model a "worst case"

[1] Research performed in part while this author was visiting the Systems Research Center, University of Maryland, College Park, where he was supported in part through NSF Grant NSF ECS-83-51836.

analysis (in terms of a performance vector) of a single-user system, where any dynamic uncertainties or time variations are modeled as control variables chosen by "nature".

A similar problem has been addressed by Blackwell [4] where an "approachability-excludability theory" has been introduced for infinitely repeated matrix games with vector payoffs. A matrix game involves two players, where a payoff $m_{i,j}$ is generated whenever player 1 chooses action i while player 2 chooses action j. Thus, in a repeated matrix game the players face exactly the same situation at each decision epoch. Blackwell's model is therefore a special case of our model, where the state space is reduced to a single state. Let us briefly review the main ideas of Blackwell's theory. Consider a two-person matrix game, where the elements of the payoff matrix $M = (m_{i,j})$ are *vectors* in $\mathbb{R}^q$, $q \geq 2$. Blackwell addressed the following question: if the game is repeated infinitely in time, with both players observing and recalling the evolution of the game, can player 1 guarantee that the time-average payoff will asymptotically approach some desired set (in $\mathbb{R}^q$), no matter what the other player may do? Conversely, can player 2 exclude the average payoff from this set?

For an arbitrary (closed) set B, a sufficient condition for approachability was given, based on the following idea. Player 1 monitors at each stage n the current average payoff. For each possible value $\overline{m}$ of the average payoff which is outside B, consider the hyperplane which passes through C, a closest point in B to $\overline{m}$, and which is perpendicular to the line segment $C - \overline{m}$. Suppose that player 1 has a strategy (i.e., a randomized action) in the matrix game such that, for every possible strategy of player 2, the expected one stage payoff is separated from $\overline{m}$ by this hyperplane. Then, by using such a strategy whenever the average payoff is outside B, the average payoff is constantly driven in the direction of B, and finally converges to it.

For convex sets, a complete solution was given. A set is obviously excludable by player 2 in the infinitely repeated game if it is excludable by him in the one-shot matrix game. For convex sets, this condition turns out to be both necessary and sufficient for excludability, and its negation is necessary and sufficient for approachability. Further results on approachability in repeated matrix games may be found in [13, 14, 15, 21, 27]. For some applications, mostly game-theoretical, see [3, 5, 6, 11, 12, 16, 25].

In this paper, the basic ideas of [4] are applied to obtain approachability results for multi-user controlled Markov processes with vector payoffs. Following terminology from game theory we shall refer to this model as a stochastic game with vector payoff. Here we consider a class of two-person stochastic games, with countable state space, finite action spaces and a (not necessarily bounded) vector payoff function; the formal setup is given in Section 2. A basic assumption which underlies the approach of this paper is the existence of a fixed state, say state 0, for which certain uniform recurrence-type properties hold. It is then possible to obtain results which are similar to those described above for repeated matrix games, except that the strategies of the players in the one-shot matrix game are replaced by (possibly stationary) strategies which are played between subsequent visits to state 0. Thus, a basic idea in the construction of approaching strategies is to play a fixed strategy on each interval between visits to 0, which may be modified (according to the current average payoff) only when state 0 is reached. (See [1, 2] for a similar approach in Markov decisions problems).

The paper is organized as follows. In section 2 the stochastic game model is formally defined. Section 3 contains the main results, and the proof of the basic Theorem 3.1 is presented in section 4. In section 5 a simple queueing example is given to illustrate the applicability of these concepts, followed by some concluding remarks.

Notation: We denote by $|\cdot|$, $\langle\cdot,\cdot\rangle$ and $d(\cdot,\cdot)$ the Euclidean norm, inner product and distance in $\mathbb{R}^q$. Let U denote the set of unit vectors in $\mathbb{R}^q$.

2   The Model

We consider a controlled Markov process with two independent decision makers, 'player 1' and 'player 2'. The model is specified by the following objects: a countable state-space S, finite action spaces A_1 and A_2, a state transition law p, and an $\mathbb{R}^q$-valued payoff function r (where $q \geq 2$).

At stage (or time instant) $n = 0, 1, 2, \ldots$ the players observe the current state s, and each player i $(i = 1, 2)$ chooses independently an action $a^i \in A_i$. As a result, the payoff vector $r(s, a^1, a^2)$ is collected, and the next state s' is chosen according to the probability distribution $p(\cdot|s, a^1, a^2)$ on S. The state and action pair at stage n will be denoted by s_n and $a_n = (a_n^1, a_n^2)$, respectively. Let $r_n = r(s_n, a_n)$ be the payoff vector at stage n, and let

$$\bar{r}_n = \frac{1}{n} \sum_{m=0}^{n-1} r_m \tag{2.1}$$

denote the time-average payoff vector up to stage n.

Note that we have yet to specify the players' goals. This will be done in the next section. For now, the payoff should just be considered as some vector which measures system performance.

A (randomized, past-dependent) strategy π_i for player i is a sequence

$$\pi_i = \{\pi_0^i, \pi_1^i, \ldots\}, \quad \pi_n^i : H_n \to \mathcal{P}(A_i)$$

where $\mathcal{P}(A_i)$ is the set of probability vectors over A_i, and $H_n = S \times (A_1 \times A_2 \times S)^{n-1}$ is the set of possible "histories" up to stage n. Thus, given $h_n = (s_0; a_0, s_1; \ldots a_{n-1}, s_n)$, the action a_n^i is chosen according to the probability vector $\pi_n^i(h_n)$. Let Π_i denote the class of strategies for player i. A *stationary* strategy for player 1 is specified by a single function $f : S \to \mathcal{P}(A_1)$, so that $\pi_n^1(h_n) = f(s_n)$, $n \geq 0$. The class of stationary strategies for player 1 will be denoted by F, and that of player 2 (defined similarly) by G.

Given the strategy pair $\pi = (\pi_1, \pi_2)$ and an initial state $s_0 = s$, the above description induces a unique probability measure P_π^s and expectation operator E_π^s on the product space $S \times (S \times A_1 \times A_2)^\infty$. When s_0 and π are determined by the context, we just write P and E for the corresponding measure and expectation.

Some definitions from game theory will be required in the sequel. For every vector $u \in \mathbb{R}^q$ and initial state s, consider the case where at each stage, player 2 pays player 1 an amount which is specified by the *scalar* payoff function $r^u = \langle r, u \rangle$. If the objective of player 1 (resp. player 2) is to maximize (resp. minimize) the average expected payoff, then the model becomes a zero-sum stochastic game, which we will denote by $\Gamma_s(u)$. Stochastic games have been extensively studied; for a survey the reader is referred to [18, 19]. The connection between our model (with vector payoff) and this family of zero-sum stochastic game will be clarified below. We say that $\Gamma_s(u)$ has a value $\text{val}\,\Gamma_s(u)$ if

$$\text{val}\,\Gamma_s(u) = \sup_{\pi_1} \inf_{\pi_2} \liminf_{n\to\infty} E^s_{\pi_1,\pi_2}(\langle \bar{r}_n, u\rangle) \tag{2.2}$$

$$= \inf_{\pi_2} \sup_{\pi_1} \limsup_{n\to\infty} E^s_{\pi_1,\pi_2}(\langle \bar{r}_n, u\rangle). \tag{2.3}$$

A strategy $\pi_1 \in \Pi_1$ $[\pi_2 \in \Pi_2]$ is *optimal* in $\Gamma_s(u)$ if it satisfies the sup in (2.2) [the inf in (2.3), respectively].

The basic assumptions made in this paper will involve certain recurrence conditions for some fixed state, say state 0. Let τ denote the first passage time to state 0:

$$\tau = \inf\{n \geq 1 : s_n = 0\}.$$

A strategy $\pi_1 \in \Pi_1$ is said to be *stable* if there exist positive constants M_2 and R_2 such that:

$$E^0_{\pi_1,\pi_2}(\tau^2) \leq M_2 \quad \forall \pi_2 \in \Pi_2, \tag{2.4}$$

$$E^0_{\pi_1,\pi_2}\left(\sum_{n=0}^{\tau-1} |r_n|\right)^2 \leq R_2 \quad \forall \pi_2 \in \Pi_2. \tag{2.5}$$

Note that (2.5) is redundant in case the payoff function r is bounded. A set $\Pi'_1 \subset \Pi_1$ is *uniformly stable* if (2.4) and (2.5) are satisfied for every $\pi_1 \in \Pi'_1$ with the same constants M_2 and R_2. Stability of player 2's strategies is defined symmetrically.

We introduce now some conditions on the model. Reference to these conditions will be made explicitly when required.

C1: For every unit vector $u \in U$, the game $\Gamma_0(u)$ has a value, and player 1 has a stationary optimal strategy $f^*(u)$ in this game. Moreover, the set $\{f^*(u) : u \in U\}$ is uniformly stable.

C2: Condition C1 holds. Furthermore, for each $u \in U$ player 2 has an optimal strategy $g^*(u)$ in $\Gamma_0(u)$ which is stationary and stable.

Existence of stationary optimal strategies is stochastic games has been established under various assumptions (cf. [18, 19]). For concreteness, a particular set of assumptions which imply C1–C2 (and which are similar to those employed in [20, Ch. 6] in the context of Markov decision processes) is specified in the full version [24] of this paper, Lemma 2.1.

3 Approachability: Definitions and Results

Let us define first the concept of *uniform* almost sure (a.s.) convergence which will be used here. Let $(X_n, n \geq 0)$ be a sequence of random variables over some measurable space $(\Omega, \mathcal{F})$, and let $\{P_\lambda, \lambda \in \Lambda\}$

be a collection of probability measures on $(\Omega, \mathcal{F})$. It is well known ([23]) that, for a fixed $\lambda \in \Lambda$, $X_n \to 0$ P_λ-a.s. is equivalent to

$$\lim_{N \to \infty} P_\lambda(\sup_{n \geq N} |X_n| > \varepsilon) = 0 \qquad \forall \varepsilon > 0 . \tag{3.1}$$

Now, we say that $X_n \to 0$ P_λ-a.s., *at a uniform rate over* Λ, if the convergence in (3.1) is uniform over Λ, that is

$$\lim_{N \to \infty} \sup_{\lambda \in \Lambda} P_\lambda(\sup_{n \geq N} |X_n| > \varepsilon) = 0 . \tag{3.2}$$

The basic concepts of this paper, namely approachability and the dual concept of excludability, are introduced in the following definition.

Definition 3.1 *Let the initial state s be fixed. A set $B \subset \mathbb{R}^q$ is* approachable *(from s by player 1) if there exists a B-approaching strategy $\pi_1^* \in \Pi_1$ such that*

$$d(\bar{r}_n, B) \to 0 \quad P\text{-a.s. for every } \pi_2 \in \Pi_2 ,$$

at a uniform rate over Π_2.
B is excludable *(from s by player 2) if there exists a B-excluding strategy $\pi_2^* \in \Pi_2$ such that, for some $\delta > 0$,*

$$d(\bar{r}_n, B_\delta^c) \to 0 \quad P\text{-a.s. for every } \pi_1 \in \Pi_1 ,$$

at a uniform rate over Π_1, *where $B_\delta^c = \{\beta \in \mathbb{R}^q : d(\beta, B) \geq \beta\}$.*

Remarks:

1. The convergence $d(\bar{r}_n, B_\delta^c) \to 0$ in the definition of excludability may be equivalently written as: $\liminf_n d(\bar{r}_n, B) \geq \delta$. Thus, loosely speaking, a set B is approachable if player 1 can guarantee (irrespective of the other's strategy) that the long-term average payoff vector is in B, and B is excludable if player 2 can guarantee (irrespective of player 1's strategy) that the long-term average payoff is at least a distance $\delta > 0$ away from B.

2. It is obvious that approachability and excludability are contradictory, in the sense that a given set cannot be both approachable by player 1 and excludable by player 2. However, these concepts are not exact opposites of each other. Indeed, it was demonstrated in [4] that, even in repeated matrix games, some (non-convex) sets may be neither approachable nor excludable.

3. In the sequel, it will be convenient to assume that the set B is closed. This involves no loss of generality, since approachability (and excludability) of a set and its closure are plainly the same.

4. An important aspect of the definition is the uniform rate of convergence. This requirement is essential if the infinite stage model is considered as an approximation to the model with very long, but finite, time horizon.

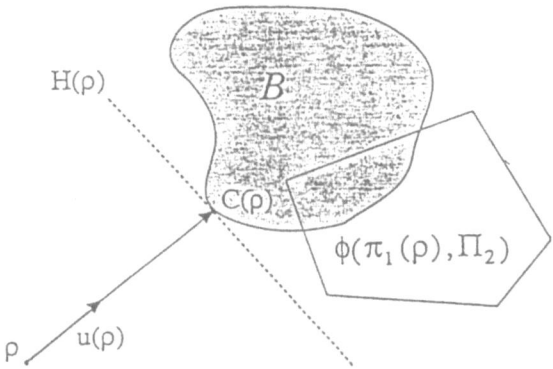

Figure 1: Geometric interpretation of Theorem 3.1

We proceed to formulate the key result, which presents a sufficient condition for approachability. To this end, let

$$\phi(\pi_1, \pi_2) = \frac{E^0_{\pi_1, \pi_2} \left(\sum_{n=0}^{\tau-1} r_n \right)}{E^0_{\pi_1, \pi_2}(\tau)} \tag{3.3}$$

denote the averaged payoff per "cycle" from state 0 and back. Note that $\phi(\pi_1, \pi_2)$ is well defined if either π_1 or π_2 is a stable strategy. We say that a strategy $\pi_1 \in \Pi_1$ is *started at stage* T if at stage $n = T$ (possibly random) player 1 resets an internal clock to 0 and starts playing according to π_1 as if the state s_T is the initial state.

Let B be a closed set in $\mathbb{R}^q$. For any point $\rho \notin B$ let $C(\rho)$ denote a closest point in B to ρ. Let $H(\rho)$ be the hyperplane through $C(\rho)$ which is perpendicular to $(C(\rho) - \rho)$, and let $u(\rho)$ be a unit vector in the direction of $(C(\rho) - \rho)$ (see Fig. 1).

Theorem 3.1
Assume that the following condition is satisfied:
SC1:   For every $\rho \notin B$, there exists a stable strategy $\pi_1(\rho) \in \Pi_1$ such that

$$\langle \phi(\pi_1(\rho), \pi_2) - C(\rho), u(\rho) \rangle \geq 0 \qquad \forall \pi_2 \in \Pi_2, \tag{3.4}$$

(equivalently, $\phi(\pi_1(\rho), \pi_2)$ is weakly separated by $H(\rho)$ from ρ, for all $\pi_2 \in \Pi_2$). Furthermore, the set $\{\pi_1(\rho) : \rho \notin B\}$ is uniformly stable.
Then B is approachable from state 0 by player 1, and a B-approaching strategy is given as follows.
Let $0 < T(1) < T(2) < \cdots$ be the subsequent arrival instants to state 0. Let π_1' be some fixed stable strategy for player 1. Then:

- *at stages $0 \leq n < T(1)$: play π_1'.*

- *at stages* $T(k) \leq n < T(k+1), \quad k \geq 1$:

    if $\bar{r}_{T(k)} \notin B$, *then play* $\pi_1(\bar{r}_{T(k)})$, *started at* $T(k)$.

    if $\bar{r}_{T(k)} \in B$, *then play* π'_1, *started at* $T(k)$.

Proof: The proof of this theorem, as well as the proofs of the following results, may be found in [24].

The sufficient condition SC1 and the approaching strategy of Theorem 3.1 admit an intuitively appealing geometric interpretation ([4], [16]). As already noted, (3.4) simply means that $H(\rho)$ separates ρ from the set $\phi(\pi_1(\rho), \Pi_2)$ (cf. Fig. 1). Consider then the approaching strategy suggested above: whenever state 0 is reached and the average payoff $\bar{r}$ is outside B, player 1 employs $\pi_1(\bar{r})$ for the next cycle (i.e., up to the next time when 0 is reached). Thus, the averaged payoff in that cycle, as defined in (3.3), will belong to the set $\phi(\pi_1(\bar{r}), \Pi_2)$, and will therefore cause the average payoff $\bar{r}$ to advance towards that set (in some probabilistic sense). Now, the stability conditions imposed on $\{\pi_1(\rho)\}$ imply that, as time progresses, the effect of one cycle on the average payoff becomes small; therefore the average payoff actually moves closer to B on each cycle. This suggests that the average payoff will converge to B in the long run.

In Theorem 3.1 it was assumed that the initial state is 0. If not, the conditions of Theorem 3.1 may still be applied provided player 1 can guarantee that state 0 is reached "fast enough"; See [24, Corollary 3.2]. From here on, we shall always assume that the initial state is 0, while keeping in mind that the results may be extended to other initial states in a similar way.

Under assumption C1 (defined in section 2), the sufficient condition of Theorem 3.1 may be expressed in terms the values of the games $\Gamma_0(u)$. Furthermore, the implied approaching strategy is 'piecewise stationary', as specified in the following:

Corollary 3.1 *Assume* C1. *Let* B, ρ, $C(\rho)$ *and* $u(\rho)$ *be as in Theorem 3.1. Then* B *is approachable from state 0 if*

$$\text{val}\,\Gamma_0(u(\rho)) \geq \langle C(\rho), u(\rho) \rangle \quad , \qquad \forall \rho \notin B . \tag{3.5}$$

An approaching strategy is then as specified in Theorem 3.1, with $\pi_1(\rho) \triangleq f^*(u(\rho))$, *the stationary* $\Gamma_0(u(\rho))$- *optimal strategy specified in* C1.

We consider next the important special case where the set B is convex. It is then possible to obtain (under C2) a complete characterization of approachability.

For every stable $g \in G$, the long-run average expected payoff:

$$R(f,g) = \lim_{n \to \infty} E^0_{f,g}(\bar{r}_n) \quad , \qquad f \in F , \tag{3.6}$$

is well defined, and (as noted in the proof of Cor. 3.3) equals $\phi(f,g)$. Furthermore, define the following bounded subsets of $\mathbb{R}^q$:

$$R(F,g) = \{R(f,g) : f \in F\} \tag{3.7}$$

$$\bar{R}(F,g) = co\{R(f,g)\} , \tag{3.8}$$

where 'co' denotes the closed convex hull. (In fact, it may be established as in [2] or [8, p. 95] that $R(F, g)$ is convex, so that $\bar{R}(F, g)$ is just its closure). Note that boundedness of $R(F, g)$ follows from property (2.5) in the definition of a stable strategy. The sets $R(f, G)$ and $\bar{R}(f, G)$ are similarly defined for any stable $f \in F$, and the same comments apply.

Theorem 3.4

Assume C2. Let B be a closed convex set in $\mathbb{R}^q$, and let the initial state $s_0 = 0$.

(i) B is approachable if and only if either one of the following equivalent conditions is satisfied:

NSC1: *For every $\rho \notin B$, the hyperplane $H(\rho)$ separates ρ from $\bar{R}(f^*(u(\rho)), G)$, i.e.,*

$$\langle R(f^*(u(\rho)), g) - C(\rho), u(\rho) \rangle \geq 0 \qquad , \qquad g \in G.$$

NSC2: *For every unit vector $u \in U$:*

$$\operatorname{val} \Gamma_0(u) \geq \inf_{\beta \in B} \langle \beta, u \rangle.$$

NSC3: *$\bar{R}(F, g)$ intersects B for every stable $g \in G$.*

(ii) If B is not approachable, then it is excludable by player 2 with a stationary strategy.

Remark: Condition NSC3 is obviously a natural candidate for a *necessary* condition for approachability. We note, however, that it may not be generally true that any stable $g \in G$ for which $\bar{R}(F, g) \cap B = \emptyset$ is an excluding strategy unless we know that $\bar{\phi}(\Pi_1, g) = \bar{R}(F, g)$. This equality is obviously a matter for Markov-decision-processes investigation, and requires some (fairly mild) additional model assumptions; some conditions in this direction may be found, e.g., in [2].

Finally, we note that condition C2, under which the last result was established, may be somewhat weakened when the convex set B is unbounded. For example, one may wish to keep a pair of performance measures below some prescribed thresholds (α_1, α_2), so that $B = \{b \in \mathbb{R}^2 : b_1 \leq \alpha_1, b_2 \leq \alpha_2\}$. Applying NSC1 of the last theorem to this case, it is easily seen that the unit vector $u(\rho)$ will always be in the (closed) negative quadrant, i.e., will have non–positive components. Recalling that, loosely speaking, $u(\rho)$ represents a possible direction in which player 1 should try to 'advance' the average payoff, this conforms well with the form of B, which implies that smaller payoffs, in both components, are preferable. The restriction to directions in the negative quadrant is also reflected in NSC2, where the right–hand–side equals $(-\infty)$ whenever u is not in that quadrant; thus this inequality needs to be checked only for vectors u in the negative quadrant.

More generally, given an (unbounded) closed convex set $B \subset \mathbb{R}^q$, define (compare NSC2):

$$U(B) = \{u \in U : \inf_{\beta \in B} \langle \beta, u \rangle > -\infty\}.$$

It is not hard to show that $U(B)$ in fact represents all directions in which a point outside B might be projected orthogonally onto B, i.e., $U(B) = \{u(\rho) : \rho \notin B\}$. Consider the following condition:

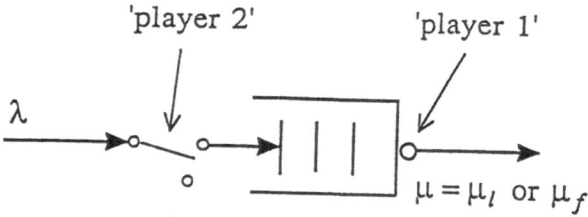

Figure 2: The queueing system

C2(B):   For every $u \in U(B)$, the game $\Gamma_0(u)$ has a value and both players have stationary stable optimal strategies in $\Gamma_0(u)$. Furthermore, the set $\{f^*(u) : u \in U(B)\}$ is uniformly stable.

Thus, in C2(B) we consider only the games $\Gamma_0(u)$ where u is in $U(B)$, instead of the whole U. Recalling that u determines the payoff function $r^u = \langle r, u \rangle$ of $\Gamma_0(u)$, this reduction may be significant when existence of optimal strategies depend on properties of the payoff functions, such as positiveness or one-side boundedness; cf. [17] and also, e.g., [2, 22] for some relevant results in the MDP case. It is easily seen that the proof of Theorem 3.4 implies also the following:

Corollary 3.3 *Let B be a closed convex set in $\mathbb{R}^q$. Then Theorem 3.4 holds with C2 replaced by C2(B), and U replaced by $U(B)$ in NSC2.*

4   A Queueing Example

In this section we apply the previous results to a simple discrete-time queueing system. Dynamic control of admission, routing and service in queueing systems has been extensively studied in the past decade; see, e.g., [28] and its references. Here we consider the case of service-rate control (by 'player 1'), while the arrival process is not completely specified (or, alternatively, controlled by 'player 2'). For illustrative purposes, the problem was simplified as much as possible so that, while not being trivial, it lends itself to a simple analytic treatment.

Consider the queueing system illustrated in Fig. 2. The time axis is divided into "slots" $n = 0, 1, 2, \ldots$. Only one customer may arrive during each time slot, and the arrival probability is λ. If the queue is empty, he joins the queue and then enters service at the beginning of the next time slot. Otherwise, he may choose either to join the queue, or to leave the system and never return.

Service is applied to one customer at each time slot (provided the queue is not empty at the beginning of the slot). The server ('player 1') may choose between a slow service mode, where the probability of successful service on each time slot is μ_l, and a fast service mode with success probability μ_f, where $1 \geq \mu_f > \mu_s > \lambda$. If the service is successful the customer leaves the system, otherwise he remains for (at least) another try. We assume that the server may switch service mode *only when the queue is empty.* A fixed cost (which is assumed for convenience to be of μ_f units) is incurred for each service attempt in the fast mode, while slow service is costless.

To fit the game model of the previous sections, we consider all arrival decisions as being made by 'player 2'. The system may be formally described as follows. Let the state $s = (x, M)$, where $x \in \{0, 1, \ldots\}$ is the number of customers in the queue at the beginning of a time slot and $M \in \{$slow, fast, empty$\}$ is the service mode in that slot. Thus,

$$
x_{n+1} = \begin{cases} A_n & : \text{ if } x_n = 0 \\ x_n + A_n a_n^2 - U_n & : \text{ if } x_n > 0 \end{cases}
$$

$$
M_{n+1} = \begin{cases} \text{empty} & : \text{ if } x_{n+1} = 0 \\ M_n & : \text{ if } x_{n+1} > 0, \ x_n > 0 \\ a_n^1 & : \text{ if } x_{n+1} > 0, \ x_n = 0 \end{cases}
$$

where $a_n^1 \in \{$slow, fast$\}$ and $a_n^2 \in \{0, 1\}$ are the players' choices, $A_n \sim Bern(\lambda)$ (i.e., $A_n = 1$ w.p. λ and $A_n = 0$ otherwise), $U_n|(M_n = $ slow$) \sim Bern(\mu_l)$ and $U_n|(M_n = $ fast$) \sim Bern(\mu_f)$.

Note that the game described above is a stochastic game with *perfect information* ([10]), which means that in each state one of the players is restricted to a single (effective) action. It is well known that in zero-sum, *finite-state* stochastic games with perfect information the players have optimal *non-randomized* stationary strategies ([10, 9]); it will be argued below that the same holds in the present (countable state) case. For now, we note that player 1 has only two such strategies, which we denote by $\mu_l^{(\infty)}$ and $\mu_f^{(\infty)}$: in the former, slow service mode is always chosen, and the latter chooses fast service always.

The performance measures which will be considered are the throughput $\bar{\Lambda}$ and the average service cost $\bar{C}$, i.e.,

$$
\bar{\Lambda}_n = \frac{1}{n} \sum_{m=0}^{n-1} 1\{U_m = 1, \ x_m > 0\} = \frac{1}{n} \sum_{m=0}^{n-1} 1\{ \text{ successful service occurred at slot } m\}
$$

$$
\bar{C}_n = \frac{1}{n} \sum_{m=0}^{n-1} 1\{M_m = \text{fast}\} = \frac{1}{n} \sum_{m=0}^{n-1} 1\{ \text{ a customer was given fast service at slot } m \}
$$

$$
\bar{r}_n = (\bar{\Lambda}_n, \bar{C}_n) .
$$

The following problem will be considered: given thresholds (Λ_0, C_0), can the server (player 1) guarantee a long-term throughput of Λ_0 or more, while the average service cost does not exceed C_0? Thus, we define the set

$$
B_0 = \{(\bar{\Lambda}, \bar{C}) \in \mathbb{R}^2 : \bar{\Lambda} \geq \Lambda_0, \ \bar{C} \leq C_0\}
$$

and look for the thresholds for which this set is approachable by player 1.

To exhibit the answer, we note first that conditions C1–C2 of section 2 are satisfied: since $\mu_f < \lambda$, it is easily seen that the entire strategy set of either player is uniformly stable (with $s = (0, \text{empty})$

defined as the '0' state). Next, we note that player 1's effective decisions are made only in state 0, and the game is of perfect information; it then follows by elementary considerations that either $\mu_l^{(\infty)}$ or $\mu_f^{(\infty)}$ is optimal in each game $\Gamma_0(u)$, $u \in \mathbb{R}^2$. Now, given that player 1 plays $\mu_l^{(\infty)}$ or $\mu_f^{(\infty)}$ in $\Gamma_0(u)$, player 2 is facing a Markov decision process, and by standard results (see, e.g., [2, 22]) there exist optimal stationary strategies (which minimize the average expected cost), say g_l^* and g_f^*, in either case. It follows that an optimal strategy for player 2 in $\Gamma_0(u)$ is to play g_l^* if $M_n = $ slow, and g_f^* if $M_n = $ fast.

Let us now calculate the sets $R(\mu_l^{(\infty)}, G)$ and $R(\mu_f^{(\infty)}, G)$ (defined below (3.7)), i.e. the range of the expected average payoff which is induced by $\mu_l^{(\infty)}$ or $\mu_f^{(\infty)}$.

Assume that $\mu_l^{(\infty)}$ is played. Obviously, the service cost is identically zero, i.e. $\bar{C}_n \equiv 0$. Now, the maximal throughput is clearly λ, while the minimal is achieved when the customers never choose to join the system (unless they have to, when the queue is empty). By considering the induced Markov chain (with the state space reduced to the two states $x = 0$ and $x = 1$), this minimal throughput is easily seen to be

$$\Lambda_l \triangleq \frac{\mu_l \lambda}{\mu_l + \lambda},$$

and thus

$$R(l) \triangleq R(\mu_l^{(\infty)}, G) = \{(\bar{\Lambda}, 0) : \Lambda_l \leq \bar{\Lambda} \leq \lambda\}.$$

Assume now that $\mu_f^{(\infty)}$ is played. It follows similarly that the throughput is in $[\Lambda_f, \lambda]$, where

$$\Lambda_f \triangleq \frac{\mu_f \lambda}{\mu_f + \lambda}.$$

Moreover, noting that a cost of μ_f units is incurred for each service attempt, it follows that the average expected cost equals the throughput; thus,

$$R(f) \triangleq R(\mu_f^{(\infty)}, G) = \{(\bar{\Lambda}, \bar{C}) : \Lambda_f \leq \bar{\Lambda} \leq \lambda, \quad \bar{C} = \bar{\Lambda}\}.$$

Let JK be the line segment between the points $J = (\Lambda_f, \Lambda_f)$ and $K = (\Lambda_l, 0)$, (see Fig. 3). We then have the following result:

Proposition 4.1 *The set B_0 is approachable by player 1 if and only if it intersects JK; otherwise it is excludable by player 2.*

Thus, the line segment KJ represents the set of Pareto-optimal performance vectors which may be secured by the server.

Proof Recalling that $f^*(u) \in \{\mu_l^{(\infty)}, \mu_f^{(\infty)}\}$ for every $u \in U$, the proof follows from Theorem 3.4, condition NSC1, by simple geometric considerations:

Assume first that $P = (\Lambda_0, C_0)$ is a point on JK. Then it is easily seen that for any $\rho \notin B$, either $R(l)$ or $R(f)$ is separated from ρ by $H(\rho)$. Thus B_0 (and any set which contains it) is approachable.

Conversely, if (Λ_0, C_0) are such that B_0 does not intersect JK, then there exists a point $\rho \notin B$ (for example, the closest point on JK to (Λ_0, C_0)) such that the required separation does not occur neither

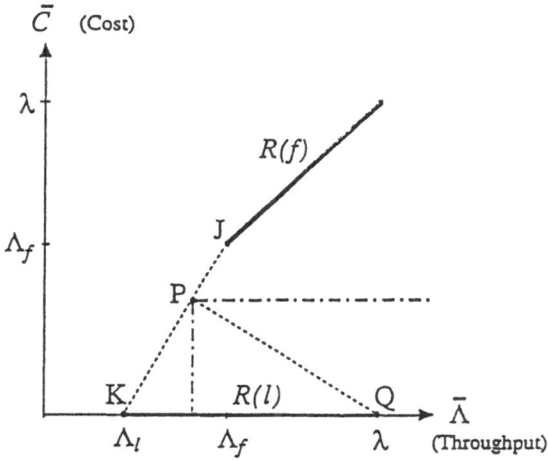

Figure 3: The payoff space

for $R(l)$ nor for $R(f)$. Thus, B_0 is not approachable, and by Theorem 3.4(ii) is therefore excludable. (In fact, an excluding strategy may always be taken as the strategy which never chooses to join a non-empty queue). □

The approaching strategy suggested by Theorem 3.1 (say, for B_0 such that $(\Lambda_0, C_0) \in JK$) is obviously non-stationary, since it switches between slow and fast service modes (when in state 0) according to the current average payoff vector. It is important to note (and may be easily shown) that there does *not* exist a stationary approaching strategy for B_0, so that dependence on the history is crucial.

The 'Pareto optimal' sets considered above, i.e. B_0 with $(\Lambda_0, C_0) \in JK$, are by no means 'minimal', in the sense that some proper subsets thereof may still be approachable. An interesting example is the following: let P be a point on JK, and let $Q = (\lambda, 0)$. Then exactly the same (geometric) arguments as in Prop. 5.1 imply that $B \triangleq \{$ the line segment $PQ\}$ is approachable by player 1 (and this set is 'minimal' in the above sense). That may seem somewhat peculiar at first glance, since the cost on PQ is decreasing with increasing throughput. However, this dependence actually reveals an 'adaptive' property inherent in the associated approaching strategy. To clarify this point, consider the extreme case where the customers always decide to join the queue. The throughput will obviously be λ, independently of the service mode. Thus, the approaching strategy adapts to this situation (without prior knowledge of the arrival policy) by adhering to the slow service mode, thereby reducing the service cost to 0 (the point Q).

5 Conclusion

The main purpose of this paper is to present a usable analytic tool for the evaluation of attainable performance, from the single user's viewpoint, in dynamically controlled multi-user systems. It should be stressed that the term 'user' here applies to any single decision maker — for example, a central system controller. Our approach is characterized by a 'worst-case' view of other users, whose actions may be chosen arbitrarily from their specified sets. Another important feature is the simultaneous consideration of several performance measures, rather than the single 'figure of merit' approach which is often considered in optimization problems.

The worst-case approach should be contrasted with the 'statistical' approach, where a certain (statistical, and usually simplified) model is imposed on the behavior of other users, thus incorporating their actions into the system dynamics. While each approach has its advantages, it is important to note that the two may be combined to yield a more realistic model. This may generally be accomplished by 'splitting' (possibly stochastically) the users' actions into two groups, where the first is incorporated into the system dynamics, while the second (which may also include model uncertainties, possible time variations, etc.) is left unrestricted.

The main results of this paper are Theorem 3.1 and its Corollary 3.3, which give a sufficient condition for any given set to be approachable, and Theorem 3.4 which gives necessary and sufficient conditions for approachability of a *convex* set; in either case, the approaching strategy is explicitly specified. These results depend in an essential way on some recurrence properties of a single fixed state, which must hold at least for certain strategies. We conclude with a few comments regarding the applicability and possible extensions of these results.

In the simple queueing example considered here, it was possible to apply directly the geometric condition NSC1 of Theorem 3.4 to get a complete solution (for convex sets). In more complicated situations, it may be more convenient to apply conditions NSC2 or NSC3 (for convex sets), or Corollary 3.3 (for general sets). Note that, in order to apply these conditions, it is sufficient to compute (possibly numerically) the values of the stochastic games $\Gamma_0(u)$ for a sufficiently dense set of unit vectors u. It should also be noted that these computations do not depend on the specific set B considered; thus, once they have been carried out, it is relatively easy to check approachability of any given set.

The approaching strategies which we considered here are adapted to the history of the process (i.e., to the relative position of the average payoff compared to the set to be approached) only when the fixed recurrent state is hit. While this is convenient for analysis and easy to implement, it may have the undesirable effect of increasing the 'variance' of the payoff if these recurrence times are far apart. Thus, it should be of interest to construct approaching strategies which adapt to the current payoff more frequently.

It is quite obvious that some sort of recurrence conditions are required to preserve the basic approach and results of this paper. However, the ones assumed here (namely recurrence of a single fixed state for all relevant strategies) are not the only possibility. Specifically, it is *conjectured* (at this point) that the basic results hold under the various recurrence conditions considered in [9].

Finally, it should be noted that the main definitions and results of this paper may be straightforwardly generalized to semi-Markov models.

References

[1] E. Altman and A. Shwartz, "Non stationary policies for controlled Markov chains," EE Pub. 633, Technion, June 1987.

[2] E. Altman and A. Shwartz, "Markov decision problems and state-action frequencies," *SIAM J. Control and Optimization* 29 No. 4, July 1991.

[3] R. J. Aumann and M. Maschler, *Game theoretic aspects of gradual disarmament.* Chapter V, Report to the U.S. Arms Control and Disarmament Agency, Contract S.T.80, prepared by Mathematica, Inc., Princeton, N.J., 1966.

[4] D. Blackwell, "An analogue for the minimax theorem for vector payoffs," *Pacific J. Math.*, 6, pp. 1–8, 1956.

[5] D. Blackwell, "Controlled random walks," *Proc. Internat. Congress Math.*, 3, pp. 336–338, 1954.

[6] D. Blackwell, "On multi-component attrition games," *Naval Res. Log. Quart.*, pp. 210–216, 1954.

[7] V. S. Borkar, "Control of Markov chains with long-run average cost criterion," in *Stochastic Differential Systems, Stochastic Control and Application*, W. Fleming and P.L. Lions, eds., IMA Vol. 10, Springer-Verlag, pp. 57–77, 1988.

[8] C. Derman, *Finite State Markovian Decision Processes*, Academic Press, New-York, 1970.

[9] A. Federgruen, "On N-person games with denumerable state space," *Adv. Appl. Prob.* 10, pp. 452–471, 1978.

[10] D. Gillette, "Stochastic games with zero stop probabilities," in *Contributions to the Theory of Games, III*, (Annals of Math. Studies 39), M. Dresher et al., editors, Princeton Univ. Press, pp. 179–188, 1957.

[11] S. Hart, "Nonzero-sum two-person repeated games with incomplete information," *Math. of Oper. Res.*, 10, pp. 117–153, 1985.

[12] J. F. Hannan, "Approximation to Bayes risk in repeated play," in *Contributions to the Theory of Games, III*, (Annals of Math. Studies 39), M. Dresher et al., editors, Princeton Univ. Press, pp. 97–139, 1957.

[13] T. F. Hou, "Weak approachability in a two-person game," *Ann. Math. Statist.*, 40, pp. 789–813, 1969.

[14] T. F. Hou, "Approachability in a two-person game," *Ann. Math. Stat.*, 42, pp. 735–744, 1971.

[15] M. Katz, "Infinitely repeatable games," *Pac. J. Math.*, 10, pp. 879–885, 1960.

[16] R. D. Luce and H. Raiffa, *Games and Decisions*, Wiley, New-York, 1957.

[17] A. Maitra and T. Parthasarathy, "On stochastic games, II," *Journal of Optimization Theory and Applications*, Vol. 8, pp. 155–160, 1971.

[18] T. Parthasarathy and M. Stern, "Markov games: a survey," in *Differential Games and Control Theory*, P.L.E. Roxin and R. Sternberg, eds., Marcel Dekker, 1977.

[19] T. E. S. Raghavan and J.A. Filar, "Algorithms for stochastic games — a survey," Preprint, June 1989.

[20] S. M. Ross, *Applied Probability Models with Optimization Applications*, Holden-Day, San Francisco, 1970.

[21] H. Sackrowitz, "A note on approachability in a two-person game," *Ann. Math. Stat.*, 43, pp. 1017–1019, 1972.

[22] L. I. Sennott, "Average cost optimal stationary policies in infinite state Markov decision processes with unbounded costs," Operations Research, Vol. 37, pp. 626–633, 1989.

[23] A. N. Shiryayev, *Probability*, Springer-Verlag, 1984.

[24] N. Shimkin and A. Shwartz, "Guaranteed performance regions for multi-use Markov models," submitted to the *IEEE Trans. Automat. Contr.*, 1991.

[25] J. Sorin, *An Introduction to Two-Person-Zero-Sum Repeated Games with Incomplete Information*. IMSSS–Economics TR–312, Stanford University, memo.

[26] M. A. Stern, *On Stochastic Games with Limiting Average Pay-Off*. Ph.D. dissertation, submitted to the University of Illinois, Circle Campus, Chicago, 1975.

[27] N. Vieille, "Weak approachability," Preprint, 1991.

[28] J. Walrand, *An Introduction to Queueing Networks*, Prentice-Hall, New Jersey, 1988.

Stability of Slowly
Time-Varying Linear Systems

by

Victor Solo[†*]

Abstract

New conditions are given in both deterministic and stochastic settings for the stability of the system $\dot{x} = A(t)x$ when $A(t)$ is slowly varying. Roughly speaking, the eigenvalues of $A(t)$ are allowed to "wander" into the right half plane so long as "on average " they are strictly in the left half plane.

†Department of Statistics

Macquarie University

Sydney NSW 2109

Australia
*This work was completed at

Johns Hopkins University, ECE department

and funded by

the NSF under grant

ECS - 8806063

1. Introduction

Recent work in adaptive control [11], [15], [21] has focussed attention on conditions for (exponential) stability of the time-varying linear system

$$\dot{x}(t) = A(t)x(t) \quad t \geq 0 \tag{1}$$

where $x(t)$ is a p-vector and $A(t)$ varies slowly. The earliest result is due to [19] while [5] found a general result. His assumptions have become more or less standard;

(I) $\|A(t)\| \leq A < \infty$ ($\|.\|$ is induced Euclidean norm)

(II) $\|\dot{A}(t)\| \leq \dot{A} < \infty$

(III) All eigenvalues of $A(t)$ have real part $\leq -\sigma < 0$.

The results of [4] are perhaps the best; he shows exponential stability of (1) if

$$\dot{A} < \sigma^2/(4M\ln M)^2 \tag{2}$$

Our aim is to relax (III) by allowing the eigenvalues to "wander" into the right half plane while "on average" remaining in the left half plane.

If $A(t)$ is periodic then Floquet theory [18] provides necessary and sufficient conditions for stability, but the theory is generally intractable [18, p57] except in the second order case.

In the stochastic case, there is little. Let us suppose

(1') $A(t)$ is strictly stationary ergodic; $E\|A(t)\| = A < \infty$.

If we denote the system transition matrix by $\Phi(t,0)$ then the Furstenburg-Kesten Theorem [12] tells us that', as $t \to \infty$,

$$t^{-1} \log \|\Phi(t,0)\| \to \theta < \infty \quad a.s.$$

If we knew $\theta < 0$ a.s. then we would have a form of stochastic stability. But there seems to be no known way to calculate θ: see [6] for some recent material.

Aside from the recently developed discrete time result in [21] there do not appear to be any stochastic stability results for slowly time varying linear systems. Below we develop a stochastic stability result simliar to the deterministic one in which the eigenvalues of $A(t)$ need only be in the left half plane "on average". There is some related work of [9] , but it is hard to interpret and has only been made to work in the second order case.

2. Deterministic Stability

We consider a perturbed form of (1)

$$\dot{x}(t) = [A(t) + P(t)]x(t) \tag{3}$$

where $P(t)$ will be a small amplitude perturbation. Introduce the following assumptions:

(A1) $\overline{\lim}_{L \to \infty} L^{-1} \int_{t_0}^{t_0+L} ||A(s)||ds \le A < \infty$, for all t_0. Equivalently, their exists $a < \infty$ such that

$$\int_{t_0}^{t_0+L} ||A(s)||ds \le a + LA < \infty, \text{ for all } t_0$$

(A2') $||A(t+h) - A(t)|| \le \beta h^\gamma$ for all t and some γ, $0 < \gamma \le 1$

(A3') Let $\alpha(t)$ be the real part of the eigenvalue of $A(t)$ whose real part is greatest.
$\overline{\lim}_{L \to \infty} L^{-1} \int_{t_0}^{t_0+L} \alpha(u)du \le \overline{\alpha}' < 0$

(A4) $\overline{\lim}_{L \to \infty} L^{-1} \int_{t_0}^{t_0+L} ||P(s)||ds \le \delta < \infty$, for all t_0

Theorem 1. Under (A1), (A2'), (A3'), (A4), (3) is exponentially stable provided we choose $\epsilon > 0$ so small that

$$\overline{\alpha}' + 2\epsilon < 0 \tag{4}$$

with δ so small that (with $M_\epsilon = \frac{3}{2}(2A/\epsilon + 1)$)

$$\overline{\alpha}' + 2\epsilon + M_\epsilon \delta < 0 \tag{5}$$

with β so small that

$$\overline{\alpha}' + 2\epsilon + M_\epsilon \delta + 2(lnM_\epsilon)^{\frac{\gamma}{\gamma+1}}[\beta(M_\epsilon + \epsilon/A)]^{\frac{1}{\gamma+1}} < 0 \tag{6}$$

and also so small that

$$\sigma(\beta^{\frac{1}{\gamma+1}}(lnM_\epsilon/(M_\epsilon + \epsilon/A))^{\frac{1}{\gamma+1}}) < \epsilon \tag{7}$$

where $\sigma(.)$ is a certain nonincreasing function.

3. Stochastic Stability

Let us consider the forced system

$$\dot{x}(t) = G(t)x(t) + v(t) \tag{8}$$

In the deterministic case, if $v(t)$ is uniformly bounded then $x(t)$ is uniformly bounded if the homogeneous system

$$\dot{x}(t) = G(t)x(t) \qquad (9)$$

is exponentially stable. In the stochastic case the situation is not as simple as this.

Let $\Phi(t,s)$ be the transition matrix of (9) and introduce the following assumptions.

(s1) $G(t)$ is strictly stationary ergodic, $-\infty \le t \le \infty$.

(s2) $E\|v(t)\| \le c < \infty$, for all t

(s3) $\overline{\lim}_{t->\infty} t^{-1} \log \|\Phi(t,0)\| \le -\gamma < 0$, a.s.

Lemma For (8) with assumptions (s1)-(s3), $x(t)$ is tight i.e.

$$\lim_{B->\infty} \sup_{0 \le t \le \infty} P(\|x(t)\| > B) = 0 \qquad (10)$$

Tightness [3], [13] ensures the existence of an invariant measure [3, p 290].

To make the Lemma useful we need a means of checking (s3). As before we allow a small amplitude perturbation and write

$$G(t) = A(t) + P(t) \qquad (11)$$

and introduce the following assumptions:

(S1) $A(t)$ is strictly stationary ergodic, $E\|A(t)\| = A < \infty$

(S2) $E\|A(t+h) - A(t)\| \le \beta|h|^\gamma$, $0 < \gamma \le 1$, h small enough.

(S3) Let $\alpha(t)$ be the real part of the eigenvalue of $A(t)$ where real part is greatest. Then $\alpha(t)$ is strictly stationary ergodic and by the elementary bound $|\alpha(t)| \le \|A(t)\|$ we deduce from (S1) that $E|\alpha(t)| < \infty$, we then suppose

$$E[\alpha(t)] = \overline{\alpha} < 0$$

(S4) $P(t)$ is strictly stationary ergodic, $E\|P(t)\| = \delta < \infty$.

Theorem 2. Consider the system (8), (11) with assumptions (S1)-(S4). Choose $\epsilon > 0$ so that (6) holds and require δ, β be small that (7), (8) hold (but with $M_\epsilon = 3(2A(\epsilon+1))^{p-1}/2$) then (s3) holds.

We now obtain our main result.

Theorem 3. For the system (8), with assumptions (S1)-(S4), (s2) and ϵ, δ, β verifying (6), (7), (8) then $x(t)$ is stable in probability.

References

[1] T. Apostol, *Mathematical Analysis,* Addison-Wesley, 1977.

[2] S.T. Ariaratnam, *Almost-Sure Stability of Some Linear Stochastic Systems,* Jl. Appl. Mech., 56, pp175-178, 1989.

[3] P. Billingsley, *Probability and Measure,* J. Wiley, 1979.

[4] W. A. Coppel, *Dichotomies in Stability Theory,* Lecture Notes in Mathematics, No. 629, Springer, Berlin, 1978.

[5] C.A. Desoer, *slowly varing system* $\dot{x} = Ax$, IEEE Trans. Automat. Control, AC-14, pp339-340, 1970.

[6] X. Feng and K.A. Loparo, *Stability of Linear Markovian Jump Systems,* 29th IEEE Conference on Decision and Control, Honolulu, Hawaii, pp xx, 1990.

[7] G.C. Goodwin, K.S. Sin, *Adaptive Filtering Prediction and Control,* Prentice-Hall, 1984.

[8] L. Guo, *On Adaptive Stabilization of Time-Varying Stochastic Systems,* Tech. Rep. Dept. Sys. Eng. Australian National University, 1989.

[9] E.F. Infante, *On the Stability of some Linear Nonautonomous Random Systems,* Jl. Appl. Mech., Vol. 35. p7-12, 1968.

[10] R.Z. Khasminski, *Stochastic Stability of Differential Equations,* Sijthoff and Nordhoff, Maryland, 1980.

[11] G. Kreisselmeier, *Adaptive Control of a Class of Slowly Time Varying Plants,* Sys. Control. Lett., Vol.8, p97-103, 1986.

[12] U. Krengel, *Ergodic Theorems,* W. de Gruyter, New York, 1985.

[13] H. Kushner, *Weak Convergence Methods and Singularly Perturbed Stochastic Control and Filtering Problems,* Birkhauser, Berlin, 1990.

[14] S.P. Meyn, L. Guo, *Adaptive Control of Time Varying Stochastic Systems* Proc. 11th IFAC World Congress, 1990.

[15] R.H. Middleton and G.C. Goodwin, *Adaptive Control of the Linear Time Varying Plants,* IEEE Trans. Automat. Control AC-33, pp 150-155, 1988.

[16] R.H. Middleton, G.C. Goodwin, D.J. Hill and D.Q. Mayne, *Design Issues in Adaptive Control,* 33, pp50-57, 1988.

[17] K.S. Narendra, A.M. Annaswamy, *Stable Adaptive Systems,* Prentice-Hall, 1989.

[18] J.A. Richards, *Analysis of Periodically Time-Varying Systems,* Springer, Berlin, 1983.

[19] H.H. Rosenbrook, *The Stability of Linear Time-Dependent Control Systems,* J. Electron and Control, pp73-80. 1963.

[20] S. Sastry and M. Bodson, *Adaptive Control, Stability, Convergence and Robustness,* Prentice-Hall, 1989.

[21] V. Solo, *A One Step Ahead Adaptive Controller With Slowly Time Varying Parameters,* Submitted to IEEE. Trans. Automat. Control, 1991.

[22] E.D. Sontag, *Mathematical Control Theory,* Springer, Berlin, 1990.

[23] K.S. Tsakalis, P.A. Ioannou, *Adaptive Control of Linear Time Varying Plants: a new Model Reference Controller Structure,* IEEE. Trans. Autom. Control., 34, pp1038-1046, 1989.

[24] N. Wiener, *The Fourier Integral and Certain of Its Applications,* Dover, New York, 1958.

[25] G. Zames, L.Y. Wang, *Local-Global Double Algebras for Slow H^{∞} Adaptation: Part I - Inversion and Stability,* IEEE Trans. Automat. Control, pp130-142, 1991.

On Adaptive Control of a Singularly Perturbed Diffusion Model

L. Stettner

Institute of Mathematics, Polish Academy of Sciences,
Śniadeckich 8, 00–950 Warsaw, Poland

Abstract. In the paper a control of a singularly perturbed diffusion with unknown parameter in the drift term of the equation for slow variable, is studied. It is assumed, that the perturbation parameter which is small, and also unknown. An adaptive procedure that guarantees nearly optimal value of the long run average cost functional is constructed. As an intermediate result some new facts concerning the ergodic control of singularly perturbed diffusion are shown.

1. Introduction

Assume $(x_t^\varepsilon, y_t^\varepsilon) \in R^{d_1} \times R^{d_2}$ is a weak solution to the following system of stochastic differential equations

$$dx_t^\varepsilon = [a_1(x_t^\varepsilon) + b_1(x_t^\varepsilon, y_t^\varepsilon, \alpha^0) + c_1(x_t^\varepsilon, u(x_t^\varepsilon, y_t^\varepsilon), \alpha^0)]dt + \sigma_1(x_t^\varepsilon)dw_t^1 \qquad x_0^\varepsilon = x \quad (1)$$

$$dy_t^\varepsilon = \varepsilon^{-1}[a_2(y_t^\varepsilon) + b_2(x_t^\varepsilon, y_t^\varepsilon)]dt + \varepsilon^{-\frac{1}{2}}\sigma_2(y_t^\varepsilon)dw_t^2 \qquad y_0^\varepsilon = y \qquad (2)$$

where

$(w_t^1), (w_t^2)$ are independent d_1 and d_2 dimensional standard Wiener processes on a probability space (Ω, F, P),

$\alpha^0 \in A$ a compact subset of R^k, is an unknown parameter,

$\varepsilon > 0$ is small, also unknown and plays the role of perturbation parameter,

$a_1, \sigma_1, a_2, \sigma_2$ satisfy the global Lipschitz and linear growth conditions,

$\sigma_1 \sigma_1^*(x) \geq \kappa_1 \cdot I, \quad \sigma_2 \sigma_2^*(y) \geq \kappa_2 \cdot I$ for $x \in R^{d_1}, y \in R^{d_2}$ with $\kappa_1, \kappa_2 > 0$, where I stands above either for $d_1 \times d_1$ or $d_2 \times d_2$ identity matrix,

b_1, c_1, b_2 are bounded measurable,

$u \in \mathcal{A} = B(R^d, U)$ – the set of all Borel measurable functions from R^d into U a given compact metric space of control parameters, with $d = d_1 + d_2$.

Clearly, under the above assumptions a unique weak solution to (1)–(2) exists. In what follows we will be interested in the minimization of the following pathwise long run average cost

$$J(u) = \limsup_{t \to \infty} t^{-1} \int_0^t k(x_s^\varepsilon, u(x_s^\varepsilon, y_s^\varepsilon))ds \qquad P \text{ a.e.} \qquad (3)$$

where $k \in B(R^{d_1} \times U, R^1)$ is bounded.

If ϵ and α^0 are known the above problem is closely related to the minimization of the long run expected average cost

$$\bar{J}_{xy}^{\alpha^0,\epsilon}(u) = \limsup_{t \to \infty} t^{-1} E_{xy}^{\alpha^0,\epsilon} \left\{ \int_0^t k(x_s^\epsilon, u(x_s^\epsilon, y_s^\epsilon)) ds \right\} \tag{4}$$

Below we impose another set of assumptions to guarantee (see [K1], [S1]) the existence of optimal strategies for (4) with known $\alpha^0 \in A$ and $\epsilon > 0$.
Namely,

$c_1(x,v)$ and $k(x,v)$ are continuous in $v \in U$ and satisfy so called Roxin
assumption i.e. the set of vectors $\begin{pmatrix} c_1(x,v) \\ k(x,v) \end{pmatrix} v \in U$ is convex $\tag{5}$

$$\sup_{\alpha \in A} \sup_{(x,y) \in \Gamma_1} \sup_{u \in A} \sup_{\epsilon > 0} E_{xy}^{\alpha,\epsilon} \tau^2 < \infty \tag{6}$$

$$\sup_{\alpha \in A} \sup_{u \in A} E_{xy}^{\alpha,\epsilon} T_{\Gamma_1} < \infty \quad \text{for any } (x,y) \in R^d, \epsilon > 0 \tag{7}$$

where in the last two assumptions $\tau = T_{\Gamma_2} + T_{\Gamma_1} \circ \Theta_{T_{\Gamma_2}}, \Gamma_1 = \left\{ \begin{pmatrix} x \\ y \end{pmatrix} \in R^d, \|\begin{pmatrix} x \\ y \end{pmatrix}\| = r_1 \right\}$,
$\Gamma_2 = \left\{ \begin{pmatrix} x \\ y \end{pmatrix} \in R^d, \|\begin{pmatrix} x \\ y \end{pmatrix}\| = r_2 \right\}$ with $r_1 < r_2$.

Since $\epsilon > 0$ is small we shall consider a limit equation for (x_t^ϵ) with $\epsilon \to 0$. For this purpose assume that
(A1) For fixed $x \in R_1^d$, there exists an invariant measure μ_x of the solution (y_t^x) to

$$dy_t^x = [a_2(y_t^x) + b_2(x, y_t^x)]dt + \sigma_2(y_t^x)dw_t^2 \tag{8}$$

and for compact sets $K_1 \subset R^{d_1}, K_2 \subset R^{d_2}$ we have

$$\sup_{x \in K_1} \sup_{y \in K_2} \|E_y\{y_t^x \in \cdot\} - \mu_x(\cdot)\|_{\text{var}} \to 0 \tag{9}$$

as $t \to \infty$ with $\| \ \|_{\text{var}}$ standing for variation of the measure.

In Section 5 we formulate a sufficient condition for (A1).
Define

$$\bar{b}_1(x, \alpha) = \int_{R^{d_2}} b_1(x, y, \alpha)\mu_x(dy)$$

$$\mathcal{A}_r = B(R^{d_1}, U)$$

For given $u \in \mathcal{A}_r$ let (x_t) be the weak solution to

$$dx_t = [a_1(x_t) + \bar{b}_1(x_t, \alpha^0) + c_1(x_t, u(x_t), \alpha^0)]dt + \sigma_1(x_t)dw_t^1 \tag{10}$$

It is known ([K2]), that $x_t^\epsilon \Rightarrow x_t$ as $\epsilon \to 0$, and as an approximating problem to (1), (2), (4) with $\epsilon \to 0$ one can consider (10) with

$$\bar{J}_x^{\alpha^0}(u) = \limsup_{t\to\infty} t^{-1} E_x^{\alpha^0} \left\{ \int_0^t k(x_s, u(x_s)) \right\} ds \qquad (11)$$

over $u \in \mathcal{A}_r$. The approximation mentioned above is in the sense that almost optimal control function $u \in \mathcal{A}_r$ for (10)–(11) is also nearly optimal for (1), (2)–(4). Because of its importance we shall formulate this result more explicitly and following [B–S] call "limit control principle".

Let

$$\lambda(\alpha^0) = \inf_{u\in\mathcal{A}_r} \bar{J}_x^{\alpha^0}(u) \qquad (12)$$

$$\lambda^\epsilon(\alpha^0) = \inf_{u\in\mathcal{A}} \bar{J}_{xy}^{\alpha^0,\epsilon}(u) \qquad (13)$$

If for any $\delta > 0$, there exists ϵ_0 such that for $\epsilon < \epsilon_0$ and $u \in \mathcal{A}_r$

$$\bar{J}_x^{\alpha^0}(u) \le \lambda(\alpha^0) + \delta \quad \text{for any} \quad x \in R^{d_1} \Rightarrow$$
$$\bar{J}_{xy}^{\alpha^0,\epsilon}(u) \le \lambda^\epsilon(\alpha^0) + 2\delta \quad \text{for} \quad (x,y) \in R^d \qquad (14)$$

then we say that limit control principle (LCP) holds. It is shown in [K2] Theorem 4.4.2 that the implication (14) is satisfied for those $u \in \mathcal{A}_r$ that are continuous. In [B–S] (LCP) is proved for diffusion models in which (y_t^ϵ) is independent of (x_t^ϵ). Below, in Section 2 we show (LCP) for (1), (2) and (4).

Since the parameter α^0 is unknown, for adaptive control purposes it would be important to have (14) satisfied uniformly in α. Denote by $\lambda(\alpha)$ and $\lambda^\epsilon(\alpha)$ the optimal values of the cost functionals $\bar{J}_x^\alpha(u)$ and $\bar{J}_{xy}^{\alpha,\epsilon}(u)$ respectively, corresponding to the controlled diffusion equations (10) or (1)–(2) with α^0 replaced by $\alpha \in A$.

If for any $\delta > 0$, there exists ϵ_0 such that for $\epsilon < \epsilon_0$ any $\alpha \in A$ and $u \in \mathcal{A}_r$

$$\bar{J}_x^\alpha(u) \le \lambda(\alpha) + \delta \quad \text{for any} \quad x \in R^{d_1} \Rightarrow$$
$$\bar{J}_{xy}^{\alpha,\epsilon}(u) \le \lambda^\epsilon(\alpha) + 2\delta \quad \text{for} \quad (x,y) \in R^d \qquad (15)$$

then we say that uniform (in α) limit control principle (ULCP) holds.

In Section 3 we show (ULCP) for the model studied above. Then in Section 4 we apply the adaptive control procedure introduced in [D–PD–S1] to the problem (1),(2)–(4). Namely by (ULCP) we discretize the space of almost optimal control functions. Then we estimate α^0, using the cost biased MLE, at suitably chosen random moments, and apply certainly equivalent control i.e. control that is almost optimal for a given value of the estimation. As we show in Section 4 this procedure makes cost functional (3) less than $\lambda^\epsilon(\alpha^0) + 3\delta$, $P_{xy}^{\alpha^0,\epsilon}$ a.e., for any given $\delta > 0$.

In other words the nearly selfoptimality property is guaranteed, i.e. the value of the cost functional (3) is close to the optimal, corresponding to the situation with known α^0.

In the last Section 5 we clarify the meaning of some assumptions imposed in the paper.

2. Limit Control Principle

In this section we assume that $\alpha^0 \in A$ is known. Therefore to simplify notations we shall skip α^0.

By the boundedness of b_1, c_1 and Girsanov formula it is known that under a new probability measure $P^0, (x_t^\epsilon, y_t^\epsilon)$ is a solution to

$$
\begin{aligned}
dx_t^\epsilon &= a_1(x_t^\epsilon)dt + \sigma_1(x_t^\epsilon)dw_t^1 \\
dy_t^\epsilon &= \epsilon^{-1}[a_2(y_t^\epsilon) + b_2(x_t^\epsilon, y_t^\epsilon)]dt + \epsilon^{-\frac{1}{2}}\sigma_2(y_t^\epsilon)dw_t^2
\end{aligned}
\tag{16}
$$

and the restrictions of P^0 and of P to the σ fields generated by the trajectories up to time t, are absolutely continuous.

Define for $f \in B(R^d, R)$, and any $F_t^1 = \sigma\{x_s, s \leq t\}$ measurable, finite Markov time τ

$$
m_x^\tau(f) = E_x^0 \left\{ \int\limits_{R^{d_2}} f(x_\tau, z)\mu_{x_\tau}(dz) \right\}
\tag{17}
$$

Assume $\tau(w) \geq h > 0$, P a.e. and

(A2) b_2 is Lipschitz with respect to x, uniformly in y, i.e. there exists L_{b_2} such that for $x, \bar{x} \in R^{d_1}, y \in R^{d_2}$

$$
\|b_2(x, y) - b_2(\bar{x}, y)\| \leq L_{b_2}\|x - \bar{x}\|
$$

where $\| \ \|$ stands above for Euclidean norm either in R^{d_2} or in R^{d_1}.

(A3) the trajectories of x_t^ϵ are Hölder continuous in mean square in the left neighbourhood of τ i.e. for some constant M we have

$$
E_x^0 \left\| x_\tau^\epsilon - x_{\tau-u}^\epsilon \right\|^2 \leq Mu^\beta \quad \text{with } \beta > 0, \text{ for } u \leq h
$$

with M, β uniform for x from compact subsets of R^{d_1}. Let $\bar{y}_t^\epsilon = y_{\epsilon t}^\epsilon$. Then

$$
d\bar{y}_t^\epsilon = [a_2(\bar{y}_t^\epsilon) + b_2(x_{\epsilon t}^\epsilon, \bar{y}_t^\epsilon)]dt + \sigma_2(\bar{y}_t^\epsilon)dw_t^2
\tag{18}
$$

We shall also assume

(A4) for each $\delta > 0$, there exists compact set $K_\delta \subset R^d$ such that for each $t > t_0 > 0$ and $(x, y) \in K_\delta$ we have

$$
E_{xy}^0 \left\{ \bar{y}^\epsilon(t) \in K_\delta \mid x_u, u \leq \frac{t}{\epsilon} \right\} \geq 1 - \delta.
$$

The fundamental result of this section is

Proposition 1. *Under (A1), (A2), (A3) and (A4)*

$$E^0_{xy} \{(x^\epsilon_\tau, y^\epsilon_\tau) \in \cdot\} \to m^\tau_x(\cdot) \tag{19}$$

as $\epsilon \to 0$, *in variation norm, uniformly for* (x, y) *from compact subsets of* R^d.

P r o o f. Define $\bar{y}^\epsilon_t = y^\epsilon_{\epsilon t}$. Then

$$d\bar{y}^\epsilon_t = [a_2(\bar{y}^\epsilon_t) + b_2(x^\epsilon_{\epsilon t}, \bar{y}^\epsilon_t)]dt + \sigma_2(\bar{y}^\epsilon_t)dw^2_t$$

Given $h \geq r_\epsilon > 0$ such that $r_\epsilon \to 0, r_\epsilon \cdot \epsilon^{-1} \to \infty, r^{1+\beta}_\epsilon \cdot \epsilon^{-1} \to 0$ as $\epsilon \to 0$, define

$$\hat{y}^{\epsilon,\tau}(t) = \bar{y}^\epsilon \left(\frac{\tau - r_\epsilon}{\epsilon}\right) + \int\limits_{(\tau - r_\epsilon)\epsilon^{-1}}^{t} [a_2(\hat{y}^{\epsilon,\tau}(u)) + b_2(x^\epsilon_\tau, \hat{y}^{\epsilon,\tau}(u))]du$$

$$+ \int\limits_{(\tau - r_\epsilon)\epsilon^{-1}}^{t} \sigma_2(\hat{y}^\epsilon(u))dw_2(u) \quad \text{for } t \geq \frac{\tau - r_\epsilon}{\epsilon} \tag{20}$$

Let

$$\Lambda_t = \exp\left[\int\limits_0^t (b_2(x^\epsilon(\tau - r_\epsilon + \epsilon u), \hat{y}^\epsilon(u)) - b_2(x^\epsilon(\tau), \hat{y}^\epsilon(u)))^*\right.$$

$$\sigma_2^{-1*}(\hat{y}^\epsilon(u))dw_2(u) - \frac{1}{2}\int\limits_0^t (b_2(x^\epsilon(\tau - r_\epsilon + \epsilon u), \hat{y}^\epsilon(u)) - b_2(x^\epsilon(\tau), \hat{y}^\epsilon(u)))^* \tag{21}$$

$$\left.\sigma_2^{-1*}(\hat{y}^\epsilon(u))\sigma_2^{-1}(\hat{y}^\epsilon(u))(b_2(x^\epsilon(\tau - r_\epsilon + \epsilon u), \hat{y}^\epsilon(u)) - b_2(x^\epsilon(\tau), \hat{y}^\epsilon(u)))du\right]$$

and define measure $\hat{P}$ such that restrictions $\hat{P}_{|t}$ and $P^0_{|t}$ of $\hat{P}$ and P^0 to $F^2_t = \sigma\{y_s, s \leq t\}$ satisfy

$$dP^0_{|t} = \Lambda_t d\hat{P}_{|t}. \tag{22}$$

We have, for $B \in B(R^{d_2})$

$$\left| P^0_{xy}\left\{ \bar{y}^\varepsilon\left(\frac{\tau}{\varepsilon}\right) \in B \mid F^1_\tau \right\} - P^0_{xy}\left\{ \hat{y}^\varepsilon\left(\frac{\tau}{\varepsilon}\right) \in B \mid F^1_\tau \right\} \right| \le$$

$$\le E^0_{xy}\left| \left\{ \hat{E}_{\bar{y}^\varepsilon((\tau - r_\varepsilon)\varepsilon^{-1})}\left\{ \chi_B\left(\hat{y}^\varepsilon\left(\frac{r_\varepsilon}{\varepsilon}\right)\right)\left(\Lambda_{r_\varepsilon \varepsilon^{-1}} - 1\right) \mid F^1_\tau \right\} \mid F^1_\tau \right\} \right.$$

$$\le E^0_{xy}\left\{ \hat{E}_{\bar{y}^\varepsilon((\tau - r_\varepsilon)\varepsilon^{-1})}\left\{ \frac{1}{2} \int_0^{r_\varepsilon \varepsilon^{-1}} (b_2(x^\varepsilon(\tau - r_\varepsilon + \varepsilon u)), \hat{y}^\varepsilon(u)) \right. \right.$$

$$- b_2(x^\varepsilon(\tau), \hat{y}^\varepsilon(u))^* \sigma_2^{-1}(\hat{y}^\varepsilon(u))^* \sigma_2^{-1}(\hat{y}^\varepsilon(u))(b_2(x^\varepsilon(\tau - r_\varepsilon + \varepsilon u), \hat{y}^\varepsilon(u))$$

$$\left. \left. - b_2(x^\varepsilon(\tau), \hat{y}^\varepsilon(u)))du(\Lambda_{r_\varepsilon \varepsilon^{-1}} + 1) \mid F^1_\tau \right\} \mid F^1_\tau \right\} +$$

$$+ E^0_{xy}\left\{ \hat{E}_{\bar{y}^\varepsilon((\tau - r_\varepsilon)\varepsilon^{-1})}\left\{ \int_0^{r_\varepsilon \varepsilon^{-1}} |(b_2(x^\varepsilon(\tau - r_\varepsilon + \varepsilon u), \hat{y}^\varepsilon(u)) - b_2(x^\varepsilon(\tau), \hat{y}^\varepsilon(u)))^* \right. \right.$$

$$\left. \left. \sigma_2^{-1^*}(\hat{y}^\varepsilon(u))dw_2(u)|(\Lambda_{r_\varepsilon \varepsilon^{-1}} + 1) \mid F^1_\tau \right\} \mid F^1_\tau \right\} \le \kappa_2^{-1} L^2_{b_2}$$

$$E^0_x\left\{ \int_0^{r_\varepsilon \varepsilon^{-1}} \|x^\varepsilon(\tau - r_\varepsilon + \varepsilon u) - x^\varepsilon(\tau)\|^2 du \mid F^1_\tau \right\} + \tag{23}$$

$$+ E^0_{xy}\left\{ E^0_{\bar{y}^\varepsilon(\tau - r_\varepsilon)\varepsilon^{-1}}\left\{ \int_0^{r_\varepsilon \varepsilon^{-1}} (b_2(x^\varepsilon(\tau - r_\varepsilon + \varepsilon u), \bar{y}^\varepsilon(u)) - b_2(x^\varepsilon(\tau), \bar{y}^\varepsilon(u)))^* \right. \right.$$

$$\left. \sigma_2^{-1^*}(\bar{y}^\varepsilon(u))dw_2(u) \mid F^1_\tau \right\} \mid F^1_\tau \right\} + E^0_{xy}\left\{ \hat{E}_{\bar{y}^\varepsilon((\tau - r_\varepsilon)\varepsilon^{-1})}\left\{ \int_0^{r_\varepsilon \varepsilon^{-1}} (b_2(x^\varepsilon(\tau - r_\varepsilon + \varepsilon u), \right.$$

$$\left. \left. \hat{y}^\varepsilon(u)) - b_2(x^\varepsilon(\tau), \hat{y}^\varepsilon(u)))^* \sigma_2^{-1^*}(y^\varepsilon(u))dw_2(u) \mid F^1_\tau \right\} \mid F^1_\tau \right\} \le$$

$$\le \kappa_2^{-1} L^2_{b_2} E^0_x\left\{ \int_0^{r_\varepsilon \varepsilon^{-1}} \|x^\varepsilon(\tau - r_\varepsilon + \varepsilon u) - x^\varepsilon(\tau)\|^2 du \mid F^1_\tau \right\} +$$

$$+ 2\kappa_2^{-1/2} L_{b_2}\left(E^0_x\left\{ \int_0^{r_\varepsilon \varepsilon^{-1}} \|x^\varepsilon(\tau - r_\varepsilon + \varepsilon u) - x^\varepsilon(\tau)\|^2 du \mid F^1_\tau \right\} \right)^{\frac{1}{2}}$$

$$\le \kappa_2^{-1} L^2_{b_2} \frac{r_\varepsilon}{\varepsilon} \cdot M \cdot r^\beta_\varepsilon + 2\kappa_2^{-1/2} L_{b_2}\left(\frac{r_\varepsilon}{\varepsilon} \cdot M \cdot r^\beta_\varepsilon \right)^{\frac{1}{2}} \to 0$$

as $\varepsilon \to 0$, uniformly in $B \in B(R^{d_2})$.

Now,

$$
|E^0_{xy}\{f(x^\epsilon_r, y^\epsilon_r)\} - E^0_x\Big\{\int_{R^{d_2}} f(x_r, z)\mu_{x_r}(dz)\Big\}| \le
$$

$$
|E^0_{xy}\Big\{f\Big(x_r, \bar{y}^\epsilon\Big(\frac{\tau}{\epsilon}\Big)\Big)\Big\} - E^0_{xy}\Big\{f\Big(x_r, \hat{y}^\epsilon\Big(\frac{\tau}{\epsilon}\Big)\Big)\Big\}|+ \tag{24}
$$

$$
+ |E^0_{xy}\Big\{f\Big(x_r, \hat{y}^\epsilon\Big(\frac{\tau}{\epsilon}\Big)\Big)\Big\} - E^0_x\Big\{\int_{R^{d_2}} f(x_r, z)\mu_{x_r}(dz)\Big\}| = I + II
$$

By (23) $I \to 0$ uniformly in $f \in B(R^d), 0 \le f \le 1$, uniformly in (x, y) from compact subsets of R^d.

By (A4) for a sufficiently large compact set K_1

$$
II \le E^0_{xy}\{E^0_{xy}\{\chi_{\hat{y}^\epsilon((\tau-r_\epsilon)\epsilon^{-1})\in K_\delta}\chi_{K_1}(x_r)\big|E^0_{xy}\Big\{f\Big(x_r, \hat{y}^\epsilon\Big(\frac{\tau}{\epsilon}\Big)\Big)|F^1_r\Big\}
$$

$$
- E^0_x\Big\{\int_{R^{d_2}} f(x_r, z)\mu_{x_r}(dz)|F^1_r\Big\}\big||F^1_r\} + 4\delta \|f\| \tag{25}
$$

Therefore by (A1), letting $\epsilon \to 0, II \le 4\delta \|f\|$ uniformly in (x, y) from compact subsets of R^d.

Finally, since the convergence in (24) and (25) was uniform in $f \in B(R^d), 0 \le f \le 1$ we obtain (19).

The proof of Proposition 1 is complete.

In the remaining part of this section we shall follow Section 3.3 of [B–S]. First we show an analog of Proposition 6 of [B–S].

Proposition 2. Assume $L_{\epsilon_n}(x, y) \to L(x, y)$ as $\epsilon_n \to 0$ in weak * topology of $L^\infty(R^d)$. Then for $T > 0$, under (A1)–(A4) we have

$$
E^0_{xy}\Big\{|\int_0^T (L_{\epsilon_n}(x_s, y^{\epsilon_n}_s) - I(x_s))ds|^2\Big\} \to 0 \quad \text{as } \epsilon_n \to 0 \tag{26}
$$

with

$$
I(x) = \int_{R^{d_2}} L(x, y)\mu_x(dy)
$$

P r o o f. Let

$$
V^t_\epsilon(x, y) = E^0_{xy}\Big\{\int_0^t L_\epsilon(x_s, y^\epsilon_s)ds\Big\}
$$

$$
V^t(x) = E^0_x\Big\{\int_0^t I(x_s)ds\Big\}
$$

For compact subset $K \subset R^{d_2}$ we have

$$
\sup_{y \in K} |V_{\epsilon_n}^t(x, y) - V^t(x)| \leq \sup_{y \in K} \Big| \int_0^t \Big\{ L_{\epsilon_n}(x_s, y_s^{\epsilon_n})
$$

$$
- \int_{R^{d_2}} L_{\epsilon_n}(x_s, y)\mu_{x_s}(dy) \Big\} ds \Big| + \Big| \int_0^t E_x^0 \Big\{ \int_{R^{d_2}} L_{\epsilon_n}(x_s, y)\mu_{x_s}(dy) \tag{27}
$$

$$
- \int_{R^{d_2}} L(x_s, y)\mu_{x_s}(dy) \Big\} ds \Big| = I + II
$$

$I \to 0$ by Proposition 1 for $\tau = s$, as $\varepsilon_n \to 0$. Denote by $p_s(x, z), n(z, y)$ the densities of the transition kernel $P_x\{x_s \in \cdot\}$ and invariant measure $\mu_x(\cdot)$, with respect to Lebesque measures l_1, l_2 on R^{d_1}, R^{d_2} respectively.

We have

$$
II = \Big| \int_0^t p_s(x, z)[L_{\epsilon_n}(z, y) - L(z, y)]n(z, y)l_1(dz)l_2(dy) \Big| \to 0 \tag{28}
$$

by the weak convergence.

Therefore

$$
V_{\epsilon_n}^t(x, y) \to V^t(x) \quad \text{as } \varepsilon_n \to 0, \tag{29}
$$

uniformly for y from compact subsets of R^{d_2}.

Now,

$$
E_{xy}^0 \Big\{ \Big| \int_0^t (L_{\epsilon_n}(x_s, y_s^{\epsilon_n})ds - I(x_s))ds \Big|^2 \Big\} =
$$

$$
= 2E_{xy}^0 \Big\{ \int_0^t (L_{\epsilon_n}(x_s, y_s^{\epsilon_n}) - I(x_s))(V_{\epsilon_n}^{t-s}(x_s, y_s^{\epsilon_n}) - V^{t-s}(x_s))ds \Big\} \tag{30}
$$

$$
\leq 4M E_{xy}^0 \Big\{ \int_0^t \sup_{y \in K} |V_{\epsilon_n}^{t-s}(x_s, y) - V^{t-s}(x_s)|ds \Big\} + 4M\delta
$$

where $|L_{\epsilon_n}| \leq M$, and compact set $K \subset R^{d_2}$ is such that $E_{xy}^0\{y_s^{\epsilon_n} \in K\} \geq 1 - \delta$ for $n = 1, 2, \ldots$

Thus, from (30), (29), since δ could be chosen arbitrarily small we obtain the assertion of Proposition 2.

Once we have shown an analog of Proposition 6 of [B–S] we can prove for (1), (2), (4) all results obtained for the model of [B–S], adapting the majority of considerations in [B–S], assuming that additionaly we impose the following conditions on the limit equation (10)

(A5)

$$\sup_{\alpha \in A} \sup_{x \in \gamma_1} \sup_{u \in A_r} E^\alpha_x \, \tau^2 < \infty$$

$$\sup_{\alpha \in A} E^\alpha_x \, T_{\gamma_1} < \infty \quad \text{for } x \in R^{d_1}$$

where this time $\tau = T_{\gamma_2} + T_{\gamma_1} \circ \Theta_{T_{\gamma_2}}$ with $\gamma_1 = \{x \in R^{d_1}, \|x\| = r_1\}$, $\gamma_2 = \{x \in R^{d_1}, \|x\| = r_2\}$ and $r_1 < r_2$.
This way we obtain a version of Theorem 5 of [B–S]

Theorem 1. *Under (A1)–(A5)*
(i)
$$\lambda^\epsilon(x^0) \to \lambda(\alpha^0) \quad \text{as } \epsilon \to 0 \tag{31}$$

(ii) *(LCP) holds.*

3. Uniform limit control principle

The adaptive control procedure introduced in [D–PD–S1] is based on the construction of a finite net of δ-optimal controls, for given $\delta > 0$. Since our real model depends on the pair α^0, ϵ of unknown parameters, we shall need such property uniformly for $\alpha \in A$ and ϵ sufficiently small. This result will be obtained as Corollary from the following theorem.

Theorem 2. *Under (A1)–(A5) we have*
(i)
$$\sup_{\alpha \in A} |\lambda^\epsilon(\alpha) - \lambda(\alpha)| \to 0 \quad \text{as } \epsilon \to 0 \tag{32}$$

(ii) *if $u_\alpha \in A_r$ is δ optimal for (10),(11) with α^0 replaced by α, then*
$$\sup_\alpha |\bar{J}^{\alpha,\epsilon}_{xy}(u_\alpha) - \bar{J}^\alpha_x(u_\alpha)| \to 0 \quad \text{as } \epsilon \to 0 \tag{33}$$

P r o o f. By [K1] and [S1] there exists $u^\epsilon_\alpha \in A$ such that
$$\lambda^\epsilon(\alpha) = \bar{J}^{\alpha,\epsilon}_{xy}(u^\epsilon_\alpha) \tag{34}$$

If (i) does not hold then there exist subsequences $\alpha_n \to \alpha, \epsilon_n \to 0$ such that
$$|\lambda_{\epsilon_n}(\alpha_n) - \lambda(\alpha_n)| > \eta > 0 \quad \text{for } n = 1, 2, \ldots \tag{35}$$

Since by Proposition 2.3 of [D–PD–S1], $\lambda(\alpha_n) \to \lambda(\alpha)$, we have
$$|\lambda_{\epsilon_n}(\alpha_n) - \lambda(\alpha)| > \eta/2 > 0 \quad \text{for } n = 1, 2, \ldots \tag{36}$$

By Proposition 7 of [B–S] and the first part of the proof of Theorem 5 of [B–S], there exists subsequence n_k, for simplicity further denoted by n, and $u \in \mathcal{A}^r$ such that

$$\lambda^{\varepsilon_n}(\alpha_n) = \bar{J}_{xy}^{\alpha_n, \varepsilon_n}(u_{\alpha_n}^{\varepsilon_n}) \to \bar{J}_x^\alpha(u) \geq \lambda(\alpha)$$

Consequently

$$\liminf_{n \to \infty} \lambda^{\varepsilon_n}(\alpha_n) \geq \lambda(\alpha) \tag{37}$$

For $u_\alpha \in \mathcal{A}$ such that

$$\lambda_\alpha = \bar{J}_x^\alpha(u_\alpha) \tag{38}$$

we have by the first part of the proof of Theorem 5 of [B–S] again

$$\bar{J}_{xy}^{\alpha_n, \varepsilon_n}(u_{\alpha_n}) \to \bar{J}_x^\alpha(u_\alpha) \quad \text{as } n \to \infty \tag{39}$$

Comparing (37), (38) and (39) we obtain

$$\lim_{n \to \infty} \lambda^{\varepsilon_n}(\alpha_n) = \lambda(\alpha) \tag{40}$$

a contradiction to (36).

It remains to show (ii). Assume (33) does not hold. Then for some $\delta > 0, \varepsilon_n \to 0, \alpha_n \to \alpha$ we have

$$|\bar{J}_{xy}^{\alpha_n, \varepsilon_n}(u_{\alpha_n}) - \bar{J}^{\alpha_n}(u_{\alpha_n})| > \delta \quad \text{for } n = 1, 2, \ldots \tag{41}$$

By Proposition 7 of [B–S], proof of Theorem 5 of [B–S] again and (39)–(40) of [S1], there exists $u \in \mathcal{A}_r$ for which

$$\bar{J}_{xy}^{\alpha_n, \varepsilon_n}(u_{\alpha_n}) \to \bar{J}_x^\alpha(u)$$
$$\bar{J}_x^{\alpha_n}(u_{\alpha_n}) \to \bar{J}_x^\alpha(u)$$

a contradiction to (41).

Corollary 1. *(ULCP) holds.*

P r o o f. In fact, for $\varepsilon < \varepsilon_0$

$$\sup_\alpha |\lambda^\varepsilon(\alpha) - \lambda(\alpha)| < \frac{\delta}{2}, \quad \sup_\alpha |\bar{J}_{xy}^{\alpha, \varepsilon}(u_\alpha) - \bar{J}_x^\alpha(u_\alpha)| \leq \frac{\delta}{2} \tag{41}$$

where u_δ is any δ–optimal control for $\bar{J}_x^\alpha$.

Therefore if $\bar{J}_x^\alpha(u) \leq \lambda(\alpha) + \delta$, by (41) we have

$$\bar{J}_{xy}^{\alpha, \varepsilon}(u) \leq \bar{J}_x^\alpha(u) + \frac{\delta}{2} \leq \lambda(\alpha) + \frac{3}{2}\delta \leq \lambda^\varepsilon(\alpha) + 2\delta$$

and (15) is satisfied.

Corollary 2. *For each $\delta > 0$ there exists $\rho > 0$ such that for $\varepsilon < \varepsilon_0$ if $\|\alpha - \alpha'\| < \rho$ then any $\frac{\delta}{2}$–optimal control $u \in \mathcal{A}_r$ for $\bar{J}_x^\alpha$ is 2δ optimal for $\bar{J}_{xy}^{\alpha', \varepsilon}$. There exists a finite set $U_\delta = \{u_1, \ldots, u_r\}, u_i \in \mathcal{A}_r$ of δ-optimal controls i.e. for each $\alpha \in A, \varepsilon < \varepsilon_0$, there is i such that*

$$\bar{J}_{xy}^{\alpha, \varepsilon}(u_i) \leq \lambda^\varepsilon(\alpha) + \delta \tag{42}$$

P r o o f. By Proposition 2.4 of [D–PD–S1] one can find ρ such that $\|\alpha - \alpha'\| < \rho$ implies that any $\frac{\delta}{2}$ optimal control $u \in \mathcal{A}_r$ for $\bar{J}_x^\alpha$ is δ-optimal for $\bar{J}_x^{\alpha'}$. Then we apply (ULCP). The second assertion is an immediate consequence of the first and the compactness of A.

4. Adaptive control

Denote by $\Pi_u^{\alpha,\epsilon}$ an invariant measure of (1),(2) with α^0 replaced by α and $u \in \mathcal{A}_r$. Let

$$\tau_1 = T_{\gamma_2} + T_{\gamma_1} \circ \Theta_{T_{\gamma_2}}$$

$$\tau_{n+1} = \tau_n + \tau_1 \circ \Theta_{\tau_n} \quad \text{with} \quad \gamma_1 = \{x \in R^{d_1}, \|x\| = r_1\}$$

$$\gamma_2 = \{x \in R^{d_1}, \|x\| = r_2\}$$

We assume:

(A6)

$$\sup_{\alpha \in A} \sup_{\epsilon > 0} E_x^{\alpha,\epsilon}\{T_{\gamma_1}\} < \infty$$

$$\sup_{\alpha \in A} \sup_{\epsilon > 0} \sup_{x \in \gamma_1} \sup_{y \in R^{d_2}} \sup_{u \in \mathcal{A}_r} E_{xy}^{\alpha,\epsilon}\{\tau_1^2\} < \infty \tag{43}$$

In addition for $u \in \mathcal{A}_r$

(A7)

$$w_u^{\alpha\epsilon}(x,y) = \lim_{n \to \infty} \inf \ E_{xy}^{\epsilon}\left\{\int_0^{\tau_n}\left(k(x_s, u(x_s)) - \int_{R^d} k(z_1, u(z_1))\Pi_u^{\alpha,\epsilon}(dz_1, dz_2)\right) ds\right\} \tag{44}$$

is well defined and

$$\sup_{\alpha \in A} \sup_{\epsilon > 0} \sup_{u \in \mathcal{A}_r} \sup_{x \in \gamma_1} \sup_{y \in R^{d_2}} |w_u^{\alpha,\epsilon}(x,y)| = K < \infty \tag{45}$$

For fixed $\delta > 0$ choose $\rho > 0$ as in Corollary 2 such that $|\alpha - \alpha'| < \rho \Rightarrow \sup_{u \in \mathcal{A}_r} |\bar{J}_x^\alpha(u) - \bar{J}_x^{\alpha'}(u)| \leq \frac{\delta}{6}$ (what is possible in view of Theorem 2.1 of [D–PD–S1]).

Define a disjoint cover of A by

$$A_1(\delta) = B(\alpha_1, \rho) \cap A$$
$$A_2(\delta) = (B(\alpha_2, \rho) \setminus A_1(\delta)) \cap A$$

$$\vdots$$

$$A_r(\delta) = \left(B(\alpha_r, \rho) \setminus \bigcup_{j=1}^{r-1} A_j(\delta)\right) \cap A \quad \text{with} \quad B(\alpha, \rho) = \{\alpha' \in A, \|\alpha - \alpha'\| \leq \rho\}$$

$$A = \bigcup_{j=1}^{r-1} A_j(\delta)$$

and set $U = \{u_1, \ldots, u_r\}, u_i \in \mathcal{A}_r$ where u_i are $\frac{\delta}{2}$–optimal controls for $\bar{J}_x^{\alpha_i}$.

Let $e : A \to \{\alpha_1, \ldots, \alpha_r\}$ and $\bar{e} : A \to \{1, \ldots, r\}$ be defined by

$$e(\alpha) = \alpha_j \quad \text{if} \quad \alpha \in A_j, \quad \bar{e}(\alpha) = j \quad \text{if} \quad \alpha \in A_j$$

and

$$\bar{\lambda}(\alpha) = \lambda(e(\alpha)) \tag{46}$$

By a suitable modification of $A_i(\delta)$ on the boundaries, we can assume that $\bar{\lambda}$ is lower semicontinuous.

Choose positive integer N such that

$$N \geq 8K \cdot \delta^{-1} m \tag{47}$$

where $m > 0$ satisfies

$$m \leq \inf_{x \in \gamma_1} \inf_{y \in R^{d_2}} \inf_{u \in A_r} \inf_{\alpha \in A} E_{xy}^{\alpha,\epsilon} \tau \tag{48}$$

Define a sequence of stopping times $\sigma_n, n = 1, 2, \ldots$ as

$$\sigma_1 = \tau_N, \sigma_2 = \sigma_1 + \tau_N \circ \Theta_{\sigma_1}, \ldots, \sigma_{n+1} = \sigma_n + \tau_N \circ \Theta_{\tau_n} \tag{49}$$

We estimate the unknown parameter α^0 at random times σ_n by the biased maximum likelihood method that is we take $\hat{\alpha}(\sigma_n)$ which is a maximizer of

$$L_n(\alpha) = \ln \bar{M}(\sigma_n, \varepsilon, \alpha, \alpha^0) + z(\sigma_n) \ln(\frac{\bar{\lambda}(\alpha^0)}{\bar{\lambda}(\alpha)}) \tag{50}$$

where $\bar{M}(\sigma_n, \varepsilon, \alpha, \alpha^0) = \dfrac{dP^{\alpha,\epsilon}}{dP^{\alpha^0,\epsilon}}$ is the likelihood evaluated at time σ_n and $z(t) > 0$, $t^{-1}z(t) \to 0, t^{-\beta}z(t) \to \infty$ for some $\beta \in (\frac{1}{2}, 1)$ as $t \to \infty$.

Let

$$\hat{\alpha}(t) = \hat{\alpha}(\sigma_n) \qquad \text{for} \quad \sigma_n \leq t < \sigma_{n+1}$$

Using the family of estimates $\hat{\alpha}(t)$ by an approximate certainty equivalence principle we define adaptive control as

$$\eta(s) = u_{\bar{e}(\hat{\alpha}(s))}(x(s)) \tag{51}$$

where $u_i \in U$.

Following the consideration of [D-PD-S1] and [S2] we finally obtain

Theorem 3. Under (A1) –(A7) there is ε_0 such that for $\varepsilon < \varepsilon_0$ we have

$$J(\eta(s)) \leq \lambda^\epsilon(\alpha^0) + 2\delta \qquad P \text{ a.e.} \tag{52}$$

5. Remarks on assumptions

In this section we provide sufficient conditions and comments on the assumptions imposed in the paper. We start with the assumption (A1). Let

$$\eta_1 = \{y \in R^{d_2} : \|y\| = r_1\} \qquad \eta_2 = \{y \in R^{d_2} : \|y\| = r_2\} \qquad r_1 < r_2$$

Define $\tau^z = T_{\eta_2}^z + T_{\eta_1}^z \circ \Theta_{T_{\eta_2}^z}$ with $T_\eta^z = \inf\{s \geq 0 : y_s^z \in \eta\}$.
    We have

Proposition 3. *If*

$$E_y T_{\eta_1}^z < \infty \qquad \text{for} \quad y \in R^{d_1}$$

and for compact set $K_1 \subset R^{d_1}$

$$\sup_{z \in K_1} \sup_{y \in \eta_1} E_y(\tau^z)^2 < \infty,$$

then under (A2), (9) is satisfied.

P r o o f. (Sketch). We point out only main steps.
    Step 1. By Theorem 2.1 of [D–PD–S1]

$$\underset{\varepsilon > 0}{\forall} \; \underset{\delta > 0}{\exists} \; \|y - y'\| < \delta \Rightarrow \|\mu_x(\cdot) - \mu_y(\cdot)\|_{\text{var}} < \varepsilon \tag{53}$$

Step 2. Suppose contrary to (9) for some $x_n \in K_1, y_n \in K_2, x_n \to x, y_n \to y$, $A_n \in B(R^{d_2}), t_n \to \infty$

$$|E_{y_n}\{y_{t_n}^{x_n} \in A_n\} - \mu_{x_n}(A_n)| > \beta > 0 \tag{54}$$

We use the fact that the transition kernel $P^x(y, t, A)$ of (y_t^x) is continuous with respect to x, y uniformly in $A \in B(R^{d_1})$ for $t > t_0 > 0$ ([Kh] Theorem 4.6.2), and

$$\|E_{\bar{y}}\{y_{t_n}^{\bar{x}} \in \cdot\} - \mu_{\bar{x}}\{\cdot\}\|_{\text{var}} \to 0$$

and then (53) obtain a contradiction to (54).
    The assumption (A2) appears in Proposition 1 and seems to be fundamental for our approach.
    The assumption (A3) is clearly satisfied when a_1 is bounded.
    To have assumptions (A4), (A5) or (A6) satisfied we can impose Lyapunov type conditions (for details see [BO], [K1], [S1]). The assumption (A7) is crucial for the adaptive procedure introduced in Section 5. Unfortunately, the author although believes that (44) is true, cannot prove this without extra restriction to (y_t^ε) being a diffusion with reflection in a bounded regular domain. In what follows we assume that (y_t^ε) is diffusion with reflection in a bounded domain D with regular boundary. One can easily see that in this case all conclusions of Sections 2,3,4 hold, and (A4) is trivially satisfied.

Proposition 4. *If*

$$\sup_{x \in \gamma_1} \sup_{y \in D} \sup_{\varepsilon > 0} \sup_{\alpha \in A} \sup_{u \in A_r} E_{xy}^{2,\varepsilon}\{\tau_1^4\} < \infty \tag{55}$$

then the pair $(x_{\bar{\tau}_n}, y_{\bar{\tau}_n})$ *where*

$$\bar{\tau}_1 = T_{\gamma_2} + T_{\gamma_1}^h \circ \Theta_{T_{\gamma_2}} \qquad T_{\gamma_1}^h = \inf\{s \geq h, x_s \in \gamma_1\}$$
$$\bar{\tau}_{n+1} = \bar{\tau}_n + \bar{\tau}_1 \circ \Theta_{\bar{\tau}_n}$$

is uniformly in $\varepsilon > 0, \alpha \in A, u \in A_r$ *ergodic i.e. there exist probability measures* $\varphi_u^{\alpha,\varepsilon}$ *on* $\gamma_1 \times D$, *and constants* $0 < \gamma < 1, 0 < K$ *independent on* α, ε, n *such that*

$$\|P_{xy}^{\alpha\varepsilon}\{(x_{\bar{\tau}_n}, y_{\bar{\tau}_n}) \in \cdot\} - \varphi_u^{\alpha\varepsilon}(\cdot)\|_{\text{var}} \leq K \cdot \gamma^n \tag{56}$$

P r o o f. (Sketch)

By Proposition of the Appendix of [D–PD–S2] from (55) we have

$$\sup_{x \in \gamma_1} \sup_{y \in D} \sup_{\varepsilon > 0} \sup_{\alpha \in A} \sup_{u \in A_r} E_{xy}^{\alpha,\varepsilon}\{\bar{\tau}_1^2\} < \infty \tag{57}$$

By Theorem 4.1 of [B] it is sufficient to show that

$$\sup_{x,x' \in \gamma_1} \sup_{y,y' \in D} \sup_{\varepsilon > 0} \sup_{\alpha \in A} \sup_{u \in A_r} \sup_{A \in B(\gamma_1 \times D)}$$

$$|P_{xy}^{\alpha,\varepsilon}\{(x_{\bar{\tau}_1}^\varepsilon, y_{\bar{\tau}_1}^\varepsilon) \in A\} - P_{x'y'}^{\alpha,\varepsilon}\{(x_{\bar{\tau}_1}^\varepsilon, y_{\bar{\tau}_1}^\varepsilon) \in A\}| < 1 \tag{58}$$

Suppose for $\alpha_n, \varepsilon \to 0, x_n, x_n', y_n, y_n', A_n$ we have

$$P_{x_n y_n}^{\alpha_n, \varepsilon_n}\{(x_{\bar{\tau}_1}^{\varepsilon_n}, y_{\bar{\tau}_1}^{\varepsilon_n}) \in A_n\} \to 1 \tag{59}$$

and

$$P_{x_n' y_n'}^{\alpha_n, \varepsilon_n}\{(x_{\bar{\tau}_1}^{\varepsilon_n}, y_{\bar{\tau}_1}^{\varepsilon_n}) \in A_n\} \to 0$$

Then by (57) for a new measure P^0 corresponding to the solution of (16)

$$P_{x_n' y_n'}^0\{(x_{\bar{\tau}_1}^{\varepsilon_n}, y_{\bar{\tau}_1}^{\varepsilon_n}) \in A_n\} \to 0$$

Since from Proposition 1

$$|P_{x_n' y_n'}^0\{(x_{\bar{\tau}_1}^{\varepsilon_n}, y_{\bar{\tau}_1}^{\varepsilon_n}) \in A_n\} - E_{x_n'}^0\left\{\int_{R^{d_2}} \chi_{A_n}(x_{\bar{\tau}_1}, z)\mu_{\bar{\tau}_1}(ds)\right\}| \to 0,$$

we have

$$E_{x_n'}^0\left\{\int_{R^{d_2}} \chi_{A_n}(x_{\bar{\tau}_1}, z)\mu_{\bar{\tau}_1}(dz)\right\} \to 0$$

Therefore

$$\int_{\gamma_2}\int_{R^{d_2}} \chi_{A_n}(z_1,z)\mu_{z_1}(dz)\sigma(dz_1) \to 0$$

where σ is surface measure on γ_2.

Consequently

$$E^0_{x_n}\left\{\int_{R^{d_2}} \chi_{A_n}(x_{\bar\tau_1},z)\mu_{\bar\tau_1}(dz)\right\} \to 0$$

and by Proposition 1 again

$$P^0_{x_n y_n}\{(x^{\epsilon_n}_{\bar\tau_1}, y^{\epsilon_n}_{\bar\tau_1}) \in A_n\} \to 0$$

a contradiction to (59). Thus (58) holds and $(x_{\bar\tau_n}, y_{\bar\tau_n})$ is uniformly ergodic.

From (56) in the same way as in Lemma 2.3 of [D–PD–S1] we obtain

Corollary 3. *Under (55) and the first part of (A6), for $u \in \mathcal{A}_r$*

$$\bar{w}^{\alpha\epsilon}_u(x,y) = \liminf_{n\to\infty} E^\epsilon_{xy}\left\{\int_0^{\bar\tau_n}\left(k(x_s,u(x_s)) - \int_{R^d} k(z_1,u(z_1))\Pi^{\alpha\epsilon}_u(dz_1,dz_2)\right)ds\right\}$$

is well defined and

$$\sup_{\alpha\in A}\sup_{u\in\mathcal{A}_r}\sup_{x\in\gamma_1}\sup_{y\in R^{d_2}} |\bar{w}^{\alpha\epsilon}_u(x,y)| = K < \infty$$

and the adaptive procedure introduced in Section 4 with τ_n replaced by $\bar\tau_n$ works.

References

[B] A. Bensoussan, *Perturbation Methods in Optimal Control*, J. Wiley, 1988

[BO] V. Borkar, *Optimal Control of Diffusion Processes*, Longman, 1990

[B–S] T. Bielecki, L. Stettner, *On Ergodic Control Problems for Singularly Perturbed Markov Processes*, JAMO 20 (1989), 131–161

[D–PD–S1] T. E. Duncan, B. Pasik–Duncan, L. Stettner, *Almost Self-Optimizing Strategies for the Adaptive Control of Diffusion Processes*, submitted for publication

[D–PD–S2] T. E. Duncan, B. Pasik–Duncan, L. Stettner, *On Ergodic and Adaptive Control Problems for Stochastic Differential Delay Equations*, in preparation

[K1] H. J. Kushner, *Optimality conditions for the average cost per unit time problem with a diffusion model*, SIAM J. Control Optimiz. 16 (1978), 330–346

[K2] H. J. Kushner, *Weak Convergence Methods and Singularly Perturbed Stochastic Control and Filtering Problems*, Birkhäuser, 1990

[Kh] R. Z. Khasminskii, *Stochastic Stability of Differential Equations*, Sigthoff and Noordhoff, Alphen van den Rijn, 1980

[S1] L. Stettner, *On the Existence of an Optimal per Unit Time Control for a Degenerate Diffusion Model*, Bull. Pol. Acad. Sci. 34 (1986), 749–769

[S2] L. Stettner, *On nearly Selfoptimizing Strategies for a Discrete Time Uniformly Ergodic Adaptive Model*, JAMO to appear

Recent Results of Finite Dimensional Estimation Algebras

Wing Shing Wong

AT&T Bell Laboratories

Holmdel, NJ 07733

Abstract

In this note we provide a perspective and a brief summary of some recent advances on the topic of estimation algebras.

1. Introduction

When the idea of an estimation algebra was first introduced [1-3], it had two popular applications. A good amount of efforts were spent using it to discover new finite dimensional nonlinear filters or to enhance our understanding of known finite dimensional filters (see [4] for a good reference up to 1981.) An equally popular application was to use it to prove the non-existence of finite dimensional filters for certain systems, most notably, the so called *cubic sensor problem* [5, 6]. Recently, a number of new results on estimation algebras were obtained. In a nutshell, these new results characterize when an estimation algebra has finite dimension and shed light on its structure. The objective of this note to provide a perspective and a brief summary of these results.

To facilitate our subsequent presentation, let us note that the filtering problem considered here is based on the following signal observation model:

$$
\begin{cases}
dx(t) = f(x(t))\,dt + g(x(t))\,dv(t) & x(0) = x_0 , \\
dy(t) = h(x(t))\,dt + dw(t) & y(0) = 0 ,
\end{cases}
\tag{1}
$$

in which x, v, y, and w, are respectively, $I\!R^n$, $I\!R^p$, $I\!R^m$, and $I\!R^m$ valued processes, and v and w have components which are independent, standard Brownian processes. We further assume that $n = p$, f, h are C^∞ smooth, and that g is an orthogonal matrix.

Let $\rho(t, x)$ denote the conditional probability density of the state given the observation $\{y(s): 0 \leq s \leq t\}$. It is well known that $\rho(t, x)$ is given by normalizing a function, $\sigma(t, x)$, which satisfies the following Duncan-Mortensen-Zakai equation:

$$d\sigma(t, x) = L_0\sigma(t, x)dt + \sum_{i=1}^{m} L_i\sigma(t, x)dy_i(t) , \qquad \sigma(0, x) = \sigma_0 , \qquad (2)$$

where

$$L_0 = \frac{1}{2} \sum_{i=1}^{n} \frac{\partial^2}{\partial x_i^2} - \sum_{i=1}^{n} f_i \frac{\partial}{\partial x_i} - \sum_{i=1}^{n} \frac{\partial f_i}{\partial x_i} - \frac{1}{2} \sum_{i=1}^{m} h_i^2$$

and for $i = 1, \ldots, m$, L_i is the zero degree differential operator of multiplication by h_i.[†] σ_0 is the probability density of the initial point, x_0.

The estimation algebra, E, of the filtering problem defined by (1), is the Lie algebra generated by $\{L_0, L_1, \ldots, L_m\}$.

2. Finite Dimensional Estimation Algebra

It was shown very early on by Ocone [7] that in order for the system defined by (1) to have a finite dimensional estimation algebra, the functions, h_i, must be polynomials with degree less than or equal to 2. This result was sharpened by Wong in [8] by showing that the h_i's must be affine functions, provided that f and h are assumed to be real analytic and f satisfies some simple growth conditions. It was further shown in [8] that a finite dimensional estimation algebra must have a very special structure. In particular, it has a basis consisting of one second degree differential operator, L_0, first degree differential operator(s) of a certain type, and zero degree differential operators affine in x.

----

† If p is a vector, we use the notation p_i to represent the i-th component of p.

A direct consequence of this structure theorem is that all such finite-dimensional estimation algebras are solvable, partially settling a question raised by Brockett about whether there exist simple or semi-simple finite-dimensional estimation algebras.

More recently, Yau and Chiou [9] partially extended these results by removing the growth conditions. They showed that under some non-degenerate assumptions, these structure results are valid without assuming the growth conditions for systems with state dimension, n, less than or equal to 4. It is conjectured that under some non-degeneracy conditions and without any growth assumption, these results should still hold. One implication of all these theorems is that finite dimensional estimation algebras form a very restrictive class of systems and that other than the linear and Benes filters, it is unlikely that there exist other significant classes of systems with finite dimensional estimation algebras. Of course, the question whether there exists a finite dimensional filtering system with an infinite dimensional estimation algebra remains an open problem.

The definition of Ω was introduced in [10], which is defined as the matrix whose i,j-element is $\dfrac{\partial f_j}{\partial x_i} - \dfrac{\partial f_i}{\partial x_j}$. Ω turns out to be a useful parameter to characterize the estimation algebra of a filtering system. In [11], systems with Ω equals to zero were studied. The estimation algebra of such systems are said to be *exact*. If $\Omega = 0$, then by the Poincare Lemma, f is a gradient vector field. This latter statement provides an equivalent description of these systems. For such a system, there exists a simple algebraic condition that is both necessary and sufficient for its estimation algebra to be finite dimensional. Define

$$\eta = \sum_{i=1}^{n} \frac{\partial f_i}{\partial x_i} + \sum_{i=1}^{n} f_i^2 + \sum_{i=1}^{m} h_i^2 \tag{3}$$

It was shown in [11] that an exact estimation algebra is finite-dimensional if and only if

$$\nabla h_i^T J_\eta^j \text{ is a constant vector for } \leq i \leq m \text{ and all } j = 0, 1, \cdots, \tag{4}$$

where $J_\eta = \left[\dfrac{\partial^2 \eta}{\partial x_i \partial x_j} \right]$, denote the Hessian matrix of η.

Notice that this theorem contains the result mentioned earlier which requires the h_i's for a finite-dimensional estimation algebra to be affine. This is proved for an exact estimation algebra without any growth or non-degenerate assumptions. We can also sharpen the structure theorem for an exact estimation algebra. In particular, it was proved in [11] that an exact estimation algebra has a basis consisting of L_0, first degree differential operator(s) with constant coefficients for the $\dfrac{\partial}{\partial x_i}$ terms, and zero degree differential operator(s) affine in x. Moreover, if X and Y are in E with degree less than or equal to 1, then $[X, Y]$ is a constant.

One of the original motivations for introducing the concept of an estimation algebra is to extend the Wei-Norman approach for solving time varying ordinary differential equations to the setting of robust Duncan-Mortensen-Zakai equations, in another word, to the setting of time varying partial differential equations. This extension is highly non-trivial for a couple of reasons. First of all, the Baker-Campbell-Hausdorff formula which is the crux of the Wei-Norman approach, does not hold in general for unbounded operators. Secondly, the Wei-Norman solution, even if it holds, is generally valid for only a finite time duration, except when the all the operators involved form a solvable Lie algebra; in that case the Wei-Norman solution is valid for all time.

The fact that an exact estimation algebra is solvable implies that if the Wei-Norman approach works at all, one can expect that it works globally in time. The remaining difficult step then is to show that the Baker-Campbell-Hausdorff formulae hold for all the operator pairs in an exact estimation algebra. Thanks to the structure theorem quoted above, one can prove, somewhat tediously, that this is correct. Hence, one shows that for all such systems the Wei-Norman approach does provide a viable program for constructing finite dimensional filters.

Recently, Yau has extended these results for exact estimation algebras to the case where Ω is a constant [12], this class of systems clearly includes both the linear and Benes cases.

The necessary and sufficient condition stated in (4) can be restated slightly differently as:

If E is finite dimensional, then the matrix,

$$M = \left[\nabla h_1, \ldots, \nabla h_m, J_\eta \nabla h_1, \ldots, J_\eta \nabla h_m, J_\eta^2 \nabla h_1, \ldots, J_\eta^2 \nabla h_m, \cdots \right] \tag{5}$$

is a constant matrix.

If one makes a non-degeneracy type of assumption that the rank of M is full, which holds for a generic system, then it is possible to provide an interesting interpretation of (5). The following result was proved to hold in [13]:

If M has full rank, then E is a real vector space of dimension $2n + 2$ with a basis given by 1, $x_1, x_2, \ldots, x_n, \frac{\partial}{\partial x_1} - f_1, \ldots, \frac{\partial}{\partial x_n} - f_n$, and L_0. Moreover η is a polynomial of degree at most two and the quadratic part of $\eta - \sum_{i=1}^{m} h_i^2$ is positive semi-definite.

The latter statement says that the drift term, f, for all systems with finite dimensional exact estimation algebra must be a solution to equation (3) with $\eta - \sum_{i=1}^{m} h_i^2$ being a polynomial of degree of at most two and a positive semidefinite quadratic part. This is the well known condition defining a Benes filter [14]. So all exact finite dimensional estimation algebras with a full rank M come from Benes filters.

Interestingly, even though (3) can be viewed as a straight forward extension of the Riccati equation, very little is known about it in the literature until recently. In [13], the existence and uniqueness properties of such a equation were examined. By using an technique pioneered by Li and Yau [15], the following theorem was proved in [13]:

Theorem: Consider the following equation

$$\sum_{i=1}^{n} \frac{\partial f_i}{\partial x_i} + \sum_{i=1}^{n} f_i^2 = \sum_{i,j=1}^{n} a_{ij} x_i x_j - c \tag{6}$$

where $(x_1, \ldots, x_n) \in I\!R^n$, $c \in I\!R$ and the constant matrix $A = (a_{ij})$ is positive semidefinite.

Let $\{\lambda_1, \ldots, \lambda_n\}$ be the eigenvalues of A and $c_0 = \sum_{i=1}^{n} \sqrt{\lambda_i}$. Then we have the following

I) (Existence). When $c < c_0$, there is a family of C^∞ solution of Equation (6) with $2n$ parameters such that f has at most linear growth at ∞ , namely

$$|f(x)| \leq C(1 + |x|) , \tag{7}$$

for some constant C.

II) (Uniqueness). When $c = c_0$, there is a quadratic polynomial, uniquely determined up to a constant, which satisfies Equation (6). Moreover, this is the unique solution up to a constant if either one of the following conditions holds.

    (i) rank $A = 0$ (namely $A = 0$)

    (ii) rank $A \geq n - 2$

III) (Non-existence). When $c > c_0$, there is no smooth solution to Equation (6).

This theorem sheds some light on the issue of how general is the class of Benes filters. It is interesting to note that the growth condition (7) is typically assumed for the Ito equation defined in (1) to have a well defined solution. It is conjectured that the uniqueness property should hold regardless of the rank of the matrix A. It is also natural to ask whether one can classify all systems with a finite dimensional exact estimation algebra. This latter question is obviously a very challenging problem.

REFERENCES

[1] Brockett, R. W., and Clark, J. M. C., The geometry of the conditional density functions, in: O. L. R. Jacobs *et al.* Eds., *Analysis and Optimization of Stochastic Systems* (Academic Press, New York, 1980), 299-309.

[2] Brockett, R. W., Nonlinear systems and nonlinear estimation theory, in: M. Hazewinkel and J. S. Willems, Eds., *The Mathematics of Filtering and Identification and Applications* (Reidel, Dordrecht, 1981).

[3] Mitter, S. K., On the analogy between mathematical problems of non-linear filtering and quantum physics, *Ricerche di Automatica* 10(2) (1979) 163-216.

[4] Marcus, S. I., Algebraic and geometric methods in nonlinear filtering, *SIAM J. Control* 22 (1984) 817-844.

[5] Hazewinkel, M. and Marcus, S. I. On Lie algebras and finite dimensional filtering, *Stochastics* 7 (1982) 29-62.

[6] Hazewinkel, M., Marcus, S. I. and Sussman, H. J., Nonexistence of finite dimensional filters for conditional statistics of the cubic sensor problem, *Syst. Contr. Lett.* 3 (1983) 331-340.

[7] Ocone, D. L., Finite dimensional estimation algebras in nonlinear filtering, in:M. Hazewinkel and J. S. Willems, Eds., *The Mathematics of Filtering and Identification and Applications* (Reidel, Dordrecht, 1981).

[8] Wong, W. S., Theorems on the structure of finite dimensional estimation algebras, *Syst. Contr. Lett.* 9 (1987) 117-124.

[9] Yau, S. S.-T. and Chiou, W. L., "Recent Results on Classification of Finite Dimensional Estimation Algebras, Dimension of State Space Less Than 2," *Proc. of the 30th IEEE Conference on Decision and Control,* Brighton, England, (1991).

[10] Wong, W. S., On a new class of finite dimensional estimation algebras, *Syst. Contr. Lett.* 9 (1987) 79-83.

[11] Tam, L., Wong, W. S. and Yau, S. S.-T., On a Necessary and Sufficient Condition for Finite Dimensionality of Estimation Algebras, *SIAM J. of Control and Optimization* 28(1) (1990) 173-185.

[12] Yau, S. S.-T., "Finite Dimensional Filters with Nonlinear Drift I: A Class of Filters Including Both Kalman-Bucy Filters and Benes Filters," preprint.

[13] Dong, R.-T., Tam, L., Wong, W. S. and Yau, S. S.-T., Structure and classification theorems of finite-dimensional exact estimation algebras, *SIAM J. of Control and Optimization* 29(4) (1991) 866-877. LI Benes, V., Exact finite dimensional filters for certain diffusions with nonlinear drift, *Stochastics* 5 (1981) 65-92.

[14] Benes, V., Exact finite dimensional filters for certain diffusions with nonlinear drift, *Stochastics* 5 (1981) 65-92.

[15] Li, P. and Yau, S. T., On the parabolic kernel of the Schrödinger operator, *Acta Math.* 156(3-4) (1986), 153-201.

ASYMPTOTIC OPTIMAL RATE OF CONVERGENCE FOR AN ADAPTIVE ESTIMATION PROCEDURE

G. Yin*

Department of Mathematics
Wayne State University, Detroit, MI 48202

Abstract. Stochastic recursive estimation algorithms for adaptive filtering is considered and a procedure with averaging of the iterates is suggested. In contrast with the traditional approach, two sequences are constructed. One is the iterates of the estimates obtained from an ordinary adaptive filtering algorithm with slowly varying gains, and the other one is the arithmetic average of the aforementioned iterates. It is shown that under weak conditions, the algorithm with averaging has asymptotic optimal rate of convergence.

Keywords. Adaptive filtering, averaged iterates, optimal rate of convergence.

AMS subject classification. 93C40, 93E11.

1. Introduction

Adaptive filtering consists of recursively updating an approximating sequence to the vector $x^* \in \mathbb{R}^r$ that minimizes the estimation error of a random signal, $\psi \in \mathbb{R}$, from an observation vector $y \in \mathbb{R}^r$. One of the crucial points is that the calculations are done without knowing the statistics of ψ and y, on the basis of a sequence of realizations $\{(y_n, \psi_n)\}$. Throughout the paper, we shall assume that the sequence $\{(y_n, \psi_n)\}$ to be stationary and

$$Ey_n y_n' = R > 0, \quad Ey_n \psi_n = q, \tag{1.0}$$

where $R > 0$ means that the matrix is symmetric positive definite. By virtue of (1.0), it is easily seen that x^* is the unique solution of the Wiener-Hopf equation $Rx^* = q$.

A standard algorithm for approximating x^* is of the following form:

$$x_{n+1} = x_n + a_n y_n (\psi_n - y_n' x_n), \tag{1.1}$$

where $\{a_n\}$ is a sequence of positive scalars satisfying $\sum_n a_n = \infty$, $a_n \xrightarrow{n} 0$, and z' denotes the transpose of z.

Algorithm (1.1) and its variations have been studied extensively for many years, various results of convergence and rates of convergence have emerged, and numerous successful applications have been reported.

Analyzing adaptive filtering algorithms by means of ODE (ordinary differential equation) methods was considered in Ljung [1] and Kushner and Clark [2]. Correlated

* This research was supported in part by the National Science Foundation under grant DMS-9022139.

signals with finite memory and finite moments were discussed in Macchi and Eweda [3]. Rates of convergence were obtained in Kushner and Shwartz [4] via the weak convergence methods. Algorithms with adaptation and multi-step procedures were studied in Benveniste, Metivier and Priouret [5]. Related problems in adaptive beam forming was studied in [6] among others.

In contrast with these developments, the efficiency issue (asymptotic optimality) is the main focus here. Our primary concern is to design asymptotically efficient and easily implementible algorithms with asymptotic optimal convergence speed so as to improve the performance of the algorithm.

Let $a_n = n^{-\gamma}$. Then, under appropriate conditions, $n^{\gamma/2}(x_n - x^*)$ converges in distribution to a normal random variable with 0 mean and some covariance matrix Σ. Σ together with the scale factor $n^{\gamma/2}$ is often considered to be a measure of rate of convergence. It is clear that among these γ with $1/2 < \gamma \le 1$, $\gamma = 1$ leads to the highest order of convergence. If two algorithms have the same order of convergence, i.e., they have the same scale factor, the 'smaller' the covariance is, the better the rate of convergence is. In fact, much effort has been devoted to the improvement of the rate of convergence issues. Optimal design problems for adaptive filtering and a more general version of stochastic recursive algorithms were considered in the work of Benveniste, Metivier and Priouret [5]. Related problems in adaptive stochastic approximation type of algorithms were treated in Chung [7], Venter [8], Lai and Robbins [9], Wei [10] among others.

To proceed, examine a related problem in stochastic approximation

$$x_{n+1} = x_n + a_n f(x_n) + a_n \zeta_n, \tag{1.2}$$

where $f(\cdot)$ is a suitable function, $\{\zeta_n\}$ is a sequence of random disturbances (observation noise) and $\{a_n\}$ is the step-size or gain sequence. Suppose that $a_n \to 0$ slower than $1/n$. Define

$$\bar{x}_n = \frac{1}{n} \sum_{j=1}^{n} x_j. \tag{1.3}$$

It was shown by Polyak in [11], this averaging algorithm has asymptotic optimal rate of convergence in that

$$E(\bar{x}_n - \theta)(\bar{x}_n - \theta)' = \frac{1}{n}\Sigma^* + o(1/n), \tag{1.4}$$

where θ is the true parameter to be sought, and Σ^* is the optimal asymptotic covariance matrix. The results in [11] are similar in spirit to the approach of Ruppert [12] for a scaler problem. The main assumptions in [11] require the noise processes be i.i.d. It was mentioned that if correlated noise of moving average type is encountered, prewhitening filters might be needed. These assumptions were relaxed, φ-mixing (cf. [13]) type of noise processes were considered in Yin [14] and state-dependent noise processes were dealt with in Kushner and Yang [15]. Direct constructions were carried out in [14], whereas previous results in stochastic approximation, in particular the weak convergence methods were utilized to obtain the desired asymptotic optimality in [15]. In both papers, non-additive noise were treated. It was shown that no prewhitening filters are needed even correlated noise is encountered. Simulation results support these assertions.

Motivated by the approach suggested in [11], the idea of averaging was adopted to treat the adaptive filtering problem in [16]. It was proposed to use the following procedure in lieu of (1.1):

$$x_{n+1} = x_n + \frac{1}{n^\gamma} y_n(\psi_n - y'_n x_n), \quad \text{for some } \frac{1}{2} < \gamma < 1 \tag{1.5}$$

$$\bar{x}_{n+1} = \bar{x}_n - \frac{1}{n+1}\bar{x}_n + \frac{1}{n+1}x_{n+1}. \tag{1.6}$$

Note that $\bar{x}_n$ is the average of the iterates x_i, $i \leq n$ as in (1.3).

The averaging procedure has certain advantages. First, to implement stochastic recursive algorithms, such as adaptive filtering procedures, one would like to have the iterates get to a neighborhood of the true parameter x^* reasonably fast. The rapid decreasing sequence a_n would often lead poor results in the first a few iterations, i.e., the iterates get away from the neighborhood of the true parameter. Therefore, one may wish to choose large step size a_n, i.e., $\gamma < 1$. Nevertheless, from previous discussions, larger step size will result in slower rate of convergence. The use of the averaging approach allows one to get to a vicinity of x^* faster. Mean while, it keeps the asymptotic covariance to be the optimal one. To some extent, this can be thought of as a multiple scaling approach. It produces a "squeezing effect" which forces the iterates get to a vicinity of x^* without paying the price of increasing the asymptotic covariance matrix or slowing down the convergence speed. Secondly, unless the usual adaptive schemes (cf. Section 2), the algorithm with averaged iterates can easily be implemented with virtually no additional computational burdens. This should particularly be pronounced for large dimensional systems.

The remainder of the paper is arranged as follows. Asymptotic optimality issue is discussed in the next Section. Convergence and rate of convergence will then be addressed in Section 3. Asymptotic results of averaging with "moving windows" will be given. The idea of proofs which is akin to that of [16] will be explained. Finally, some further discussions are made at the last section.

2. Asymptotic optimality

To proceed, we assume the following conditions hold.

Assumption: $\{(y_n, \psi_n)\}$ is a stationary φ-mixing sequence such that

(C1) $E|y_n|^{4+\alpha} < \infty$, for some $\alpha > 0$.

(C2) $E|\psi_n|^{4+\beta} < \infty$ for some $\beta > 0$.

(C3) For $l > 0$, let

$$\varphi(l) = \sup_{A \in \mathcal{F}^{n+l}} E^{\frac{1}{1+\delta}}[P(A|\mathcal{F}_n) - P(A)]^{\frac{2+\delta}{1+\delta}}$$

and assume

$$\sum_l \varphi^{\frac{\delta}{1+\delta}}(l) < \infty, \quad \text{where } \delta = \min(\frac{\alpha}{2}, \frac{\beta}{2}),$$

and

$$\mathcal{F}_n = \sigma\{(y_k, \psi_k); k \leq n\}, \quad \mathcal{F}^n = \sigma\{(y_k, \psi_k); k \geq n\}.$$

To study the asymptotic optimality, we begin with the following algorithm:

$$x_{n+1} = x_n + \frac{\Gamma}{n} y_n \psi_n - \frac{\Gamma}{n} y_n y_n' x_n \tag{2.1}$$

where $\Gamma \in \mathbf{R}^{r \times r}$ is a non-singular matrix. We use the asymptotic covariance matrix as a performance index. The matrix Γ is introduced as a matrix-valued parameter. To evaluate the asymptotic performance, consider the following optimization problem. Choose an appropriate Γ so that the asymptotic covariance matrix is minimized.

It can be shown that $x_n \to x^*$ w.p.1 as $n \to \infty$, and

$$\sqrt{n}(x_n - x^*) = M_n(1) + o(1)$$

where $o(1) \xrightarrow{n} 0$ in probability. We let

$$C_{jk} = \begin{cases} \prod_{l=k+1}^{j}(I - \frac{\Gamma R}{l}), & j \geq k+1; \\ I, & j = k, \end{cases}$$

and define

$$M_n(t) = \frac{[nt]}{\sqrt{n}} \sum_{k=1}^{[nt]} \frac{1}{k} C_{[nt]k} \Gamma \xi_k, \quad t \in [0, 1],$$

where

$$\xi_k = y_k \psi_k - y_k y_k' x^*, \tag{2.2}$$

and $[s]$ denotes the largest integral part of s. It then can be proved (cf. [17, Section 3]) that $M_n(\cdot)$ converges weakly to a Gauss-Markov process $M(\cdot)$ which satisfies the following equation

$$M(t) = \int_0^t e^{-(I-\Gamma R)(\ln u - \ln t)} d\hat{B}(u),$$

where $\hat{B}(\cdot)$ is a Brownian motion with an appropriate covariance matrix. As a result, we arrive at $\sqrt{n}(x_n - x^*) \xrightarrow{n} N(0, \Sigma)$ in distribution, with Σ given by

$$\Sigma = \int_0^\infty e^{(\frac{I}{2} - \Gamma R)u} \Gamma S \Gamma' e^{(\frac{I}{2} - \Gamma R)'u} du,$$

where S is the error covariance to be specified later. It turns out that when $\Gamma = R^{-1}$, the asymptotic covariance is the optimal one given by $\Sigma^* = R^{-1} S R^{-1}$. This can be demonstrated by use of either the Crámer-Rao bound, or algebraic computations.

Since R is normally unavailable, a sequence of approximations $\{\Gamma_n\}$ is constructed. By means of the inversion lemma, Γ_n can be computed recursively. Γ in (2.1) is then replaced by Γ_n, and a pair of iterates is generated.

Such an algorithm preserves the consistency and asymptotic normality, and the asymptotic covariance matrix is the optimal one given by $\Sigma^* = R^{-1}SR^{-1}$. Some discussions on the optimal design issues can be found in the reference [5, Chapter 4] and the references therein.

Although the asymptotic optimality is achieved by using the above adaptive estimation scheme, the computation is rather intensive. A sequence of matrix-valued estimates has to be given. The computational complexity becomes a real issue, especially for large dimensional problems since each entry of the matrix needs to be estimated.

3. Results of the averaging algorithm

In this section, we shall present some asymptotic results concerning the algorithm (1.5)-(1.6). These results are collected into two theorems. The first one gives the convergence of the iterates $\{\bar{x}_n\}$ in the sense of with probability one, and the second one concentrates on the estimation error $\{x_n - x^*\}$.

Theorem 3.1. *Under (1.0) and (C1)-(C3), for any initial condition x_1, the iterates $\bar{x}_n$ defined by (1.5) and (1.6) converges to x^* w.p.1.*

Proof: See the proof of [16, Theorem 3.1]. □

Theorem 3.2. *Let $\{\kappa_n\}$ be a sequence of positive integers and $\{\nu_n\}$ a sequences of increasing, positive integers such that $\nu_n \xrightarrow{n} \infty$. Define*

$$\tilde{B}_n(t) = \frac{1}{\sqrt{\nu_n}} \sum_{j=\kappa_n}^{\kappa_n+[\nu_n t]+1} (x_j - x^*), \quad \text{for } t \in [0,1],$$

where $[s]$ denotes the largest integral part of s. Under the conditions of Theorem 3.1, $\tilde{B}_n(\cdot)$ converges weakly to a Brownian motion $\bar{B}(\cdot)$, with mean 0 and covariance matrix

$$\Sigma^* = R^{-1}SR^{-1} \tag{3.1}$$

where S is given by

$$S = E(\xi_1\xi_1') + \sum_{k=2}^{\infty} E(\xi_1\xi_k') + \sum_{k=2}^{\infty} E(\xi_k\xi_1'). \tag{3.2}$$

Remark: Theorem 3.2 is a slight generalization of the corresponding result in [16]. In lieu of a fixed window of averaging, a "moving window" is used. The sequence $\{\kappa_n\}$ can be chosen in many different ways. $\kappa_n = 1$ for all n corresponds to the case considered in [16]. $\kappa_n = \tilde{M}$ for some $\tilde{M}$ and all n, is the case that the averaging is started after the first $\tilde{M}$ iterations. In fact, our simulation results have shown that it is better to start the averaging after the iterates are "settled down" and become "stable" (cf. [14, 16]). Moreover, $\kappa_n = n$ is considered in [15]. Clearly, other sequences $\{\kappa_n\}$ can also be dealt with.

Due to the assumptions (C1)-(C3),

$$E|\xi_n|^{2+\delta} \leq 2^{2+\delta} \left(E|y_n\psi_n|^{2+\delta} + E|y_n y_n' x^*|^{2+\delta} \right)$$

$$\leq 2^{2+\delta} \left(E^{\frac{1}{2}}|y_n|^{4+2\delta} E^{\frac{1}{2}}|\psi_n|^{4+2\delta} + E|y_n|^{4+2\delta}|x^*|^{2+\delta} \right)$$

$$\leq 2^{2+\delta} \left(E^{\frac{1}{2}}|y_n|^{4+\alpha} E^{\frac{1}{2}}|\psi_n|^{4+\beta} + E|y_n|^{4+\alpha}|x^*|^{2+\delta} \right) < \infty.$$

By virtue of a basic φ-mixing inequality (cf. [13, Chapter 7]),

$$|E\xi_1\xi_k'| = |E\xi_1\xi_k' - E\xi_1 E\xi_k'| \leq K\varphi^{\frac{\delta}{1+\delta}}(k-1)|\xi_1|_{2+\delta}^2,$$

where $|z|_{2+\delta} = E^{1/(2+\delta)}|z|^{2+\delta}$ denotes the $L_{2+\delta}$ norm. Consequently,

$$\sum_{k=2}^{\infty} |E\xi_1\xi_k'| \leq K \sum_{k=1}^{\infty} \varphi^{\frac{\delta}{1+\delta}}(k) < \infty.$$

Thus, the series defined in (3.2) is absolutely summable.

In view of the definition of the interpolations, the function space we are using is $D^r[0,1]$, i.e., the space of functions defined on $[0,1]$ that are right continuous with left-hand limits and endowed with the Skorokhod topology (cf. [13, Chapter 3] and the references therein).

We outline the proof of Theorem 3.2 below. The proof consists of several steps; the details are similar to that of the corresponding theorems in [16].

Step 1: Write $z_n = x_n - x^*$, $\bar{z}_n = \bar{x}_n - x^*$ and define

$$A_{kj} = \begin{cases} \prod_{l=j+1}^{k}(I - R/l^{\gamma}), & k \geq j+1; \\ I, & k = j. \end{cases} \tag{3.3}$$

In view of (1.5), for each $m > 0$,

$$\sqrt{\nu_n + 1}\,\bar{z}_{\kappa_n+\nu_n+1} = \frac{1}{\sqrt{\nu_n+1}} \sum_{l=\kappa_n}^{\kappa_n+\nu_n+1} z_l$$

$$= \frac{1}{\sqrt{\nu_n+1}} \sum_{l=\kappa_n}^{\kappa_n+m-1} z_l$$

$$+ \frac{1}{\sqrt{\nu_n+1}} \sum_{l=\kappa_n+m}^{\kappa_n+\nu_n} A_{l,\kappa_n+m-1} z_{\kappa_n+m}$$

$$+ \frac{1}{\sqrt{\nu_n+1}} \sum_{l=\kappa_n+m}^{\kappa_n+\nu_n} \sum_{j=\kappa_n+m}^{l} \frac{1}{j^{\gamma}} A_{lj}(R - y_j y_j') z_j$$

$$+ \frac{1}{\sqrt{\nu_n+1}} \sum_{l=\kappa_n+m}^{\kappa_n+\nu_n} \sum_{j=\kappa_n+m}^{l} \frac{1}{j^{\gamma}} A_{lj}\xi_j. \tag{3.4}$$

Step 2: We show that the first three terms on the right-hand side of (3.4) contribute nothing to the limit. Denote

$$g_n = \frac{1}{\sqrt{\nu_n}} \sum_{l=\kappa_n+m(n)}^{\kappa_n+\nu_n} \sum_{j=\kappa_n+m(n)}^{l} \frac{1}{j^\gamma} A_{lj} \xi_j,$$

where $\{m(n)\}$ is a monotone increasing sequence of positive integers satisfying $m(n) \to \infty$ and $m(n)/\sqrt{n} \to 0$ as $n \to \infty$. To accomplish this goal, we show

$$\sqrt{\nu_n + 1}\,\bar{z}_{\kappa_n+\nu_n+1} = g_n + o(1), \tag{3.5}$$

where $o(1) \xrightarrow{n} 0$ in probability.

Step 3: We show that under the conditions of Theorem 3.2,

$$g_n = R^{-1} \frac{1}{\sqrt{\nu_n}} \sum_{j=\kappa_n}^{\kappa_n+\nu_n} \xi_j + o(1), \tag{3.6}$$

where $o(1) \xrightarrow{n} 0$ in probability.

Combine Step 2 and Step 3, we obtain

$$\sqrt{\nu_n + 1}\,\bar{z}_{\kappa_n+\nu_n+1} = R^{-1} \frac{1}{\sqrt{\nu_n}} \sum_{j=\kappa_n}^{\kappa_n+\nu_n} \xi_j + o(1), \tag{3.7}$$

where $o(1) \xrightarrow{n} 0$ in probability.

Step 4: To proceed, define $B_n(t)$ and $\bar{B}_n(t)$ for each $t \in [0,1]$ as:

$$B_n(t) = \frac{1}{\sqrt{\nu_n}} \sum_{j=\kappa_n}^{\kappa_n+[\nu_n t]} \xi_j$$
$$\bar{B}_n(t) = R^{-1} B_n(t). \tag{3.8}$$

We then show that $B_n(\cdot)$ converges weakly to a Brownian motion $B(\cdot)$ with mean 0 and covariance matrix S, where S is given by (3.2). $\bar{B}_n(\cdot)$ converges weakly to a Brownian motion $\bar{B}(\cdot)$ as given in Theorem 3.2.

Furthermore, notice that

$$\sqrt{\frac{[\nu_n t]}{\nu_n}} \left(\sqrt{[\nu_n t] + 1}\,\bar{z}_{\kappa_n+[\nu_n t]+1} \right) = \bar{B}_n(t) + o(1), \tag{3.9}$$

where $o(1) \xrightarrow{n} 0$ in probability. The left-hand side of the above equation is equal to

$$\sqrt{\frac{[\nu_n t]}{[\nu_n t] + 1}} \frac{1}{\sqrt{\nu_n}} \sum_{j=\kappa_n}^{\kappa_n+[\nu_n t]+1} z_j = \sqrt{\frac{[\nu_n t]}{[\nu_n t] + 1}} \tilde{B}_n(t). \tag{3.10}$$

Consequently, owing to the Slutsky's Theorem, $\tilde{B}_n(\cdot)$ and $\bar{B}_n(\cdot)$ have the same weak limit. The desired results then follows.

4. Further discussions

The essence of the approach studied in this paper is the use of averaged iterates which are computed from a usual adaptive filtering algorithm with 'slowly varying' gain $a_n = \frac{1}{n^\gamma}$, $\frac{1}{2} < \gamma < 1$. To some extent, the algorithm can be viewed as an accelerated convergence scheme. The averaging approach suggested in [11, 12] has been generalized. We demonstrate that no prewhitening filtering is needed in the presence of correlated signals of φ-mixing type. The new algorithm provides us with an efficient procedure having optimal rate of convergence and without excessive computations.

It seems that the approach discussed here can easily be adopted to the constrained adaptive array processing or adaptive beam forming algorithms (cf. [6] and the references therein). Incorporated with the averaging approach, various projection and truncation algorithms can also be developed. Moreover, such an idea could be applied to multiprocessor problems such as those discussed in [18].

Finally, a few words about algorithms with constant gain will be said below. Let $\varepsilon > 0$ be a small parameter and let the algorithm be given by:

$$x_{n+1} = x_n + \varepsilon y_n(\psi_n - y_n' x_n). \tag{4.1}$$

In this case, we shall work with $D^r[0, \infty)$. Define piecewise constant interpolations $x^\varepsilon(\cdot)$ and $u^\varepsilon(\cdot)$ by

$$x^\varepsilon(t) = x_n, \quad u^\varepsilon(t) = x_n/\sqrt{\varepsilon}, \text{ for } t \in [n\varepsilon, n\varepsilon + \varepsilon).$$

Then, under suitable conditions, it can be shown that for a sequence $t_\varepsilon \xrightarrow{\varepsilon} \infty$, $u^\varepsilon(t_\varepsilon + \cdot) \Rightarrow u(\cdot)$ a stationary process that satisfies

$$du = -Ru\,dt + S^{1/2}dw, \tag{4.2}$$

where $w(\cdot)$ is a standard Brownian motion. Let

$$\begin{aligned} z^\varepsilon(t) &= \frac{1}{\sqrt{t}} \int_0^t u^\varepsilon(s)ds, \\ \bar{z}(t) &= \frac{1}{\sqrt{t}} \int_0^t u(s)ds. \end{aligned} \tag{4.3}$$

By virtue of the weak convergence of $u^\varepsilon(\cdot) \Rightarrow u(\cdot)$, it is easily seen that $z^\varepsilon(\cdot) \Rightarrow \bar{z}(\cdot)$. Furthermore, define

$$\bar{z}^\varepsilon(t) = \sqrt{\frac{\varepsilon}{t}} \sum_{j=[t_\varepsilon]}^{[t_\varepsilon]+[t/\varepsilon]} (x_j - x^*). \tag{4.4}$$

As in Kushner and Yang [15], it can be established that for each t,

$$z^\varepsilon(t) - \bar{z}^\varepsilon(t) = o(1),$$

where $o(1) \xrightarrow{\varepsilon} 0$ in probability. In addition,

$$\text{Cov} u(s) = Eu(t)u'(t+s),$$

where $u(\cdot)$ is a stationary solution of (4.2). $\text{Cov} u(s) \to 0$ exponentially as $s \to \infty$. Therefore, there is a $\eta > 0$, such that

$$\begin{aligned}
\text{Cov}\bar{z}(t) &= \frac{1}{t} \int_0^t \int_0^t \text{Cov} u(s - \tilde{s}) ds d\tilde{s} \\
&= \int_{-\infty}^{\infty} \text{Cov} u(s) ds + O\left(e^{-\eta t}\right) \\
&= R^{-1} S R^{-1} + O\left(e^{-\eta t}\right).
\end{aligned}$$

Consequently, $\bar{z}^{\varepsilon}(t)$ converges in distribution to $\bar{z}(t)$, a normal random variable with 0 mean and covariance $R^{-1} S R^{-1} + O\left(e^{-\eta t}\right)$.

References

[1] L.Ljung, Analysis of recursive stochastic algorithms, *IEEE Trans. Automatic Control* **AC-22** (1977), 551-575.

[2] H.J. Kushner and D.S. Clark, *Stochastic Approximation Methods for Constrained and Unconstrained Systems*, Springer-Verlag, 1978.

[3] E. Eweda and O. Macchi, Quadratic and almost sure convergence of unbounded stochastic approximation algorithms with correlated observations, *Ann. Institut. Henri Poincáre* **19** (1983), 235-255.

[4] H.J. Kushner and A. Shwartz, Weak convergence and asymptotic properties of adaptive filters with constant gains, *IEEE Trans. Inform. Theory* **IT-30** (1984), 177-182.

[5] A. Benveniste, M. Metivier and P. Priouret, *Adaptive Algorithms and Stochastic Approximation*, Springer-Verlag, Berlin, 1990.

[6] G. Yin, Asymptotic properties of an adaptive beam former algorithm, *IEEE Transactions on Information Theory* **IT-35** (1989), 859-867.

[7] K.L. Chung, On a stochastic approximation method, *Ann. Math. Statist.* **25** (1954), 463-483.

[8] J.H. Venter, An extension of the Robbins-Monro procedure, *Ann. Math. Statist.* **38** (1967), 181-190.

[9] T.L. Lai and H. Robbins, Consistency and asymptotic efficiency of slope estimates in stochastic approximation schemes, *Z. Wahrsch. verw. Gebiete* **56** (1981), 329-360.

[10] C.Z. Wei, Multivariate adaptive stochastic approximation, *Ann. Statist.* **15** (1987), 1115-1130.

[11] B.T. Polyak, New method of stochastic approximation type, *Automat. Remote Control* **51** (1990), 937-946.

[12] D. Ruppert, Efficient estimations from a slowly convergent Robbins-Monro process, Technical Report, No. 781, School of Oper. Res. & Industrial Eng., Cornell Univ., 1988.

[13] S.N. Ethier and T.G. Kurtz, *Markov Processes, Characterization and Convergence*, Wiley, New York, 1986.

[14] G. Yin, On extensions of Polyak's averaging approach to stochastic approximation, *Stochastics* **36** (1991), 245-264.

[15] H.J. Kushner and J. Yang, Stochastic approximation with averaging of the iterates: optimal asymptotic rate of convergence for general processes, Technical Report, LCDS #91-9, Brown Univ., 1991.

[16] G. Yin and Y.M. Zhu, Averaging procedures in adaptive filtering: an efficient approach, *IEEE Trans. Automat. Control* **AC-37** (1992).

[17] G. Yin, A stopping rule for the Robbins-Monro method, *J. Optim. Theory Appl.* **67** (1990), 151-173.

[18] H.J. Kushner and G. Yin, Asymptotic properties of distributed and communicating stochastic approximation algorithms, *SIAM J. Control Optim.* **25** (1987), 1266-1290.

Controlled diffusions with rapidly oscillating unknown parameter processes

Qing Zhang

Department of Mathematics
University of Kentucky
Lexington, KY 40506

Abstract

This paper presents an asymptotic analysis of a controlled diffusion with an unknown system parameter process. The oscillation rate of the parameter process is assumed to be very fast. This gives rise to a limiting problem in which the unknown system parameter is replaced by its averaged mean value. A control for the original problem can be constructed from the optimal control of the limiting problem in a way which guarantees its asymptotic optimality. The convergence rate of the value function for the original problem is obtained. This helps in providing an error estimate for the constructed asymptotically optimal control.

Key words: diffusion processes, Markov chains, convex costs, asymptotically optimal controls, error estimates

1 Introduction

We consider a stochastic system with the state $x_t \in R^1$ and control $u_t \in \Gamma := [a_1, a_2] \in R^1$ that satisfy

$$dx_t = b(t, \theta_t^\varepsilon, x_t, u_t)dt + \sigma(t)dw_t, \ x_s = x, \ t \geq s \geq 0, \tag{1.1}$$

where $w_t \in R^1$ is a standard Brownian motion and $\theta_t^\varepsilon \in \Theta$ (Θ is a finite set) is an unknown parameter process with fast 'fluctuation rate'. Here $\varepsilon > 0$ is assumed to be a small number and $1/\varepsilon$ is used to stand for the fast oscillating rate of θ_t^ε.

We consider a cost function $J^\varepsilon(s, x, u.)$ with a finite time horizon $T \geq s \geq 0$ and convex functions $l(t, x)$ and $g(x)$,

$$J^\varepsilon(s, x, u.) = E[\int_s^T l(t, x_t)dt + g(x_T)]. \tag{1.2}$$

The problem is to find a control process $u_t \in \Gamma$ as a function of x_r, $s \leq r \leq t$ that minimizes the cost $J^\varepsilon(s, x, u.)$.

Theoretically, this is a stochastic adaptive control problem. Exact optimal control of this problem is very difficult to obtain (cf. Caines and Chen [2], Hijab [9], and Benes, Karatzas, and Rishel [1]). Instead of searching for exact optimal solutions, we turn to approximate optimal solutions. Let us first look at the following two cases in which either the parameter process θ_t^ε oscillates very slow or very fast, i.e., either ε is very large or very small. If ε is large, it means the process θ_t^ε will stay at each of its current state for a quite length of time before it jumps to another state. This period of staying time allows the controller to identify the value of θ_t^ε and enables him to construct approximate optimal controls accordingly. In this case, one may use the methods given in [8] together with the nonlinear filtering estimates given in [12] to derive approximate solutions. The purpose of this paper is to study the other extreme case in which ε is small. Namely, the process θ_t^ε switches in Θ more and more frequent as ε tends to zero. In this case, it will be more an more difficult for the controller to identify the value of θ_t^ε due to the fact that θ_t^ε changes its values fast. On the other hand, the value of θ_t^ε at each time t is not important anymore since it is going to switch to another value in a short period of time anyway. What we want to show in this paper is the following: when the process θ_t^ε oscillates fast, the controller may not need the information of θ_t^ε at each time t at all. In fact, the controller may actually ignore the details of θ_t^ε at each time t and average out the unknown parameter θ_t^ε. Examples of systems with fast oscillating parameters are not hard to provided. For instance, some failure-prone manufacturing systems can be formulated as a model with fast oscillating production capacities when the rate of fluctuation in part demand is small compare to the change of production capacities, see [10] and [13] for more discussion.

In this paper, we consider the system (1.1) with linear function b and convex cost functions l and g that are only functions of control t and x. A more general model of such problem is studied in [15].

The plan of this paper is as follows. In §2 we formulate the problem and make assumptions on the system under consideration. In §3 we introduce two auxiliary control problems: a control problem $(\mathcal{P}_a^\varepsilon)$ with complete information of θ_t^ε and a control problem $(\mathcal{P}_a^0)$ with only the information given by the mean value of θ_t^ε. By showing $\mathcal{P}_a^\varepsilon \to \mathcal{P}_a^0$ as $\varepsilon \to 0$ in terms of their value functions, we are able to construct feedback controls from $\mathcal{P}_a^0$ for our original problem $\mathcal{P}^\varepsilon$. In §4 we give a method to construct controls and then show that the constructed controls are asymptotically optimal. Error bounds of the constructed controls are obtained in terms of ε. Finally, we conclude the paper by making some remarks in §5.

2 Problem formulation and assumptions

We make the following assumptions on the functions b, σ, l, g, and the random processes θ_t^ε.

A1) There exist bounded functions $B_i(t)$, $i = 1, 2, 3, 4$ such that

$$b(t, \theta, x, u) = [B_1(t)x + B_2(t)] + \theta[B_3(t)u + B_4(t)].$$

Moreover, $\sigma(t)$ is bounded and there exists $c_0 > 0$ such that $\sigma(t) \geq c_0$ for all t.

A2) Let $\phi(t, x)$ denote $l(t, x)$ or $g(x)$. Then $\phi(t, \cdot)$ is convex. For all x, x', there exist constants

C_g and k_g such that

$$0 \leq \phi(t,x) \leq C_g(1+|x|^{k_g})$$

and $\quad |\phi(t,x)-\phi(t,x')| \leq C_g(1+|x|^{k_g}+|x'|^{k_g})|x-x'|.$

A3) Let $\Theta = \{1,2,\cdots,p\}$. $\theta_t^\epsilon \in \Theta$ is a finite state Markov chain governed by

$$L_\theta \psi = \epsilon^{-1} Q\psi \tag{2.1}$$

for any function ψ on Θ where Q is an $p \times p$ matrix such that $Q=(q_{ij})$ with $q_{ij} \geq 0$ if $i \neq j$ and $q_{ii} = -\sum_{i \neq j} q_{ij}$. Thus

$$L_\theta \psi(\cdot)(i) = \sum_{i \neq j}(\epsilon^{-1}q_{ij})(\psi(j)-\psi(i)) = \epsilon^{-1}Q\psi(\cdot)(i).$$

Moreover, Q is irreducible and θ_t^ϵ is independent of w_t.

Remark. In A1), we assume the function b is linear in x and u. Such assumption is used to obtain estimated error bounds (see Remark of Theorem 3.1 for further discussion). Moreover, the assumption on $\sigma(t)$ is only required in §4 for construction of optimal controls. Results up to §3 hold without this assumption.

Definition. We say that a control $u. = \{u_t : s \leq t \leq T\}$ is *admissible* if there exists a Borel function $u(t,x)$ such that $u_t = u(t,x_t) \in \Gamma = [a_1,a_2]$ and (1.1) has a unique strong solution. We use $\mathcal{A}^\epsilon$ to denote the set of all admissible controls.

We write $v^\epsilon(s,x)$ to be the corresponding value function of the system (1.1), i.e.,

$$v^\epsilon(s,x) = \inf_{u. \in \mathcal{A}^\epsilon} J^\epsilon(s,x,u.).$$

In the following, we always use $\mathcal{P}^\epsilon$ to denote our original problem, i.e.,

$$\mathcal{P}^\epsilon : \begin{cases} \text{minimize} & J^\epsilon(s,x,u.) = E[\int_s^T l(t,x_t)dt + g(x_T)], \\ \text{subject to} & dx_t = b(t,\theta_t^\epsilon,x_t,u_t)dt + \sigma(t)dw_t, \ x_s = x, \ u. \in \mathcal{A}^\epsilon, \\ \text{value function} & v^\epsilon(s,x) = \inf_{u. \in \mathcal{A}^\epsilon} J^\epsilon(s,x,u.). \end{cases} \tag{2.2}$$

Note that θ_t^ϵ is a process that is unknown to us. The usual dynamic programming method is no longer applicable here anymore. In order to handle the problem, we introduce two auxiliary control problems, which will be used to approximately characterize the dynamics of our control problem.

3 Auxiliary control problems

Let $\nu = (\nu_1,\cdots,\nu_p)$ denote the equilibrium distribution of Q. Then ν is the only solution to

$$\nu Q = 0 \text{ and } \sum_{i=1}^{p}\nu_i = 1.$$

Let $\bar{\theta}$ denote the equilibrium mean of Q, i.e., $\bar{\theta} = \sum_{i=1}^{p}\nu_i i$. In problem $\mathcal{P}^\epsilon$, we only assume the value of $\bar{\theta}$ is available to the controller.

We now introduce two related control problems. We use $\mathcal{P}_a^\epsilon$ to denote a control problem with complete information of the parameter process θ_t^ϵ. Let $\mathcal{A}_a^\epsilon$ denote the class of controls,

$$\mathcal{A}_a^\epsilon := \{u_t \in \Gamma = [a_1, a_2] : u_t \text{ is } \sigma\{\theta_r^\epsilon, w_r, s \leq r \leq t\} \text{ measurable }\}.$$

Then $\mathcal{P}_a^\epsilon$ is described as follows.

$$\mathcal{P}_a^\epsilon : \begin{cases} \text{minimize} & J^\epsilon(s, x, u.) = E[\int_s^T l(t, x_t)dt + g(x_T)], \\ \text{subject to} & dx_t = b(t, \theta_t^\epsilon, x_t, u_t)dt + \sigma(t)dw_t, \; x_s = x, \; u. \in \mathcal{A}_a^\epsilon, \\ \text{value function} & v_a^\epsilon(s, x) = \inf_{u. \in \mathcal{A}_a^\epsilon} J^\epsilon(s, x, u.). \end{cases} \quad (3.1)$$

It is clear that $\mathcal{A}^\epsilon \subset \mathcal{A}_a^\epsilon$. Therefore,

$$v_a^\epsilon(s, x) \leq v^\epsilon(s, x) \leq J^\epsilon(s, x, u.) \text{ for } u. \in \mathcal{A}^\epsilon. \quad (3.2)$$

We now define another control problem $\mathcal{P}_a^0$ with a modified system equation to 'average out' the random parameter θ_t^ϵ, i.e., to replace θ_t^ϵ by its mean value $\bar{\theta}$ in (1.1). Let $\mathcal{D}_t = \sigma\{w_r, s \leq r \leq t\}$. We consider a class of controls $\mathcal{A}_a^0$,

$$\mathcal{A}_a^0 := \{u_t \in \Gamma = [a_1, a_2] : u_t \text{ is } \mathcal{D}_t \text{ measurable }\}.$$

Let $\bar{b}(t, x, u) = [B_1(t)x + B_2(t)] + \bar{\theta}[B_3(t)u + B_4(t)]$. Then

$$\mathcal{P}_a^0 : \begin{cases} \text{minimize} & J^0(s, x, u.) = E[\int_s^T l(t, x_t)dt + g(x_T)], \\ \text{subject to} & dx_t = \bar{b}(t, x_t, u_t)dt + \sigma(t)dw_t, \; x_s = x, \; u. \in \mathcal{A}_a^0, \\ \text{value function} & v_a^0(s, x) = \inf_{u. \in \mathcal{A}_a^0} J^0(s, x, u.). \end{cases} \quad (3.3)$$

In the following, we sometimes refer the system state equation of $\mathcal{P}^\epsilon$ ($\mathcal{P}_a^\epsilon$ or $\mathcal{P}_a^0$) by saying 'system $\mathcal{P}^\epsilon$ ($\mathcal{P}_a^\epsilon$ or $\mathcal{P}_a^0$)'.

Let us now state a result regarding the convexities of the value functions. The proof of this result is standard and can be found in, e.g., [14].

Proposition 3.1. *Under the assumptions A1)-A2), the value functions $v^\epsilon(s, x)$, $v_a^\epsilon(s, x)$, and $v_a^0(s, x)$ are convex in x.*

We will show that $\mathcal{P}_a^\epsilon$ is close to $\mathcal{P}_a^0$ as ϵ is small. Then we choose a feedback control law from $\mathcal{P}_a^0$. We apply such control law to $\mathcal{P}^\epsilon$ and show that the incurred cost J^ϵ is close to v_a^0. Then by making use of (3.2), we conclude J^ϵ is close to v^ϵ. Therefore, the control constructed from $\mathcal{P}_a^0$ is nearly optimal.

We first show that $\mathcal{P}_a^\epsilon$ converges to $\mathcal{P}_a^0$ as $\epsilon \to 0$. More precisely, we have the following theorem.

Theorem 3.1. *There exists a constant C such that*

$$|v_a^\epsilon(s, x) - v_a^0(s, x)| \leq C(1 + |x|^{k_g})\sqrt{\epsilon}.$$

Remark. In general, the convergence of v_a^ϵ to v_a^0 can be proved without assuming the linearity on b in A1). However, the error bound of convergence is not available anymore. Detailed discussion concerns with this point can be found in [15]. The convergence rate of the value function v_a^ϵ is essentially determined by the convergence rate of $\theta_t^\varepsilon \to \bar\theta$, which is an order of $\sqrt\varepsilon$ as shown in Lemma A.2. In [13], it is pointed out by giving an example (with $\sigma(t) = 0$) that $\sqrt\varepsilon$ is the best convergence rate possible.

Proof. Let $\bar u_t \in \mathcal{A}_a^0 \subset \mathcal{A}_a^\varepsilon$ and let x_t and $\bar x_t$ denote, respectively, the corresponding states of the systems:

$$dx_t = b(t, \theta_t^\varepsilon, x_t, \bar u_t)dt + \sigma(t)dw_t, \quad x_s = x,$$
$$d\bar x_t = \bar b(t, x_t, \bar u_t)dt + \sigma(t)dw_t, \quad \bar x_s = x,$$

Then we have immediately, for some constants C_1, C_2,

$$
\begin{aligned}
E|x_t - \bar x_t| &\le C_1 \int_s^t E|x_r - \bar x_r|dr + E|\int_s^t (\theta_r^\varepsilon - \bar\theta)[B_3(r)\bar u_r + B_4(r)]dr| \\
&\le C_1 \int_s^t E|x_r - \bar x_r|dr + C_2\sqrt\varepsilon\sqrt{t-s}.
\end{aligned}
$$

The last inequality is due to Lemma A.2 (Appendix). From Gronwall's inequality, it follows that

$$E|x_t - \bar x_t| \le C_2\sqrt\varepsilon(\sqrt{t-s} + C_3 \int_s^t \sqrt{r-s}dr) \le C_4\sqrt\varepsilon \text{ for } s \le t \le T.$$

Then, we have

$$
\begin{aligned}
|J^\varepsilon(s, x, \bar u.) - J^0(s, x, \bar u.)| &= |E[\int_s^T [l(t, x_t) - l(t, \bar x_t)]dt + g(x_T) - g(\bar x_T)]| \\
&\le E[\int_s^T C_g(1 + |x_t|^{k_g} + |\bar x_t|^{k_g})|x_t - \bar x_t|dt \\
&\quad + C_g(1 + |x_T|^{k_g} + |\bar x_T|^{k_g})|x_T - \bar x_T|] \\
&\le C_5\sqrt\varepsilon(1 + |x|^{k_g}).
\end{aligned}
\tag{3.4}
$$

This implies

$$v_a^\varepsilon(s, x) - v_a^0(s, x) \le C_5(1 + |x|^{k_g})\sqrt\varepsilon. \tag{3.5}$$

We now show the opposite inequality. To this end, we first show that for any control $u_t \in \mathcal{A}_a^\varepsilon$ there exists a control $\bar u_t \in \mathcal{A}_a^0$ such that $E[\theta_t^\varepsilon u_t | \mathcal{D}_t] - \bar\theta \bar u_t$ is small in terms of ε. More precisely, we have the following lemma.

Lemma 3.1 *Let $u_t \in \mathcal{A}_a^\varepsilon$. Then there exist $\bar u_t \in \mathcal{A}_a^0$ and a constant C such that*

$$|E[\theta_t^\varepsilon u_t | \mathcal{D}_t] - \bar\theta \bar u_t| \le Ce^{-k_0(t-s)/\varepsilon}, \text{ a.s.}$$

Proof. For each $i \in \Theta$, let $u_t^i = E[u_t | \mathcal{D}_t, \theta_t^\varepsilon = i]$. Then u_t^i is $\mathcal{D}_t$ measurable and $u_t^i \in \Gamma$.

We define $\bar u_t = \bar\theta^{-1} \sum_{i=1}^p \nu_i i u_t^i$. Then $\bar u_t \in \mathcal{A}_a^0$. By direct computation,

$$
\begin{aligned}
E[\theta_t^\varepsilon u_t | \mathcal{D}_t] &= \sum_{i=1}^p P(\theta_t^\varepsilon = i) i u_t^i \\
&= \sum_{i=1}^p \nu_i i u_t^i + \sum_{i=1}^p (P(\theta_t^\varepsilon = i) - \nu_i) i u_t^i \\
&:= \bar\theta \bar u_t + R_t,
\end{aligned}
\tag{3.6}
$$

where $R_t = \sum_{i=1}^p (P(\theta_t^\varepsilon = i) - \nu_i) i u_t^i$. By Lemma A.1 (Appendix), there exist constants C and $k_0 > 0$ such that

$$|R_t| \le Ce^{-k_0(t-s)/\varepsilon}, \text{ a.s.} \quad \square$$

We consider the systems $\mathcal{P}_a^\varepsilon$ and $\mathcal{P}_a^0$ with controls $u_t \in \mathcal{A}_a^\varepsilon$ and $\bar{u}_t \in \mathcal{A}_a^0$ (here $\bar{u}_t$ is given by (3.6)). Then

$$dx_t = b(t, \theta_t^\varepsilon, x_t, u_t)dt + \sigma(t)dw_t, \quad x_s = x,$$
$$d\bar{x}_t = \bar{b}(t, x_t, \bar{u}_t)dt + \sigma(t)dw_t, \quad \bar{x}_s = x.$$

We now show that $\bar{x}_t$ is an approximation to $E[x_t|\mathcal{D}_t]$.

Lemma 3.2. *There exists C such that*

$$|E[x_t|\mathcal{D}_t] - \bar{x}_t| \le C\varepsilon, \text{ for } s \le t \le T.$$

Proof. Let $\Psi(t, r)$ denote the fundamental solution of

$$\frac{d}{dt}\Psi(t, r) = B_1(t)\Psi(t, r), \ \Psi(r, r) = I.$$

Then

$$x_t - \bar{x}_t = \int_s^t \Psi(t, r)[B_3(t)(\theta_r^\varepsilon u_r - \bar{\theta}\bar{u}_r) + B_4(t)(\theta_r^\varepsilon - \bar{\theta})]dr.$$

From Lemma A.3 (Appendix) and the fact that $\bar{u}_r$ is $\mathcal{D}_t$ measurable for $r \le t$, it follows

$$E[x_t|\mathcal{D}_t] - \bar{x}_t = \int_s^t \Psi(t, r)[B_3(t)(E[\theta_r^\varepsilon u_r|\mathcal{D}_r] - \bar{\theta}\bar{u}_r) + B_4(t)E(\theta_r^\varepsilon - \bar{\theta})]dr.$$

Then, applying Lemmas 3.1 and A.2, we have

$$|E[x_t|\mathcal{D}_t] - \bar{x}_t| \le C_6[\int_s^t |E[\theta_r^\varepsilon u_r|\mathcal{D}_r] - \bar{\theta}\bar{u}_r|dr + \sqrt{\varepsilon}] \le C_7\sqrt{\varepsilon}. \ \square$$

Now we continue to prove the theorem. By Jensen's inequality, we have, for any $u_t \in \mathcal{A}_a^\varepsilon$,

$$
\begin{aligned}
J^\varepsilon(s, x, u.) &\ge E[\int_s^T l(t, E[x_t|\mathcal{D}_t])dt + g(E[x_T|\mathcal{D}_T])] \\
&= E[\int_s^T l(t, \bar{x}_t)dt + g(\bar{x}_T)] \\
&\quad + E[\int_s^T (l(t, E[x_t|\mathcal{D}_t])dt - l(t, \bar{x}_t))dt + (g(E[x_T|\mathcal{D}_T]) - g(\bar{x}_T))] \\
&\ge v(x, z) + E[\int_s^T (l(t, E[x_t|\mathcal{D}_t])dt - l(t, \bar{x}_t))dt \\
&\quad + (g(E[x_T|\mathcal{D}_T]) - g(\bar{x}_T))].
\end{aligned}
$$

By Lemma 3.2 and Jensen's inequality, it follows,

$$
\begin{aligned}
&E[\int_s^T (l(t, E[x_t|\mathcal{D}_t])dt - l(t, \bar{x}_t))dt + (g(E[x_T|\mathcal{D}_T]) - g(\bar{x}_T))] \\
&\le C_g E[\int_s^T (1 + |x_t|^{k_g} + |\bar{x}_t|^{k_g})|E[x_t|\mathcal{D}_t] - \bar{x}_t|dt \\
&\quad + C_g(1 + |x_T|^{k_g} + |\bar{x}_T|^{k_g})|E[x_T|\mathcal{D}_T] - \bar{x}_T|] \\
&\le C_8(1 + |x|^{k_g})\sqrt{\varepsilon}.
\end{aligned}
$$

Thus,

$$v_a^\varepsilon(s, x) - v_a^0(s, x) \ge -C_8(1 + |x|^{k_g})\sqrt{\varepsilon}. \tag{3.7}$$

We now complete the proof of Theorem 3.1 by combining (3.5) with (3.7). $\square$

4 Asymptotically optimal controls

In this section, we use the control policy associated with the problem $\mathcal{P}_a^0$ and derive from it, a feedback control policy for the original problem $\mathcal{P}^\epsilon$. Then we study the asymptotics of such a control policy as ϵ tends to zero.

We first study the dynamic properties of the problem $\mathcal{P}_a^0$. It is well known (cf. [5]) that $v_a^0(t, x)$ satisfies the following dynamic programming equation.

$$0 = \frac{\partial}{\partial t} v_a^0(t, x) + \frac{1}{2}\sigma^2(t)\frac{\partial^2}{\partial x^2} v_a^0(t, x) + \min_{u \in \Gamma = [a_1, a_2]}[\bar{b}(t, x, u)\frac{\partial}{\partial x} v_a^0(t, x)] + l(t, x)$$

with the boundary condition $v_a^0(T, x) = g(x)$.

Let $u^*(t, x)$ denote the minimizer of the above equation. Then

$$u^*(t, x) = \begin{cases} a_2 & \text{if } \frac{\partial}{\partial x} v_a^0(t, x) \leq 0 \\ a_1 & \text{if } \frac{\partial}{\partial x} v_a^0(t, x) > 0 \end{cases} \tag{4.1}$$

Example. If $\Gamma = [-a, a]$ for some $a > 0$, then

$$u^*(t, x) = -a\,\text{sgn}[\bar{\theta}B_3(t)\frac{\partial}{\partial x} v_a^0(t, x)].$$

Moreover, suppose that the cost $l(t, x)$ and $g(x)$ are both even functions in x. Then it can be shown as in [7] that

$$\text{sgn}[\bar{\theta}B_3(t)\frac{\partial}{\partial x} v_a^0(t, x)] = \text{sgn}[\bar{\theta}B_3(t)x].$$

This yields that

$$u^*(t, x) = -a\,\text{sgn}[\bar{\theta}B_3(t)x].$$

We would like to work with the control $u_t = u^*(t, x_t)$ given in (4.1) for $\mathcal{P}_a^0$ and $\mathcal{P}^\epsilon$. But for such control, systems $\mathcal{P}_a^0$ and $\mathcal{P}^\epsilon$ may not have strong solutions. So it is natural to seek a smooth approximation to $u^*(t, x)$.

We define the kernels $q(t, x) = (2\pi)^{-1}\exp-(x^2 + t^2)/2$ and $q_\eta(t, x) = \eta^{-2}q(t/\eta, x/\eta)$ for $\eta > 0$. We consider $u_\eta(t, x)$ the convolution of $u^*(t, x)$ with the kernel $q_\eta(t, x)$. Then $u_\eta(t, x) \in C^\infty$ is smooth and $u_\eta(t, x) \in \Gamma$ for all (t, x). Moreover, $u_\eta(t, x) \to u^*(t, x)$ a.e. and for $p' \geq 1$, $\|u_\eta - u^*\|_{p'} \to 0$, as $\eta \to 0$.

We use the feedback control law $u_\eta(t, x)$ for $\mathcal{P}_a^0$. Then the following equation

$$d\bar{x}_t = \bar{b}(t, \bar{x}_t, u_\eta(t, \bar{x}_t))dt + \sigma(t)dw_t, \quad \bar{x}_s = x \tag{4.2}$$

has a unique strong solution for each $\eta > 0$. Moreover, if we let $\bar{u}_t = u_\eta(t, \bar{x}_t)$, then it can be shown (cf. [11]) with the nondegenrate assumption ($\sigma(t) \geq c_0$) that

$$J^0(s, x, \bar{u}.) \to v_a^0(s, x), \quad \text{as } \eta \to 0.$$

Now, for each $\epsilon > 0$, we take $\eta_\epsilon > 0$ such that under control $\bar{u}_t = u_{\eta_\epsilon}(t, \bar{x}_t)$,

$$|J^0(s, x, \bar{u}.) - v_a^0(s, x)| \leq \epsilon.$$

For this fixed η_ϵ, we construct a control $u_t^\epsilon = u_{\eta_\epsilon}(t, x_t)$ for the problem $\mathcal{P}^\epsilon$. Then it is easy to see that the system $\mathcal{P}^\epsilon$ has a strong solution with such control (cf. [6]) We now show u_t^ϵ is nearly optimal. Namely, $J^\epsilon(s, x, u^\epsilon.)$ is close to $v^\epsilon(s, x)$. First of all,

$$
\begin{aligned}
\frac{d}{dt}(x_t - \bar{x}_t) = \ & B_1(t)(x_t - \bar{x}_t) + B_3(t)\theta_t^\epsilon[u_{\eta_\epsilon}(t, x_t) - u_{\eta_\epsilon}(t, \bar{x}_t)] \\
& + (\theta_t^\epsilon - \bar{\theta})[B_3(t)u_{\eta_\epsilon}(t, \bar{x}_t) + B_4(t)].
\end{aligned}
$$

$$
\begin{aligned}
\frac{d}{dt}(x_t - \bar{x}_t)^2 = \ & 2(x_t - \bar{x}_t)\{B_1(t)(x_t - \bar{x}_t) + B_3(t)\theta_t^\epsilon[u_{\eta_\epsilon}(t, x_t) - u_{\eta_\epsilon}(t, \bar{x}_t)] \\
& + (\theta_t^\epsilon - \bar{\theta})[B_3(t)u_{\eta_\epsilon}(t, \bar{x}_t) + B_4(t)]\}.
\end{aligned}
$$

Note that $B_3(t)u^*(t, \cdot)$ is a nonincreasing function, so is $B_3(t)u_{\eta_\epsilon}(t, \cdot)$. It follows that

$$
(x_t - \bar{x}_t)B_3(t)\theta_t^\epsilon[u_{\eta_\epsilon}(t, x_t) - u_{\eta_\epsilon}(t, \bar{x}_t)] \leq 0. \tag{4.3}
$$

Let $\Psi_t := \int_s^t (\theta_r^\epsilon - \bar{\theta})[B_3(r)u_{\eta_\epsilon}(r, \bar{x}_r) + B_4(r)]dr$. Then by (4.3), we have

$$
\begin{aligned}
|x_t - \bar{x}_t|^2 \leq \ & C_9\{\int_s^t (x_r - \bar{x}_r)^2 dr \\
& + \int_s^t 2(x_r - \bar{x}_r)(\theta_r^\epsilon - \bar{\theta})[B_3(r)u_{\eta_\epsilon}(r, \bar{x}_r) + B_4(r)]dr \\
:= \ & C_9 \int_s^t (x_r - \bar{x}_r)^2 dr + 2\int_s^t (x_r - \bar{x}_r)d\Psi_r.
\end{aligned}
$$

By integating by parts, we obtain

$$
\begin{aligned}
\int_s^t (x_r - \bar{x}_r)d\Psi_r = \ & (x_t - \bar{x}_t)\Psi_t - \int_s^t \frac{d}{dr}(x_r - \bar{x}_r)\Psi_r dr \\
\leq \ & \frac{1}{4}|x_t - \bar{x}_t|^2 + |\Psi_t|^2 - \int_s^t \frac{d}{dr}(x_r - \bar{x}_r)\Psi_r dr \\
\leq \ & \frac{1}{4}|x_t - \bar{x}_t|^2 + |\Psi_t|^2 + C_{10}\int_s^t (x_r - \bar{x}_r)^2 + |\Psi_s|^2 + |\Psi_r|)dr.
\end{aligned}
$$

The last inequality is because $|\frac{d}{dr}(x_r - \bar{x}_s)| \leq C(1 + |x_r - \bar{x}_r|)$ for some constant C.

Let $\phi_t = E|x_t - \bar{x}_t|^2$. Then,

$$
\phi_t \leq 2C_9 \int_s^t \phi_r dr + \frac{1}{2}\phi_t + 2E|\Psi_t|^2 + 2C_{10}\int_s^t (\phi_r + E|\Psi_r|^2 + E|\Psi_r|)dr.
$$

Note that Lemma A.2 (Appendix) implies $E|\Psi_t|^2 \leq C\epsilon t$. Thus,

$$
\begin{aligned}
\phi_t \leq \ & C_{11} \int_s^t \phi_r dr + 4E|\Psi_t|^2 + 4C_{10}\int_s^t (E|\Psi_r|^2 + E|\Psi_r|)dr \\
\leq \ & C_{11} \int_s^t \phi_r dr + C_{11}\sqrt{\epsilon}(1 + (t - s)).
\end{aligned}
$$

By Gronwall's inequality, we obtain

$$
\phi_t \leq C_{12}\sqrt{\epsilon}[1 + (t - s) + C_{11}\int_s^t (1 + (r - s))\exp C_1(t - r)dr] \leq C_{13}\sqrt{\epsilon}.
$$

Therefore,

$$
E|x_t - \bar{x}_t| \leq \sqrt{C_{13}}\sqrt[4]{\epsilon}.
$$

Lemma 4.1. *There exists C such that*

$$
|J^\epsilon(s, x, u^\epsilon.) - v_a^0(s, x)| \leq C(1 + |x|^{k_\theta})\sqrt[4]{\epsilon}.
$$

Proof. Similarly as in (3.4),

$$|J^\varepsilon(s, x, u^\varepsilon.) - J^0(s, x, \bar{u}.)| \le C_{14}(1 + |x|^{k_9})\sqrt[4]{\varepsilon},$$

where $u_t^\varepsilon = u_{\eta_\varepsilon}(t, x_t)$ and $\bar{u}_t = u_{\eta_\varepsilon}(t, \bar{x}_t)$. It follows

$$
\begin{aligned}
|J^\varepsilon(s, x, u^\varepsilon.) - v_a^0(s, x)| &\le \quad |J^\varepsilon(s, x, u^\varepsilon.) - J^0(s, x, \bar{u}.)| + |J^0(s, x, \bar{u}.) - v_a^0| \\
&\le \quad \varepsilon + C_{14}(1 + |x|^{k_9})\sqrt[4]{\varepsilon} \\
&\le \quad C_{15}(1 + |x|^{k_9})\sqrt[4]{\varepsilon}. \ \square
\end{aligned}
$$

Theorem 4.1. *There exists C such that for each $\varepsilon > 0$ there exists $\eta_\varepsilon > 0$ for $u_t^\varepsilon = u_{\eta_\varepsilon}(t, x_t) \in \mathcal{A}^\varepsilon$ (here $u_{\eta_\varepsilon}(t, x) = (\rho_{\eta_\varepsilon} * u^*)(t, x)$ with u^* given in (4.1)),*

$$|J^\varepsilon(s, x, u^\varepsilon.) - v^\varepsilon(s, x)| \le C(1 + |x|^{k_9})\sqrt[4]{\varepsilon}.$$

That is, $u_t^\varepsilon = u_{\eta_\varepsilon}(t, x_t)$ is an asymptotically optimal control with error bound $\sqrt[4]{\varepsilon}$.

Proof.

$$
\begin{aligned}
0 \le \quad & J^\varepsilon(s, x, u^\varepsilon.) - v^\varepsilon(s, x) \\
\le \quad & J^\varepsilon(s, x, u^\varepsilon.) - v_a^\varepsilon(s, x) \\
= \quad & (J^\varepsilon(s, x, u^\varepsilon.) - v_a^0(s, x)) + (v_a^0(s, x) - v_a^\varepsilon(s, x)) \\
\le \quad & C_{16}(1 + |x|^{k_9})\sqrt[4]{\varepsilon} + C_{17}(1 + |x|^{k_9})\sqrt{\varepsilon} \\
\le \quad & (C_{16} + C_{17})(1 + |x|^{k_9})\sqrt[4]{\varepsilon}. \ \square
\end{aligned}
$$

Remark. Note that the control we used in our analysis is $u_{\eta_\varepsilon}(t, x_t)$, a smooth approximation of $u^*(t, x)$. The purpose of smoothing $u^*(t, x)$ is to ensure the systems $\mathcal{P}_a^\varepsilon$ and $\mathcal{P}_a^0$ have unique strong solutions. For numerical simulation purpose, such smooth approximation will not be necessary since the system $\mathcal{P}_a^\varepsilon$ and $\mathcal{P}_a^0$ always have solutions under their discrete time version; see [7] explanation of this point.

5 Concluding remarks

In this paper, we have presented an asymptotic result for a controlled diffusion with a fast oscillating parameter. We introduce two auxiliary problems and make use of the singular perturbation methods to reduce the problem $\mathcal{P}^\varepsilon$ into a simpler problem $\mathcal{P}_a^0$. We then describe a method to construct a control for the system by using the optimal control of the simpler problem $\mathcal{P}_a^0$. Therefore, by introducing a problem $\mathcal{P}_a^\varepsilon$ (a control problem with complete information of θ_t^ε) and showing that the associated value functions for $\mathcal{P}_a^\varepsilon$ converges to the value function of $\mathcal{P}_a^0$, we can construct a control for $\mathcal{P}^\varepsilon$ from the optimal control of $\mathcal{P}_a^0$. It turns out that the controls so constructed are asymptotically optimal as the oscillation rate of the parameter goes to infinity, i.e., $\varepsilon \to 0$. Furthermore, error estimates of the asymptotic optimality are provided in terms of their corresponding cost functions.

6 Appendix

We give three technical lemmas, concerning the asymptotics of the capacity process θ_t^ε, which are required to prove the results the previous sections.

Lemma A.1. *There exist constants C and $k_0 > 0$ such that, for all $t \geq s$,*

$$|P(\theta_t^\epsilon = i) - \nu_i| \leq Ce^{-k_0(t-s)/\epsilon}.$$

Proof. See [4, Theorem II.10.1] and [4, Theorem II.12.8].

Lemma A.2. *Let $\beta_t = (\beta_1(t), \cdots, \beta_m(t))$ be a bounded process on $[s, \infty)$ independent of θ_t^ϵ. Then, there exists a constant C such that*

$$\Lambda_t := E| \int_s^t (\theta_r^\epsilon - \bar{\theta}) \beta_r dr |^2 \leq C\epsilon(t-s).$$

Proof. We may take $s = 0$ for brevity. First of all,

$$
\begin{aligned}
\Lambda_t &= E \int_0^t \int_0^t (\theta_r^\epsilon - \bar{\theta})(\theta_{r'}^\epsilon - \bar{\theta}) \beta_r \beta_{r'} dr dr' \\
&\leq C \int_0^t \int_0^t |[E(\theta_r^\epsilon - \bar{\theta})(\theta_{r'}^\epsilon - \bar{\theta})]| dr dr'.
\end{aligned}
$$

Applying Lemma A.1, we have, for $\tau \geq 0$ and $k = 1, \cdots, p$,

$$E\theta_\tau^\epsilon = \sum_{i=1}^p iP(\theta_\tau^\epsilon = i) = \sum_{i=1}^p i\nu_i + O(e^{-k_0\tau/\epsilon}) = \bar{\theta} + O(e^{-k_0\tau/\epsilon}).$$

Moreover, for all $\tau' \geq \tau \geq 0$,

$$
\begin{aligned}
E\theta_\tau^\epsilon \theta_{\tau'}^\epsilon &= \sum_{i,j=1}^p ijP(\theta_\tau^\epsilon = i, \theta_{\tau'}^\epsilon = j) \\
&= \sum_{i,j=1}^p ijP(\theta_{\tau'}^\epsilon = j|\theta_\tau^\epsilon = i)P(\theta_\tau^\epsilon = i) \\
&= \sum_{i,j=1}^p ijP(\theta_{\tau'-\tau}^\epsilon) = j|\theta_0^\epsilon = i)P(\theta_\tau^\epsilon = i) \\
&= \bar{\theta}^2 + \sum_{i,j=1}^p ij\nu_j(P(\theta_\tau^\epsilon = i) - \nu_i) \\
&\quad + \sum_{i,j=1}^p ij(P(\theta_{\tau'-\tau}^\epsilon) = j|\theta_0^\epsilon = i) - \nu_j)P(\theta_\tau^\epsilon = i) \\
&= \bar{\theta}^2 + O(e^{-k_0\tau/\epsilon} + e^{-k_0|\tau'-\tau|/\epsilon}).
\end{aligned}
$$

Thus,

$$|E(\theta_r^\epsilon - \bar{\theta})(\theta_{r'}^\epsilon - \bar{\theta})| \leq C(e^{-k_0|r-r'|/\epsilon} + e^{-k_0r/\epsilon} + e^{-k_0r'/\epsilon}).$$

Therefore,

$$
\begin{aligned}
\Lambda_t &\leq Cp \int_0^t \int_0^t [e^{-k_0|r-r'|/\epsilon} + e^{-k_0r/\epsilon} + e^{-k_0r'/\epsilon}] dr dr' \\
&\leq Cp\epsilon[4k_0^{-1}t]. \quad \square
\end{aligned}
$$

Lemma A.3. *Let u_t denote an $\mathcal{F}_t := \sigma\{\theta_s^\epsilon, w_s : s \leq t\}$ adapted process and let $\mathcal{D} = \vee_{t \geq s} \mathcal{D}_t$. Then $E[u_t|\mathcal{D}_t] = E[u_t|\mathcal{D}]$. In particular, $E[u_t|\mathcal{D}_t] = E[u_t|\mathcal{D}_{t'}]$ for all $t' \geq t$.*

Proof. Define $\mathcal{G}_t = \sigma\{\theta_r^\epsilon : s \leq r \leq t\}$. Let $\mathcal{C}_t = \{A \cap B : A \in \mathcal{D}_t, B \in \mathcal{G}_t\}$ and let

$$\mathcal{H}_t = \{\xi : \mathcal{F}_t \text{ measurable, such that } E[\xi|\mathcal{D}_t] = E[\xi|\mathcal{D}]\}$$

denote the set of random variables ξ. Then $\mathcal{H}_t$ is a λ-system and $\mathcal{C}_t$ is a π-class (see [3] for definitions). Moreover, $\sigma(\mathcal{C}_t) = \mathcal{D}_t \vee \mathcal{G}_t = \mathcal{F}_t$. Furthermore, it can be shown that $\mathcal{H}_t$ contains all the indicator functions of $\mathcal{C}_t$, i.e.,

$$\mathcal{H}_t \supset \{\chi_K : K \in \mathcal{C}_t\}.$$

Then, by applying [3, Theorem 1.3], we can conclude that $\mathcal{H}_t$ contains all $\mathcal{F}_t$ measurable random variables. This proves the lemma. $\square$

References

[1] Benes, V.E., Karatzas, I., Rishel, R.W., The separation principle for a Bayesian adaptive control poblem with no strict-sense optimal law, preprint, (1991).

[2] Caines, P.E. and Chen, H.F., Optimal adaptive LQG control for systems with finite state process parameters, *IEEE Trans, Automat. Contr.* 30, No. 2, pp. 185-189, (1985).

[3] Chow, Y.S., *Probability theory*, Springer-Verlag, New York, (1978).

[4] Chung, K.L., *Markov chains with stationary transition probabilities*, Springer, Berlin, (1960).

[5] Fleming, W.H. and Rishel, R.W. (1975). *Deterministic and Stochastic Optimal Control.* Springer-Verlag, New York.

[6] Ghosh, M.K., Arapostathis, A., and Marcus, S.I., Optimal control of switching diffusions with applications to flexible manufacturing systems, *Working paper*, University of Texas at Austin, Department of Electrical and Computer Engineering, (1991).

[7] Haussmann, U. and Zhang, Q., Optimal control of diffusions with small observation noise, *Proceedings of the Imperial College Workshop on Applied Stochastic Analysis*, M.H. Davis and R.J. Elliott ed., Stochastics monograph, pp. 237-263, (1989).

[8] Haussmann, U. and Zhang, Q., Stochastic adaptive control with small observation noise, *Stochastics and Stochastics Reports*, Vol. 32, pp. 109-144, (1990).

[9] Hijab, O., The adaptive LQG problem - Part I, *IEEE, Trans. Automat. Contr. 28*, No. 2, pp. 171-178, (1983).

[10] Lehoczky, J., Sethi, S.P., Soner, H.M., and Taksar, M., An asymptotic analysis of hierarchical control of manufacturing systems under uncertainty, forthcoming in *Mathematics of Operations Research*, (1990).

[11] Krylov, N.V., *Controlled Diffusion Processes*, Springer-Verlag, New York, (1980).

[12] Roubaud, M.C., Suboptimal filters for piecewise linear systems with small observation noise, *Working paper*, INRIA Sophia-Antipolis, France, (1991).

[13] Sethi, S. and Zhang, Q., Hierarchical production planning in dynamic stochastic manufacturing systems: asymptotic optimality and error bounds, submitted to *SIAM Journal on Control and Optimization*, (1990).

[14] Sethi, S.P., Soner, H.M., Zhang, Q., and Jiang, J., Turnpike sets and their analysis in stochastic production problems, *Mathematics of Operations Research*, to appear, (1991).

[15] Zhang, Q., An asymptotic analysis of controlled diffusions with fast oscillating system parameters, submitted to *Stochastics and Stochastics Reports*, (1991).

Lecture Notes in Control and Information Sciences

Edited by M. Thoma and A. Wyner

Lecture Notes in Control and Information Sciences

Edited by M. Thoma and A. Wyner

Lecture Notes in Control and Information Sciences

Edited by M. Thoma and A. Wyner